Structural Modeling
and
Experimental Techniques

Prentice-Hall Civil Engineering and Engineering Mechanics Series

N. M. NEWMARK AND W. J. HALL, *Editors*

Structural Modeling and Experimental Techniques

Gajanan M. Sabnis
Professor of Civil Engineering
Howard University, Washington, DC

Harry G. Harris
Professor of Civil Engineering
Drexel University, Philadelphia, PA

Richard N. White
Professor of Structural Engineering
Cornell University, Ithaca, NY

M. Saeed Mirza
Professor of Civil Engineering and Applied Mechanics
McGill University, Montreal, Canada

PRENTICE-HALL, INC., Englewood Cliffs, N.J. 07632

Library of Congress Cataloging in Publication Data
Main entry under title:

Structural modeling and experimental techniques.

    Bibliography: p.
    Includes index.
    1. Structural frames—Models—Testing.
2. Structural design. I. Sabnis, Gajanan M.
TA643.S87   624.1'7'0724   82-481
ISBN 0-13-853960-X   AACR2

With special contributions by:
    Prof. Robert J. Hansen, MIT
    Prof. William A. Litle, University of California—Irvine

Editorial/production supervision by Karen Skrable
Manufacturing buyers: Joyce Levatino and Anthony Caruso
Art production by Peter J. Ticola, Jr.
Cover design by Marvin Warshaw

©1983 by Prentice-Hall, Inc., Englewood Cliffs, N.J. 07632

Printed in the United States of America

10 9 8 7 6 5 4 3 2 1

ISBN 0-13-853960-X

Prentice-Hall International, Inc., *London*
Prentice-Hall of Australia Pty. Limited, *Sydney*
Prentice-Hall Canada Inc., *Toronto*
Prentice-Hall of India Private Limited, *New Delhi*
Prentice-Hall of Japan, Inc., *Tokyo*
Prentice-Hall of Southeast Asia Pte. Ltd., *Singapore*
Whitehall Books Limited, *Wellington, New Zealand*

TO OUR CHILDREN

Rahul
Madhavi
Maria
George
Constantine
Basil
Julia
Barbara
David
Arif
Nasim

Contents

∿∿∿∿∿∿∿∿∿∿∿∿∿∿

2   THE THEORY OF STRUCTURAL MODELS   26

3   ELASTIC MODELS—MATERIALS AND TECHNIQUES   63

Preface

Structural models have always played a significant role in structural engineering research and design. Since 1950 a number of outstanding models laboratories were developed in Europe; this fact has greatly stimulated the usage of the modeling process by European engineers. More recently, substantial improvements in modeling techniques for reinforced concrete structures have been made in the United States and Canada. The purpose of this book is to present a current, up-to-date treatment of structural modeling for applications in design, research, and education. Primary emphasis is given to modeling the behavior of reinforced and prestressed concrete structures.

The authors view the structural model as a complement to the mathematical model and not as a competitor nor a replacement for analysis. If an appropriate, well-tested analytical approach exists for a given situation, it will usually be less expensive and quicker than an experimental approach. If analysis is not feasible, or if boundary conditions are poorly defined or highly variable, then the model test may be the only solution to the problem. Models must be used selectively, and their range of application is constantly changing as analytical methods get increasingly more powerful. This book should prepare the reader to form the proper perspective as to when and where models should be utilized.

The considerable recent developments in structural modeling are reflected by the amount of literature published since 1960 and in the well-attended symposia sponsored by the American Concrete Institute in 1968 and 1972, by McGill University in 1972, and by the University of Sydney in 1972. These symposium volumes brought together most of the new literature, but they are more useful as reference books for people already involved in the field rather than as the principal introductory books for those not familiar with this important tool in structural engineering.

Models for determining the elastic response of structures have been used for many years, and considerable information is available in the form of research reports and books. Many elastic modeling techniques have been replaced by computer-based analysis methods, and therefore they are treated here in a relatively brief manner.

The major emphasis is on the modeling of the true inelastic behavior of structures. Compared with elastic models, the problems in inelastic modeling are considerably greater—from the development of materials to be used in the models, the techniques of fabrication, and testing to the interpretation of model results to predict the behavior of the simulated prototype.

Applications of the modeling techniques to real structures help one to better understand the actual process of model analysis. They also assist in forming a perspective on the types of structures for which physical modeling is important. There are some types of special structures where the models approach plays an essential role in design. These topics are given detailed treatment in the form of case histories.

Chapters 1 and 2 discuss the historical background of model analysis and similitude principles that govern the design, testing, and interpretation of models. Chapter 3 deals with the various aspects of elastic models, pointing out their limitations and usefulness in certain situations, and also the problems associated with them. Chapters 4 and 5 treat materials for reinforced concrete models. Accurate modeling of the properties of both concrete and steel is absolutely essential and is one of the more difficult parts of the modeling process. Chapter 6 treats the problem of scale effects of model testing. An understanding of the nature and source of scale effects is crucial in understanding the capabilities and limitations of the modeling technique. Chapter 7 deals with selected laboratory techniques and loading methods. Developing a sound familiarity with these techniques is an integral part of model analysis.

Instrumentation techniques are discussed in Chapter 8, with a strong emphasis on strain measurement and interpretation. Modern automated data acquisition systems are treated in sufficient depth to enable the reader to help plan new systems for his or her own laboratory. Chapter 9 contains material on errors and the accuracy and reliability of physical modeling. Random errors are discussed, and their propagation in experimental work is introduced. Reliability of results is extremely important to any reader, and size effects still bother those who have already used the models technique. The detailed treatment of these topics should remove some of the "misapprehensions" of structural modeling.

Chapter 10 covers actual applications of structural modeling. Case studies of a number of important uses of modeling in design and research are presented. Time requirements and costs are given when the information is available. These cases, which are drawn from the experience of the authors and their colleagues, cover a wide spectrum of applications.

Chapter 11 deals with dynamic similitude and modeling techniques used in studying dynamic loading of structures. The problems discussed in the earlier chapters become more difficult in dynamic studies. Additional similitude requirements and new experimental techniques used in studying the dynamic behavior of models are introduced, and several examples of different types of structures under dynamic loading are presented.

We believe that this book will be of substantial assistance not only to students in model analysis and experimental methods but also to those involved at the professional level in manufacturing and testing structural models. It should also be useful to those engaged in testing large or full-scale structures since the instrumentation techniques and overall approaches used in testing large structures are very similar to those utilized in small-scale modeling work.

GAJANAN M. SABNIS
HARRY G. HARRIS
RICHARD N. WHITE
M. SAEED MIRZA

Acknowledgments

∿∿∿∿∿∿∿∿∿∿∿∿

The authors would like to thank a number of individuals who helped at one stage or another of this book. Notable contributions are as follows:

- Professors William A. Litle of the University of California (Irvine) and Robert Hansen of the Massachusetts Institute of Technology, whose earlier draft in coauthorship with Professor Richard N. White was utilized in portions of several chapters, particularly for Chapters 2 and 9, and for Appendix A.
- Professors Alex Haney of the University of New South Wales and C. Douglas Sutton of Purdue University reviewed the entire manuscript with extreme thoroughness and care; their comments were most useful.
- Professor William G. Godden of the University of California (Berkeley), who helped considerably with the practice problems in several chapters.

A number of companies and individuals donated photographs and are credited for them. They are as follows:

- Wiss Janney Elstner and Associates, Northbrook, Illinois
- Portland Cement Association, Skokie, Illinois
- Measurements Group, Raleigh, North Carolina
- BLH Corporation, Waltham, Massachusetts

• Roberto Riccioni, ISMES Laboratories, Bergamo, Italy
• Professor Henry Cowan, University of Sidney, Australia

Finally, the graduate students of Cornell and Drexel Universities, who assisted in reviewing several chapters.

Introduction to Physical Modeling in Structural Engineering

1

1.1 INTRODUCTION

A perspective on physical modeling of structures is presented in this chapter, including classification of the various types of physical models that have evolved over the years and some comments on the general role of these categories of models in design, development, and research in North America and elsewhere.

Structural models (and reduced-scale structures) have always played a significant role in structural engineering research and design. Experiments on reduced-scale structures and specimens have also been important in teaching structural mechanics and structural engineering. A wide range of problems is met in planning, conducting, and interpreting an experimental study of structural behavior. Each of these areas, which range from theoretical similitude requirements to the rather extensive discipline of experimental stress analysis, will be treated in detail in subsequent chapters. This chapter gives the reader an appreciation of structural modeling in its broadest physical sense, and how it is used in the profession. Purely architectural models are also important in planning new construction and correlating spaces, but this type of model will not be considered here because its role is completely different from the structural model.

1.2 STRUCTURAL MODELS—
DEFINITIONS AND CLASSIFICATIONS

A structural model is defined as "any physical representation of a structure or a portion of a structure. Most commonly, the model will be constructed at a reduced scale."* It applies equally well to models of structures made of any material, of course. A second definition given by Janney et al. (1970) is:

> A structural model is any structural element or assembly of structural elements built to a reduced scale (in comparison with full size structures) which is to be tested, and for which laws of similitude must be employed to interpret test results.

Both definitions encompass a broad class of modeling studies on prototype (full-size) structures such as bridges, buildings, dams, towers, reactor vessels, shells, aerospace and mechanical engineering structures, undersea structures, etc. Loadings include static, simulated seismic, thermal, and wind effects.

Many reduced-size structural elements are customarily used in research studies; some investigators also class these structures as models even though similitude conditions are not normally applied to the large-scale research models. Instead, design methods and equations are based directly upon the observed behavior of these research models and are accordingly given full acceptance by the design profession.

This book contains extensive material on models of reinforced concrete structures with a geometric scale factor of about $\frac{1}{10}$. However, much of the material to be covered is applicable to models of other types of structures and to different scale factors.

1.2.1 Models Classification

Structural models can be defined and classified in a variety of ways. The definitions adopted here relate to the intended function of the model. That is, what do we expect to achieve from the tests? Do we want only *elastic response*, or do we expect to load the model up to failure to observe its complete behavior, including the *failure mode and capacity*? Are we content to work with *influence lines* determined from the models, or do we need actual *strain measurements* for prescribed loadings? The models needed in each of these applications have been given well-accepted names.

Elastic Model.   This type of model has a direct geometric resemblance to the prototype but is made of a homogeneous, elastic material that does not necessarily resemble the prototype material. The elastic model is restricted to the elastic range of behavior of the prototype and cannot predict postcracking behavior of concrete, postyield behavior of steel, nor the many other inelastic behavior

---

*As evolved by ACI Committee 444, Models of Concrete Structures.

modes that develop in actual structures when they are loaded. Chapter 3 treats elastic models in detail, including selection of materials. Plastics such as methyl methacrylate (Plexiglas, Lucite, Perspex) and polyvinylchloride (PVC) are most widely used in constructing elastic models, even though their time-dependent properties present difficulties.

Indirect Model. An indirect model (Fig. 1.1) is a special form of the elastic model that is used to obtain *influence diagrams* for reactions and for internal stress resultants such as shearing forces, bending moments, and axial forces. The loading applied to indirect models has no correspondence to the actual loads expected on the prototype structure since load effects are obtained from superposition of the influence values. An indirect model often does not have a direct physical resemblance to the prototype; for example, a frame whose behavior is controlled by its flexural stiffness properties (*EI*) can be modeled with an indirect model that correctly reproduces the relative stiffness values. The latter can be done without precise scaling of the cross-sectional shape (circular shapes in the indirect model can represent prototype wide-flange sections), and the element areas may be grossly distorted without affecting the results.

Figure 1.1 Influence line for right base moment for knee-braced frame (note rotation applied to right base to produce deflected shape).

In the past most applications of the indirect model have been for non-uniform members in indeterminate frames, but now this type of model finds very little use because these purely elastic calculations are better done by computer.

Direct Model. A direct model is geometrically similar to the prototype in all respects, and the loads are applied to it in the same manner as to the prototype. Strains, deformations, and stresses in the model for each loading condition are representative of similar quantities in the prototype for the corresponding loading condition. Thus an elastic model can also be a direct model.

Strength Model. This type of model is also called *ultimate strength* or *realistic model* and is a direct model that is made of materials that are similar to the prototype materials such that the model will predict prototype behavior for all loads up to failure (see Figs. 1.5, 1.7, 1.9, and 1.14). A strength model of a reinforced concrete element or structure must be made from model concrete and model reinforcing elements where each of the materials satisfies the similitude conditions for the prototype materials. The latter represents the most difficult problem met in strength models for concrete structures; Chapters 4 and 5 cover these topics in depth. Strength models can also be made for steel structures, timber structures, etc., and in each case the major problem is in finding the proper materials and fabrication techniques for the models.

A strength model must be a direct model, by definition. To use the results of indirect models, one must rely on superposition of results, and the superposition principle is not valid for the postlinear response found in all strength models. It is not economical to build strength models and use them only in the elastic range of behavior.

Wind Effects Model. There are various ways of classifying wind effects modeling. We can utilize *shape models*, where either total forces or the wind pressures on the structure may be measured (Fig. 1.3), and *aeroelastic models*, where both the shape and stiffness properties of the prototype structure are modeled in order to measure the wind-induced stresses and deformations and the dynamic interaction of the structure with the wind.

Research, Design, and Instructional Models. One often sees the classifications of *research models, instructional models,* and *design models.* While the use of each is obvious, it is worthwhile to point out that the degree of sophistication needed in each may be markedly different. Instructional models should be made as simple as possible to demonstrate the concepts under study—any similitude distortion that does not markedly influence the desired behavior is permissible. Research models, from which theories may be substantiated and generalizations made for a class of structures, usually must be made with as much accuracy as the laboratory technicians can muster. Design models may range in accuracy requirements from the instructional model to the research model, depending upon the desired results. Some design models may be used only as a conceptual tool to get a better idea of how a proposed structure deforms under load; others may be expected to predict possible instability modes or to predict the true load capacity of the structure.

The use of structural models as direct aids in design is one of the most powerful applications of structural models. An engineer is often called upon to design structures such as the free-form shell of the Kodak Pavilion in the 1965 New York World Fair, which could not be solved using the existing available analysis tools. In this instance a series of plastic models were used as the main approach in the final design.

Another major physical modeling application in design is to help verify calculations for very large and monumental structures where failure consequences could be extremely serious (such as heavy loss of life or capital investment, or disruption of essential lifeline services). A nuclear reactor structure is a good example of this application.

Perhaps the major disadvantages of using a model for design purposes, from the standpoint of the consulting engineer, are the time and money involved in the modeling study. This topic is given further attention in Section 1.7.

Other Models.   Other classifications of models include *dynamic models*, which are usually elastic, direct models, and *thermal models*, where effects of temperature gradients are studied. Thermally loaded models are also usually elastic, direct models, although some attempts have been made to combine mechanical loads and thermal loads for strength models. There is also a group of *photomechanical* models that utilize optical effects, such as the *photoelastic effect* for stress intensities and directions and *interference effects* from grids to measure plate displacements, internal strain fields, and deflections of framed structures. *Construction procedure* models are used to help plan the building of very complex structures, such as in reinforcement placement in nuclear reactor containments, and in cantilever bridges.

1.3 A BRIEF HISTORICAL PERSPECTIVE ON MODELING

The use of small-scale models by engineers and builders dates back many hundreds and even thousands of years. However, these early models were primarily aids for planning and constructing structures and were not useful for predicting deformations and strengths of prototypes. They more nearly resemble the modern architectural model and should not be thought of in the same context as structural models.

Most models used to predict structural behavior require measurement of strains, displacements, and forces. Thus the development of modeling as a practical tool has been sharply influenced by our abilities in experimental stress analysis. The most-used techniques in experimental stress analysis have been established only since the turn of the century. They include:

1. Photoelasticity for elastic stress analysis of complex geometries
2. Deformeters (Fig. 1.1) developed by Beggs, Eney, Gottschalk, and others for introducing deformations into indirect models and then determining influence lines by use of the Müller-Breslau principle
3. Mechanical and optical strain gages for measurement of surface strain
4. Electrical resistance strain gages

5. Linear variable differential transformers (LVDT), linear potentiometers, and similar devices for electrical recording of displacements

6. Brittle coatings, Moire and interference fringe methods, and photoelastic coatings for "full-field" strain measurements on the surface of a structure or model

7. Automated data acquisition systems that use a minicomputer to control and process many channels of data

Item (4), the electrical resistance strain gage, is perhaps the single most important development in terms of providing an easily used method for determining either static or dynamic strains in a structure. The same gage forms the sensing device in commonly used load cells and transducers. Thus its introduction in the 1940s can be considered to be the basis for modern experimental stress analysis and structural model analysis.

Relatively little model analysis other than photoelastic studies and indirect models was done prior to 1940. The $\frac{1}{240}$-scale Hoover Dam model built in 1930 was a notable exception. Since that time the technology needed for rapid construction, instrumentation, and testing of structural models has continued to develop. The current use of structural modeling is introduced in this chapter, and the full range of applicability will become apparent as the reader progresses through the book.

1.4 STRUCTURAL MODELS AND CODES OF PRACTICE

Modeling has received relatively little attention in most North American building codes and specifications. However, most codes do contain special provisions that permit the engineer to make rather substantial use of models in the design process. For example, the 1969 City of New York Building Code contains the following paragraph:

> (5) MODEL TESTS.—Tests on models less than full size may be used to determine the relative intensity, direction, and distribution of stresses and applied loads, but shall not be considered as a proper method for evaluating stresses in, nor the strength of, individual members unless approved by the commissioner for this purpose. Where model analysis is proposed as a means of establishing the structural design, the following conditions shall be met:
>
> a. Analysis shall be made by a firm or corporation satisfactory to the commissioner.
>
> b. The similitude, scaling, and validity of the analysis shall be attested to by an officer or principal of the firm or corporation making the analysis.
>
> c. A report on the analysis shall be submitted showing test set-ups, equipment, and readings.

ACI 318, Building Code Requirements for Reinforced Concrete, permits model analysis for shell structures in Section 19.3.3:

19.3.3—Analysis based on the results of elastic model tests approved by the Building Official shall be considered as valid elastic analyses. When such model analysis is used, only those portions which significantly affect the items under study need be simulated. Every attempt shall be made to insure that these tests reveal the quantitative behavior of the prototype structure.

The same code has a general clause in Section 1.2.2 that reads: "Calculations may be supplemented by model analysis." This clause is further elaborated on in the Code Commentary.

Some countries, such as Australia, permit complete designs of certain types of structures by model analysis alone. Thus there is a relatively healthy potential usage of modeling in design codes. Many engineers would be even more receptive of modeling for design if they would only realize that many of the code provisions that they apply analytically every day are in fact derived mainly from tests on reduced-scale models. The engineer who wishes to use models should not hesitate to contact the responsible building official to seek approval and should also seek proper assistance from an expert in model analysis.

There are numerous situations in which these code provisions might be applied in practice; in most cases it is where the analytical approach is not fully adequate. Basic doubts may arise in applying existing analytical techniques to new and complex structural forms. Analytical methods are not yet developed to handle the extremely complicated behavior of reinforced concrete structures loaded to near-failure or certain other limit-state conditions. This is why modeling is often used by engineers studying the failure of structures.

Types of structures suitable for possible structural model studies during the design phase include:

1. Shell roof forms of complex configuration and boundary conditions
2. Tall structures and other wind-sensitive structures for which wind tunnel modeling is indicated
3. New building structural systems involving the interaction of many components
4. Complex bridge configurations such as multicell prestressed concrete box girder highway bridges
5. Nuclear reactor vessels and other reinforced and prestressed concrete pressure vessels
6. Ordinary framed structures subjected to complex loads and load histories such as wind and earthquake forces
7. Structural slabs with unusual boundary or loading conditions, or with irregular geometry produced by cutouts and thickness changes

8. Dams
9. Undersea stuctures
10. Detailing

Item (10) points out an important use of models: for studying problems that arise in only a limited region of a structure, such as an involved connection detail or localized stresses due to large prestressing forces. Carefully designed and tested, the partial model can be extremely important in clarifying these situations and in leading to an improved design. The major difficulty with the partial model is in providing proper boundary conditions: an inadequate boundary condition in a physical model may produce even worse results than a poor boundary condition in an analytical solution.

1.5 CHOICE OF GEOMETRIC SCALE

Any given model being built in a given laboratory has an optimum geometric scale factor. Very small models require light loads but can present great difficulties in fabrication and instrumentation. Large models are easier to build but require much heavier loading equipment. The latter requirement is not serious in a laboratory that is fully equipped to conduct tests on large structures, but it is a severe handicap in a smaller laboratory. Typical scale factors for several classes of structures are:

Type of Structure	Elastic Models	Strength Models
Shell roof	$\frac{1}{200}$ to $\frac{1}{50}$	$\frac{1}{30}$ to $\frac{1}{10}$
Highway bridge	$\frac{1}{25}$	$\frac{1}{20}$ to $\frac{1}{4}$
Reactor vessel	$\frac{1}{100}$ to $\frac{1}{50}$	$\frac{1}{20}$ to $\frac{1}{4}$
Slab structures	$\frac{1}{25}$	$\frac{1}{10}$ to $\frac{1}{4}$
Dams	$\frac{1}{400}$	$\frac{1}{75}$
Wind effects	$\frac{1}{300}$ to $\frac{1}{50}$	Not applicable

The rationale behind this table should become more apparent as we progress through the chapters of this book. Strength models of concrete structures have many practical dimensional limitations such as minimum feasible thickness, bar spacing, cover, etc. Maintaining materials similitude requirements is a crucial problem in this category of models.

1.6 THE MODELING PROCESS

The successful modeling study is one that is characterized by careful *planning* of the many diverse steps in the physical modeling process. An experimental study of an engineering structure is a small engineering project in itself, and as in any engineering venture, a logical and careful sequencing of events is an absolute requirement.

Detailed planning of an experiment is even more essential than planning of an analytical approach because refinement of a structural model halfway through the modeling process is usually impossible. A major aspect of planning is deciding what is expected from the model. Do we need only elastic stresses and displacements, or do we want to see how the structure behaves at overloads leading up to failure? Is instability a possible failure mode? Are thermal stresses involved? Do we have to simulate dynamic effects? The time required to complete the model study can range from perhaps a week or two for a very limited elastic model of a portion of a structure to six months or more for a detailed, ultimate strength reinforced mortar model for predicting failure behavior of a complete structure. We obviously must guard against "overdoing" the model study just as we have to avoid excessive analysis of a structure. The engineer who bears final responsibility for the project must be the key person in prescribing precisely what the model is supposed to accomplish.

A typical modeling study can be broken into the following multistep process:

1. Define the *scope* of the problem, deciding what is needed from the model and what is not needed.

2. Decide on the required level of *reliability* or *accuracy*. If $\pm 30\%$ is adequate for design purposes, then an attempt to achieve $\pm 10\%$ accuracy is wasted effort and time (see Chapter 9).

3. Specify *similitude* requirements for geometry, materials, loading, and interpretation of results. Pay particular attention to those similitude requirements that cannot be met, such as the desired equality of Poisson's ratio for concrete and plastics when doing elastic modeling of shell and slab structures (Chapter 2).

4. Select model *materials* with proper attention to steps (1), (2), and (3) above (Chapters 3, 4, and 5).

5. Plan the *fabrication* phase in consultation with the technicians who will be constructing the model, and follow the fabrication activities closely. This can be a frustrating part of modeling because it is often quite time-consuming.

6. Design and prepare the *loading equipment*; new systems should be thoroughly checked out and calibrated before use on a model (Chapter 7).

7. Select *instrumentation* and recording equipment for strains, displacements, forces, and other quantities. This step must be closely coordinated with steps (5) and (6), particularly if embedded strain gages are to be used in concrete models. Special strain gages and other equipment must be ordered well in advance of the actual time of usage (Chapter 8).

8. Observe the *response* of the model during loading, taking complete notes and photographic records of the behavior. Do not rush through a test, and never leave anything to memory. Some investigators use tape or video

recorders in this phase to record detailed comments on load history, cracking development, instability modes, and other information that may be difficult to describe numerically. Approximate calculations should be done before the experiment to estimate expected levels of response. Equilibrium checks should be done on results obtained early in the test.

Because of the great importance in properly recording data when it must be done manually, a few specific comments are in order here:

(a) Prepare a ruled sheet with columns; put the date, names of test personnel, and the model designation on the first sheet.
(b) Record the readings directly, and do not attempt to reduce the data in your head.
(c) Record the zero readings, allowing at least two lines of space since zero readings often must be taken more than once.
(d) Allow adjacent columns for reduction of results.
(e) Take readings at lower load increments as failure is approached.
(f) Take readings as yielding or failure actually occurs, even if the level of accuracy achieved is not high—approximate readings can give a better idea of behavior.
(g) Take final readings when the load is removed.

9. *Analyze the data and write the report* as soon as possible, while the entire test is still fresh in your mind. In addition to reporting the results, suggestions for improvement in techniques should be recorded to facilitate better modeling results in subsequent experiments.

Most of these steps are merely statements of common sense, but it is surprising how often common sense is ignored or left out of a crucial step in an experimental study. Several "laws" should be kept in mind when thinking about the difficulties of experimental work:

Murphy's law: If something can go wrong, it will.

O'Toole's law: Murphy is wildly optimistic.

SHWM law:* Things are never as bad as they turn out to be.

There is a moral in these tongue-in-cheek laws, and that is the fact that experiments must be carefully planned, controlled, and interpreted if they are to succeed.

1.7 ADVANTAGES AND LIMITATIONS OF MODEL ANALYSIS

The main advantage of a physical model over an analytical model is that it portrays behavior of a complete structure loaded to the collapse stage. Although substantial progress is continually made in computer-based procedures for

*Sabnis-Harris-White-Mirza law.

analysis of structures, we still cannot predict analytically the failure capacity of, say, a three-dimensional assemblage of reinforced concrete elements.

The prime motivation to conduct experiments on structures at reduced scales is to reduce the cost. Cost reductions come about from two areas: reduction of loading equipment and associated restraint frames, etc., and a reduction in cost of test-structure fabrication, preparation, and disposal after testing. The load-reduction factor is most dramatic since the concentrated load on a prototype is reduced in proportion to the square of the geometric scale factor of the model (a 100-kN prototype load is 0.25 kN on a $\frac{1}{20}$-scale model). This reduction is even more dramatic when a low-modulus material such as plastic is used in the model.

The major limitations of using structural models in a design environment are those of time and expense. In comparing physical models with analytical models, we find that the latter are normally less expensive and faster, and we cannot expect physical models to supplant or replace analytical modeling of structures when the latter procedure leads to acceptable definition of behavior of the prototype structure. Thus physical models are almost always confined to situations where the mathematical analysis is not adequate or not feasible. Another limiting factor is that changes in the prototype design resulting from the results of a model study may require a second model to check the design. Practical considerations therefore often dictate that the model will be used to verify a "nearly finalized" design.

The time involved in modeling is often subjected to further pressures because the decision to go to a physical study is often made at the last minute, after more conventional approaches are proven inadequate. An engineer who is accustomed to getting all answers by analytical means is naturally hesitant to admit that the analysis is insufficient and that a physical model is needed. Suitable efforts must be made to predict earlier in the design process that a test is needed. This would enable earlier planning and a smoother, less hectic approach to the model study.

Design applications of structural models have been outlined earlier in this chapter. Structural models are also widely employed in research programs in such applications as:

1. Development of experimental data for verification of the adequacy of proposed analytical methods.
2. Study of basic behavior of complex structural forms such as shells.
3. Parametric studies on member behavior. Much of our basic research on reinforced concrete flexural members has been done on large-scale models.
4. Behavior of complete structural systems subjected to complex loading histories, such as coupled shear walls and connecting beams.

Many of these areas of research modeling will be explored through examples in subsequent chapters. It is well recognized that research models play an invaluable role in improving our knowledge of structural behavior and thereby pave the way for new and improved design methods. This role will always be important in structural engineering because it is a discipline founded strongly on physical behavior of real systems made of ordinary materials of construction.

1.8 ACCURACY OF STRUCTURAL MODELS

The reliability of the results from a given physical modeling study is perhaps the single most important factor to the user of the modeling approach. This topic is explored in depth in Chapter 9, and only a few general comments will be given here to stimulate the reader into thinking about this important topic. Adequate definitions of reliability and accuracy are difficult to formulate. One obvious measure is the degree to which a model can duplicate the response of a prototype structure. The problem met in such a comparison is the inherent variability in the prototype itself, particularly if it is a reinforced concrete structure. Two supposedly identical reinforced concrete elements or structures will normally show differences, sometimes as high as 20% or more, and when one must compare a model to a single prototype, the difficulty in making a firm conclusion on accuracy becomes rather apparent. We need multiple prototypes and multiple models in order to treat the results statistically, but the expense of even a single test structure is usually high, and the availability of sufficient data for application of statistical tests of significance is severely limited.

The factors affecting model accuracy include model material properties, fabrication accuracy, loading techniques, measurement methods, and interpretation of results. Elastic models can be built to give extremely high correlation with detailed computer-based results—the only limitation is in the cost of properly fabricating and loading the model. Elastic models of reinforced concrete structures can only predict elastic response and thus will have high accuracy (errors on the order of less than 5 to 10%) for structures with minimal cracking, such as shells.

Carefully designed and tested strength models of reinforced concrete beams, frames, shells, and other structures normally have maximum errors on the order of less than 15% for prediction of postcracking displacements and ultimate load capacity of the structure, provided that bond between steel reinforcement and model concrete is not the governing factor in behavior. As will be shown later in the book, the results of a multistory frame modeling study (Case Study C in Section 1.10.3) showed excellent agreement between model and prototype beam-column joint behavior, including cracking, stiffness, failure mode, and ultimate load capacity. The frame models provided behavior predictions that cannot be supplied by currently available computer analysis programs.

A better perspective on the degree of reliability to be expected in any particular model testing program can be achieved only by careful study of a large number of individual cases. The material presented in Chapter 9 as well as in other chapters and in the cited references will provide the reader with much of the material needed in studying model reliability.

1.9 MODEL LABORATORIES

There are a number of outstanding laboratories in Europe that have developed excellent reputations in physical modeling. There are no similar commercial laboratories in North America, but there are excellent structural modeling facilities at a number of private and institutional laboratories, including Wiss Janney Elstner and Associates in Northbrook, Illinois, the Portland Cement Association in Skokie, Illinois, and educational/research laboratories at Cornell University, McGill University, the University of Texas, Drexel University, and elsewhere. The Boundary Layer Wind Tunnel at the University of Western Ontario and a similar facility at Colorado State University are widely used for both research and investigation of wind effects on actual structures. Dynamic tests on shake tables may be done at laboratories located at the University of California (in Berkeley and in Los Angeles), the University of Illinois, Stanford University, the University of Calgary, Drexel University, and elsewhere.

The many and diverse problems associated with structural modeling make it evident that high-quality structural modeling is best done by skilled engineers and technicians at established laboratories. This statement is not made to discourage newcomers; instead, it is merely a realistic comment on the difficulties of good experimental work. Considerable time and patience are required in establishing a diverse structural testing laboratory, whether it be at full scale or at the greatly reduced small model scale. One particularly important point to be made is that considerable amounts of trial-and-error approaches to materials development are now in the literature, and the careful individual can take full advantage of this material in setting up a laboratory. Thus advances in structural modeling since 1960, coupled with the simplicity and reliability of modern instrumentation, have at least partially eased many of the difficulties to be faced in beginning a new laboratory operation.

1.10 MODELING CASE STUDIES

Several structural modeling studies will be used in this text to help illustrate the many facets of modeling. Two design model studies (designated A and B) and two research projects (C and D) are utilized. General descriptions are given in this chapter, and the rest of the material is given in Chapter 10. Other modeling studies related to more specialized topics such as prestressed concrete pressure vessels, dynamic response, and shell stability will also be given subsequently.

1.10.1 Case Study A, TWA Hangar Structures

The Trans-World Airlines Maintenance Hangar Facility located in Kansas City, Missouri, was designed with the aid of a series of structural models. The total facility consists of four long-span shell structures, each located at the corner of a cruciform-shaped framed building. Two completed shells are shown in Fig. 1.2. Each shell spans about 320 ft (96 m) and consists of hyperbolic paraboloid (abbreviated here as *hypar*) surfaces spanning between hollow triangular edge members and an arch formed at the intersection of the two hypar surfaces. The shell varies in thickness from 3 to 6 in. (75 to 150 mm). Four separate model studies were used for the TWA hangar to resolve a series of design questions. They are referred to as models A1, A2, A3, and A4.

Figure 1.2   Completed Trans-World Airlines maintenance facilities at Kansas City International Airport. (Courtesy of Wiss, Janney, Elstner & Associates, Inc., Northbrook, Illinois.)

Model A1 was a $\frac{1}{300}$-scale shape model of the entire complex of four shells. It was placed in a wind tunnel and loaded to investigate the general character of wind pressures and wind flow over the site. Only the shape of the model is needed in this investigation; the model can be built of any convenient solid material such as softwood.

Model A2 (Fig. 1.3) was another wind tunnel shape model of a single hangar shell structure, built at a scale of $\frac{1}{100}$ and tested to obtain a more detailed and accurate representation of local wind pressures and flow patterns with doors opened, closed, etc.

Figure 1.3 Detailed wind model of hangar with doors closed. (Courtesy of Wiss, Janney, Elstner & Associates, Inc., Northbrook, Illinois.)

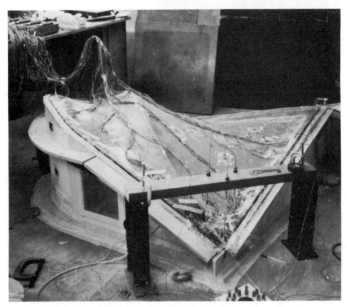

Figure 1.4 Elastic model enclosed in vacuum frame. (Courtesy of Wiss, Janney, Elstner & Associates, Inc., Northbrook, Illinois.)

The wind tunnel model results were utilized in the analytical design phase to help generate the proposed final design. The third model was then constructed to determine elastic behavior under many different load conditions. Model A3 (Fig. 1.4) was a $\frac{1}{50}$-scale structure made from plastic. It was loaded with discrete concentrated loads and with a vacuum loading. The evaluation of bending effects produced by the heavy maintenance equipment suspended from the shell was of particular importance in this model study. The stress resultants in the shell near the stiff edge members were also given considerable attention.

Model A4 (Fig. 1.5) was a realistic or strength model. Built at a scale of $\frac{1}{10}$, the model was designed to simulate the true behavior of a reinforced concrete shell structure loaded to failure. This strength model required duplication of both concrete and reinforcement at small scale. The load-deflection behavior of the

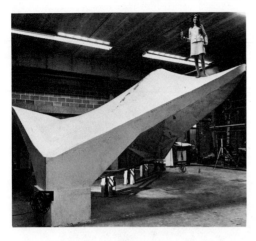

Figure 1.5 Completed micro-concrete model and hostess. (Courtesy of Wiss, Janney, Elstner & Associates, Inc., Northbrook, Illinois.)

shell was studied for several loadings before the model was loaded to failure to determine its failure mode and load capacity and factor of safety against collapse.

The TWA hangar models were done by the firm of Wiss Janney Elstner and Associates of Northbrook, Illinois.

1.10.2 Case Study B, Three Sisters Bridge

The proposed Three Sisters Bridge across the Potomac River in Washington, D.C., is shown in Fig. 1.6. The prototype structure was 110 ft wide and had three continuous spans of 440, 750, and 440 ft (134-228-134 m). The side spans were curved in plan, the deck contained a superelevation transition, and the profile was on a vertical curve. The bridge cross section was a four-cell box with vertical interior webs and curved exterior fascia. The deck was posttensioned longitudinally and transversely, the webs were posttensioned diagonally, the fascia was posttensioned vertically, the pier and abutment diaphragms were posttensioned transversely, and a portion of the soffit was posttensioned longitudinally. This design called for very deep and thin concrete webs, and it represented a substantial step beyond the normal state of the art of prestressed concrete bridge construction in North America. The modeling study was undertaken to determine the true factor of safety of the structure.

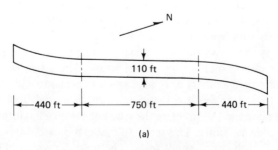

(a)

Figure 1.6 Three Sisters Bridge. (a) Plan.

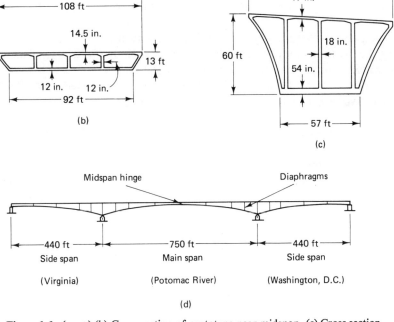

(b)

(c)

Midspan hinge   Diaphragms

|◄— 440 ft —►|◄——— 750 ft ———►|◄— 440 ft —►|
Side span   Main span   Side span

(Virginia)   (Potomac River)   (Washington, D.C.)

(d)

Figure 1.6 (*cont.*) (b) Cross section of prototype near midspan. (c) Cross section of prototype near pier. (d) Elevation.

The $\frac{1}{10}$-scale reinforced microconcrete model B1 of the Three Sisters Bridge is shown during a load test in Fig. 1.7. Since the prototype structure had a hinged condition at midspan, only half the structure had to be modeled, but even this partial model was nearly 80 ft (24 m) long.

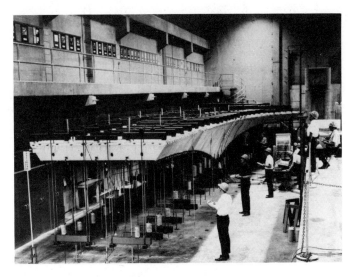

Figure 1.7 Three Sisters Bridge model under load. (Courtesy of Portland Cement Association.)

A plastic cross-sectional model B2 was also used in this study to help assess the load-transfer characteristics of the torsionally stiff bridge. This model represented a slice through the bridge cross section.

Models B1 and B2 were designed, constructed, and tested at the Portland Cement Association Laboratories in Skokie, Illinois.

1.10.3 Case Study C, Multistory Reinforced Concrete Frames

This research program on behavior of multistory reinforced concrete frames subjected to simulated seismic forces utilized a series of structural models at a scale of about $\frac{1}{10}$. Since the postcracking behavior of reinforced

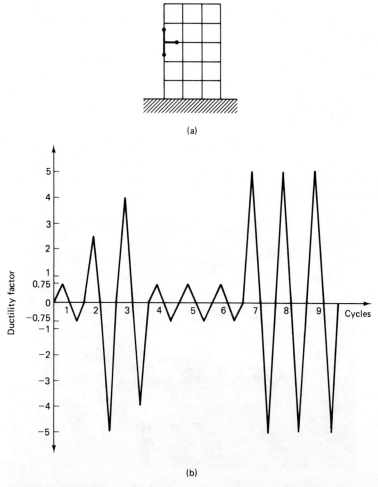

(a)

(b)

Figure 1.8 Beam-column joint area and loading history. (a) Frame and portion subjected to reversing loads. (b) Loading cycles for beam-column joint area.

concrete was being studied under rather severe reversed loading conditions, the model materials (concrete and reinforcing) had to be chosen with extreme care.

The reliability of the modeling approach was established with seven models (designated C1a, C1b, etc.) of the exterior beam-column region of a multi-story frame (Fig. 1.8a). Each $\frac{1}{10}$-scale model (Fig. 1.9) represented the exterior beam-column joint of a high-rise building between column inflection points and the beam inflection point for the condition of high lateral force on the frame.

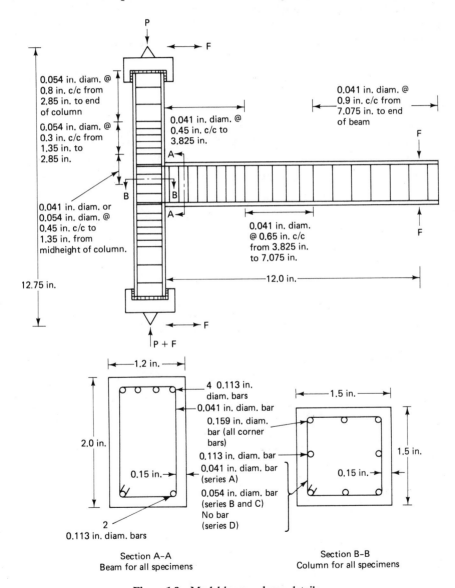

Figure 1.9 Model beam-column details.

Full-scale results were available from four prototype tests reported by Hanson and Conner (1967).

The models were loaded with combined axial force in the column and a reversing load at the end of the beam section. The load history for models and prototype is shown in Fig. 1.8b, where the ductility factor is defined as the ratio of the total rotation in the beam at the face of the column (measured over a length of a half-beam depth) to the rotation that exists at yield load.

The C1 models accurately duplicated prototype results. Thus the same modeling techniques were utilized in two C2 models of three-story, two-bay reinforced concrete frames subjected to combined constant gravity load and

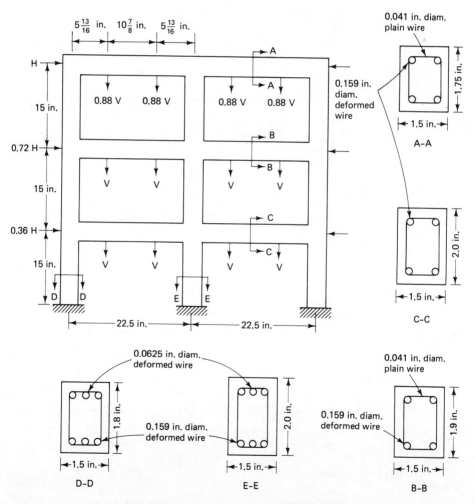

Figure 1.10   Multistory frame model details.

gradually increasing lateral forces. The $\frac{1}{10}$-scale frame design is shown in Fig. 1.10. The frames were designed to conform in all details to the requirements of the Structural Engineers Association of California Recommendations for Seismic Design. Frame C2a was loaded with 1.4 gravity design load and a monotonically increasing lateral load up to failure. Frame C2b had the same gravity load, but the lateral loads were fully reversing. Defining the lateral load factor (LLF) as the ratio of applied lateral load to the design lateral load, the loading history was:

Cycle: 1   2   3   4   5   6   7   8   9
LLF: 1   3   3   3   3   5   5   5.5   6

The frame modeling program was conducted in the Structural Models Laboratory at Cornell University, Ithaca, New York.

1.10.4 Case Study D, Precast Concrete Large-Panel Buildings

This research investigation, dealing with the behavior of precast concrete large-panel (LP) buildings under simulated progressive collapse conditions, utilized three types of ultimate strength models: joint details, planar components, and three-dimensional models.

Progressive collapse, usually initiated by an abnormal load such as a blast or impact, may be defined as a chain reaction of failures following damage to only a relatively small part of the structure. Large-panel (LP) buildings, which consist of large numbers of precast floor and wall panels, are basically bearing-wall structures arranged in box-type layouts (Fig. 1.11). They have inherent lines of weakness in their many cast-in-place joints and are susceptible to progressive collapse [Breen (1975)] if not properly tied together. A system of ties in the three orthogonal directions has been recommended by the Portland Cement Association to enable the LP structure to bridge over local damage. Direct models were used in the study to check the adequacy of the proposed tie system.

The first group of models, designated D1, consisted of thirty $\frac{1}{4}$-scale horizontal joints between floor slabs and bearing-wall panels (Fig. 1.12). These were relatively large-scale models, tested in triplicate, of 10 different prototype joints studied at the Portland Cement Association, Skokie, Illinois, and helped to establish the reliability of the modeling technique. The prestressed hollow-core slabs, the joint grout and drypacking material, and the bearing-wall panels were all properly simulated, and the joint was loaded to failure under increasing vertical load. Correlation between model and prototype joints in both the ultimate load and the mode of failure was very good to excellent.

A second group of three models (D2a, D2b, and D2c), at $\frac{3}{32}$-scale, consisted of planar wall and floor slab assemblies with one wall missing. This test

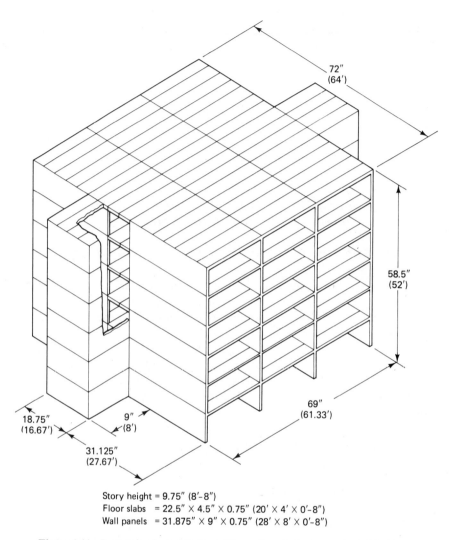

Story height = 9.75" (8'-8")
Floor slabs = 22.5" × 4.5" × 0.75" (20' × 4' × 0'-8")
Wall panels = 31.875" × 9" × 0.75" (28' × 8' × 0'-8")

Figure 1.11 Isometric view of 6-story, 3-bay, $\frac{3}{32}$-scale large panel model showing model (prototype) dimensions.

simulates the condition of a loss of a critical load-bearing member (Fig. 1.13). Its main objective is to see if the damaged structure can absorb the dead load plus one-half the live load used in design by means of a cantilever action of the wall panels above the damage. Two models were constructed with the same amount of transverse reinforcing in the horizontal joints, to demonstrate the repeatability of two identical models. A third contained one-half the amount of the transverse reinforcing to determine its effect on the behavior.

Overall, the model cantilever walls exhibited a nonlinear trend toward

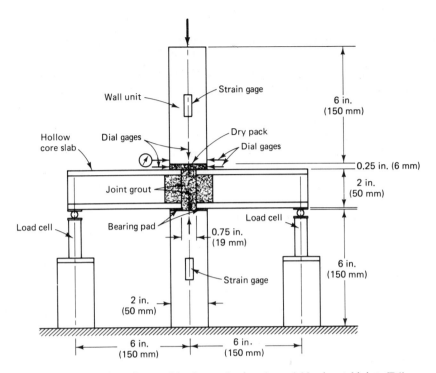

Figure 1.12 Dimensions and load setup for ¼-scale model horizontal joints (D1).

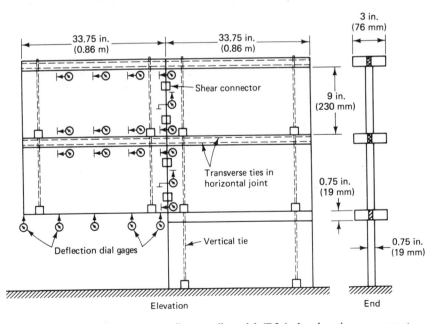

Figure 1.13 Two-story cantilever wall model (D2a) showing tie arrangement and instrumentation at $\frac{3}{32}$ scale.

increasing deflection with increasing load similar to that found in the prototype structure also tested at the Portland Cement Association. All models, irrespective of their failure mode, exhibited the same ability to undergo large deflections. The ultimate load was also found to be directly related to the amount of transverse joint reinforcement.

A final and three-dimensional model, designated D3, was constructed in the same way as a typical precast concrete LP building with American-type details (Fig. 1.11). The properties of the model material were chosen to be related as closely as possible to those used in typical prototype construction. Mild steel was used for the reinforcement of the wall panels and floor slabs and continuous-thread steel-threaded rod and stainless steel cables for the system of vertical and horizontal ties, respectively. The model was built with four removable wall panels at different critical locations. Thus, a damaged configuration could be studied by removal of a panel without hampering the loading operation. Assembly of the model from the various precast components followed the procedures used for full-scale structures of this type. The elevation views of the assembled model with the exterior removable steel-frame wall panels clearly visible are shown in Fig. 1.14. Load was applied through a system of whiffle trees and loading beams connected to a system of hydraulic tension jacks.

The results of the several tests of model D3 support the recommendations of the Portland Cement Association for the use of a three-dimensional tensile tie system to enable a large-panel structure to bridge over local damage. Specifically, crack patterns observed in the joints indicated that the tie system was

(a)

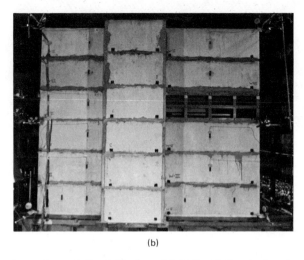

(b)

Figure 1.14 Two views of the finished model (D3). (a) South elevation. (b) North elevation.

successfully utilized to provide the necessary support and continuity for both the cantilever and floor suspension mechanisms to develop.

The large-panel model research was conducted in the Structural Models Laboratory, Drexel University, Philadelphia, Pennsylvania, under contract to the U.S. Department of Housing and Urban Development.

1.11 SUMMARY

Physical models of structures are used in education, in research and development, and in design. A model can be built and tested at a small fraction of the cost of a prototype (full-scale) structure because of the great reductions in loading magnitudes and in construction costs. Many different types of models are used, and the cost and time requirements vary widely for each type.

The applicability of models in design applications changes almost continuously as improved analytical capabilities permit the engineer to mathematically model increasingly complex structures, but the development of new combinations of structural forms and materials will most likely always be one step ahead of analysis.

Experimental stress analysis and the general principles of experimentation form integral parts of the modeling process. The execution of the many steps involved in a structural model study is an engineering project in itself; hence a considerable degree of "art" is involved along with the rather well-developed technology given in this book.

The Theory
of Structural Models

∿∿∿∿∿∿∿∿∿∿∿∿

2

2.1 INTRODUCTION

Any structural model must be designed, loaded, and interpreted according to a set of *similitude requirements* that relate the model to the prototype structure. These similitude requirements are based upon the theory of modeling, which can be derived from a dimensional analysis of the physical phenomena involved in the behavior of the structure. Accordingly, this chapter examines two distinct topics:

1. Dimensional analysis and similitude theory.
2. Actual similitude requirements for different types of structural models, aimed at studying their response under elastic and ultimate load conditions as well as under thermal loadings. Similitude requirements for dynamic loadings are given in Chapter 11.

It must be emphasized that a strictly formal application of modeling theory to a structural problem, without at least some understanding of the expected structural behavior, can lead to an inadequate and even incorrect modeling program. Similitude theory must be viewed simply as one aspect of the total modeling problem.

2.2 DIMENSIONS AND DIMENSIONAL HOMOGENEITY

The use of dimensions dates from early history when human beings first attempted to define and measure physical quantities. It was essential for these descriptions to have two general characteristics: qualitative and quantitative.

The *qualitative* characteristic enables physical phenomena to be expressed in certain *fundamental measures* of nature. The three general classes of physical problems, namely, mechanical (static and dynamic), thermodynamic, and electrical, are conveniently described qualitatively in terms of the following fundamental measures:

1. Length
2. Force (or mass)
3. Time
4. Temperature
5. Electric charge

These fundamental measures are commonly referred to as *dimensions*. Full chapters of books [such as those by Ipsen (1960) and Bridgman (1922)] are devoted to establishing and categorizing the fundamental measures.

Most structural modeling problems are mechanical; thus the measures of length, force, and time are most important in structural work. Thermal problems require the additional measure of temperature.

The *quantitative* characteristic is made up of both a *number* and a standard of *comparison*. The standard of comparison, also called the standard *unit*, was often established rather arbitrarily from traditional usage (such as the inch). Each of the fundamental measures, or dimensions, thus has its associated standard units in the several different unit systems in use today (U.S. Customary, SI, metric, etc.). Dimensions and units are such logical quantities that we now take them completely for granted. It is difficult to realize that the present state of physical description of occurrences did not always exist.

Keeping the above definitions of dimensions and units in mind, the theory of dimensions can be summarized in two essential facts:

1. Any mathematical description (i.e., equation) that describes some aspect of nature must be in a dimensionally homogeneous form. That is, the governing equation must be valid regardless of the choice of dimensional units in which the physical variables are measured. As an example, the equation for bending stress, $\sigma = Mc/I$, is correct regardless of whether force and length are measured in newtons and meters, pounds and inches, or other consistent units.
2. As a consequence of the fact that all governing equations must be dimensionally homogeneous, it can be shown that any equation of the form

$$F(X_1, X_2, \ldots, X_n) = 0 \qquad (2.1)$$

can be expressed in the form

$$G(\pi_1, \pi_2, \ldots, \pi_m) = 0 \tag{2.2}$$

where the π (Pi) terms are dimensionless products of the n physical variables $(X_1, X_2, \ldots, X_n)$, and $m = n - r$, where r is the number of fundamental dimensions that are involved in the physical variables.

This second fact, that any equation of the form $F(X_1, X_2, \ldots, X_n) = 0$ is expressible as $G(\pi_1, \pi_2, \ldots, \pi_m) = 0$, has two very important implications:

1. The form of a physical occurrence may be partially deduced by proper consideration of the dimensions of the n physical quantities X_i involved. The deductions are made by *dimensional analysis*, which is discussed at length in Section 2.3.
2. Physical systems that differ only in the magnitudes of the units used to measure the n quantities X_i, such as the quantities for a prototype structure and its reduced-scale model, will have identical functionals G. *Similitude requirements* for modeling result from forcing the Pi terms $(\pi_1, \pi_2, \ldots, \pi_m)$ to be equal in model and prototype, which is a necessary condition for the full functional relationships to be equal. Section 2.4 expands upon this concept.

Example 2.1

Hooke's law furnishes a good example of a dimensionally homogeneous relation. For a stretched bar of a perfectly elastic material, the equation which describes the stress-strain relation is

$$\sigma - E\epsilon = 0$$

It is important to notice that the fundamental measures that describe the physical quantities in this equation combine in a certain fashion. Thus:

$$k_1\frac{\text{force unit}}{(\text{length unit})^2} - \left[k_2\frac{\text{force unit}}{(\text{length unit})^2} \cdot k_3\frac{\text{length unit}}{\text{length unit}}\right] = 0$$

or

$$k_1\frac{\text{force unit}}{(\text{length unit})^2} - k_2 k_3\frac{\text{force unit}}{(\text{length unit})^2} = 0$$

and the equation holds for any system of force and length units. Such an equation is said to be *dimensionally homogeneous*.

Consider the special case of a bar made of steel with modulus of elasticity $E = 200 \text{ N/mm}^2$. Then the equation is

$$\sigma - 200\epsilon = 0$$

The equation is no longer dimensionally homogeneous. It is obviously correct only when σ and E are measured in terms of newtons and millimeters.

The following examples further verify the fact that physical phenomena are always expressible in dimensionally homogenous form.

Example 2.2

An algebraic equation describing the deflection of a simple prismatic elastic beam that is subjected to the triangularly distributed total load W (Fig. 2.1a) is

$$y(x) - \frac{Wx}{180EIl^2}(3x^4 - 10l^2x^2 + 7l^4) = 0$$

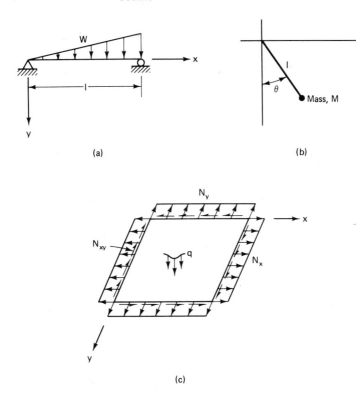

(a)   (b)

(c)

Figure 2.1

All quantities may be expressed in the fundamental measures of length (L) and force (F); thus the dimensions of the left side of the equation are

$$L - \frac{FL}{(F/L^2)L^4L^2}(L^4 - L^2L^2 + L^4)$$

Example 2.3

A general nonlinear ordinary differential equation describing the undamped motion of the simple pendulum (Fig. 2.1b) is

$$l\frac{d^2\theta}{dt^2} + g \sin \theta = 0$$

Dimensionally, the left side of the equation is

$$L\frac{1}{T^2} + \frac{L}{T^2}$$

Example 2.4

A partial differential equation describing small deflections of a flat plate subjected to lateral and edge loads (Fig. 2.1c) is

$$\frac{\partial^4 w}{\partial x^4} + 2\frac{\partial^4 w}{\partial x^2\,\partial y^2} + \frac{\partial^4 w}{\partial y^4} - \frac{12(1-\nu^2)}{Eh^3}\left(q + N_x\frac{\partial^2 w}{\partial x^2} + N_y\frac{\partial^2 w}{\partial y^2} + 2N_{xy}\frac{\partial^2 w}{\partial x\,\partial y}\right) = 0$$

and the dimensions of the left side become

$$\frac{L}{L^4} + \frac{L}{L^2 L^2} + \frac{L}{L^4} - \frac{1}{(F/L^2)L^3}\left(\frac{F}{L^2} + \frac{F}{L}\frac{L}{L^2} + \frac{F}{L}\frac{L}{L^2} + \frac{F}{L}\frac{L}{LL}\right)$$

Note in each example that the right-hand side of the governing equation is zero. The fact that zero is a constant and thus has no dimensions would seem to indicate a contradiction. Of course, there is no contradiction. Zero is a very special constant as far as dimensional considerations are concerned, and one could either transpose one quantity from the left to the right or nondimensionalize the equation.

2.3 DIMENSIONAL ANALYSIS

Dimensional analysis is of substantial benefit in any investigation of physical behavior because it permits the experimenter to combine the variables into convenient groupings (Pi terms) with a subsequent reduction of unknown quantities.

Example 2.5

As an introduction, consider the problem of experimentally determining the maximum stress at a section of a multispan girder subjected to a known distributed loading q per unit length. The analytical equation for stress has the form of Eq. (2.1). Assuming that one had a good insight into the nature of this problem, it would be apparent that the stress σ is a function of loading q and a representative length l, or

$$F(q, l, \sigma) = 0 \tag{2.3}$$

Buckingham (1914) proves that any equation of this form can be represented as a product of powers in the form below:

$$\sigma = Kq^a l^b \tag{2.4}$$

where K is a dimensionless parameter that may itself be a function of dimensionless groupings of the pertinent physical quantities, but is more often simply a constant.

In dimensional terms, Eq. (2.4) takes the form*

$$\frac{F}{L^2} \doteq \left(\frac{F}{L}\right)^a L^b$$

or

$$FL^{-2} \doteq F^a L^{-a+b} \tag{2.5}$$

---

*The symbol $\doteq$ will be used where dimensional equivalence is meant, and the symbol $=$ will be left for those equations where numerical equivalence is also maintained.

The dimensional homogeneity requirement forces equal dimensions on each side of the equation, or the exponents on each of the fundamental measures must be equal for the two sides of the equation. Thus we write exponential equalities for both F and L, or

$$F: \quad 1 = a$$
$$L: \quad -2 = -a + b \tag{2.6}$$

from which

$$a = 1$$
$$b = -1$$

and thus

$$\sigma = K\left(\frac{q}{l}\right) \tag{2.7a}$$

in which K can be determined experimentally. Note that dimensional analysis alone has shown σ to be a linear function of (q/l).

Equation (2.7a) can also be cast in the form of Eq. (2.2):

$$G\left(\frac{l\sigma}{q}\right) = 0 \tag{2.7b}$$

which tells us that the problem is formulated in terms of a single dimensionless ratio and an unknown functional G.

Recourse to a mathematical solution of the problem would lead to

$$\sigma = \frac{Mc}{I} = \frac{a_1 q l^2 (a_2 l)}{a_3 l^4} = \frac{a_1 a_2}{a_3} \frac{q}{l} \tag{2.8}$$

where a_1, a_2, and a_3 are constants that depend upon the geometry of the girder and taken together are the constant K in Eq. (2.7a). Of course, the dimensional analysis of the problem could not have determined the magnitude of the constant $a_1 a_2 / a_3$.

Example 2.6

Now suppose that the load in the previous example is of a dynamic nature and, therefore, has a variation with time. If one is interested in determining the maximum elastic displacement, then a logical set of pertinent physical parameters would include modulus of elasticity E, geometric length l, time t, loading q, density of steel ρ, and the acceleration of gravity g. The last two quantities are needed to account for inertia forces of the girder. The functional relationship implied by Eq. (2.1) becomes

$$F(u, E, l, t, q, \rho, g) = 0 \tag{2.9}$$

or writing the displacement u as a function of the other variables,

$$u = F'(E, l, t, q, \rho, g)$$

which can also be expressed in the continued product form

$$u = K E^a l^b t^c q^d \rho^e g^f \tag{2.10}$$

The dimensional equation for this expression is

$$L \doteq (FL^{-2})^a L^b T^c (FL^{-1})^d (FL^{-3})^e (LT^{-2})^f$$

Forcing this expression to be dimensionally homogeneous, we then have three equations for the three fundamental measures of force, length, and time, or

$$
\begin{aligned}
F&: \quad 0 = a + d + e \\
L&: \quad 1 = -2a + b - d - 3e + f \\
T&: \quad 0 = c - 2f
\end{aligned}
\tag{2.11}
$$

The three equations (2.11) have six unknowns and thus allow a threefold infinity of solutions. Selecting d, e, and f "arbitrarily,"* the equations can be solved for a, b, and c in terms of d, e, and f. The solutions are

$$
\begin{aligned}
c &= 2f \\
a &= -d - e \\
b &= -d + e - f + 1
\end{aligned}
$$

Then

$$
u = K(E^{-d-e}l^{-d+e-f+1}t^{2f}q^d p^e g^f)
$$

or

$$
\left(\frac{u}{l}\right) = K\left[\left(\frac{q}{El}\right)^d\left(\frac{pl}{E}\right)^e\left(\frac{t^2 g}{l}\right)^f\right]
\tag{2.12}
$$

Equation 2.12 can also be written in the alternate form

$$
G\left(\frac{u}{l}, \frac{q}{El}, \frac{pl}{E}, \frac{t^2 g}{l}\right) = 0
\tag{2.13}
$$

The important fact that can be drawn from this analysis is that the dimensionless ratio involving the desired displacement u can be expressed as a function of a set of dimensionless ratios, or

$$
\frac{u}{l} = \phi\left(\frac{q}{El}, \frac{pl}{E}, \frac{t^2 g}{E}\right)
\tag{2.14}
$$

Thus we have been able to use dimensional considerations alone to deduce a substantial insight into the problem and to eliminate a number of the exponents involved in the original formulation. The form of the functional relationship would have to be determined by experiments in which the several dimensionless ratios were systematically varied. Murphy (1950) prescribes tests that can be made to determine if the parameter K in Eq. (2.12) is merely a constant or if it is a function of the dimensionless parameters.

A fully equivalent formulation of this problem that involves one less variable can be obtained by realizing that the acceleration of gravity g enters the problem only indirectly in converting the specific weight to a mass density, and that if mass density M had been used instead of specific weight p, then g would not have been needed. This alternate formulation leads to

$$
G\left(\frac{u}{l}, \frac{q}{El}, \frac{Ml^2}{Et^2}\right) = 0
\tag{2.15}
$$

*Later in the chapter it will become evident that this choice is really not arbitrary; the three chosen exponents must embrace all three fundamental measures.

and

$$\left(\frac{u}{l}\right) = \phi\left(\frac{q}{El}, \frac{Ml^2}{Et^2}\right) \tag{2.16}$$

2.3.1 Buckingham's Pi Theorem

The discerning reader will have noticed that an analysis of dimensions has led in the first example from

$$\left.\begin{array}{c} F(q, l, \sigma) = 0 \\ \\ G\!\left(\dfrac{l\sigma}{q}\right) = 0 \end{array}\right\} \tag{2.17}$$

and in the second example from

$$\left.\begin{array}{c} F(u, E, l, t, q, M) = 0 \\ \\ G\!\left(\dfrac{u}{l}, \dfrac{q}{El}, \dfrac{Ml^2}{Et^2}\right) = 0 \end{array}\right\} \tag{2.18}$$

These two examples are illustrations of a general theorem stated by Buckingham (1914). This theorem states that *any dimensionally homogeneous equation involving certain physical quantities can be reduced to an equivalent equation involving a complete set of dimensionless products*. For the structural models engineer, this theorem states that the solution equation for some physical quantity of interest, i.e.,

$$F(X_1, X_2, \ldots, X_n) = 0 \tag{2.19}$$

can equivalently be expressed in the form

$$G(\pi_1, \pi_2, \ldots, \pi_m) = 0 \tag{2.20}$$

The Pi terms are dimensionless products of the physical quantities $X_1, X_2, \ldots, X_n$. A complete set of dimensionless products are the $m = n - r$ independent products that can be formed from the physical quantities $X_1, X_2, \ldots, X_n$. In the previous two illustrations it turned out that three and six physical variables reduced to one and three dimensionless products, respectively. Generally, it can be stated that the number of dimensionless products (m) is equal to the difference between the number of physical variables (n) and the number of fundamental measures (r) that are involved. The first example was a static mechanical problem; the fundamental dimensions were force and length, or $m = n - r = 3 - 2 = 1$. The second problem was a dynamic mechanical problem; the fundamental dimensions were force, length, and time, and $m = 6 - 3 = 3$.

Buckingham's Pi theorem occupies a very important place in the theory of dimensional analysis. Before proceeding to the applications of dimensional analysis, let us consider in some detail the procedures used in obtaining the dimensionless products that go into Buckingham's Pi equations.

2.3.2 Dimensional Independence and Formation of Pi Terms

Table 2.1 presents a sample of physical quantities that might be involved in structural problems, and the dimensional measures required to describe them. Some structural problems will involve temperature and heat considerations; the extension of Table 2.1 to cover these quantities is done in Section 2.5.3.

TABLE 2.1   Typical List of Physical Quantities

	Quantity	Units
l	Length	L
Q	Force	F
M	Mass	$FL^{-1}T^2$
σ	Stress	FL^{-2}
ϵ	Strain	—
a	Acceleration	LT^{-2}
δ	Displacement	L
ν	Poisson's ratio	—
E	Modulus of elasticity	FL^{-2}

An examination of the dimensional measures in Table 2.1 makes it obvious that the dimensions that are required to describe any physical quantity occur in the form of a single product. The fact that all such dimensional descriptions occur in the product form is a direct result of the nature of dimensions and the basic foundations upon which scientific measurements were first established.

Now, from any set of physical quantities, such as that listed in Table 2.1, it is possible *dimensionally* to form certain quantities in the set by combining others in the form of products. Thus, the dimensions of length (l) divided by the dimensions of time (t) yield the dimensions of velocity (v). Also, the dimensions of stress (σ) divided by the product of the dimensions of acceleration (a) and of time (t) squared, yield the dimensions of specific weight (ρ). Consequently, it is seen that the products vt/l and $\rho at^2/\sigma$ are dimensionless. In any set of physical quantities there is a limited number of quantities that cannot themselves be combined with other quantities in the set to yield a dimensionless product. The quantities involved in the limited set are said to be dimensionally independent, while the other quantities are dimensionally dependent upon the special limited set. The main question met in applying the Buckingham Pi theorem pertains to the formation of appropriate Pi terms. The following simple points are the only guidelines needed in the formation process:

1. All variables must be included.
2. The m terms must be independent.
3. In general, there is no unique set of Pi terms for a given problem; alternate formulations are possible either by forming the Pi terms in several different ways or by suitable transformations of one set of Pi terms. Thus it is not possible to state that a set of complete, independent Pi terms is either "right" or "wrong" for a given problem.

The best method for arriving at the groupings of Pi terms is open to personal preference; there are a number of rather formal methods which involve setting up the appropriate dimensional equations. One less formal approach involves the following steps:

1. Choose r variables that embrace the r dimensions (fundamental measures) required in expressing all variables of the problem, and that are dimensionally independent. This means that if a problem involves the dimensions of force F, length L, and time T, then the three variables chosen must collectively have dimensions which include F, L, and T, but no two variables can have the same dimensions. Variables that are in themselves dimensionless (strain, Poisson's ratio, angles) cannot be chosen in the set of r variables.
2. Form the m Pi terms by taking the remaining $(n - r)$ variables and grouping them with the r variables in such a fashion that all groups are dimensionless. This procedure will guarantee a set of independent, dimensionless terms. It should be noted that the r variables chosen in step (1) above will in general appear more than once in the total set of Pi terms while the remaining $(n - r)$ variables will each appear only once.

Example 2.7
How does the elastic stiffness of a rectangular cantilever beam depend on its properties? Assume the beam is loaded at its end, as shown in Fig. 2.2, and that stiffness is defined as the force per unit displacement measured at the location of the load. With no knowledge of the actual relationship between beam properties and stiffness, one might choose the stiffness S to be a function of beam length l, depth h, width w, elastic modulus E, and Poisson's ratio v.

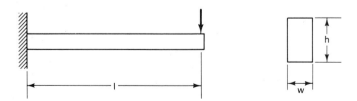

Figure 2.2   Beam stiffness example.

The variables and their dimensions are conveniently represented in tabular form:

	S	l	w	h	E	v
F	1	0	0	0	1	0
L	−1	1	1	1	−2	0

All six variables are expressible in two dimensions: force and length. Thus there will be $6 - 2 = 4$ Pi terms. Since the desired relationship is for the stiffness S, it is best not to include it in the two variables that may appear more than once. A convenient choice for the two multiple variables is span length l and modulus of elasticity E. These variables embrace both dimensions (F and L) and do not have identical dimensions.

The Pi terms may be formed by inspection by appropriate grouping of l and E with the other variables:

$$\pi_1 = \frac{S}{El}, \qquad \pi_2 = \frac{w}{l}$$
$$\pi_3 = \frac{h}{l}, \qquad \pi_4 = v \tag{2.21}$$

The behavior of the beam can then be summarized as

$$G\left(\frac{S}{El}, \frac{w}{l}, \frac{h}{l}, v\right) = 0$$

or, alternately,

$$\frac{S}{El} = G'\left(\frac{w}{l}, \frac{h}{l}, v\right) \tag{2.22}$$

Example 2.8

Form a complete, independent set of dimensionless Pi terms from the quantities listed in Table 2.1.

The quantities and their dimensions are listed in array form as:

	l	Q	M	σ	ϵ	a	δ	v	E
F	0	1	1	1	0	0	0	0	1
L	1	0	−1	−2	0	1	1	0	−2
T	0	0	2	0	0	−2	0	0	0

In selecting the three independent quantities that will appear at least once in the $9 - 3 = 6$ Pi terms, it is evident that either mass M or acceleration a must be included, as these are the only two quantities which possess the dimension of time. The three quantities chosen here are length l, modulus E, and acceleration a. The six Pi terms are then formed by inspection:

$$\pi_1 = \frac{Q}{EI^2}, \qquad \pi_2 = \frac{Ma}{EI^2}$$

$$\pi_3 = \frac{\sigma}{E}, \qquad \pi_4 = \epsilon \qquad\qquad (2.23)$$

$$\pi_5 = \frac{\delta}{l}, \qquad \pi_6 = \nu$$

A more rigorous treatment of dimensional independence is given in Appendix A.

2.3.3 Uses of Dimensional Analysis

Dimensional analysis can be used by the engineer in two separate ways. First, it can be useful in deducing, from experimental observations, certain theoretical results regarding the behavior of a physical phenomenon. Such a situation could arise if one knew the relevant physical variables that affected the state of some other physical variable but did not know the mathematical relationship that connected these variables. For example, if only three or four physical quantities are involved, dimensional analysis may reveal the solution to within some constant value or some unknown function of one or two variables. Rather simple experiments can then be performed to determine the constant value or the functional relationship. Of course, if there are 10 or 20 physical variables to begin with, there will still be so many dimensionless products remaining as to make experimental analysis difficult or impractical.

The second use in the area of structural design work has been stated very succinctly by Bridgman (1922):

> There are in engineering practice a large number of problems so complicated that the exact solution is not obtainable. Under these conditions dimensional analysis enables us to obtain certain information about the form of the result which could be obtained in practice only by experiments with an impossibly wide variation of the arguments of the unknown function. In order to apply dimensional analysis we merely have to know what kind of a physical system it is that we are dealing with, and what the variables are which enter into the equation; we do not even have to write the equations down explicitly, much less solve them.
>
> Suppose that the variables of the problem are denoted by X_1, X_2, etc., and that the dimensionless products are found, and that the result is thrown into the form
>
> $$X_1 = X_2^{a_1} X_3^{b_1} \ldots \phi(X_2^{a_2} X_3^{b_2} \ldots, X_2^{a_3} X_3^{b_3}, \ldots) \qquad (2.24)$$
>
> where the arguments of the function and the factor outside embrace all the dimensionless products, so that the result as shown is general. Now in passing from one physical system to another, the arbitrary function ϕ will in general change in an unknown way, so that little if any useful information could be obtained by indiscriminate model experiments. But if the models are chosen in such a restricted way that all the arguments of the unknown function have the

same value for the model as for the full scale example, then the only variable in passing from model to full scale is in the factors outside the functional sign, and the manner of variation of these factors is known from the dimensional analysis.

Stated in this way it would appear that the structural design engineer's model problems are solved, and, in fact, they are if the dimensionless products that are the arguments of the unknown functions have the same value for the model as for the prototype. As will be seen later, technological problems may make it impossible to strictly satisfy this condition, particularly for models that are intended to reproduce the inelastic response of reinforced concrete structures.

2.3.4 Additional Considerations in Using Dimensional Analysis

It is essential that the experimenter who wishes to use dimensional analysis have sufficient insight into the problem to choose the proper physical quantities in the formulation. The multispan beam problem of Example 2.5 is a convenient example for discussion here. To the person who does not have a keen insight into beam bending, it might not be immediately apparent that the modulus of elasticity E does not enter into the stress problem. If E was added to the problem, then the governing relationship would be

$$F(\sigma, q, l, E) = 0 \tag{2.25}$$

which is expressible in the two fundamental measures F and L. By following the procedures set down in Section 2.3.2, it is seen that q and l, q and σ, q and E, l and σ, and l and E are all dimensionally independent. Taking q and l as the dimensionally independent physical quantities, the complete set of dimensionless products is

$$\pi_1 = \frac{\sigma l}{q}, \qquad \pi_2 = \frac{El}{q} \tag{2.26}$$

According to Buckingham's Pi theorem, Eq. (2.26) can be reduced to

$$G\left(\frac{\sigma l}{q}, \frac{El}{q}\right) = 0 \tag{2.27}$$

or

$$\sigma = \left(\frac{q}{l}\right)\phi_1\left(\frac{El}{q}\right) \tag{2.28}$$

There was nothing unique about the selection of q and l as the independent physical quantities. It is seen that q and E could just as easily have been chosen, and a procedure identical to the above would have led to

$$\sigma = E\phi_2\left(\frac{El}{q}\right) \tag{2.29}$$

Equation 2.27 should be compared with Eq. (2.7a). The unnecessary inclusion of E as a relevant physical variable does not make Eq. (2.27) incorrect.

However, it is needlessly complicated, and it would require additional experimental work to find out that the beam stress was in fact independent of E. The experimental design engineer thus is faced with the following dilemma:

1. If a relevant physical quantity is omitted from a dimensional analysis of a problem, the investigation either meets an impasse or leads to erroneous results. This statement is not rigorously proved here, but reason alone indicates its validity.

2. If not only the relevant physical quantities but also some irrelevant ones are included, the dimensional analysis will lead to a result that will make the experimental investigation much more difficult than need be. In fact, it may eliminate a model study as a practical means of obtaining the desired information.

In this light Langhaar (1951) has stated,

Frequently the question arises: How do we know that a certain variable affects a phenomenon? To answer this question, one must understand enough about the problem to explain why and how the variable influences the phenomenon. Before one undertakes the dimensional analysis of a problem, he should try to form a theory of the mechanism of the phenomenon. Even a crude theory usually discloses the actions of the more important variables. If the differential equations that govern the phenomenon are available, they show directly which variables are significant.

If they are not available, then the engineer must have some other insight into the phenomenon, because it is clear that dimensional analysis can be of no use unless one can identify the relevant physical variables.

Example 2.9

Free transverse vibrations of a flat elastic plate are known to be governed by the partial differential equation

$$\frac{\partial^4 w}{\partial x^4} + 2\frac{\partial^4 w}{\partial x^2 \, \partial y^2} + \frac{\partial^4 w}{\partial y^4} + \frac{Mh}{Eh^3/12(1-v^2)} \frac{\partial^2 w}{\partial t^2} = 0 \qquad (2.30)$$

subject to certain prescribed boundary and initial conditions, with the pertinent variables being:

$$x, y = \text{coordinates of points on the surface of the plate}$$
$$h = \text{plate thickness}$$
$$E = \text{modulus of elasticity}$$
$$v = \text{Poisson's ratio}$$
$$t = \text{time}$$
$$M = \text{mass density}$$
$$w = \text{out-of-plane displacement of plate middle surface}$$

It is desired to make a model study of a large irregularly shaped plate in order to determine the lowest natural frequency of vibration. What are the relevant physical variables? In the light of a dimensional analysis of the problem, how should a model study be conducted, and is it likely that useful results will be obtained from a model investigation?

Assume that the solution equation for the natural frequency f is of the form

$$F(f, l, M, E, v, h) = 0 \tag{2.31}$$

where l is a characteristic plate dimension for either plan or thickness dimensions. Taking l, M, and E to be dimensionally independent, Eq. (2.31) can be reduced to

$$G\left(\frac{flM^{1/2}}{E^{1/2}}, v\right) = 0 \tag{2.32}$$

or in the solved form

$$f = \frac{1}{l}\sqrt{\frac{E}{M}}\,\phi(v) \tag{2.33}$$

Equation 2.32, which is the most that can be obtained from a dimensional analysis, can be compared with the mathematical solution of the simply supported rectangular plate. Such a solution yields for the fundamental frequencies

$$f = \left(\frac{k^2}{a^2} + \frac{j^2}{b^2}\right)\pi\sqrt{\frac{Eh^2}{12M(1 - v^2)}}$$

where a and b are the side lengths and m and n are integers indicating the number of half sine waves in the deflected shape of the plate. For a square plate with j and k equal to 1,

$$f = \frac{h}{a^2}\sqrt{\frac{E}{M}}\sqrt{\frac{\pi^2}{3(1 - v^2)}}$$

Returning to the model problem, Eq. (2.33) can be written once for the prototype and once for the model. Dividing the prototype equation by the model equation, one obtains

$$\frac{f_{\text{prototype}}}{f_{\text{model}}} = \frac{(1/l_p)\sqrt{(E_p/M_p)}}{(1/l_m)\sqrt{(E_m/M_m)}}\frac{\phi(v)_p}{\phi(v)_m}$$

If Poisson's ratio in the model and the prototype are equal, then the magnitude of the function is an identical constant for each. In that case

$$f_{\text{prototype}} = f_{\text{model}}\frac{l_m}{l_p}\sqrt{\frac{E_p M_m}{E_m M_p}} \tag{2.34}$$

In fact, it might be reasoned that Poisson's ratio does not greatly influence this physical phenomenon, and the experimenter might then allow the model material to have a Poisson's ratio different from that of the prototype. The experimenter must be aware, of course, that the final result would be subject to the error associated with the departure from the dictates of the dimensional analysis.

The boundary conditions also must be modeled. Clearly, if the edges of prototype plate are rigidly fixed or simply supported, then the edges of the model plate must also be rigidly fixed or simply supported. Similarly, elastically restrained edges in the prototype must be modeled. As a matter of interest, it should be expected that the technological problems associated with providing similar boundary conditions in this problem would be of considerably more concern than the effect of Poisson's ratio.

2.4 STRUCTURAL MODELS

It is a relatively simple matter to apply dimensional analysis principles to the structural model. As the discussion is developed, three types of structural models will be described. These are:

1. The *true model*, which maintains *complete similarity*. Any model that satisfies each and every stipulation set forth by a proper dimensional analysis would be said to have complete similarity.

2. The *adequate model*, which maintains *"first-order" similarity*. If an engineer has a special insight into a problem, then it may be possible to reason that some of the stipulations set forth by proper dimensional analysis are of "second-order" importance. For example, in rigid-frame problems it is known that axial and shearing forces are of second-order importance relative to bending moments insofar as deformations are concerned. Thus it may be adequate to model the moment of inertia but not the cross-sectional areas of members. Thus, any model which satisfies each and every first-order stipulation which is set forth by a proper dimensional analysis but which may not satisfy certain second-order stipulations would be said to have first-order similarity.

3. The *distorted model*, which *fails to satisfy* one or more of the first-order stipulations as set forth by proper dimensional analysis.

Of course, complete similarity is desirable in all structural models, but usually the economic and technological conditions preclude a model study that maintains complete similarity with the prototype. By neglecting certain second-order effects, it is usually possible to make an *adequate model* study to obtain results to accurately predict the behavior of a prototype structure.

2.4.1 Models with Complete Similarity

It has been seen from Buckingham's theorem that the mathematical formulation of any physical phenomenon can be reduced to an equation involving a complete set of dimensionless products.

$$\pi_1 = \phi(\pi_2, \pi_3, \ldots, \pi_n) \tag{2.35}$$

If Eq. (2.35) is written once for the prototype and once for the model, the following quotient can be formed:

$$\frac{\pi_{1p}}{\pi_{1m}} = \frac{\phi(\pi_{2p}, \pi_{3p}, \ldots, \pi_{np})}{\phi(\pi_{2m}, \pi_{3m}, \ldots, \pi_{nm})} \tag{2.36}$$

where π_{1m} refers to π_1 in the model and π_{1p} refers to π_1 in the prototype, etc. Complete similarity is defined to be that condition in which all of the dimensionless products are the same in both model and prototype. When complete similarity is maintained,

$$\pi_{2m} = \pi_{2p}$$

$$\pi_{3m} = \pi_{3p}$$
$$\dots \tag{2.37}$$
$$\pi_{nm} = \pi_{np}$$

so that Eq. (2.36) may be written

$$\frac{\pi_{1p}}{\pi_{1m}} = \frac{\phi(\pi_{2p}, \pi_{3p}, \dots, \pi_{np})}{\phi(\pi_{2m}, \pi_{3m}, \dots, \pi_{nm})} = 1 \tag{2.38}$$

or

$$\pi_{1p} = \pi_{1m} \tag{2.39}$$

Equations (2.37) and (2.39) are the basis for the model method. The relations between model and prototype quantities as implied in Eq. (2.37) are called the *design and operating conditions*; the single Eq. (2.39) is the *prediction equation* for the dependent variable of the problem. Thus, in Example 2.9,

$$\pi_1 = \frac{flM^{1/2}}{E^{1/2}}$$

$$\pi_2 = v$$

Equation (2.39) took the form

$$f_p l_p \sqrt{\frac{M_p}{E_p}} = f_m l_m \sqrt{\frac{M_m}{E_m}}$$

which yielded Eq. (2.34).

The similitude relations corresponding to the Pi terms of Example 2.8 can be derived by equating $\pi_m = \pi_p$ for each of the six terms and solving for the scale factor $s_i = i_p/i_m$. s_i is defined as the scale factor for the quantity i, and the subscripts p and m denote prototype and model, respectively.

There are three dimensionally independent quantities (l, E, and a) appearing in the six Pi terms of Example 2.8. Accordingly, scale factors may be arbitrarily chosen for only three quantities. It is logical to choose S_l, S_E, and S_a. The remaining scale factors are then:

$$S_Q = S_l^2 S_E, \qquad S_\varepsilon = 1$$

$$S_M = \frac{S_l^2 S_E}{S_a}, \qquad S_\delta = S_l \tag{2.40}$$

$$S_\sigma = S_E, \qquad S_v = 1$$

It should be noted that model and prototype strains must be identical for true models. Stresses will be identical *only* if $S_E = 1$. This results automatically when the same material is used in model and prototype.

The formulation of scaling relations for any true modeling problem can be established quite easily by the simple translation of π terms into required scale factors. The reader is encouraged to select a complete set of quantities for some more complicated problem, derive a set of π terms, establish the similitude relations, and examine them carefully.

2.4.2 Technological Difficulties Associated with Complete Similarity

Several types of departures from complete similarity can occur, including:

1. Accidental overlooking of a pertinent variable
2. Deliberate violation of a similitude requirement that is considered to be not critical, such as using a model material with Poisson's ratio different from that of the prototype material
3. Necessary deviations from true modeling, such as using a discrete load system to replace a continuous load

Lack of complete similarity in the model and prototype dimensionless ratios means that the ratio ϕ_p/ϕ_m in Eq. (2.38) is no longer unity. While it is not feasible in most instances to make an evaluation of the true value of the ratio, it is essential to realize that this departure from true similarity affects the model result to the degree that the ratio departs from unity. Lack of similarity, either known or unknown, often leads to differences that are misleadingly called "size effects." Size effects as such do not exist if complete similarity is maintained in geometry, materials properties, and loading (Chapters 3 to 7 deal with the latter two aspects in depth).

Example 2.10   Gravity Load Simulation
This example illustrates a departure from similitude of the second type mentioned above. Corresponding to the fact that the maximum number of dimensionally independent quantities equals the number of fundamental dimensions involved in the quantities, the model engineer can select only a restricted number of model quantities without regard for the prototype. Thus, in static and dynamic problems only two and three model quantities, respectively, can be arbitrarily selected. Practical considerations generally demand that the model geometric scale and certain model material properties be selected to be compatible with the available equipment and materials.

If the problem involves only static response and the dead weight of the structure exerts an important influence, then one of the dimensionless products in the problem will be $\rho l/E$, where ρ is the specific weight of the material, E is the modulus of elasticity of the material (or some equivalent quantity which represents the stress-strain characteristics of the material), and l is a representative length. Now only two model quantities can be selected arbitrarily, and thereafter the preceding dimensionless product must have the same magnitude in model and in prototype. Thus

$$\left(\frac{\rho l}{E}\right)_{\text{model}} = \left(\frac{\rho l}{E}\right)_{\text{prototype}}$$

or

$$\rho_{\text{model}} = \rho_{\text{prototype}} \frac{l_p E_m}{l_m E_p} \tag{2.41}$$

As a typical illustration, consider a reinforced concrete prototype structure simulated with a $\frac{1}{10}$-scale polyvinylchloride plastic model. Then

$$\rho_{\text{model}} = 90 \text{ lb/ft}^3 \text{ (1440 kg/m}^3)$$

but the dead-load similitude requirement says that the density of the model material should be

$$150(10)\left(\frac{450,000}{3,000,000}\right) = 225 \text{ lb/ft}^3 \text{ (3600 kg/m}^3)$$

Thus Eq. (2.41) is not satisfied. A departure from complete similarity is necessary. If the structure is "slender," the difference between the dead weight dictated by similitude considerations and that furnished by the model material may be added at the surface of the structure. On the other hand, a "massive" structure may require that the additional mass be dispersed throughout the volume of the structure. Thus, many of the model studies of large concrete dams have required such an approach.

Three model quantities can be arbitrarily selected in dynamic problems. This fact should not lead one to believe that the specific weight could be that additional quantity. Length, stress, and specific weight are not dimensionally independent quantities. Thus, dead-weight stresses constitute a static phenomenon, and therefore arti-

Figure 2.3 Model for continuous bridge under dynamic loading. (Courtesy of ISMES, Bergamo, Italy.)

ficial means must ordinarily be used to provide the simulation. Of course, if the structural response that is of interest does not depend significantly upon the dead weight, then the most appropriate procedure may be simply to neglect the discrepancy completely.

Figure 2.3 shows how additional artificial mass was attached to a bridge vibration model to simulate prototype dead-weight effects.

2.4.2.1 Other Types of Distortion

There are a number of distortions that are met frequently in modeling, including:

1. Discrete loading in place of distributed loading
2. Lack of similitude in bond strength of reinforcing wires for reinforced concrete models
3. Model concrete with failure criteria different from that of prototype concrete
4. Time scaling in dynamic loadings
5. Violation of $v_m = v_p$ when using elastic models for concrete structures

Several of these situations will be discussed subsequently.

2.4.3 Models with First-Order Similarity

Some of the practical difficulties of obtaining complete similarity in the model were discussed in the preceding section. If the experimental method is to be used for both analysis and design, it may be necessary to relax the restriction that the ratio of the ϕ functions in Eq. (2.38) be identically 1. First-order similarity is herein defined to be that degree of model-to-prototype similarity such that the engineer is willing to neglect the difference between the actual value of the ratio ϕ_p/ϕ_m and unity, or to neglect an error introduced by incomplete similarity of quantities outside the ϕ function. In other words first-order similarity is when ϕ_p/ϕ_m is approximately equal to 1, or such that, for example, the behavioral difference between a uniformly applied load and a discrete pattern of loads can be neglected.

In line with the examples that were discussed in the preceding section, the bridge vibration model that incorporated a series of additional masses along the length of the bridge and a thin-shell buckling model that is loaded by a grid of concentrated loads instead of a distributed load could be considered to be models with first-order similarity.

In another vein, there are certain types of structural problems that are special situations. The nature of these problems can best be understood by considering first the nature of all structural response. In this regard, the deformations in any structure are dependent upon: (1) the force and displacement boundary and initial conditions imposed on the structure, (2) the geometry of the structure, and (3) the materials of which the structure is composed. In the determination of these deformations (and hence stresses, etc.) from a mathematical point of view, certain special types of structural behavior have been categorized. Thus, from an analytical point of view, any structural response is said to arise from either (1) axial, (2) shearing, (3) bending, or (4) torsional deformation or *any* combination of the four. These categories have been invented in an attempt to surmount certain of the difficulties associated with analytical stress analysis. Now, the greatest asset of the experimental method of stress analysis lies in the fact that the model, like the prototype, is unaware of these four categories, and hence simply yields the complete result. Nevertheless, certain special problems may arise in which our knowledge of these four types of response can be usefully applied in a model problem.

For example, consider a planar rigid-frame structure that is very highly statically indeterminate. The component members, even with a variable moment of inertia, are all relatively long in comparison with their cross-sectional configuration. If the external loads are all applied in the plane of the frame and the connections of the members do not involve any lateral eccentricities, then it is known that the resulting deformations are primarily the result of bending. Axial and shearing deformations are of second order, and torsional deformations are nonexistent. If the problem is to determine the bending moment at a certain cross section and the behavior is known to be elastic, it would suffice to maintain similarity with respect to the moment of inertia of any particular cross section rather than to maintain similarity with respect to shape as well. At the price of not being able to properly account for axial and shear force effects, the fabrication problems associated with the model study may have been reduced significantly. However, suitable care must be exercised when taking such liberties. Thus, for example, the fact that bending moments are similar does not ensure that stresses are similar. Further, one might ask whether model and prototype bending moments would be similar if the material stress-strain characteristics were nonlinear.

2.4.4 Distorted Models

The question now arises whether or not it is possible that deviations from complete similarity can be permitted that will lead to gross dissimilarities between model and prototype behavior. The only answer that can be given is that any deviations can be permitted as long as it is somehow possible to determine the influence of such deviations. It has been said that complete similarity demands that the ϕ_p/ϕ_m ratio in Eq. (2.38) be identically 1. When certain second-order deviations from complete similarity are permitted, the ratio might only be approximately equal to 1; if first-order deviations from complete similarity are permitted, the resulting ϕ_p/ϕ_m ratio will, in general, be unknown. Accordingly, the model-to-prototype extrapolation equation given by Eq. (2.39) would no longer be correct. Models with such first-order deviations are said to be *distorted*.

Distortion can arise through a dissimilarity in boundary and initial conditions, geometry, or material properties. In structural problems a distortion in boundary and initial conditions or in geometry is seldom necessary or, for that matter, practically advantageous. Murphy (1950) discusses structural models of eccentrically loaded compression blocks that involve geometric distortion, but in these cases the analytical solution is available to enable one to evaluate the ϕ_p/ϕ_m ratio. Of course, if one knows this, the physical model may be less useful than the analytical method.

Geometric distortion has been used in hydraulic models of tidal basins. In these cases the effect of surface tension on the water, which can be neglected in real life, may become significant if the geometric scale is faithfully reduced in the model. If the height dimension of the model is distorted, however, the water is deeper and surface tension again can be neglected. Means exist to account for this distortion.

Of greater significance to the structural model engineer is the possibility of permitting distortion in the reproduction of the prototype material stress-strain characteristics. For complete similarity with regard to a uniaxial stress state, the model material will have to behave according to Fig. 2.4. Suppose that such a model material is simply not available but materials are available which follow the stress-strain laws shown in Fig. 2.5. Certainly, if a model

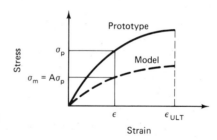

Figure 2.4 Completely similar model material.

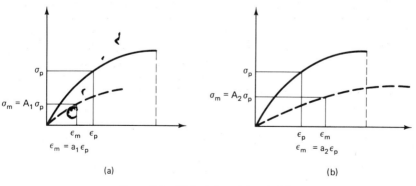

Figure 2.5 Distorted model materials.

material conforming to Fig. 2.5b were to be used, the strains in the model would be larger than those in the prototype. As a result the model displacements, which are a function of strain times length, would not be similar to the displacements of the prototype. If the structural behavior is dependent upon the displacements (e.g., the stresses in a beam-column or the critical buckling pressure in a thin-shell roof), such a distortion cannot be permitted. However, if the displacements are sufficiently small so as not to disturb the conditions of equilibrium, then such a distortion can be permitted. The corresponding strains, displacements, velocities, accelerations, etc., which would be set up in the model would not be similar to those of the prototype, but would be known to err by the factor that indicates the magnitude of the strain distortion (a_1 and a_2 in Fig. 2.5). Beaujoint (1960) gives an extensive discussion of strain distortion.

Another type of material property distortion may arise when the Poisson's ratio of the model material does not equal that of prototype material. If the structural behavior is known to be characterized by plane stress (e.g., the bending of a common beam), a Poisson's ratio discrepancy may very well distort the model strains but not the model stresses, reactions, bending moments, etc. In such cases proper interpretation of the model results can preserve the integrity of the model study. In other instances, analytical or experimental results from a related problem may substantiate that Poisson's ratio v is not of predominant importance. Many of our analytical expressions contain terms like $\sqrt{1 - v^2}$. Such knowledge leads one to allow certain models to be fabricated from a material having a dissimilar Poisson's ratio with only a small percentage of error. On the other hand, if the structural behavior is not well understood, then a Poisson's ratio distortion can lead to incorrect results.

2.5 SIMILITUDE REQUIREMENTS

Similitude requirements for static, elastic modeling are summarized in Table 2.2. This is followed by requirements for static inelastic modeling of reinforced and prestressed concrete structures. Special cases of thermal structural modeling are treated, and general remarks about dynamic modeling are presented. A more detailed treatment of dynamic modeling is presented in Chapter 11.

In Table 2.2 the independent scale factors chosen are those for modulus of elasticity and length; all remaining scale factors are either unity or functions of s_E and s_l. The material for an elastic model of a prototype need only satisfy the requirement that it must remain elastic within the model loading range, and that it have the same Poisson's ratio as the prototype material. Assuming that loadings are then scaled by the factors given in Table 2.2, which can easily be derived by forming the appropriate dimensionless ratios, the model stresses are s_E times as small as those in the prototype, while model strains are identical to prototype strains. This type of similitude preserves the correct state of

TABLE 2.2   Similitude Requirements, Static Elastic Modeling

Quantities	Dimensions	Scale Factor
Material-related properties		
Stress	FL^{-2}	S_E
Modulus of elasticity	FL^{-2}	S_E
Poisson's ratio	—	1
Mass density	FL^{-3}	S_E/S_l
Strain	—	1
Geometry		
Linear dimension	L	S_l
Linear displacement	L	S_l
Angular displacement	—	1
Area	L^2	S_l^2
Moment of inertia	L^4	S_l^4
Loading		
Concentrated load Q	F	$S_E S_l^2$
Line load w	FL^{-1}	$S_E S_l$
Pressure or uniformly		
distributed load q	FL^{-2}	S_E
Moment M or torque T	FL	$S_E S_l^3$
Shear force V	F	$S_E S_l^2$

strain for beam-column effects and other geometry-dependent phenomena. If the latter type of behavior is not present, then in many modeling applications the loading is increased by some "slicing" factor that has the advantage of increasing the deformations and displacements of the model. This permits more accurate measurement of the response of the model. The same slicing factor must be taken out again in projecting the model results to the prototype.

The load similitude requirements demonstrate one of the major advantages of reduced-scale elastic modeling in structures. Loads are reduced from prototype loads by the factor $s_E s_l^2$, which is a very large number for a small-scale plastic model of a concrete or steel structure. For typical plastics, E is about 400,000 pounds per square inch (psi) (2,720 MPa); thus s_E is about 8 for a concrete prototype and about 75 for a steel prototype. The above product of $s_E s_l^2$ automatically becomes large, thereby resulting in extremely small loads in comparison with those of the prototype.

2.5.1 Reinforced Concrete Models

It is not easy to model the complete inelastic behavior of a reinforced or prestressed concrete structure, including the proper failure mode and capacity. The highly inelastic nature of concrete under both tensile and compressive stress states is a substantial problem in itself. The other major difficulty is in the reinforcing phase of this two-component material. The strength properties and surface roughness characteristics (bond capacity) of ordinary reinforcement must be given very careful attention if successful models are to be realized. In prestressed models the stress-strain characteristics of the prestressing steel are crucial, as is the anchorage detail.

Because the modeling of reinforced and prestressed concrete structures normally includes loading to failure, the failure criteria for model concrete subjected to multiaxial stresses should be identical with that of the prototype concrete. The lack of a well-defined failure criterion normally leads one to relax this requirement, as outlined below, even for true models:

1. Stress-strain curves must be geometrically similar in model and prototype concrete for both uniaxial tension and compression.
2. $\epsilon_m = \epsilon_p$ at failure under uniaxial tension and compression.

These requirements are summarized in Fig. 2.6. The corresponding similitude requirements are given in column (4) of Table 2.3.

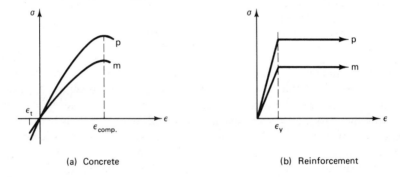

(a) Concrete   (b) Reinforcement

Figure 2.6 Similitude requirements.

The use of a nonunity stress scale factor (s_σ) in true models is justified from dimensional analysis. If $s_\sigma \neq 1$ is used for the model concrete, the same nonunity s_σ must be applied to the reinforcing steel, or $s_\sigma = s'_\sigma$, where the primed quantity refers to the reinforcing. The problem of using this type of scaling arises from the fact that s_E must equal s_σ, and the model reinforcing must have a modulus differing from steel by the factor $s'_E = s_E = s_\sigma$. The extensive discussion on materials in Chapters 4 and 5 will show that steel is the only feasible material for reinforced concrete models. Thus one is led to the conclusion that the only practical way to conduct true modeling of reinforced concrete structures is to use $s_\sigma = s_E = 1$; the corresponding similitude requirements are given in column (5) of Table 2.3. The rather stringent requirement of forcing model concrete to have a stress-strain curve identical to the prototype concrete may be impossible to meet. Finally, it should be noted that the density requirement cannot be satisfied.

TABLE 2.3 Summary of Scale Factors for Reinforced Concrete Models

(1)	Quantity (2)	Dimension (3)	True Model (4)	Practical True Model (5)	Distorted Model, Case 1 (Fig. 2.7) (6)	Distorted Model, Case 3 (Fig. 2.8) (7)
Material Related Property	Concrete stress, σ_c	FL^{-2}	S_σ	1	S_σ	S_σ
	Concrete strain, ϵ_c	—	1	1	S_ϵ	S_ϵ
	Modulus of concrete, E_c	FL^{-2}	S_σ	1	S_σ/S_ϵ	S_σ/S_ϵ
	Poisson's ratio, ν_c	—	1	1	1	1
	Mass density, ρ_c	FL^{-3}	S_σ/S_l	$1/S_l$	S_σ/S_l	S_σ/S_l
	Reinforcing stress, σ_r	FL^{-2}	S_σ	1	S_σ	S_σ
	Reinforcing strain, ϵ_r	—	1	1	S_ϵ	S_ϵ
	Modulus of reinforcing, E_r	FL^{-2}	S_σ	1	S_σ	1
	Bond stress, u	FL^{-2}	S_σ	1	S_σ	*
Geometry	Linear dimension, l	L	S_l	S_l	S_l	S_l
	Displacement, δ	L	S_l	S_l	$S_\epsilon S_l$	$S_\epsilon S_l$
	Angular displacement, β	—	1	1	S_ϵ	S_ϵ
	Area of reinforcement, A_r	L^2	S_l^2	S_l^2	S_l^2	$S_\sigma S_l^2/S_\epsilon$
Loading	Concentrated load, Q	F	$S_\sigma S_l^2$	S_l^2	$S_\sigma S_l^2$	$S_\sigma S_l^2$
	Line load, w	FL^{-1}	$S_\sigma S_l$	S_l	$S_\sigma S_l$	$S_\sigma S_l$
	Pressure, q	FL^{-2}	S_σ	1	S_σ	S_σ
	Moment, M	FL	$S_\sigma S_l^3$	S_l^3	$S_\sigma S_l^3$	$S_\sigma S_l^3$

*Function of choice of distorted reinforcing area.

51

It is necessary to utilize a distorted model approach when the available model concrete does not have $s_\epsilon = s_\sigma = 1$. A number of possible distortions are discussed by Zia, White, and Van Horn (1970) and are summarized in Table 2.4.

TABLE 2.4   Possible Distortions in Reinforced Concrete Models

	Concrete			Reinforcement		
Case	S_ϵ	S_σ	S_E	S'_ϵ	S'_σ	S'_E
1	$\neq 1$	S_ϵ	1	S_ϵ	S_ϵ	1
2	$\neq 1$	1	$1/S_\epsilon$	S_ϵ	1	$1/S_\epsilon$
3	$\neq 1$	$\neq 1$	$\neq 1$	S_ϵ	S_ϵ	1
4	$\neq 1$	$\neq 1$	$\neq 1$	S_ϵ	S_σ	$\neq S_E$

Only cases 1 and 3 are of interest because the others require reinforcement made of a material other than steel. Note that both cases utilize a distortion in strain, and thus they should not be used when the structural response is sensitive to absolute magnitude of strain (such as in a beam-column). The case 1 distortion would have material properties as shown in Fig. 2.7.

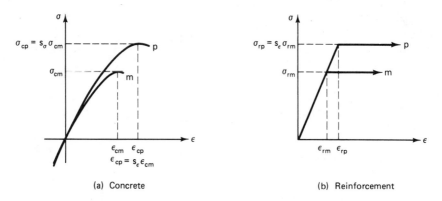

(a) Concrete   (b) Reinforcement

Figure 2.7   Case 1: distortion similitude requirements.

The effects of the strain distortion on the scale factors are given in column (6) of Table 2.3. Loading similitude is not affected by strain distortion; this statement is true for any type of model with a strain distortion. This type of similitude still requires geometrically similar concrete stress-strain curves but has the flexibility of not requiring $s_\sigma = s_\epsilon = 1$.

Case 3 distortion treats the combination of $s'_E = 1$ and $s_E \neq 1$ (Fig. 2.8). In order to have the steel strains distorted in the same fashion as the concrete strains, $s'_\epsilon = s_\epsilon$ and with $s'_E = 1$, s'_σ must equal s'_ϵ and the model steel yield strength must satisfy the requirement ($\sigma_{rp} = s_\epsilon \sigma_{rm}$). By equating the

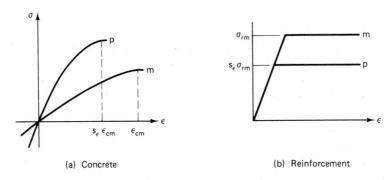

Figure 2.8   Case 3: distorted models.

scaled steel force in the model to that in the prototype ($F_p = s_\sigma s_l^2 F_m$) and using the relation $F = A\sigma$ for model and prototype steels, it follows that

$$\frac{A_p}{A_m} = s_A = \frac{s_\sigma s_l^2}{s_\epsilon}$$

is the appropriate scaling factor for model reinforcing areas.

In conclusion, it must be emphasized that the tensile strength properties of the model concrete must be properly scaled if such phenomena as shear (diagonal tension) strength and cracking, and even deflections, are to be modeled. Detailed physical properties of appropriate model concretes and reinforcements are given in Chapters 4 and 5.

Concrete stresses are normally transmitted to reinforcing elements through the actions at the steel-concrete interface (pullout and dowel action). The character of the bond (pullout) action is very important in many structural elements, and cannot be ignored in modeling. The following discussion from Zia, White, and Van Horn (1970) is intended to cast light on bond similitude requirements in reinforced concrete models.

The basic similitude requirement for bond between concrete and its reinforcement, for true models, is that bond stresses developed by the model reinforcing be identical to those of the prototype reinforcing. It also requires that ultimate bond strength of model and prototype reinforcement be identical. This would be possible only when $s_\sigma = 1$.

The scale of the model is an important factor in the bond similitude problem. For large models, #2 and #3 deformed bars which satisfy ASTM requirements may be within the desired range of size as model steel. For smaller models, one is forced to utilize either wire (smooth, rusted or deformed), threaded rod, or stranded cables (for prestressed models). It should be noted, however, that even the use of standard deformed bars does not necessarily ensure true bond modeling because of the difference in bond strength of small and large size bars.

Modeling of bond is seriously complicated by our limited knowledge of the bond mechanism in prototype concrete. According to the current notion of the bond behavior of prototype reinforcing [see Lutz and Gergely (1967)] one must attribute bond strength mainly to the mechanical wedging action afforded by the protruding deformations of the bars. Based upon the ACI Code, the ultimate bond strength per inch of bar length, for normal size bars, is:

$$U_\mu = 30\sqrt{f'_c} < 2500D \tag{2.42}$$

where f'_c = concrete compressive strength in psi and D = bar diameter in inches. It would be grossly misleading to attempt to model this relationship with the types of small reinforcement mentioned previously. In the first place, the above expression is dimensionally inhomogeneous. Secondly, it was not intended to be applied to the small size reinforcement as it was derived experimentally from tests on ordinary prototype bars. If the above expression were valid for all size bars, the ultimate bond strength of small size reinforcement [0.25 in. (6 mm) dia. or below] would always be dictated by the $2500D$ limitation. However, from the work of Lutz and others, it can be stated that the ultimate bond strength is not independent of bar size for any range of diameters. Thirdly, the geometric dissimilarity between the wire, threaded rod, stranded cable and the prototype deformed bars is so great that their bonding action is undoubtedly different. For wires and stranded cables, the lack of protruding lugs greatly reduces the tendency to produce a splitting failure in concrete. On the other hand, pullout tests on threaded rods often produce a complete shearing of the surrounding concrete at the top of the threads, also with little or no splitting accompanying the failure. The effect of concrete cover upon bond strength, and the increased bond strength afforded by stirrups and other steel which tends to prevent splitting from developing adjacent to bars under high bond stress, are also difficult to predict and to model.

In spite of these difficulties, it should be noted that it may be possible to provide model steel which will duplicate local bond behavior of the prototype steel even though its ultimate bond strength is not identical to that of the latter. Cracks normal to the direction of the reinforcement can open through the mechanism of localized slipping of reinforcing without necessarily having longitudinal splitting cracks. Therefore, partial satisfaction of similitude would be provided by this type of model steel.

Another method to improve modeling is strictly dependent upon a detailed knowledge of the bond behavior of model reinforcements. If this information is available, then the bond strength of a certain combination of prototype bars may be modeled by using the proper number and size of model bars. The philosophy of this type of modeling is to provide the correct amount of total bond strength instead of being concerned with scaling unit bond stress. It is felt that this approach has considerable merit and will provide the best answer once a detailed knowledge of both prototype and model reinforcing bond behavior is at hand. True bond stress distribution bears little resemblance to the computed average unit bond stress; it will always be difficult to attempt to model such a poorly defined quantity.

The bond characteristics of several model reinforcements will be presented in Chapter 5. Another problem often associated with concrete model similitude, that of increasing scaled strength with decreasing size of structural element, will also be dealt with in Chapter 6. Results of cracking similitude investigations are summarized in Chapter 10.

2.5.2 Concrete Masonry Models

An early attempt to model masonry structures was made in England [Vogt (1956)] using $\frac{1}{4}$-scale bricks and later $\frac{1}{10}$-scale bricks. Engineering feasibility of model brickwork investigations of structures was established [Hendry and Murthy (1965), Murthy and Hendry (1966), Sinha (1967), Sinha and Hendry (1969)] by direct comparison of prototype brickwork and $\frac{1}{3}$- and $\frac{1}{6}$-scale models. It was concluded from these early investigations that the strength of full-size brickwork can be reproduced by means of model tests. Extension of the basic modeling techniques developed by Hendry and his co-workers at the University of Edinburgh was made in studies of deflections and stresses in multistory brick structures under lateral loads [Sinha et al. (1970), Kalita and Hendry (1970), Haselite and Fisher (1973)]. Test on model masonry structures have also been successfully conducted in Australia [Mohr (1971), Baker (1972)] on axially loaded and laterally loaded brick walls at $\frac{1}{3}$ and $\frac{1}{6}$ scales.

In the United States, the earliest reported tests using model masonry were conducted in studies dealing with the shear resistance of infilled frames [Benjamin and Williams (1958), Yorulmaz and Sozen (1968)]. An extensive experimental study of multistory and multibay reinforced concrete masonry infilled frames using $\frac{1}{4}$-scale clay bricks [Fiorato et al. (1970)] has provided considerable insight of the interaction of the masonry with the boundary frame in such systems.

The modeling of concrete masonry structures has not received as much attention as those of clay brick construction. Early attempts at the National Bureau of Standards (NBS) in the late sixties to model concrete masonry structures using carefully fabricated $\frac{1}{4}$-scale masonry blocks made from Ottawa sand were not conclusive. The earliest reported work on the direct modeling of concrete masonry using the same $\frac{1}{4}$-scale units, manufactured for NBS by the National Concrete Masonry Association, was conducted at Drexel University [Harris and Becica (1977), Becica and Harris (1977), Harris and Becica (1978)]. This work on the modeling of concrete masonry structures is reported in some detail in Chapters 4 and 10 of the text.

The most general and useful modeling techniques used in the design and analysis of masonry structures subjected to static and dynamic loads are those which can predict inelastic as well as elastic behavior and have the ability to study with confidence the mode of failure of the structure. These techniques are, however, very restrictive on the choice of model materials and their methods of fabrication. Let us consider first the case of static loading on the structure, which usually consists of the dead and live loads but could also consist of equivalent static loads for the dynamic effects of wind or earthquake and even abnormal loads such as explosions or impact. Under the assumption that there are no significant time-dependent effects that influence the structural behavior, the pertinent parameters that enter the modeling process are listed in Table 2.5, column (2). For complete similarity of the structural behavior including the inelastic effects of cracking and yielding, a dimensional analysis will give the scale factors shown in Table 2.5, column (4). If it is assumed that the stresses

TABLE 2.5   Summary of Scale Factors for Masonry

| | | | Static Loading | |
| | | | | |
Group (1)	Quantity (2)	Dimension (3)	True Model (4)	Practical True Model (5)
Loading	Concentrated load, Q	F	$S_\sigma S_l^2$	S_l^2
	Line load, w	FL^{-1}	$S_\sigma S_l$	S_l
	Pressure, q	FL^{-2}	S_σ	1
	Moment, M	FL	$S_\sigma S_l^3$	S_l^3
Geometry	Linear dimension, l	L	S_l	S_l
	Displacement, δ	L	S_l	S_l
	Angular displacement, β		1	1
	Area, A	L^2	S_l^2	S_l^2
Material properties	Masonry unit stress, σ_m	FL^{-2}	S_σ	1
	Masonry unit strain, ϵ_m		1	1
	Modulus of masonry unit, E_m	FL^{-2}	S_σ	1
	Masonry unit Poisson's Ratio, v_m		1	1
	Specific mass, ρ_m	FL^{-3}	S_σ/S_l	$1/S_l$
	Mortar stress, σ'_m	FL^{-2}	S_σ	1
	Mortar strain, ϵ'_m		1	1
	Modulus of mortar, E'_m	FL^{-2}	S_σ	1
	Mortar Poisson's ratio, v'_m		1	1
	Reinf. stress, σ_{rm}	FL^{-2}	S_σ	1
	Reinf. strain, ϵ_{rm}		1	1
	Modulus of reinf., E_{rm}	FL^{-2}	S_σ	1

caused by the self-weight of the structure are not significant, as is usually the case in most masonry buildings, the scale factors given in Table 2.5, column (5) will be adequate for modeling masonry structures. For this latter, "practically true" modeling approach, the stress-strain curves of both model and prototype masonry must be the same, presenting a very difficult challenge to the model analyst. The reason for this is that since masonry is a composite material, one has to model all of its constituents: block or brick, mortar, and reinforcements. In addition, fabrication difficulties arise because of the small size of the individual units.

2.5.3 Structures Subjected to Thermal Loadings

Modeling of thermal effects has proven useful for several classes of structural problems. Examples include studies of thermal stress in arch dams [Rocha (1961)] and in nuclear reactor vessels, and temperature distribution and temperature stresses in spacecraft structures [Katzoff (1963)]. Issen (1966) has studied the problem of modeling the effects of fire on concrete structures. Thermal model analysis should also be considered for other thermostructural problems for which analytical methods are very difficult.

The elastic response of a structure built of a homogeneous, isotropic material will be treated in the initial discussion. Transient heat conduction with no internal (to the material) heat generation will be assumed. With these limitations in mind, the only thermal properties needed are the coefficient of linear expansion, α, and the thermal diffusivity, D. It is further assumed that these thermal properties and the elastic properties of the model material are independent of temperature.

The 10 quantities involved in the analysis are defined in Table 2.6. The thermal diffusivity D is equal to $k/(c\gamma)$ where k = thermal conductivity, c = specific heat per unit weight, and γ = specific weight of the material. Four fundamental measures are needed to express the 10 quantities: force F or mass M, length L, time T, and temperature θ. Choosing the four independent quantities as l, E, D, and θ, a suitable set of Pi terms is:

$$\pi_1 = \frac{\sigma}{E}, \qquad \pi_2 = \epsilon$$

$$\pi_3 = \nu, \qquad \pi_4 = \frac{\delta}{l}$$

$$\pi_5 = \alpha\theta, \qquad \pi_6 = \frac{tD}{l^2}$$

The scale factors for a true thermal model are summarized in column 3 of Table 2.6. The term π_6, known as *Fourier's number* in heat transfer, leads to the time scaling factor of $s_t = s_l^2/s_D$. With model time inversely proportional to the square of the geometrical scale factor, it is possible to model long-time thermal effects in a greatly reduced time.

Whenever identical materials are used in model and prototype, the scale factors shown in column (4) of Table 2.6 should be followed. Taking this approach on a true modeling basis compels one to use the same temperature in model and prototype and eliminates the possible problem of having a distortion introduced because of dependence of material properties on temperature level.

Distortion of strain in thermal modeling is often possible; the criteria for deciding on acceptability of strain distortion is the same as for mechanically

TABLE 2.6 Summary of Scale Factors for Thermal Modeling

Quantity (1)	Dimension (2)	True Model (3)	Same Materials in Model and Prototype, and Same Temperatures (4)	Distorted Strain Scaling (5)
Stress, σ	FL^{-2}	S_E	1	$S_\alpha S_\theta S_E$
Strain, ϵ	—	1	1	$S_\alpha S_\theta$
Elastic modulus, E	FL^{-2}	S_E	1	S_E
Poisson's ratio, ν	—	1	1	1
Coefficient of linear expansion, α	θ^{-1}	S_α	1	S_α
Thermal diffusivity, D	$L^2 T^{-1}$	S_D	1	S_D
Linear dimension, l	L	S_l	S_l	S_l
Displacement, δ	L	S_l	S_l	$S_\alpha S_\theta S_l$
Temperature, θ	θ	$1/S_\alpha$	1	S_θ
Time, t	T	S_l^2/S_D	S_l^2	S_l^2/S_D

loaded models. If thermal actions and mechanical actions are being modeled simultaneously, the same degree of strain distortion must be followed for both actions.

Using a distorted strain scale and different materials in model and prototype leads to the scaling shown in column (5) of Table 2.6. The strain is distorted because the temperature scaling is distorted from its desired true value of $s_\theta = 1/s_\alpha$. It can be reasoned that a distortion of temperature will have a linear effect on strains, stresses, and displacements in the model. Accordingly, a temperature distortion factor d_θ is defined as follows. In the true model, $s_\theta = 1/s_\alpha$; in the distorted model, $s_\theta = d_\theta(1/s_\alpha)$. Solving for the distortion factor, $d_\theta = s_\theta s_\alpha$. The distorted scaling relations for strain, stress, and displacement are formed by multiplying the distortion factor d_θ by the true scaling factors for the affected quantities, or

$$s_\epsilon = 1 \cdot d_\theta = s_\alpha s_\theta$$
$$s_\sigma = s_E \cdot d_\theta = s_E s_\alpha s_\theta$$
$$s_\delta = s_l \cdot d_\theta = s_l s_\alpha s_\theta$$

Properties of materials suitable for studying thermal stresses in dams at a scale of $\frac{1}{100}$ to $\frac{1}{500}$ will be given in the next chapter. The time scaling factor allows annual temperature waves on the prototype to be modeled in a matter of minutes. Models of this type must be planned very carefully to ensure temperatures and temperature strains large enough to be measured accurately and yet be within the temperature ranges allowed by the model material.

Transfer of heat at a boundary is a function of the surface coefficient of heat transfer, h. The corresponding dimensionless parameter is hl/k, where k = coefficient of thermal conductivity and l = characteristic length. This parameter, called *Nusselt's number* (N), should be equal in model and prototype in order to have properly scaled surface temperature gradients, but a direct contradiction arises when the same material is used in model and prototype. With $h_m = h_p$ and $k_m = k_p$, Nusselt's number can be satisfied only when $l_m = l_p$ or a full-sized model is used. Lack of similitude in temperature gradients at and near the boundary layer in a model has not been a serious factor in either high-temperature fire tests or long-time dam studies [Rocha (1961) and Issen (1966)].

Another type of thermal loading is where the heat transfer is by radiation, such as in a spacecraft subjected to the heat of the sun. No additional similitude requirements are placed on the model itself, but the intensity of radiation H (dimensions of Q/L^2T, where Q is a measure of heat) must be scaled in accordance with the parameter $\sigma\theta^4/H$, where σ is the Stefan-Boltzmann constant and θ is the absolute temperature. Radiation scaling may lead to substantial difficulties; the reader is referred to Katzoff (1963) for further discussion of thermal radiation modeling.

Thermal effects leading to extensive inelastic actions and structural failures can be modeled, but only with extreme care being given to all material properties. Full discussion of the inherent problems is beyond the scope of this text.

2.5.4 Structures Subjected to Dynamic Loadings

The modeling of the effects of dynamic loads on structures has been utilized in a wide variety of problems, including:

1. Aerodynamic response of suspension bridges and cable-suspended roof systems
2. Behavior of buildings and dams subjected to earthquakes and wind
3. Machinery-induced vibrations of complex structures
4. Blast loading of above-ground and buried structures
5. Wave-impact loadings on ship structures
6. Dynamic and aerodynamic response characteristics of aircraft and spacecraft, including lateral, longitudinal, and local vibration, and fuel sloshing, buffeting, flutter, and ground wind effects

Dynamic modeling requirements are given in Chapter 11, along with several examples in this increasingly important area of structural modeling.

PROBLEMS

2.1. In studying the effect of the size of a submarine on the depth to which it can safely dive, a single design is considered and then the size effect is studied by scaling all linear dimensions (including plate thicknesses) in the same proportion. The critical depth of dive is one where the structure fails in elastic buckling. What can you say about the critical depth as related to the size of the hull and its material? (Courtesy: W. Godden, U. C., Berkeley)

2.2. Derive the similitude relation for displacements in a flexural member,

$$U_m = \frac{1}{s_l s_\epsilon} U_p$$

from the fact that the displacement can be expressed as

$$\delta = k\frac{ML^2}{EI}$$

where k is a geometric constant, and the model and prototype values of $M, L, E,$ and I are related by scale factors that are readily derived (or recognized by inspection).

2.3. For the case of $s_\epsilon \neq 1$, $s_\sigma \neq 1$, and $E_{rm} \neq E_{rp}$, derive expressions for a reinforced concrete beam for:
(a) Required area of model reinforcing A_{rm}
(b) Yield strength of model reinforcing as a function of yield strength of prototype steel
(*Hint*: This can be accomplished by considering the basic flexural mechanics of a simple reinforced concrete beam.)

2.4. (a) Using a dimensional analysis, develop a general expression for the distance that a freely falling object will drop in time t, neglecting resistance afforded by the air.
(b) What happens in the analysis if we erroneously assume that the weight of the object is a variable?
(c) Assume the dropped object starts with an initial velocity v_0. Develop an expression for distance traveled in time t.

2.5. Determine the π terms for the beam-stiffness problem discussed in this chapter, expressing the parameters in terms of the fundamental quantities of mass, length, and time. What is the basic difficulty met in this formulation of the problem?

2.6. The following equations are to be checked for dimensional homogeneity. Give values for the dimensions attached to any constants in the nonhomogeneous equations.
(a) Column strength formula:

$$P = A_g(0.25f'_c + f_s p_g)$$

(b) Bending of a transversely loaded plate:

$$\frac{\partial^4 w}{\partial x^4} + 2\frac{\partial^4 w}{\partial x^2 \partial y^2} + \frac{\partial^4 w}{\partial y^4} - \frac{q}{D} = 0$$

(c) AISC column stress formula:

$$F_a = \left[1 - \frac{(kl/r)^2}{2C_c^2}\right] f_y/\text{factor of safety}$$

(d) Manning's formula for open-channel flow:

$$V = \frac{1.486}{n} R^{2/3} S^{1/2}$$

2.7. You are working in a design office and have just completed the calculation to determine the maximum bending moment in a three-span continuous girder carrying a uniform load of 3.0 kips/ft. Your supervisor then informs you that the spans have been increased from 60, 80, and 60 feet to 72, 96, and 72 ft. Can you, without going through the calculations again, quickly find the new maximum bending moment?

2.8. Given

$$\sigma_{yp} = s_\epsilon \sigma_{ym}$$

$$\text{Steel } A_p = \frac{s_\sigma s_l^2}{s_\epsilon} A_m$$

$$f'_{cp} = s_\sigma f'_{cm}$$

$$d_p = s_l d_m$$

$$b_p = s_l b_m$$

Show that $M_p = s_\sigma s_l^3 M_m$ (ultimate moment capacities) by applying conventional ultimate strength analysis

$$M = A_s f_y\left(d - \frac{a}{2}\right)$$

and forming the ratio of M_p to M_m.

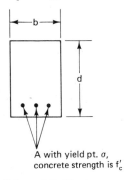

A with yield pt. σ,
concrete strength is f'_c

2.9. Prove or disprove the validity of Galileo's statement (given below) by applying similitude theory to the problem of dead-weight stresses in a structure of constantly increasing size.

... it would be impossible to build up the bony structures of men, horses, or other animals so as to hold together and perform their normal functions if these animals were to be increased enormously in height; for this increase in height can be accomplished only by employing a material which is harder and stronger than usual, or by enlarging the size of the bones, thus changing their shape until the form and appearance of the animals suggest a monstrosity. If the size of a body be diminished, the strength of that body is not diminished in the same proportion; indeed the smaller the body the greater its relative strength. Thus a small dog could probably carry on his back two or three dogs of his own size, but I believe a horse could not carry even one of his own size.

2.10. A rectangular tubular steel section is subject to the loading shown below. Use the Pi theorem to develop a general expression for the deflection of point a with respect to point b.

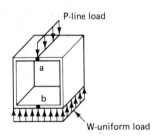

Elastic Models— Materials and Techniques

$\wedge\wedge\wedge\wedge\wedge\wedge\wedge\wedge\wedge\wedge\wedge\wedge\wedge\wedge\wedge\wedge$

3

3.1 INTRODUCTION

An elastic model can be used to study the behavior of the prototype *only* in the linear elastic range; it cannot be used to predict any inelastic behavior of a loaded structure resulting from material nonlinearities such as the postcracking behavior of concrete, postyielding behavior of steel, or the postbuckling behavior of a plate or column. Elastic models have been used extensively to study the response of multistory buildings, bridges, nuclear reactor pressure vessels, dams, and other types of structures subjected to static, dynamic (including earthquake, blast, and wind loads), and thermal loadings. Elastic models of structural components such as columns, frames, slabs, and shells have been used in elastic stability studies.

Similitude requirements for static elastic modeling were presented in Table 2.2. Independent scaling factors were chosen for modulus of elasticity and length, and all remaining scale factors were established as functions of s_E and s_l. It follows from Table 2.2 that the material for an elastic model of a prototype structure need only satisfy the requirements that it must not be loaded beyond the linear elastic range and that it must have the same Poisson's ratio as the prototype material, although, this latter requirement can be waived in one-dimensional structures. Elastic models can be used as direct models (the pattern of model loading is similar to that of the prototype) or as indirect models (to derive influence lines and influence surfaces, where the pattern of prototype loading does not have to be reproduced). In indirect models, it is not necessary to satisfy the condition of equality of strains at corresponding (homologous) points in the prototype and the model ($\epsilon_p = \epsilon_m$). It follows that the intensity of applied loading can be varied at will; the resulting deformations will be proportional to the applied loads provided the linear elastic range of the material is not exceeded. Conceptually, such experimental models are very similar to the available procedures for elastic analysis of structures and account for the stiffnesses of members and joints; in some ways, especially with respect to boundary conditions, the physical model represents a better idealization of the structure than the mathematical model.

3.2 MATERIALS FOR ELASTIC MODELS

The choice of materials available for the construction of any structural model is perhaps even greater than that for prototype structures. The advantages, disadvantages, and limitations of commonly used model materials have been examined by several investigators. The selected material must satisfy the laws of similitude presented in Chapter 2, and besides having reproducible mechanical properties and geometric stability, it should be readily available, easily fabricated, and inexpensive. A detailed discussion of the physical and chemical properties of all materials suitable for the construction of models is beyond the scope of this text, and only a few relevant properties of the more commonly used model materials are presented to assist with the selection of suitable material. The most significant model material properties are:

1. Proportional (limit) stress
2. Stiffness
3. Failure mechanism
4. Influence of temperature and humidity on material properties
5. Creep characteristics
6. Load rate and strain rate effects
7. Effect of size and shape on material properties

All material properties should be confirmed by appropriate tests, as data given in the various handbooks or manufacturer's catalogs represent average or "obtainable" values and are often unreliable.

The properties of some materials such as plastics and concrete not only show large variability from one sample to another, but are also significantly dependent on the type, shape, and size of the specimen and the rate of loading. Three major types of materials suitable for the construction of elastic models are presented in this text: plastics (Chapter 3), cementitious materials (Chapter 4), and metals (Chapter 5).

The only condition required to be satisfied for *indirect models* is that the material exhibit a linear elastic stress-strain relationship; in general, plastics or metals are used for constructing indirect models. Plastics have low elastic moduli, and this leads to small load requirements and measurable large deformations in small-scale models. Using structural mechanics, it can be shown that the principles of superposition and reciprocity are valid within the linear elastic range of behavior for the prototype or model structure (this also implies that there are no stability or catenary effects, the deformations are small and the limit of proportionality of the material is not exceeded). These two principles are useful for determining deformations and stresses in direct models as a result of combinations of a variety of loading conditions such as dead, live, and wind loads. For a general case, to simulate a given state of deformation in the prototype, the following similitude relationship for Poisson's ratio must be satisfied (see Table 2.2):

$$v_p = v_m \tag{3.1}$$

In planar skeletal structures, such as frames, trusses, arches, and cables, torsion is nonexistent and shearing effects are not as predominant as the flexural and axial effects; therefore the similitude requirement (3.1) can be relaxed for these structures. However, if the response of the structure is not independent of Poisson's ratio, as in grids, three-dimensional frames, plates, and shells, then Eq. (3.1) should be satisfied.

3.3 PLASTICS

The term *plastic* is normally used for a material derived by chemical synthesis and containing carbon compounds [Preece and Davies (1964)]. A wide variety of plastic materials with varying chemical compositions and mechanical properties is available under specific trade names from manufacturers and local suppliers. Only those properties of commonly used plastics that are relevant to the construction and testing of elastic models are presented in this section.

Beggs (1932) is perhaps the first known investigator to use plastic models to solve statically indeterminate structural problems. Since the early sixties, plastics have been used effectively for the construction of direct and indirect models designed to simulate the linear elastic or linear viscoelastic response of the prototype. Comprehensive studies of structural models constructed from plastics have been reported by Fialho (1962), Litle and Hansen (1963, 1965), and Carpenter, Magura, and Hanson (1964). Mechanical properties of plastics directly relevant to structural modeling were reviewed by Rowe (1960), Preece and Davies (1964), and Roll (1968).

3.3.1 Thermoplastics and Thermosetting Plastics

Plastics can be classified into two categories: thermoplastics and thermosetting plastics.

Thermoplastics become progressively softer at temperatures between 200 and 300°F and can be formed into complex shapes such as shells of complex

TABLE 3.1   **Properties of Some Plastics**

Plastic	Thermal Charac- teristic	Available Shapes	Tensile Strength, psi	Compressive Strength, psi	Flexural Strength, psi	Modulus of Elasticity, psi
Cellulose nitrates (celluloid)	Thermo- plastic	Sheets, rods, and tubes	3000–7000	3000–30,000	3000–17,000	$65–400 \times 10^3$
Cellulose acetates (plasticele)	Thermo- plastic	Sheets, rods, and tubes	2250–11,000	2200–10,900	2200–11,500	$65–260 \times 10^3$
Methyl methacry- lates* (Plexiglas, Lucite Perspex)	Thermo- plastic	Sheets, rods, and tubes	7000–11,000	12,000–20,000	3000–17,000	$420–500 \times 10^3$
Polyvinyl chlorides (Boltaron)	Thermo- plastic	Sheets, rods, and tubes	5000–10,000	8000–13,000	13,500–	$350–600 \times 10^3$
Polyethylenes (Alkathene)	Thermo- plastic	Sheets, rods, tubes, mold- ing powders	1000–5000	—	2000–7000	$17–80 \times 10^3$
Natural or synthe- tic rubber	Thermo- plastic	Sheets and ex- truded shapes	1000–4000	—	—	200–350
Polyester resins (Marco, Palatal)	Thermo- setting	Casting resins	5000–6000	12,000–20,000	—	$300–400 \times 10^3$
Epoxy resins (Epon, Araldite)	Thermo- setting	Casting resins	5000–12,000	15,000–30,000	—	$430–600 \times 10^3$

*Acrylic resins.
Note: All plastics with specific gravity of less than 1.0 will float in water.

geometry with little or no pressure and yet retain their shape upon cooling. If a thermoplastic material is reheated, it can be remolded into another shape, although some types have a "memory." This characteristic of thermoplastics is extremely useful for the vacuum-forming process, which will be described later. The commonly used thermoplastics [namely, acrylic plastics and polyvinyl-chloride (PVC)] are generally available in the form of sheets, rods, and tubes. It must be noted that several types and grades are commercially available under a single trade name; for example, there are many different grades of Plexiglas. Plexiglas G, commonly used for many applications, will shrink about 2.2% during the heating process, with a corresponding increase in thickness. By contrast, Plexiglas II UVA, used extensively for research work (e.g., at Cornell University), is preshrunk and is manufactured to more exacting standards of optical quality, surface quality, and thickness tolerances. Other preshrunk types include I-A UVA, 55, and 5009.

The acrylic plastics (Plexiglas, Perspex, Lucite) can be easily machined and cemented, and accurate models can be rapidly assembled. Sheets can be softened by heating and formed into shells of single or double curvature using the vacuum-forming process. Comprehensive studies of acrylic plastics used for models have been reported by Fialho (1962), by Carpenter, Magura, and Hanson (1964),

Suitable for Structural Models

Poisson's Ratio	Elongation at Rupture Percent	Specific Gravity	Softening Temperature, °C	Coefficient of Expansion, in./in./°C	Machinability	Jointing Characteristics
0.40–0.42	40–90	1.35–1.70	70–100	11–17×10^{-5}	Excellent	Can be cemented with solvent cements and ethyl acetate (acetone)
0.40		1.24–1.32	250–350	8–16×10^{-5}		Can be cemented with solvent cements and ethyl acetate (acetone)
0.35–0.38	3–10	1.17–1.20	80–160	5–9×10^{-5}	Excellent	Can be cemented with commercial adhesives or a solution of the plastic and chloroform
0.38–0.40	85–100	1.38–1.40	80–105	5–8×10^{-5}	Excellent	Can be welded with PVC rods or cemented with epoxy cements
0.45–0.50		0.91–0.96	85–127	9–18×10^{-5}		Can be welded but difficult to cement
0.5	300–800	0.95–1.20	70–75	9–12×10^{-5}		Can be cemented with rubber cements
0.35–0.45	2	1.20–1.35	80–90	3–6×10^{-5}	Fair	(solvents are acetone, and cellosolve)
0.33–0.40	5–10	1.20	—	3–9×10^{-5}	Good	Can be cemented with epoxy cements

and later summarized by Roll (1968). Fumagalli (1973) has reported on the use of Lucite for several elastic model studies at the Istituto Sperimentale Modelli e Strutture (ISMES) in Italy.

Polyvinylchloride (PVC) is normally available in thinner sheets and sheets of more uniform thickness than acrylic plastics and are specially suited for model studies of various kinds of shells. A series of vacuum-formed shells of PVC have been tested at Cornell University and the Massachusetts Institute of Technology (MIT). Litle (1964) has reviewed the use of PVC for model testing and has reported a considerable amount of strength and stiffness data.

Thermosetting plastics differ from thermoplastics in that they cannot be remolded by heating once they have been cast into their original shapes. Thermosetting plastics such as the epoxies (Araldite) or polyesters (Marco, Palatal) can be used for casting models at room temperature with upper and lower molds without the use of pressure or ovens [Roll (1968)]. Since these casting resins are in liquid form prior to polymerization, they are frequently used for casting very intricate models; however, considerable skill and experience are prerequisites to achieve successful results. Thermosetting plastics are preferred to the thermoplastics in the manufacture of shell models with varying thickness. Any complex curved surface with any desired thickness variation can be cast conveniently using thermosetting plastics. One must remember that the thickness of such shells is affected by the variations in thickness due to the forming process. This variation is governed by dissipation of the heat of polymerization, and therefore care must be exercised in using thermosetting plastics.

The advantage of using epoxy resins compared with thermoplastics is that the limited development of the heat of polymerization assures a more homogeneous hardening process, which results in a constant elastic modulus throughout the mass [Fumagalli (1973)]. Also, the relatively lower shrinkage that occurs in epoxy resins after casting results in a significant decrease in the internal stresses. These are particularly useful in models of varying thickness where the internal stresses can even lead to fracture of the model. Properties of some plastics commonly used for structural models are described in Table 3.1. It must be noted that *only* typical values of the properties of plastics used by some investigators are listed. There are other available plastics that may be suitable for model work.

Epoxy resins also offer the possibility of modifying their physical properties by adjusting the quantity of hardener or by adding an inert material such as a filler dispersed homogeneously throughout the mass and/or reinforcements consisting of inorganic or organic fibers. Silica sand, powdered metal (aluminum or iron), cork, lead shot, polystyrene granules, and other ingredients have been successfully used as fillers. The addition of fillers alters the material density and modifies the modulus of elasticity within wide limits. It also decreases the temperature rise due to the generated heat of polymerization and reduces shrinkage and the associated internal stresses. Powdered cork, sand, and polystyrene help reduce the values of the Poisson's ratio, while the use of an aluminum powder increases the thermal conductivity, which helps disperse the heat generated from the electrical resistance strain gages later applied to the surface. Properties of some epoxy resin mixes with varying amounts of selected fillers are shown in Table 3.2.

3.3.2 Tension, Compression, and Flexural Characteristics of Plastics

The strength and stress-strain characteristics of plastics are dependent on a number of factors, such as the type of test (tension, compression, or flexure), the specimen size, the rate of loading, and the previous stress history in terms of creep and relaxation. The mechanical properties of plastics are also significantly influenced by temperature and relative humidity, which are discussed in more detail in Section 3.5. The measured properties vary not only from batch to batch but also from one sheet thicknesss to another within the same batch. Reasonable care must therefore be exercised to determine the properties in the laboratory under conditions of temperature and relative humidity similar to those in which the model will be both cast and tested. Also, it is important that if the model is subjected principally to direct stresses, the modulus of elasticity in

TABLE 3.2 Composition and Properties of Some Epoxy Resin Mixes with Added Fillers*

Resin Base, % by Weight		Filler Materials, % by Weight				Physicomechanical Properties	
Epoxy Resin (Alraldite M)	Hardener	Cork Dust	Aluminum Powder	Polystyrene Granules	Silica Sand 0.5–1 mm	Elastic Modulus E, kg/cm^{-2}	Poisson's Ratio
83.3	16.7	—	—	—	—	26,000– 32,000	0.38
71.5	14.3	—	14.2	—	—	25,000– 30,000	0.28
38.5	7.7	38.4	15.4	—	—	5,000– 7,000	0.27
23.7	4.8	—	—	71.5	—	18,000– 25,000	0.34
15.4	3.0	—	—	4.6	77.0	80,000– 90,000	0.27
9.8	2.0	—	—	—	88.2	140,000–160,000	0.26

*After Fumagalli, 1973.

direct tension or compression should be determined from tension or compression tests on suitable specimens. Similarly if the model is subjected principally to bending, the modulus of elasticity in flexure should be determined using a cantilever beam test or a similar flexure test.

It is recommended that the tension and flexural specimens should be at least 8 in. (200 mm) long, randomly selected from the material for model construction. A brief description of the specimens recommended by the American Society for Testing and Materials (ASTM) Standards and of specimens used by some investigators is presented in Table 3.3.

TABLE 3.3   Test Specimens Used by Various Researchers
for Tests on Plastics

Researcher Reference	Test	Specimen Size	Comments
Struminsky (1971)	Tension	ASTM Standard Specimen	Tensile properties of fiberglass-reinforced polyester resin
Carpenter et al. (1964)	Tension	1.5 × 0.25 × 16 in. (37 × 6 × 400 mm)	Tensile properties of Plexiglas; specimen ends reinforced
Balint and Shaw (1965)	Tension	5/8 in. diam. × 8 in. (16 mm diam. × 200 mm)	Plastrene 97 with varying amounts of calcite filler used for the Australia Square models; enlarged specimen ends
Balint and Shaw (1975)	Compression	2 in. diam. × 4 in. (50 mm diam. × 100 mm)	Compressive properties of the above mixes
Struminsky (1971)	Flexure	ASTM Standard Specimen	Flexural properties of fiberglass-reinforced polyester resin
Tabba (1972) Fam (1973)	Flexure	1 in. × sheet thickness × 8 in. (25 mm × sheet thickness × 200 mm)	Flexural properties of Plexiglas

Details of the tension specimen Type I recommended by ASTM Standard D638-77a (1978) for determining tensile properties of plastics are shown in Fig. 3-1. The specimen has a uniform cross section (12.5 mm × sheet thickness) over a length of 57 mm and a gradual transition to the two enlarged ends to prevent failure at the grips. It is recommended that this specimen be used in model material evaluations.

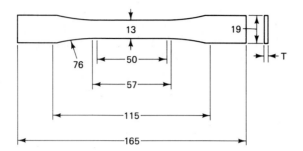

Figure 3.1 Tension test specimen dimensions in millimeters (for sheet, plate, and molded plastics.) (Courtesy of ASTM.)

The ASTM Standard 695-77 (1978) for determining compressive properties of plastics recommends that a right cylinder or prism whose length is twice its diameter or its principal width be used to determine the compressive properties of plastics; the preferred specimen size is $0.5 \times 0.5 \times 1.0$ in. ($12 \times 12 \times 25$ mm) for prisms and 0.5 in. diameter $\times$ 1.0 in. high (12 mm diameter $\times$ 25 mm high) for cylinders. For determination of elastic modulus and the yield stress, the test specimen should be of such dimensions to avoid the slenderness or instability problems. The slenderness ratio generally used is between 11 and 15. The preferred specimen size for prisms is $0.5 \times 0.5 \times 2.0$ in. ($12 \times 12 \times 50$ mm) and for cylinders the recommended dimensions are 0.5 in. diameter and 2.0 in. height (12 mm diameter and 50 mm height). It is not necessary to machine the test specimen cross sections to the preferred sizes; compression specimens of suitable height can be cut from the material used for the model construction according to the ASTM Standard 695-77.

Several investigators have used ASTM standard specimen to determine the flexural properties of different plastics. According to the ASTM Standard D790-71 (1978) for plastic materials $\frac{1}{16}$ in. (1.5 mm) or greater in thickness, the depth of the specimen for flatwise tests shall be the material thickness and for edgewide tests the depth shall not exceed the width; for all tests the support span shall be 16 times the depth with sufficient overhangs to prevent the specimen from slipping through the supports. Dimensions of the roller supports and the rounded loading nose, along with other details, are also specified.

3.3.3 Viscoelastic Behavior of Plastics

Most of the commonly used plastics exhibit linear viscoelastic properties; that is, the stress-strain relationship at any particular time after loading is linear although the modulus of elasticity varies with time. In other words, the stress-strain relationship does not conform to Hooke's law, and the strain is a function of time after loading, the loading history, and the stress level. Thus, if a plastic specimen is subjected to stress σ that is maintained constant for a duration of time t (Fig. 3.2a), the strain response of the specimen will be as shown in Fig. 3.2b. There will be an instantaneous strain ϵ_1 in the plastic immediately upon the application of stress. Under constant stress the strain increases further with time, and this rate of strain increase $(d\epsilon/dt)$ is dependent on the stress intensity. According to Preece and Davies (1964), in most model studies the maximum stress intensity must be controlled so that $(d^2\epsilon/dt^2)$ is negative. If the stress intensity is above the creep strength, the value of $(d^2\epsilon/dt^2)$ will be positive and the strain will proceed to increase at an increasing rate until the specimen fractures. This problem can normally be avoided by limiting the value of the maximum stress in the plastic well below the creep strength of the plastic.

Under a constant stress σ, the total strain ϵ_t at any time interval t (Fig. 3.2b) consists of two parts: the instantaneous elastic strain, ϵ_i and the time-dependent creep strain ϵ_{cr}. When the stress is removed, there will be an instantaneous recovery of strain approximately equal to ϵ_0 followed by a slower strain recovery that gradually disappears with time. If the stress intensity is doubled to 2σ, the corresponding total strain at the same time t will be $2\epsilon_t$, with the total instantaneous elastic and the creep strains being $2\epsilon_i$ and $2\epsilon_{cr}$, respectively, if the

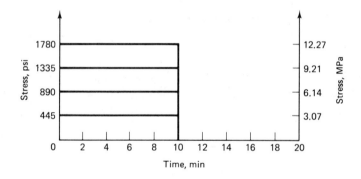

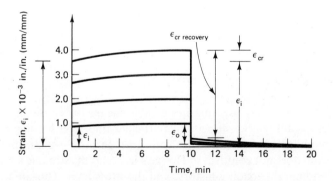

Figure 3.2 Stress-strain-time characteristics of a linear viscoelastic material.

plastic exhibits linear viscoelasticity. It follows that at any given time interval, the ratios (σ/ϵ_t), $(2\sigma/2\epsilon_t)$, ... will be constant, and thus the *effective* elastic modulus

$$E_t = \frac{\sigma}{\epsilon_t} \tag{3.2}$$

is instantaneously constant with respect to time.

3.3.4 Mechanical Properties of Plastics

The typical mechanical properties of commonly used plastics are summarized in Table 3.1. The moduli can vary from about 6000 to 8000 psi (42 to 55 N/mm²) for plastic foams to values between 1×10^6 and 2×10^6 psi (6900 and 13,800 MPa) for some glass-reinforced plastics. For common acrylics, polyesters, celluloids, epoxies, and other plastics, the moduli values range between 3×10^5 to 6×10^5 psi (2070 to 4140 MPa). These relatively low moduli values (compared with steel and concrete) result in measurable strains and deflections in plastic models without requiring large loads. Typical tension and compression stress-strain curves for Plexiglas (grade G and grade II UVA used at MIT) and a typical stress-strain curve for PVC [normal-impact grade Type I used at MIT.; sheet thickness of 0.030 in. (0.8 mm) and 0.060 in. (1.6 mm)] are shown in Fig. 3.3. A series of Cornell University tests for tension on instrumented Plexiglas coupons showed that after about 10 min of loading, the modulus of elasticity decreases to about 90% of the instantaneous value. The

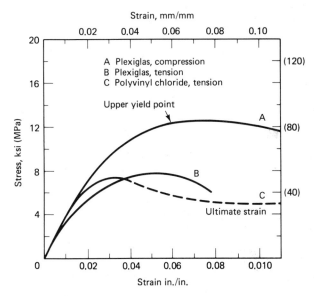

Figure 3.3   Stress-strain curves for PVC and Plexiglas. [After Pahl and Soosaar (1964).]

specimens ceased to remain linear viscoelastic after a strain of about 1700 μin./ in., which is in agreement with the range of 1500 to 2000 μin./in. observed by other investigators. Although these Plexiglas specimens underwent large total deformations, they were not ductile and exhibited little warning of failure. However, some of the softer PVCs are more ductile. Unlike other plastics, ethyl cellulose and polycarbonate (thermoplastic kind) have stress-strain curves in tension similar in shape to that of steel (Fig. 3.4) but without strain-hardening and have been used by Harris et al. (1962) for model studies of welded steel frame structures subjected to blast-type dynamic loadings. It should be noted that at small strains, the stress-strain curves for most plastics in tension and compression and the corresponding moduli of elasticity are almost identical. Thus, the modulus of elasticity in flexure will be equal to the modulus of elasticity in tension or compression. However, there is a variation between the compressive and the tensile yield strength, the former being greater.

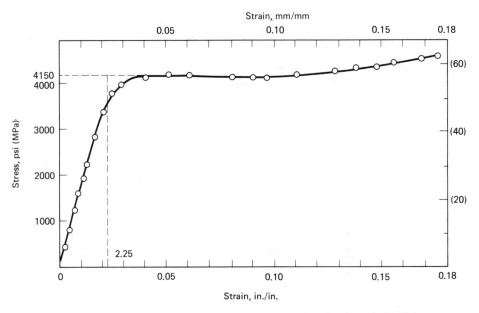

Figure 3.4 Stress-strain curve of ethyl cellulose. [After Harris et al. (1962).]

3.3.5 Mechanical Properties of Polyester Resin Combined
with Calcite Filler

Polyester casting resin with the trade name of Plastrene 97 [Balint and Shaw (1965)] is another material used for elastic models. The properties of the resin are varied by using different amounts of calcite filler; thus it is possible to match the moduli of elasticity of the dense and lightweight concretes of the prototype structure with that of the model in order to fulfill the similitude

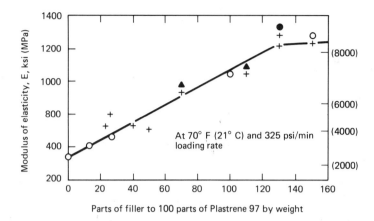

Figure 3.5 Variation of modulus of elasticity of Plastrene 97 with calcite filler. [After Balint and Shaw (1965).]

requirements (see Fig. 3.5). The influence of variation of the calcite filler on material shrinkage, coefficient of linear expansion, strength, creep characteristics, and glueability have also been investigated.

This material was used in the model studies of the circular 600-ft (183-m) high Australia Square, Sydney, which is constructed with lightweight concrete in the upper stories [Gero and Cowan (1970)]. Plastrene 97 is a moldable material and was used for this model because of the intricate shapes involved, such as recesses around doors, tapered columns and beams, and varied slab thicknesses; the change in the modulus of elasticity of the concrete through the building height; and the need for a low Poisson's ratio. At room temperature, complete polymerization of Plastrene 97 normally took place 90 days after the addition of the catalyst and the accelerator. However, this process was accelerated at elevated temperatures, and curing for about 3 hr at 150°F (65°C) resulted in a fully polymerized product [Balint and Shaw (1965)]. Best results were obtained by slow and even cooling of the heat-cured components, guarding them against warping. This slow cooling process resulted in a practically stress-free material. Reheating the cured product to 180°F softened the material and easily eliminated any unwanted warps.

Uniaxial tests on various mixes indicated linear viscoelastic behavior and equal moduli of elasticity in tension and compression. A 5-min test versus a 3-hr test showed the presence of creep strains and a varying modulus of elasticity. Typical creep curves for Mix 23 (23% calcite filler by weight) showing variation of strain with time and temperature at two stress levels (250 and 715 psi) (1.7 and 5 MPa) are shown in Fig. 3.6. The tensile strength of Mix 23 and Mix 130 were 5200 and 3300 psi (36 and 23 MPa), respectively. Variation of the modulus of elasticity of Mix 23 with time and temperature is shown in Fig. 3.7. Testing these fully cured specimens at elevated temperatures resulted in a loss of tensile strength.

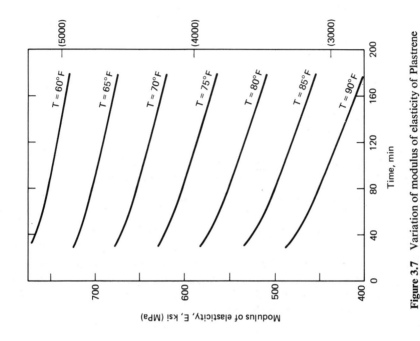

Figure 3.6 Typical creep curves for Plastrene 97 mix (23% calcite filler by weight). [After Balint and Shaw (1965).]

Figure 3.7 Variation of modulus of elasticity of Plastrene 97 mix (23% calcite by weight) with time and temperature. [After Balint and Shaw (1965).]

3.4 TIME EFFECTS IN PLASTICS—EVALUATION
AND COMPENSATION

The time-dependent strain or creep should be accounted for in interpreting the experimental data, especially if several strain gages are to be read and there is a time delay in readings. Because of creep, strain readings under a given applied load are changing with time; therefore no unique value of stress is associated with a measured strain value, and the time-dependent modulus of elasticity E_t must be used to evaluate the stress values from the corresponding strain values ϵ_t at a specified time t.

In an indirect model test one derives the influence lines or surfaces from the distorted shape of the model, and time-dependent properties do not influence the results. In these cases, the change in the stress with time at any point, that is, relaxation, is inconsequential. However in a direct model one cannot ignore the creep of the plastic, and proper care must be used in interpreting the test data.

If the stress values are low (about 20% of the material yield strength), then creep and creep recovery are linear functions of stress and all creep is eventually recovered upon the removal of stress. Similarly if the material is subjected to a strain, there is a stress corresponding to the initial strain, and this stress decreases with time under a sustained strain. This relaxation is also a linear function of strain if the imposed strain is small. Basic creep and relaxation behavior of plastics are normally determined by simple uniaxial or flexural tests, and it has been observed that in general the creep modulus of elasticity and the relaxation modulus are about equal for a specified time t. Hence only one time-dependent modulus, called the *apparent modulus*, is necessary.

3.4.1 Determination of the Time-Dependent Modulus
of Elasticity and Poisson's Ratio

Roll (1968) suggests that the time-dependent stress-strain curve or the effective modulus of elasticity can be easily determined by conducting creep tests in tension or flexure for as few as two values of stress. It is recommended that for better accuracy these tests be conducted for three different stress values σ_1, σ_2, and σ_3. As shown in Fig. 3.8a, at the time t_0 when the stresses are first applied the corresponding initial strains are ϵ_{i1}, ϵ_{i2}, and ϵ_{i3}, respectively. For stress values below the proportional limit, these points $(\sigma_1, \epsilon_{i1})$, $(\sigma_2, \epsilon_{i2})$, and $(\sigma_3, \epsilon_{i3})$ lie on a straight line corresponding to time t_0 (Fig. 3.8b), the slope of which is the initial tangent modulus of elasticity E_i (Fig. 3.8c). At some time t_1, the corresponding total strain values are ϵ_{t11}, ϵ_{t12}, and ϵ_{t13}, respectively (Fig. 3.8a), which result in the stress-strain curve (Fig. 3.8b) and the modulus of elasticity E_{t1} (Fig. 3.8c). In a similar manner, we can determine the stress-strain curves for times t_2 and t_3 and the corresponding moduli of elasticity E_{t2} and E_{t3}, respectively. Thus the variation of E_t with time can be determined (Fig. 3.8c).

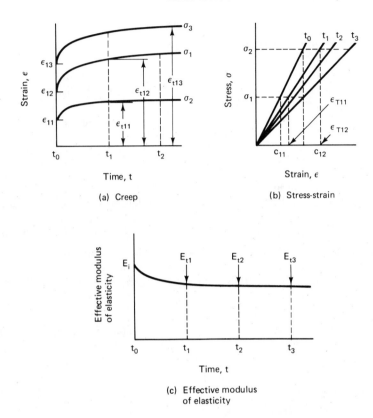

Figure 3.8 Time-dependent behavior of plastic. [After Roll (1968).]

The curve in Fig. 3.8c may be represented by the equation:

$$E_t = \frac{E_i}{1 + c(t)E_i} \tag{3.3}$$

where $c(t)$ is the unit creep function (i.e., the creep due to a unit stress). It must be remembered that Eq. 3.3 is based on the assumption that the material is strain-free when the stresses are applied. However, like all viscoelastic materials, plastics exhibit a "memory" effect that is a function of the stress history, including the currently applied stress. As for elastic materials, the principle of superposition is applicable for viscoelastic behavior provided that the governing differential equations are linear and that the resulting deformations do not significantly change the boundary conditions. In such a case, it can be shown that for a linear viscoelastic material, the total strain ϵ_t at any time is given by

$$\epsilon_t = \frac{\sigma_t}{E_i} + \sum_i \int_{t_i}^{t} \sigma_i \frac{dc(t)}{dt}\, dt \tag{3.4}$$

where $\sigma_t =$ the currently applied stress
$\sigma_i =$ the stress previously applied or removed at time t_i
$c(t) =$ the unit creep function
$E_i =$ the initial modulus of elasticity

It can be noted that if the unit creep function $c(t)$ is independent of stress at low stress levels and if the material is strain-free when stress is applied, Eq. (3.4) reduces to Eq. (3.3).

It is clearly advantageous to eliminate the complications resulting from the memory effects by loading the model in a *strain-free* condition. This can be easily achieved by removing the load after the strain readings have been taken and then allowing enough time for creep recovery before reloading the model. Because of the principle of superposition, one can expect the model to recover completely if it is left unloaded for as long a duration as for which it was loaded.

The Poisson's ratio v_t can be easily determined by using additional strain gages installed perpendicular to the longitudinal gages in a uniaxial tension or flexural test. These gages are used to measure the lateral strains. The value of the Poisson's ratio at time t, is given by

$$v_t = \frac{\text{lateral strain}}{\text{longitudinal strain}} \tag{3.5}$$

In general, the Poisson's ratio is not as sensitive to time effects as is E_t.

Another indirect method to determine the Poisson's ratio is to determine the value of the time-dependent shear modulus of elasticity G_t from a torsion test [Litle, (1964)]. The Poisson's ratio can then be calculated from the relationship

$$v_t = \frac{E_t}{2G_t} - 1 \tag{3.5a}$$

This test is cumbersome compared with the flexural test; however it is a useful and independent check when accurate values of the elastic constants are required.

For models made from plastics stressed in the range 0 to 2000 psi, (0 to 14 MPa) the Poisson's ratio normally varies from 0.3 for the thermosetting plastics and some of the more brittle thermoplastics to about 0.35 to 0.45 for most of the thermoplastics.

3.4.2 Loading Techniques to Account for Time-Dependent Effects

The difficulties caused by the time-dependent effects in plastics can be overcome by several techniques, including deferring measurements until creep has terminated, that is, until the additional increase of strain with time has become negligible. This time interval depends on the creep characteristics of the material, but in almost all cases it is also dependent on the magnitude of the applied stress. Low stress levels result in a lower creep rate and a smaller required time interval for creep termination. Preece and Davies (1964) suggest a waiting time of 6 hr; however Roll (1968) recommends a far shorter period of 20 to 30 min for low stresses. In any case, creep tests must be conducted at the highest stress expected in the model, so that the asymptotic time for creep termination can be determined for the model material. The stress at any point in the model can then be determined using the known value of the effective modulus of elasticity E_t.

Rocha (1961) has suggested a method of obtaining a sufficient number of readings from a viscoelastic model by cycling the load until the observed strain (or deflection) difference between the loading and unloading cycles becomes constant (Fig. 3.9). Under these conditions, the material behaves as if it were elastic, and the effective elastic modulus E_t is dependent upon the strain difference between the commencement of loading and the start of the following unloading cycle.

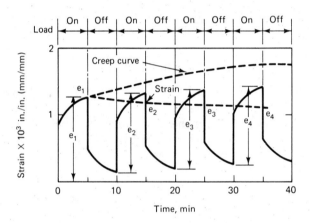

Figure 3.9 Repeated loading of a viscoeleastic material showing constant strain increment after first loading cycle. [After Rocha (1961).]

Another technique to handle the time-dependent effects in plastic models is the *constant strain method*, described by Carpenter (1963). This method has previously been used primarily for indirect tests, such as with the Beggs deformeter system. For tests on direct models, the method consists of imposing and maintaining a displacement at some point on the model and reading the resulting strains and deflections that do not vary with time. Thus the difficulties of obtaining time-dependent data from a deforming model are avoided. However, although the strains and deformations are not changing with time, the corresponding loads and stresses decrease with time, and some auxiliary method must be used to convert the measured strains into stresses. This is the major disadvantage in the use of the constant strain method.

Using the "spring balance" concept of Wilbur and Norris (1950) to circumvent this difficulty, Carpenter (1963) used a calibrated "strain cell" as the auxiliary structure in his tests. The function of the strain cell is illustrated in Fig. 3.10. The strain cell is made of the same material as the model, and its axis is oriented along the chosen direction of the applied deflection Δ. The strain cell can be instrumented as a load cell with a suitable number of strain gages and a circuitry to achieve maximum sensitivity.

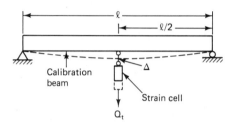

Figure 3.10 Calibration beam and strain cell. [After Carpenter (1963).]

The fixed displacement Δ imposed on the load cell results in a force Q_t that is a function of time. Since the displacement Δ is maintained constant, the strain in the load cell, ϵ_{cell}, and the strain at any point in the structure, ϵ_p, remain constant. Hence, the strain in the load cell is given by

$$\epsilon_{\text{cell}} = D\frac{Q_t}{E_t} \tag{3.6}$$

where D is the calibration constant for the cell. For the model structure,

$$\epsilon_p = \frac{\sigma_p}{E_t} \tag{3.7}$$

where σ_p and ϵ_p are the stress and strain, respectively, at some selected point, and for convenience they are assumed to be uniaxial.

Since the model material and the cell material are the same, by eliminating E_t between Eqs. (3.6) and (3.7), we get

$$\frac{\sigma_p}{Q_t} = D\frac{\epsilon_p}{\epsilon_{\text{cell}}} \tag{3.8}$$

Although the stress σ_p and the force Q_t, are changing with time, they are both proportional to the same relaxation function of time, and therefore, the stress per unit load remains unchanged with time. Thus the response of the structure is the same as that of a model constructed from a creep-free elastic material whose properties are *not* time-dependent.

The principal advantage of the constant strain method is that the strain readings do not change with time, and this eliminates the need to conduct additional creep tests to determine the effective modulus of elasticity. It is not really necessary to know the time-dependent force Q_t corresponding to the imposed displacement Δ. The construction and calibration of the strain cell is relatively simple and can be easily handled by an individual with some laboratory experience.

3.5 EFFECTS OF LOADING RATE, TEMPERATURE, AND THE ENVIRONMENT

Direct tension, compression, and flexure tests of materials at ordinary room temperatures have shown that the stress-strain relationships of most of the common metals are characterized by a typical stress-strain relationship. This is not the case for plastics that are more sensitive to the rate of testing (strain rate), temperature, and environmental effects. These effects are discussed in the following sections.

3.5.1 Influence of Strain Rate on Mechanical Properties of Plastics

The mechanical properties of plastics are influenced by the rate at which stresses or strains are applied. These effects are in addition to those due to time-dependent deformations, such as creep. ASTM D638-77a and ASTM D695-77 specify certain rates of crosshead movement for standard tension and compression tests, respectively. For tension tests, four speeds are listed, namely, 0.05, 0.20 to 0.25, 2.0, and 20.0 in./min with preference given to the 0.20 to 0.25 in./min (5 to 6 mm/min) rate. It must be noted that these speeds relate to a $4\frac{1}{2}$ in. (114 mm) grip length and may not correspond to strain rates over the gage length. For compression tests a speed of 0.05 in./min is specified between the bearing blocks supporting a 1-in. (25 mm)-long specimen.

In a model test, different parts of the structure and, for that matter, different points on the same cross section are subjected to varying rates of straining. This causes serious difficulties in assigning different values of E_t corresponding to the different strain rates for calculating stresses at various points in the structure. The influence of the different rates of straining can be minimized by applying the loads as slowly as possible and by leaving the model in a loaded state for some time before deformations are measured. Under such conditions, the creep strains will be large compared with the strain rate effects, and no significant error will be introduced by using the same value of E_t throughout the structure. It is obvious that serious difficulties can arise in interpreting test results for dynamically loaded models.

3.5.2 Effects of Temperature and Related Thermal Problems

The mechanical properties of plastics are significantly influenced by environmental changes such as changes in temperature and in some cases relative humidity. Therefore the testing environment must be carefully controlled, otherwise considerable errors can exist in correlating prototype and model behavior [Roll (1968)]. It is also important to know the effects of temperature changes on the mechanical properties of plastics. Of course, a knowledge of thermal properties of plastics is necessary in the model molding process and in the use of plastic models for thermal studies.

3.5.2.1 Temperature Effects on Elastic Constants

The values of the elastic constants are sensitive to temperature changes. For example, the tensile and shear moduli of elasticity of the acrylic Perspex, if assumed to be "standard" at 20°C, increase by about 1 % per degree Celsius fall in temperature [Preece and Davies (1968)]. These variations are practically linear at room temperatures; however, there is a marked decrease in the moduli values at temperatures greater than 100°C. The value of Poisson's ratio for Perspex remains practically unchanged (0.35 to 0.37) at room temperature, but this value increases to 0.5 if the temperature exceeds 100°C.

3.5.2.2 Temperature Effects on Strength

It should be noted that plastic models are normally used to simulate elastic behavior of a prototype; therefore it is essential to know the compressive and tensile strengths only to limit the working stresses in the model. The maximum stress in a model should be related to its type and duration. Normally, the maximum working stress should not exceed one-third of the ultimate tensile strength to ignore creep effects.

3.5.3 Coefficients of Thermal Expansion

Plastics have very high coefficients of thermal expansion relative to metals. According to Preece and Davies (1964), the coefficient of linear expansion for Perspex is 9×10^{-5} per degree Celsius over the temperature range -20 to 60°C. This is about nine times the value for steel. The coefficient of linear expansion is assumed to be one-third of the coefficient of volumetric expansion, which can be determined experimentally.

The high thermal expansion coefficient causes some problems in molding of the specimens. Because plastics either cure or are formed at high temperatures, the dimensions of any piece will be smaller than the mold dimensions after the piece has cooled to room temperature. In practice, it is difficult to separate thermal contraction and shrinkage of the resin. For example, unshrunk methyl methacrylate sheet material will shrink about $\frac{1}{4}$ in./ft (or 2 %) in the first heating cycle as a result of further polymerization; however, mold shrinkage is generally less than 1 %. Some plastics exhibit larger values. Fialho (1960) reported a 20 % shrinkage in the molding of a polyethylene model. Another direct consequence of the large thermal contractions experienced by molded parts as they cool to room temperatures is manifested as residual stresses in the molded part. This can lead to warping in the final specimen. It must be noted that careful temperature control during fabrication can reduce these effects. However, they cannot be totally eliminated.

3.5.4 Thermal Conductivity

Plastics have a very low thermal conductivity approximating 0.1 Btu/(hr) (ft) (°F), compared with 222 for copper, 28 for iron, 0.4 for asbestos, 0.2 for wood, and 0.03 for cork. Thermal conductivity is an important factor in determining stresses in thermal model studies. Also, because of the low thermal conductivity, when vacuum forming it is necessary to heat both sides of plastic sheets that are more than 0.05 in. thick. Otherwise the temperature differential through the sheet may cause problems.

The low thermal conductivity of plastics causes difficulties when strains are measured with electrical resistance strain gages (see Section 8.3.2). As the strain gage circuit is switched, the low thermal conductivity causes heating of the gage and the material under it, which in turn influences its mechanical properties. The heating of the gage itself causes a drift in the measuring circuit, resulting in a change of the output with time. Moreover, it is not possible to separate the output due to structural response and that due to drifting caused by heating of the gage. Litle et al. (1970) caution that very little quantitative data is available on the gage-heating effects and that one should be aware of this serious problem and should not "just willy-nilly" use strain gages on plastic materials. They also present some experimental data to illustrate the variation of measured strains with time as a result of gage-heating effects in a methyl methacrylate and a PVC specimen.

The following methods have been successfully used by some investigators to handle the problems resulting from gage heating:

1. Investigators at the Portland Cement Association [Carpenter et al. (1964)] simultaneously switch into the measuring circuit a cold active gage and a cold dummy gage. If both gages are in the same environment, the drift due to heating will be minimal. It is adequate to have only three or four dummy gages that can be used in turn, giving each gage sufficient time to cool off from the preceding cycle. This method has also been used at the Laboratorio Nacional de Engenharia Civil (LNEC) in Lisbon and at the University of Pennsylvania for strain measurements in tests on Perspex (Plexiglas) models of box-beam highway bridges [Roll and Aneja (1966)].

2. Johnson and Homewood (1961) have handled the gage-heating problem by having the gages heated at all times by means of a separate heating circuit that passes current through all gages not in the measuring circuit. This method has been used generally by the Cement and Concrete Association in England, and also recently by the Portland Cement Association.

3. Yates, Lucas, and Johnson (1953) used pulse excitation of the strain gage, with the measuring current passing through the gage in short pulses in the millisecond range. Since the time between two consecutive pulses is significantly longer than the pulse itself, the temperature of the gage remains low enough to prevent any significant heating effects.

The bridge voltage of the strain-indicating equipment is an important consideration in the gage-heating problem. The heating effect is proportional to the square of the bridge voltage, and therefore it is important to keep the bridge voltage as low as possible [Roll (1968)]. Carpenter et al. (1964) observed that an applied gage voltage of 3.5 V, corresponding to a bridge voltage of 7.0 V, was enough to damage a $\frac{1}{16}$-in. gage-length foil gage on Plexiglas. This is seldom a problem at present because of the availability of low voltage (1.5 V) strain-indicators. The use of gages that are compensated for plastics does not completely solve the problems; however, in general, the use of temperature-compensated foil gages along with a low-voltage strain-indicator will minimize the gage-heating effects. Regardless of the method used to handle the gage-heating problem, the mechanical properties of the plastic should be determined with the same equipment, techniques, and environment used for the model.

3.5.5 Softening and Demolding Temperatures

At the softening temperature, the shape of a plastic sheet can be easily molded to any curved profile. For acrylic plastics, such as Plexiglas, Perspex, and Lucite, this temperature is approximately 110°C [Preece and Davies (1964)]. According to Litle (1964), the vacuum forming temperature for Boltaron 6200 PVC is in the range of 85 to 127°C.

3.5.6 Influence of Relative Humidity on Elastic Properties

Like temperature, relative humidity affects the tensile strength, compressive strength, proportional limit, and yield strength of plastics. However, the elastic properties are not significantly influenced, and therefore the influence of humidity changes on the response of an elastic model can be neglected. The effects of relative humidity on the tensile and compressive strengths of methyl methacrylate (Plexiglas, Perspex, and Lucite) are shown in Fig. 3.11.

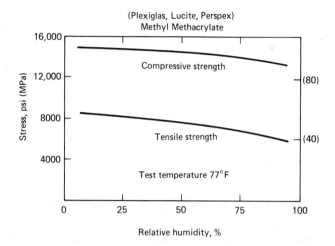

Figure 3.11   Influence of humidity on compressive and tensile strengths of methyl methacrylates. [After Manufacturing Chemists' Association (1957).]

3.6 SPECIAL PROBLEMS RELATED TO PLASTIC MODELS

There are additional characteristics peculiar to plastics; knowledge of these characteristics can be helpful in the construction and testing of the model and interpretation of the test data.

3.6.1 Modeling of Creep in Prototype Systems

It has been shown that the creep behavior of plastics can be troublesome in interpreting strain data; however it can be quite useful if the model is used to study the effects of creep on the behavior of the prototype. Ross (1946) simulated the creep behavior of reinforced concrete structures using a fabric-reinforced phenolic plastic. The unit creep of both materials was approximately equal for a time scale factor of 240; that is, the unit creep of plastic in 50 hr was equal to the unit creep of the concrete at 500 days. Using this material to construct models of reinforced concrete elements, he conducted experiments in short periods of time to determine creep-buckling in columns and to study the redistribution of reactions due to settling supports on continuous beams. Using unplasticized acrylic resin models, he also studied the effect of creep on the horizontal components of reactions in two-hinged arches. An excellent study of creep-buckling of cylindrical shell roofs under gravity loading, using Plexiglas models, has been conducted by Kordina (1964).

3.6.2 Poisson's Ratio Considerations

We noted in Section 3.2 that for skeletal-type structures the similitude requirement of equality of Poisson's ratio can be relaxed. However in grids, plates, and shell-type structures where the structural response is sensitive to Poisson's ratio, this similitude requirement must be satisfied. If the prototype structure is steel ($v_p = 0.30$), then models constructed from acrylic plastics, with Poisson's ratio of approximately 0.35, will be reasonably satisfactory, with little error in satisfying the similitude requirement. However, if it is required to model a prototype concrete structure ($v = 0.15 - 0.20$), then it may cause some discrepancy between the prototype and the model response, depending on the sensitivity of the structural response to Poisson's ratio.

3.6.3 Thickness Variations in Commercial Shapes

The forming process can cause significant thickness variations in commercial shapes of plastics that can be as large as $\pm 15\%$ of the nominal thickness. If not considered in analysis, this can seriously affect the interpretation of the test data. Therefore, if commercial shapes are used, care must be taken to use the best available grade to minimize this variation.

3.6.4 Influence of the Calendering Process on the Modulus of Elasticity

Thermoplastics such as polyvinylchloride are commonly manufactured as continuous sheets, up to 8 ft (2.5 m) in width, on rolls called *calenders*. The components are usually mixed and blended on heated rolls called *mills* or milled under pressure in a mixer and subsequently extruded through one or more sets of calenders or rolls at speeds of up to about 200 linear ft (60 m)/min. This process can produce continuous sheets ranging in thickness from 1 mil or less to about $\frac{1}{8}$ in. (3 mm) or more.

Litle (1964) examined the variability of modulus of elasticity and its sensitivity to some factors in an elastic buckling study of spherical shells, where the models were vacuum-formed from flat sheets of Boltaron 6200 PVC. A summary of the test values from four specimens from each of the 20 shells tested is shown in Fig. 3.12. This data establishes a sense of variability of the material modulus of elasticity. Also, it is important to note that although the plastic is manufactured by a calendering process, the anisotropy is not large.

Since the fabrication process for the shells [Litle (1964)] involved heat forming, the effect of annealing on the modulus of elasticity was studied. The results presented in Fig. 3.13 show the sensitivity of the modulus to annealing. It must also be noted that the vacuum forming process causes the material to

1. Four samples (two in each direction) taken for each model.
2. Modulus determination by cantilever beam test.
3. x implies parallel to calendering. Average = 454,000 psi
4. o implies perpendicular to calendering. Average = 447,000 psi

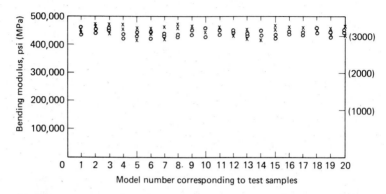

Figure 3.12   Modulus of elasticity of Boltaron 6200 PVC. [After Litle (1964).]

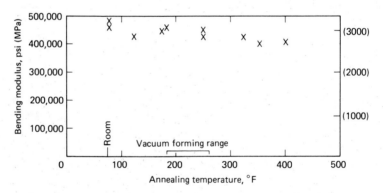

Figure 3.13   Effect of annealing temperature on bending modulus of Boltaron 6200 PVC. [After Litle (1964).]

stretch in going from the initial flat sheet to the contour of the model mold. The influence of this stretching on the material's modulus is shown in Fig. 3.14.

3.7 CONSTRUCTION OF ELASTIC MODELS

Models of buildings, bridges, and other structures are miniature structures in every sense, and as in the prototype, their construction requires careful planning and considerable skill and experience. Model construction requires good knowledge of the mechanical properties of the material selected along with its limita-

Type I PVC sample	Orientation in preformed sheet re: calendering	Linear stretch in tranverse direction, %	Linear stretch in longitudinal direction, %
1a	Perpendicular	80	−15
1b	Perpendicular	80	−15
2a	Parallel	0	0
2b	Parallel	0	0
3a	Perpendicular	0	0
3b	Perpendicular	0	0
4	Parallel	15	10
5	Perpendicular	10	15
6a	Parallel	0	80
6b	Parallel	0	80

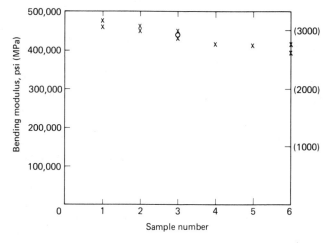

Figure 3.14 Effect of vacuum stretch on bending modulus of Boltaron 6200 PVC. [After Litle (1964).]

tions and the resulting difficulties in the construction ·process. The following factors must be given proper attention in the construction of any model: availability of proper working tools and trained personnel and limitations of available time and cost [Pahl and Soosaar (1964)].

Pahl and Soosaar (1964) and Cowan et al. (1968) classify the available methods for construction of elastic models into the following five basic techniques:

1. Cutting or carving of the model from a continuous piece
2. Assembly of two or more individual components
3. Thermal forming process
4. Casting process
5. Spin forming of soft-metal models

There are sources of systematic error and deviations inherent in any given construction process, and these can significantly influence some types of structural behavior, such as the buckling phenomenon of shells that may be very sensitive to geometrical imperfections. One must therefore carefully assess these potential sources of error before constructing the model.

3.7.1 Fabrication Considerations

The techniques of cutting, milling, gluing, welding, and drilling are normally used for relatively simple planar elastic models. Acrylic plastics possess the desirable characteristics of low elastic moduli, extensive linear ranges, good machinability, and relatively low cost and therefore are used very widely for the construction of elastic models. Plexiglas, Perspex, and Lucite, the more commonly used acrylics, are available in the form of rods, tubes, and sheets, with thickness varying from 0.03 to 4 in. (0.8 to 100 mm). The sheets are available in various sizes up to 60 × 100 in. (1.5 × 2.5 m).

Most acrylic plastics are subject to inception and growth of crazing, which is defined as visible mechanical cracks, to submicroscopic failures that result in noticeable blushing of an otherwise transparent material, or to fine cracks that may extend in a network over or under the surface of or through a plastic [Lever and Rhys, (1978)]. Crazing is normally dependent on the duration of stress application, but it can also occur under impact loads. Commonly used acrylic plastics are subject to crazing at stresses of the order of one-tenth of their tensile strength. Thus the stresses during testing must be limited to a maximum of 1500 psi at normal room temperatures. Also, proper care must be exercised in machining acrylic plastics using ordinary working tools [Rohm and Haas (1979)].

3.7.2 Assembly of Models Composed of Two or More Components

A majority of structural models are constructed from two or more individual components. In any structure, the joints are required to behave in a manner similar to the rest of the model. This may involve consideration of elastic, rheological, and ultimate load behavior.

3.7.2.1 Capillary Welding

Capillary welding for acrylic plastics is best achieved as follows: Adjoining pieces are held together firmly and solvent is applied to the joint with a hypodermic needle (Fig. 3.15). Acetone is usually used as solvent; however, any ketone or ester and some alcohols can also be used. The solvent is drawn into the joint by capillary action; it dissolves the plastic at this joint. With time the solvent evaporates, leaving a continuous and homogeneous joint that acts monolithically with the rest of the structure. This technique has an advantage in

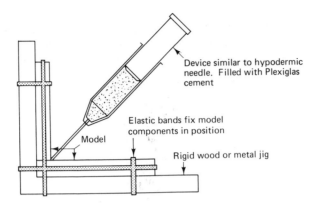

Device similar to hypodermic needle. Filled with Plexiglas cement

Elastic bands fix model components in position

Model

Rigid wood or metal jig

Figure 3.15 Capillary cementing of acrylic plastic models. [After Pahl and Soosaar (1964).]

that it requires no application of heat to aid polymerization. Also, the inspection of the joint in transparent plastics is very simple. A good continuous joint is clear and transparent, while a poor one is either cloudy or full of bubbles.

3.7.2.2 Glues or Adhesives

Some models are assembled using glues or adhesives that are normally different from the material being joined. Incompatibility problems may arise from differences that may be elastic, chemical, or viscoelastic in nature. The adhesive must be selected carefully because occasionally the solvent in the adhesive, which must evaporate to set, dissolves certain chemicals in the plastics and the joint is thus weakened. Although the areas of the model stiffened by the adhesive initially accept a greater proportion of the loads, if the adhesive is strongly viscoelastic, with time all load will be transferred to the adjoining members and the joint may possibly distort out of shape. An ideal adhesive must be capable of resisting the high level of stresses and strains that the model is expected to carry. This means that the failure criteria for the adhesive must be identical to those of the members it joins. This condition is seldom fulfilled in practice. Furthermore, the evaporation of the solvent, reduction in process temperature, or polymer network formation normally causes shrinkage in the adhesive, and parasitic stresses may result as a consequence. Therefore, special attention must be paid to the choice of an acceptable adhesive.

3.7.2.3 Epoxy Resins

Epoxy resins are used to assemble models because of their relative ease of application, their very high strengths, and the relative ease with which their elastic properties can be adjusted. By varying the proportions of the plasticizer and/or the accelerator or hardener (the two components of an epoxy resin), the elastic modulus can be adjusted from 30,000 to 600,000 psi (207 to 41401 MPa), thus permitting the joints to be compatible with the model material stiffness. Another advantage is that the setting of the adhesive does not release any volatiles, and therefore there is a minimal effect on the adjacent member material.

Considerable care must be exercised in cleaning the adherent surfaces before using any adhesives, especially for epoxy resins. Adhesives do not penetrate oil, grime, or oxidized layers, and these must be removed to obtain a good joint.

3.7.3 Thermal Forming Processes

Thermal forming techniques are used exclusively with thermoplastics which possess the basic characteristic of glass-transition temperature. Although the material does not melt, it becomes rubbery, and significant deformations can be applied to conform to any imposed shape with relative ease. It is possible to heat a thermoplastic sheet to just beyond its glass-transition temperature and force it against a prepared mold. The temperature is then reduced below the transition level, which will cause the plastic to cool in the shape of the mold.

It is important to fulfill the fundamental requirements of strength, stiffness, and creep characteristics in any thermoplastic selected for model construction. In addition, the glass-transition temperature, i.e., the temperature above which the plastic ceases to be hard and becomes rubberlike and easily deformed, must be a good deal above the room temperature to avoid accidental heating by radiators, sunlight, or electrical equipment. Moreover, this temperature must not be too high to impose unneccessarily large heating requirements. There must also be a sufficient spread between the glass-transition and melting temperatures to eliminate sudden transformations from the rubbery state to the molten state as a result of slight overheating. Extreme care must be used in selecting the model material, as not all polymers satisfy the above criteria.

It is important that the elongation required to form a given model does not exceed the ductility available in the rubbery stage between the glass-transition and melting temperatures. Any overstraining is visibly apparent as tearing and discoloration. The ductility is a function of the composition and structure of the plastic.

Polyvinylchloride (PVC) and polymethyl methacrylate are readily available commercially and have been successfully used in thermal forming of structural models. Pahl and Soosaar (1964) suggest that polymethyl methacrylate is successful in drape forming shells of single curvature; however, in vacuum forming, where stretching of the middle surface is required, the relatively low elongation available is a potential problem, and it is normally manifested in tearing of the sheets. Polyvinylchloride is more commonly used because it is easily formable, has an adequate temperature range, and has sufficient ductility. It is available as calendered sheets that have reasonably isotropic elastic properties. It is also available in several combinations with plasticizers and fillers and has many industrial applications.

3.7.4 Drape Forming or Gravity Forming of Shell Models

Gravity forming or drape forming of shells of single curvature is a relatively simple model manufacturing technique. A mold is fabricated from wood, metal, foamed plastics, or plaster to reproduce the desired geometrical configuration. It must be noted that all geometrical deviations and imperfections are reproduced in the formed shell, and therefore a high degree of precision is required in the fabrication of the mold. Wooden molds are dimensionally unstable on account of changes in humidity and temperature, and machined metal molds can be expensive and time-consuming. Quick-setting plasters can be used to fabricate accurate molds, especially for surfaces of revolution (Fig. 3.16). A skilled technician can limit the mold imperfections to about 0.001 in. (0.025 mm). Molds for hyperbolic paraboloid model structures can be formed by screeding (Fig. 3.17).

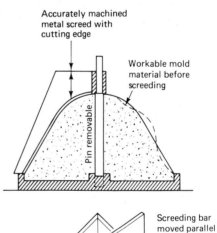

Figure 3.16 Plaster mold formation for surfaces of revolution. [After Pahl and Soosaar (1964).]

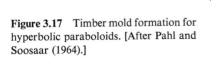

Figure 3.17 Timber mold formation for hyperbolic paraboloids. [After Pahl and Soosaar (1964).]

3.7.4.1 Drape Molding

The plastic sheet and the mold are set in controlled ovens, and the temperature is raised to just beyond the glass-transition temperature of the plastic. The plastic sheet becomes rubbery and unable to support itself and drapes onto the mold. The oven temperature is slowly lowered to eliminate any bowing of the model because of uneven cooling, and the plastic is allowed to cool in the form of the mold.

This technique can be used for forming shells of single curvature and very shallow shells of double curvature. For models with sudden changes in curvature, such as sharp corners or deep depressions, the rubbery plastic will normally not conform to sharp changes in geometry, and an additional mating mold must be used to force the rubbery plastic into the sharp corners (Fig. 3.18). Proper care must be taken to ensure that the elongation limit is not exceeded and that the plastic does not tear. Polymethyl methacrylate is well suited for the process of forming shells of single curvature and very shallow double curvature and is not suitable for forming models with sharp changes in curvature. Other plastics such as PVC are more suitable for such models.

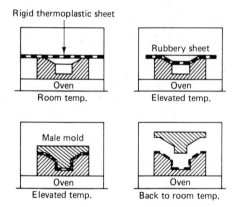

Figure 3.18 Two-mold drape forming of models with sharp corners. [After Pahl and Soosaar (1964).]

3.7.5 Vacuum Forming

Vacuum forming is used to form doubly curved models since the plastic will not stretch spontaneously in several directions under its own weight. Various techniques can be used, including blowing the rubbery sheet or vacuuming it onto a male or female mold. One technique (Fig. 3.19) consists of clamping a plastic sheet to a fixed frame and heating it to its glass-transition temperature using a movable electric coil heater. A platform with a mold is then raised and pushed against the rubbery plastic to form an airtight seal. A vacuum is applied from within the platform, forcing the plastic to conform snugly to the mold. The heating system is usually removed before applying the vacuum, and the model is allowed to cool on the mold. For models with sudden changes in curva-

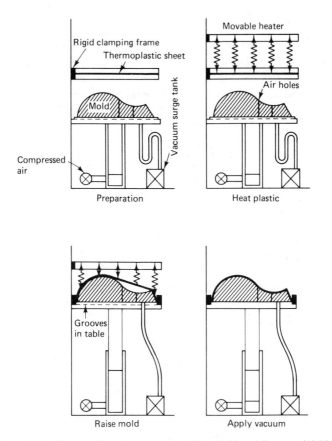

Figure 3.19   Vacuum forming sequence. [After Pahl and Soosaar (1964).]

ture, a companion mold can be used to produce local changes in geometry. Also, it may be necessary to provide additional access for the vacuum by drilling holes in the mold.

Uneven heating of the plastic sheets causes varying cooling rates in different areas of the model, which results in large residual stresses. Careful distribution of heating sources can minimize these effects. On account of the low thermal conductivity of plastics, it is possible to have a significant temperature difference between the two surfaces, which can cause tearing of the plastic sheet during forming. Pahl and Soosaar (1964) recommend that in forming models with thicknesses larger than 0.05 in. (1.3 mm) both sides of the sheet must be heated. Also, if some parts of a model are steeper than others, these parts require greater elongation during the forming process, which can be achieved by concentrated heating of these areas. A practical solution is to preheat the entire sheet with the regular heating elements, shielding the areas where the extra elongation is not required.

3.7.6 Fabrication Errors in Thermal Forming

The heating of the plastic sheet and its stretching during the forming process result in some manufacturing errors in the model. Thermal forming processes cause shrinkage in plastics upon cooling. This shrinkage has two basic components. Further heating after the primary heating cycle can cause approximately 1% contraction, and the high coefficient of thermal expansion results in another 2 to 3% contraction. In some cases, if the plastic is allowed to cool in contact with the restraining mold, the resulting contraction is usually reduced.

Vacuum forming of doubly curved shells causes the plastic sheet to stretch and become thinner in certain regions of high curvature. Litle (1964) noted that the extent of stretching and thinning are largely governed by the shape, the curvatures, and the sudden changes in the geometry of the structure. He observed very small behavioral differences, only of the order of a few percent, in some models formed using spherical and concave molds. However, structures with very pronounced geometrical changes can be expected to exhibit larger deviations in behavior.

3.7.7 Casting of Plastic Models

Some shell structures have varying thickness and thus are not amenable to precise analytical solutions. Gravity or drape and vacuum forming techniques produce shells of almost uniform thickness and cannot be used for shells of variable thickness. Although it is possible to construct a variable-thickness shell by gluing additional layers to a thermally formed model, the results from such models must be interpreted with reservation because of the problems associated with the gluing process [Pahl and Soosaar (1964)].

Variable thicknesses can be achieved by using a pair of matching and interlocking male and female molds and hand finishing with high precision to achieve good accuracy. For some structural models [Stevens (1959)] a technique of building layer upon layer of fiberglass cloth over a single mold and binding with resin is used. The dimensional control for this technique is normally not better than ± 0.01 in. (± 0.2 mm). The larger difference between the elastic moduli of the glass fibers and the resin matrix results in considerable anisotropy locally, and therefore strains measured by a short gage can be significantly different from the strains measured by a long gage. Stress concentrations also result from air bubbles trapped in the resin.

3.7.8 Spin Forming of Metal Shells

Structural models of shells can be fabricated commercially by the spin forming technique using soft metals such as aluminum alloys, copper, and some low-carbon steels. A flat disc of the selected metal is cold-formed into the desired shape, which is a surface of revolution, by spinning it at a very rapid rate while pressing a wooden or metal mold against it. This cold working does not affect the elastic properties appreciably, provided that the spinning is followed by annealing of the model. Also, curvatures in the model must be gradual relative to the material thickness to avoid local overstraining and tearing of the material.

Spin forming requires quite precise and heavy mechanical equipment, and consequently the fabrication, which can be expensive, is usually done commercially.

3.8 ELASTIC MODELS—DESIGN AND RESEARCH APPLICATIONS

For the past two decades, elastic models made from plastics have been used as a complement to a mathematical model in the design process and to obtain solutions for problems with poorly defined or very highly variable boundary conditions for which analysis is not feasible. Leading models laboratories in Europe, such as the Laboratorio Nacional de Engenharia Civil (LNEC), Portugal, Istituto Sperimentale Modelli e Strutture (ISMES), Italy, the Cement and Concrete Association (CCA), England, the Portland Cement Association (PCA), and Wiss Janney Elstner and Associates in North America have conducted model studies on many buildings, bridges, dams, and other structures. Recent development of facilities in North America and elsewhere should encourage more engineers to use models not only as an important complement to the analytical procedures but also as an integral part of research and development programs aimed at improving the current design procedures.

Elastic models have been used both to verify the designs prepared using the available analytical tools and as a partial or complete "design aid" to design several important buildings, bridges, and dams and some special-purpose structures such as aircraft hangars, prestressed concrete reactor vessels, reservoirs, and complex shell and dome roof structures. Another application of elastic models is the wind effects model that has come into prominence recently and is being used in the design of many large, exposed structures. Elastic models have also been used for studying the dynamic and seismic behavior of several important structures. These are discussed in Chapter 11.

3.9 DETERMINATION OF INFLUENCE LINES AND INFLUENCE SURFACES USING INDIRECT MODELS—MÜLLER-BRESLAU PRINCIPLE

An *influence line* is a curve the ordinate of which at any point equals the value of some particular function as a result of a unit load acting at that point. An *influence surface* can be defined similarly. The ordinates for an influence line for a particular structural action (force or moment reaction, or an internal force or moment) at a selected section in a linear elastic structure are evaluated analytically by placing a unit load successively at several possible load positions and calculating the value of the structural action at the selected station. It must be noted that the condition that the materials of the structure behave elastically is not sufficient to ensure that deformation is proportional to load. In addition, it must be ensured that there are no cable or stability effects, that is, the deformations do not influence the actions of the loads, and that the structure is sufficiently stiff so that the deformations under applied loads are small and do not change the geometry of the structure appreciably. Under such conditions, the Maxwell-Betti reciprocal theorem is valid and can be used to derive the Müller-Breslau principle, which provides a very convenient method of computing influence lines and forms the basis for certain indirect methods of model analysis [Norris and Wilbur (1976)]. This principle can be defined as follows:

> For any linear elastic structure if the restraint corresponding to any internal stress resultant (axial or shearing force, bending or twisting moment at a section) or a reaction is removed and a corresponding deformation (translation or rotation) is introduced, the ordinates of the deflected shape of the structure represent (to some scale) the influence characteristics for the stress resultant considered.

This principle is applicable to any type of structure—skeletal, surface, or solid structures—statically determinate and indeterminate alike. For statically determinate structures, the material need not be elastic for the principle to be valid; however, for statically indeterminate structures, this principle is limited to structures made of linear elastic materials.

Indirect models have been used to obtain influence lines and influence surfaces for linear elastic structures, especially for skeletal structures. The experimental procedure is quite simple. If influence characteristics are required for force, moment, or reactions at any support, the corresponding restraint (translation or rotation) is released and a prescribed deformation (translation or rotation) is introduced by a mechanical device. The results are interpreted in accordance with the Müller-Breslau principle. Similarly, if the influence characteristics are required for an internal stress resultant (axial or shearing force, bending or twist moment) at a particular section, the indirect model is cut at this section, and a prescribed deformation (translation or rotation) is imposed on this section by a mechanical device. The results are again interpreted in accordance with the Müller-Breslau principle. The normal practice is to impose a unit deformation, then the deflected shape of the indirect model becomes to some scale an influence line or an influence surface for the given reaction or internal stress resultant. For this reason, an indirect model is often also referred to as a *displacement model*. Therefore, the use of an indirect model to obtain influence characteristics for reactions or internal stress resultants requires cutting of the model and a suitable apparatus to displace the model. This technique has a significant advantage in that only deformations need be measured.

To plot the influence line for a generalized deformation (translation or rotation), a prescribed force is applied to deform the model, and the results are interpreted using the Maxwell-Betti reciprocal theorem. In this case, in addition to measuring the model deformations, it is also necessary to measure the force magnitude.

Indirect models have been used to obtain influence lines and to determine deflections in skeletal structures since the early twenties. Beggs (1932) and Eney (1939) were the first to use specially designed deformeters to obtain influence lines for skeletal structures made of plastic and other materials. The simplest application is the use of a long, flexible strip or spline of wood, brass, or steel and to measure the model deformations directly. The Gottschalk continostat (1926) is an improvement on this technique. However, large deformations must be imposed on the model, which causes other kinds of errors [Kinney, (1957)]. Bull (1930) used the brass spring model for indirect analysis of articulated structures. Development of other deformeters is reported by Ruge and Schmidt (1939) at the Massachusetts Institute of Technology (moment deformeter) and by Moakler and Hatfield (1953) at the Rensselaer Polytechnic Institute (RPI deformeter). A simple deformeter is shown in Fig. 1.1.

For reasons of space, only the Beggs deformeter is briefly discussed in the following section.

3.10 BEGG'S DEFORMETER

The first known application of an indirect model is due to Beggs, who developed a small apparatus to impose small displacements on plastic models using plugs and wedges (Beggs deformeter) as shown in Fig. 3.20. Models of planar structures are normally mounted on a sheet of smooth paper on a drawing board in a horizontal plane. An appropriate type of cut (representing a generalized force

Figure 3.20 Beggs's deformeter equipment (deformation devices, plugs, and microscope). See also Fig. 3.21 for similar devices. (Courtesy of Soiltest, Inc., Evanston, Illinois.)

release) is introduced at the section at which the reactive force is required. A specially designed deformeter is used to introduce relative axial, shear, or rotational displacements at the cut section. It is useful to drill holes along the axis of the structure, and a pin is used to mark the positions of the initial and deformed axis on the paper. These deformations are usually very small and must necessarily be measured by a micrometer microscope or by dial gages. To reduce friction, the model is often supported on ball bearings at a selected number of points. The errors arising from the deformation of the structure can be minimized and in general the accuracy of measurements can be considerably improved by applying equal deformations about both sides of the undeformed axis of the structure and by measuring the displacement between the two deflected curves [Hendry (1964)].

Pippard (1947) developed a small displacement deformeter (Fig. 3.21) for

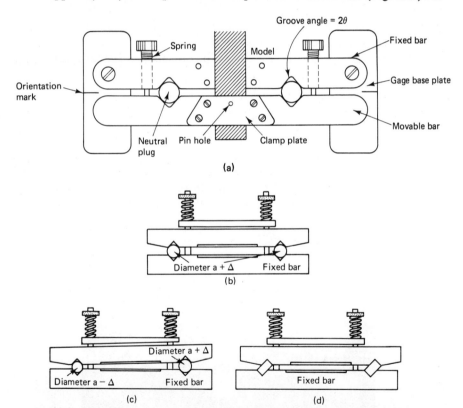

Figure 3.21 (a) Pippard's displacement deformeter for use with the Begg's deformeter: neutral or no displacement position. [After Kinney (1957).] Pippard's deformeter in position for determination of influence lines: (b) unit axial displacement—influence line for axial force; (c) unit angular displacement—influence line for bending moment; and (d) unit horizontal displacement—influence line for shearing force.

use with Begg's deformeter. The ends of the member at the cut section are attached to the two plates, which are separated by two plugs of equal diameter and located in V-shaped grooves with an angle of 2θ (Fig. 3.21a). To determine the influence characteristics for axial force, an axial displacement is introduced by replacing the original plugs by plugs of diameter $(a + \Delta)$, thus introducing an axial displacement (elongation) $\Delta \sec \theta$ along the member axis. Similarly an equal and opposite displacement (shortening) $\Delta \sec \theta$ can be introduced by using plugs of diameter $(a - \Delta)$ (Fig. 3.21b). For determining moment influence characteristics, a relative rotation $\Delta \sec \theta / L$ is introduced by using plugs of diameter $(a + \Delta)$ and $(a - \Delta)$ in the two notches a distance $2L$ apart (Fig. 3.21c). It is simple to cause a shear displacement by using rectangular plugs, as shown in (Fig. 3.21d).

Rocha and Borges (1961) developed a useful procedure for use with Beggs deformeter, which consists of photographing the model before and after the application of deformations on the same photographic plate, by "double exposure." The displacements can be measured at leisure and more accurately compared with direct measurements on the model using the permanent photographic record.

3.11 SUMMARY

Similitude relationships and materials suitable for elastic models are discussed in this chapter. Since plastics are the most commonly used material, their physical properties and uses are presented in detail. Also, some of the available fabrication techniques are presented as guidelines. Techniques are presented for the use of elastic models as indirect models to obtain influence lines and influence surfaces for any type of structure. Earlier uses of elastic models by Beggs, Eney, and others are summarized and the Beggs deformeter is discussed.

PROBLEMS

Elastic Models

3.1. The purpose of this exercise (to be done in the laboratory) is to obtain experience in the use of Plexiglas as a material for structural models. The time-dependent behavior of the material will soon become evident, and the problems of using a plastic with constant stress loading techniques will be demonstrated.

Specimen: One flexure specimen

Equipment available: Dial gages, strain gages, strain indicator, loading device with load cell, dead–load weight system.

Questions to be answered:

(a) Is Plexiglas linear viscoelastic in flexure? To what stress level?

(b) How does E vary with time?

(c) How does the behavior of the tensile side of the beam compare with that of the compressive side? Is the neutral axis of bending at middepth?

(d) How does response as measured by the dial gage compare with that measured with the strain gage?

(e) Does creep recovery appear to be the same phenomenon as creep itself?

(f) Is there a problem connected with the heating of the gages as the strain-measuring circuitry is activated?

(g) With the given equipment, what problems would you encounter with a relaxation test?

(h) For your specimen, what is the stress level at the peak deflection that you would still consider to be a "small deflection"?

(i) What is the ultimate strength of the material in flexure?

(j) Is the material ductile or brittle in its failure mode?

3.2. The lateral buckling capacity of a shallow circular timber arch is to be modeled because it may be unstable under a *dead* loading of 700 lb per foot of length until the roof is fully completed and can provide full lateral support. Using a plastic model with $E = 400,000$ psi, establish the design and operating conditions for stresses, deflections, and buckling load in a $\frac{1}{40}$-scale model. How will you load the model? How will you detect buckling?

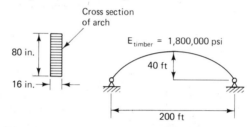

Cross section
of arch

$E_{timber} = 1,800,000$ psi

80 in.

40 ft

16 in.

200 ft

3.3. For the Plexiglas specimens tested in the laboratory:

(a) Plot strain (displacement) values as a function of time for each of the three loadings, following the format of Fig. 3.2 in the text.

(b) Plot strain (displacement) values as a function of stress and time. Compute modulus values.

(c) Is the plastic linear viscoelastic? Discuss the validity of this assumption for the two tests conducted.

(d) Suggest any improvements you would make in the experimental methods used in getting the data.

3.4. You are asked to do a stress and deflection analysis of a four-section hyperbolic paraboloid roof structure by means of model analysis. The prototype structure spans 80 ft and is built of 3000-psi concrete. It is 5 in. thick with edge beams 18 in. deep and 9 in. wide.

The model is to be no more than 3 ft square and is to be built of Plexiglas ($E = 450,000$ psi). The critical prototype loading of snow plus dead load is to be simulated by a series of uniformly spaced concentrated loads.

Design the model and determine the required model loads for a 100% overload on the prototype (basic snow load may be taken at 40 lb/ft²).

Pay particular attention to proper simulation of the dead-weight effects, including the edge members. That is, use similitude to establish the required model loads to properly simulate dead load, and specify clearly what the intensity of model loads must be.

Plexiglas sheets are available in thicknesses of 0.06, 0.08, 0.10, 0.125, 0.15, or $\frac{3}{16}$ in. for this particular project.

Specify all scale factors, including prediction equations for deflections, stresses, and strains. You may assume $E = 3,000,000$ psi for the concrete and specific weight $= 150$ lb/ft³.

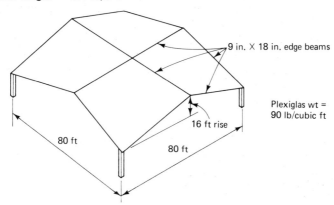

3.5. The model of Problem 3.4 is made of plastic. Why do we use plastic so often as a model material? What are the main advantages and disadvantages of using plastic for elastic models? How is the creep problem handled with respect to application of loads? (Your discussion should be complete but concise.)

3.6. It is proposed that the Plexiglas model built for the shell structure of Problem 3.4 be loaded at double the true model loading intensity in order to double the strains and make their measurement more accurate. Is this permissible? Discuss.

Inelastic Models: Materials for Concrete and Masonry Structures

〰〰〰〰〰〰〰〰〰〰〰〰〰〰

4

4.1 GENERAL

Direct models of reinforced and prestressed concrete structures require the use of materials to simulate the steel reinforcement or prestressing tendons, the prototype concrete, and the bond strength and dowel action at the steel-concrete interface. Similarly, in modeling of brick and concrete masonry structures, it is important to reproduce the tensile strength of the mortar along with the effects of the stressed volume. It must be noted that these direct models are miniature prototypes and require construction techniques that should be as close as possible to those used in the construction of the prototypes. Therefore, the success of direct models of structural concrete and masonry structures depends upon the degree of success (or accuracy) with which the models engineer can simulate the relevant prototype material properties and loading and environmental conditions.

This chapter emphasizes the modeling of prototype materials used in the construction of reinforced and prestressed concrete and brick and concrete masonry structures. Modeling of steel reinforcement along with the associated phenomena such as bond strength is discussed in Chapter 5. Principal characteristics of both prototype and model materials and their various strengths are presented along with a discussion of the influence of different parameters on these strengths.

4.2 PROTOTYPE AND MODEL CONCRETES

Prototype and model concretes consist of a mixture of inert granular substances held together by a cementing agent. More specifically, prototype concrete is a combination of cement, water, coarse and fine aggregates, and possibly admixtures, while model concrete normally consists of fine aggregates, cement, water, and possibly admixtures.

Aggregates are hard, chemically inert materials in the form of graded fragments. The most common aggregates are crushed stone, gravel, and sand, although other materials are frequently used for both prototype and model concretes. For example, expanded shales, slates, slags, and clays have been used in making lightweight prototype concrete. Gravels and sands have specific gravities of about 2.7 as compared with a specific gravity of about 1.7 or less for expanded shales. Any aggregate passing the U.S. No. 4 sieve (0.187-in. or 4-mm mesh) is designated as *fine aggregate,* and that which is retained on the sieve is known as *coarse aggregate.* The aggregate grading can influence the mechanical behavior of the resulting concrete. Well-graded aggregates yield the minimum void space and hence require the least amount of cement paste. Well-graded aggregates not only lead to economy in the use of cement, but also yield concrete having maximum strength and minimum volume change due to drying shrinkage.

The maximum aggregate size for prototype concrete is dependent upon the type of construction in which the concrete is to be used. The American Concrete Institute recommends maximum aggregate sizes from $\frac{1}{2}$ to 6 in. (10 to 150 mm); $\frac{3}{4}$ to $1\frac{1}{2}$ in. (20 to 40 mm) are the most commonly used in building construction. There are no universally accepted rules to determine the maximum aggregate size for model concrete, but it is normally established from the model geometric scale, the minimum member thickness in the model, and the reinforcement spacing. A survey of the literature shows that the maximum aggregate size used in model concrete ranges from $\frac{3}{8}$ in. (10 mm) or $\frac{1}{4}$ in. (6 mm) for $\frac{1}{2}$- to $\frac{1}{3}$-scale models to the U.S. No. 10 sieve (0.187-in. or 4-mm mesh) for $\frac{1}{6}$- to $\frac{1}{10}$-scale models.

Portland cement is used as a binder in most prototype and model concretes. It consists of four principal compounds: tricalcium silicate ($3CaO \cdot SiO_2$), dicalcium silicate ($2CaO \cdot SiO_2$), tricalcium aluminate ($3CaO \cdot Al_2O_3$), and tetracalcium aluminoferrite ($4CaO \cdot Al_2O_3 \cdot Fe_2O_3$). It has a specific gravity of about 3.15. When Portland cement is mixed with increasing amounts of water, it gradually becomes pasty and then a rather viscous liquid with considerable adhesion. After about 1 hr, the liquid begins to stiffen, and after 6 to 10 hr, it is hard or fully set. This hydration process can take place under water, and for this reason Portland cement is also known as a *hydraulic material.* No special consideration needs to be given to the water used for mixing concrete; a part of this water is used for the hydration of cement particles, while the remainder assists in generating a mix with reasonable workability for easy placing. Generally, ordinary water that is fit for drinking purposes is entirely satisfactory.

Often it is desired to improve the workability, to increase or decrease the setting time, to increase the strength, or to decrease the porosity of the cement-aggregate mixture over that which would naturally be obtained. Small quantities of materials called admixtures, such as calcium chloride, acetic acid, or an air-entraining agent, may be added in controlled quantities at the time of mixing or included in the cement itself, with the express purpose of increasing workability, etc., or bringing about other desired changes. The air-entraining agents are the most commonly used admixtures.

4.3 ENGINEERING PROPERTIES OF CONCRETE

Concrete is a unique construction material and possesses properties that are not common to other materials; for example, the tensile strength of concrete is less than its shear strength, which in turn is less than its compressive strength. Consequently, prototype concrete cannot normally be replaced by any other material in ultimate load model tests [Pahl and Soosaar (1964)]. The engineering properties of the hardened cement-aggregate mass that comprises prototype and model concretes are discussed extensively in the literature [Neville (1963)]; these are dependent on several factors, including:

1. Water-cement ratio.
2. Cement-aggregate ratio.
3. The nature of the aggregates, i.e., size, hardness, gradation, porosity, surface texture, etc.
4. Type of cement. Ordinary Portland cement Type I and rapid-hardening Portland cement Type III (high early-strength cement) are commonly used in the construction of both prototype and model structures. Rapid-hardening cement develops strength more rapidly; for the same water-cement ratio, the strength developed at the age of 3 days is of the same order as the 7-day strength of ordinary Portland cement. The increased rate of strength gain with rapid-hardening Portland cement is due to a higher tricalcium silicate content and the grinding of the cement clinker [Neville (1963)]. Use of rapid-hardening Portland cement is particularly important for model concrete as it decreases the laboratory time requirements.
5. Time history of moisture available for reaction with the cement, and time history of the temperature during this curing period.
6. Moisture content and temperature during testing.
7. Age at testing.
8. Type of stress caused by the applied loading: tension, bending, uniaxial compression, biaxial compression, triaxial compression, etc.

9. Duration of loading.
10. Strain rate of loading.

The effect of these factors and their interaction is not completely understood. In spite of significant research efforts since the early part of the century, the complete load-time-deformation behavior of concrete is one of the least well understood of all the common construction materials. These technical shortcomings are compounded by the fact that structural concrete is mixed, placed, and cured under a wide variety of conditions. Consequently, it is difficult to obtain a model material to simulate the prototype concrete by scaling the individual components according to the laws of similitude. Moreover the physical and chemical processes at the molecular level cannot be scaled down. However these limitations are not serious, as long as the physical properties of the model concrete, including its stress-strain curve and the failure criteria, are compatible with those of the prototype concrete according to the laws of similitude discussed in Chapter 2. Model concrete and gypsum mortar are commonly used in ultimate load model studies basically because of their rheological similarity to the prototype concrete. A knowledge of the relationship between the water-cement or water-plaster ratio and strength is important in the selection of a trial mix for model concrete and is discussed in Section 4.8.

4.3.1 Prototype and Model Concretes—Effect of the Microstructure

A close examination of the structure of concrete, mortar, and cement paste reveals an interesting analogy [Newman (1965)]. Model concrete is basically concrete, but on a reduced scale of at least one order of magnitude. In the concrete the stiff, coarse aggregate in a softer matrix of mortar is analogous to the hard sand particles in a softer matrix of cement paste of model concrete. A second-order magnification of the hardened cement paste structure reveals the same qualitative composition that is found in concrete and mortar of relatively hard unhydrated cement particles in a matrix of cement gel. These similarities can lead to theories of two-phase models of the behavior of these systems as a first approximation by excluding the void phase. Yet, marked differences in behavior under load are observed when cement paste, mortar, and concrete of the *same consistency* and normal compositions are compared:

1. Cement, mortar, and concrete show the same elastic portion on the stress-strain diagram.
2. The stress-strain curve for the concrete starts to deviate from the straight line at a lower stress than either mortar or paste.
3. The paste shows a linear stress-strain curve up to a very sudden and brittle failure.

4. Both mortar and concrete show gradual curving or nonlinear stress-strain curves with some ductility and warning of failure.

The addition of aggregate to the cement paste creates a heterogeneous system of a complex nature that behaves very differently from the paste under load [Hsu et al. (1963)]. Consistency is not a good basis for comparing these systems, however, and as will be shown in subsequent sections, the effect of the volume of aggregate in the system plays an important role [Ruiz (1966)]. In fact, when the volume of aggregates is less than a certain critical value, both mortars and concretes are stiffer than a paste made with the same water-cement ratio [Gilkey (1961)]. When this critical volume is exceeded, as is the case with most structural concretes, there is a very sharp decrease in strength and stiffness, with the net result that the paste is stronger than the mortar, which is in turn stronger than the concrete.

4.4 UNCONFINED COMPRESSIVE STRENGTH
AND STRESS-STRAIN RELATIONSHIP

4.4.1 Prototype Concrete

The unconfined compressive strength is considered to be the most important property of structural concrete. The stress-strain curve is reasonably linear at low stress levels and the modulus of elasticity E_c is usually taken as the slope of the tangent through the origin at the initial portion of the stress-strain curve (Fig. 4.1).

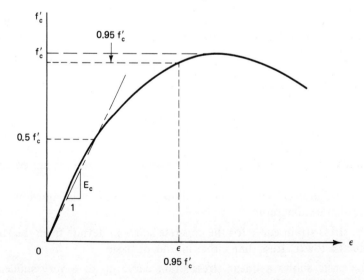

Figure 4.1   Typical stress-strain behavior of concrete in uniaxial compression.

Several different expressions are available for the modulus of elasticity. The ACI Building Code 318-77 (1977) uses Pauw's expression:

$$E_c = 33w^{1.5} \sqrt{f'_c} \qquad (4.1a)$$

where w is the unit weight of concrete in pounds per cubic foot, f'_c is the compressive strength in pounds per square inch (psi), and E_c is the modulus of elasticity in pounds per square inch. In SI units

$$E_c = 0.043w^{1.5} \sqrt{f'_c} \qquad (4.1b)$$

where w is the concrete mass density in kilograms per cubic meter (kg/m^3), f'_c is the compressive strength in megapascals (MPa), and E_c is the modulus of elasticity in megapascals.

4.4.2 Model Concrete

The response of any model concrete to external compressive load is of primary importance, because concrete as a structural material, is used mainly to carry compressive loads. The behavior of the prototype concrete under compression is normally all that is available when a model substitute is desired. For these reasons the stress-strain behavior in uniaxial compression becomes of primary importance in correlating material similitude between the prototype and the model structure. Model concrete mixes have to be designed carefully to simulate not only the strength and the stress-strain curve in compression but also some other characteristics such as the tensile strength, time-dependent behavior, etc. Design of model concrete mixes is presented in Section 4.9. The compressive strength of model concrete is determined using cylindrical specimens with a length to diameter (L/D) ratio of 2, similar to that used for prototype cylindrical specimens, although Harris et al. (1966) showed that in order to determine the true uniaxial compressive strength, an L/D ratio of at least 2.5 is needed.

The reader is cautioned that all material properties described and illustrated in the various figures should be considered to give general trends, and therefore this information can be used as a guide for initiating work on local materials to establish their properties and suitability for use as model materials. Such a study of materials must always be undertaken before initiating any model investigation.

A study of the literature on model concretes reveals that different sizes of cylinders ranging from $\frac{1}{2} \times 1$ in. (12.5 $\times$ 25 mm) to 6 $\times$ 12 in. (150 $\times$ 300 mm) have been used by various investigators. It will be shown in Chapter 6 that for the same concrete mix, the observed concrete strength increases as the test specimen size is decreased. A typical question that must be answered is: What is the strength of concrete in a $\frac{1}{2}$-in. (12.5 mm)-thick reinforced concrete shell model? One should probably attempt to obtain this strength from uniaxial compression tests on $\frac{1}{2} \times 1$ in. (12 $\times$ 25 mm) cylinders; however, because of

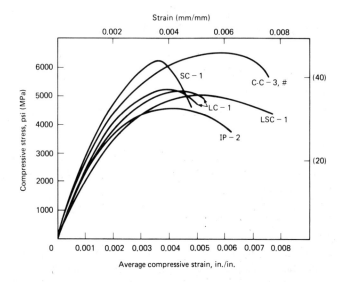

Figure 4.2   Stress-strain curves of model concrete in compression.

the very small specimen size, such tests are difficult to perform. The ACI Committee 444, Models of Concrete Structures, recommends a 2 × 4 in. (50 × 100 mm) cylinder as a standard size along with any other size cylinder that the models engineer selects. This would assist with correlation of strength data on model concretes. Typical stress-strain curves obtained with electrical resistance strain gages on 1 × 2 in. (25 × 50 mm) cylinders are shown in Fig. 4.2. Stress-strain curves obtained from tests on 2 × 4 in. (50 × 100 mm) cylinders of three model concrete mixes used at Cornell University [Harris et al. (1966)] with compressive strength ranging from $f'_c = 3200$ to 6800 psi (22 to 47 MPa) are shown in Fig. 4.3. The secant moduli ranged from $E_c = 2.1 \times 10^6$ to 3.4×10^6 psi (14480 to 23440 MPa). Several other model concretes used at Cornell University followed the same general trends.

4.4.3 Comparison of Prototype and Model Concrete Stress-Strain Characteristics

Several investigators have noted that the typical stress-strain behavior of model concrete cylinders with the same height-diameter ratio of 2 but of smaller dimensions than the standard 6 × 12 in. (150 × 300 mm) cylinders is very similar to ordinary concrete of the same ultimate strength [Sabnis and Mirza (1979)]. Figure 4.4 compares the stress-strain curve of a model concrete mix used at Massachusetts Institute of Technology and a prototype concrete mix used at the University of Illinois [Johnson (1962)]. Both mixes had an ultimate strength of 4600 psi, and the type of correlation generally expected is shown in Fig. 4.4. There are usually some minor variations in the modulus and the ultimate strain

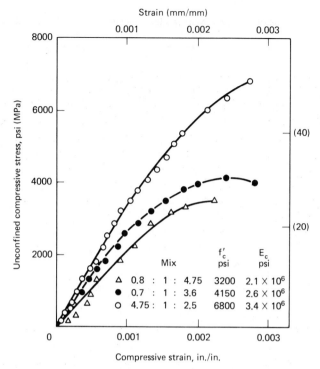

Figure 4.3 Compressive stress-strain curves for 2 × 4 in. microconcrete cylinders. [From Harris et al. (1966).]

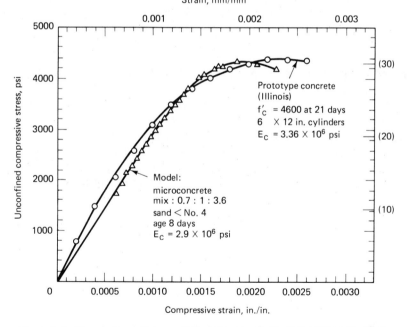

Figure 4.4 Comparison of stress-strain behavior of actual concrete and microconcrete at about 1 : 10 scale.

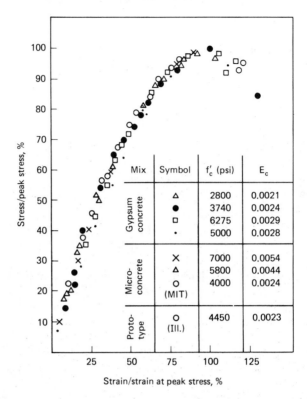

The table within the figure:

Mix	Symbol	f'_c (psi)	E_c
Gypsum concrete	△	2800	0.0021
	●	3740	0.0024
	□	6275	0.0029
	·	5000	0.0028
Micro-concrete	×	7000	0.0054
	△	5800	0.0044
	O (MIT)	4000	0.0024
Proto-type	O (Ill.)	4450	0.0023

Figure 4.5 Comparison of compressive stress-strain curves for various concretes.

with the variations in the compressive strength, but these are generally within an acceptable range.

Sabnis and White (1967) compared the stress-strain curves of prototype and model concrete and gypsum mortar mixes (see Section 4.10) using a nondimensional basis, as shown in Fig. 4.5. Excellent correlation was noted, and the nondimensional stress-strain curves can be considered to be similar for the prototype and model concretes and the gypsum mortar mix. Syamal (1969) tested prototype cylinders 6×12 in. (150×300 mm) and their $\frac{1}{2}$-, $\frac{1}{4}$-, and $\frac{1}{6}$-scale models in compression. The nondimensional stress-strain curves obtained are shown in Fig. 4.6, and it can be noted that these stress-strain curves are similar for the various sizes of cylinders, although there are slight variations in the moduli and the ultimate strains. For more details, see the paper by Mirza, White, and Roll (1972).

4.4.4 Creep and Creep Recovery of Concrete

When a concrete structure is loaded, the deformations continue to increase with time. Thus, although the initial deformations may be nearly elastic (Fig. 4.1), the strains continue to increase even under constant stress. Such additional

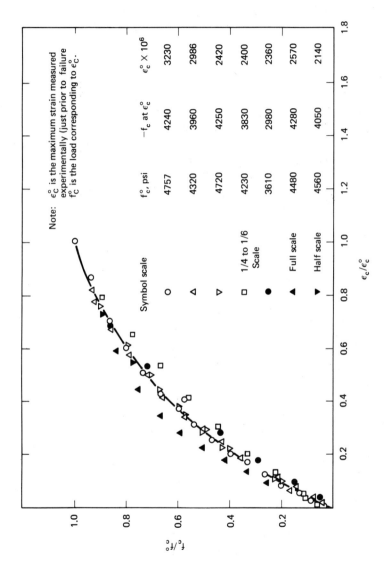

Figure 4.6 Nondimensional stress-strain curve for concrete in compression.

strains are referred to as *creep strains*. Some of the major factors that influence the creep characteristics of concrete are:

1. Age of concrete at the time of loading
2. Water-cement ratio
3. Relative humidity
4. Magnitude and time history of the applied stress system
5. Compressive strength f'_c

Results reported by Rüsch (1959) are shown in Fig. 4.7, which indicates the effect of time on the stress-strain characteristics of concrete. This shows the significance of modeling the prototype behavior for corresponding loading durations. No systematic studies have been undertaken on model concretes; however, since moisture exchange with the environment has a strong influence on creep, and since model concretes (in these small sections) can exchange moisture more readily, creep behavior of model concretes is expected to be different from that of the prototype concrete. Therefore the shape of the modeled stress-strain curve from a cylinder test must be approached by the same type of loading history that was used on the prototype.

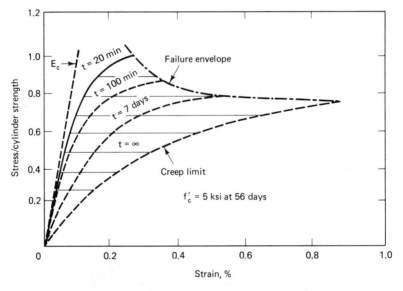

Figure 4.7   Stress-strain curves for various intensities and durations of sustained axial compressive loadings. [Rüsch (1959).]

4.4.5 Effect of Aggregate Content

Experiments at Cornell University [Harris, Sabnis, and White (1966)] showed that increasing the sand content in model concrete mixes caused the compressive strength to decrease, but the modulus of elasticity increased with

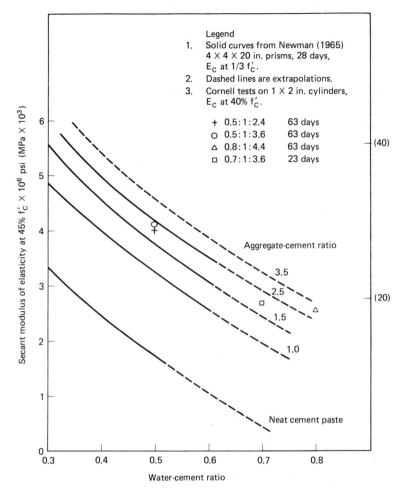

Figure 4.8   Effect of water-cement (W/C) and aggregate-cement (A/C) ratios on elastic properties of mortars. [Newman (1965).]

an increase in the aggregate-cement ratio. The Cornell University results are presented in Fig. 4.8 along with Newman's test data. This variation is due to the higher modulus of elasticity of the aggregate than that of the paste. However, the ultimate strength decreases with the increase of aggregate particles because they produce more stress concentrations and hence a greater probability of starting cracks at a given load level. The data from the Cornell tests using 1×2 in. (25×50 mm) model cylinders do not fall exactly on the curves given by Newman because his tests were on much larger specimens, $4 \times 4 \times 20$ in. ($100 \times 100 \times 500$ mm) prisms, and at a different age. The trend, however, is the same.

The effect of sand concentration on the modulus of elasticity and flexural strength of mortar beams has been studied by Ishai (1961). He found that a critical volume concentration of sand exists above which the material loses strength and stiffness very abruptly. A study of the effect of sand concentration on the modulus of elasticity, the ultimate strength, and the strain at 95% ultimate strength using square prismatic compressive specimens by Ruiz (1966) confirms Ishai's findings. The results, together with comparisons with some tests on small model cylinders, are shown in Fig. 4.9. Ruiz investigated mixes with water-cement ratios of 0.4 and 0.6. The rest of the curves are extrapolated from these findings.

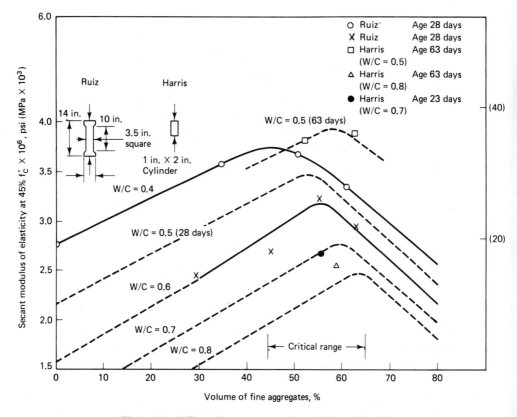

Figure 4.9 Effect of aggregate on the modulus of elasticity.

4.4.6 Effect of Strain Rate

The effect of increasing strain rate is shown in Fig. 4.10 for ordinary concrete and model concrete using 2 × 4 in. (50 × 100 mm) cylinders. The rate of increase of compressive strength with increasing strain rate seems to be greater for the model material. However, the smaller cylinder used for the model material does have a higher apparent strength than the larger cylinder at strain rates in the range of what is generally considered a "short-time" strength.

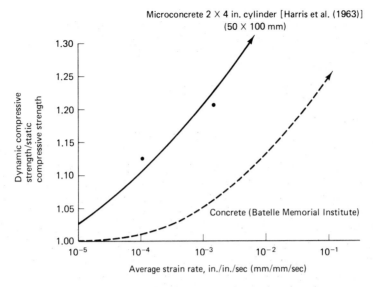

Figure 4.10   Effect of strain-rate on compressive strength.

4.4.7 Moisture Loss Effects

It has been observed by various investigators that when wet, cured concrete cylinders are allowed to lose part of their moisture, their compressive strength increases. Increases of up to 25 % have been observed when wet, cured specimens were allowed to remain at room conditions for a few days. In the case of model concrete structural elements, this partial drying during instrumentation and testing can be a cause of variations in strength. A method of coating the model that does not retard surface cracking and at the same time retains transparency is highly desirable. Shellac lacquer has reasonably good sealing qualities. It is very brittle in nature and therefore does not inhibit detection of crack formation.

An attempt to relate the amount of moisture loss with time in a controlled environment (21°C and approximately 70 % relative humidity) was made by Harris et al. (1966). Two groups of 1×2 in. (25×50 mm) compression cylinders, and one group of three $-1 \times 1\frac{1}{2} \times 3\frac{1}{2}$ in. ($25 \times 38 \times 89$ mm) prisms were used for this purpose. The results of these experiments are shown in Fig. 4.11. The behavior of cylinders and prisms was essentially the same. Of the cylinders, the densest mix had the lowest moisture loss, and the mixes that were cured for longer periods of time also had lower moisture loss.

4.4.8 Strength-Age Relations and Curing

Temperature and moisture have pronounced effects on the strength development of model and prototype concretes. The development of strength stops at an early age when the concrete specimen is exposed to dry air with no previous

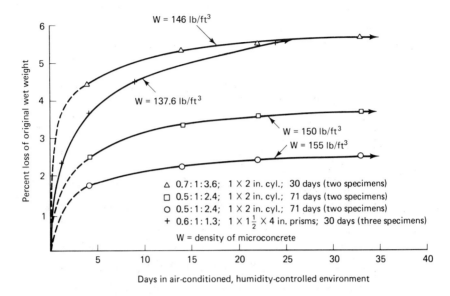

Figure 4.11   Rate of moisture loss in microconcrete in a controlled environment.

moist curing. This can have special significance in model making, where strength variations during testing need to be minimized. This effect on 6 × 12 in. (150 × 300 mm) cylinders is shown in Fig. 4.12; in the first 28 days the moist-cured specimen shows considerable increase in strength, and the air-cured specimen shows no effective increase in strength after 14 days. Figure 4.13 shows that for smaller elements used in model work the increase in strength is achieved in relatively shorter curing times, a desirable feature of the modeling technique.

Curing temperatures have a marked effect on the strength development of concrete, as shown in Fig. 4.14.

4.4.9 Statistical Variability in Compressive Strength

In order to study the statistical variation of the compressive strength of small cylinders, test series with large numbers of specimens were cast from the same mix and cured and tested in the same way at Cornell University. The standard deviation and coefficient of variation of each series was then computed. Standard deviations obtained from these model tests with those reported by Rüsch (1964) are shown in Fig. 4.15 and Fig. 4.16. For a large variety of concrete structures (Fig. 4.15a and b) the expected standard deviations show that models of very small scales are feasible (Fig. 4.16). The statistical variations of strength obtained from small model cylinders were not greater than 500 to 700 psi (3.5 to 4.8 MPa). The variation of standard deviation with mean compressive strength of 1 × 2 in. (25 × 50 mm) model concrete cylinders (Fig. 4.16) shows a trend similar to that for actual prototype structures.

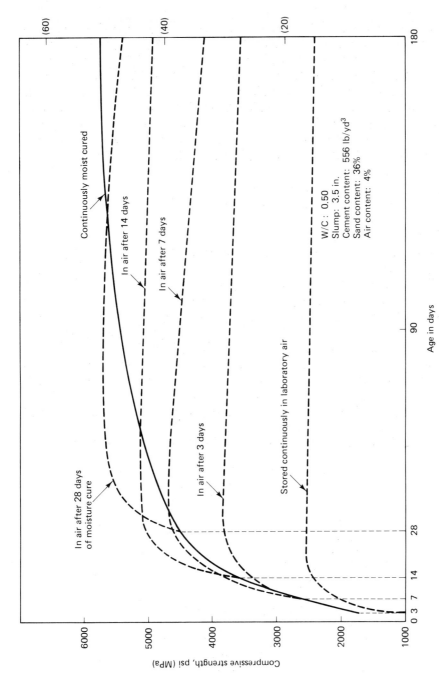

Figure 4.12 Compressive strength of concrete dried in laboratory air after preliminary moist curing. [Price (1951)].

Continuously moist cured

In air after 14 days

In air after 7 days

In air after 28 days of moisture cure

In air after 3 days

Stored continuously in laboratory air

W/C: 0.50
Slump: 3.5 in.
Cement content: 556 lb/yd^3
Sand content: 36%
Air content: 4%

Age in days

Compressive strength, psi (MPa)

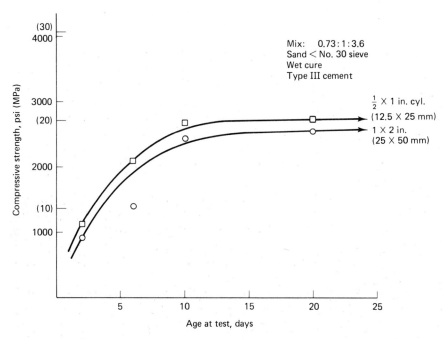

Figure 4.13 Strength-age curves of model cylinders.

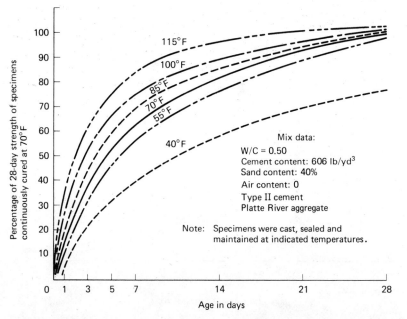

Figure 4.14 Effect of curing temperature on the compressive strength of concrete.

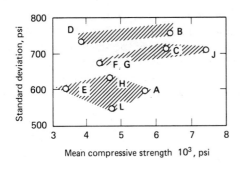

A = mass concrete
B = pavements and runways
C = bridges
D = housing, offices, schools
E = multistory structures
F, G = industrial buildings
H = tunnels
J = prefabricated members
L = ready-mixed concrete
Note: Results of an International
        questionnaire conducted
        by H. Rüsch.

(a)

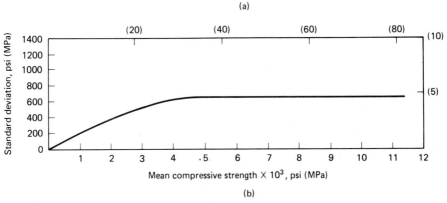

(b)

Figure 4.15 Variation of standard deviation with mean compressive strength for various types of concrete structures. [Rüsch (1964).]

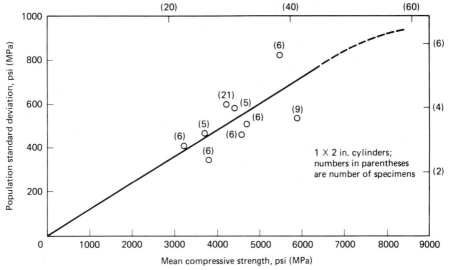

Figure 4.16 Variation of standard deviation with mean compressive strength of model specimens.

It is known that concrete cast under laboratory conditions has lower values of standard deviation than concrete on actual jobs. Therefore, it can be argued that in modeling, one should use the small size specimens to duplicate the actual prototype structure rather than those in the laboratory tests. As pointed out by Neville (1959), both the standard deviation and the mean strength decrease with increase in size of specimen, so that for the very small sizes it is to be expected that the spread in the measured strength as indicated by the standard deviation will be rather high. However, it has been shown by several investigators that with suitable care, the statistical variations of model concrete strengths can be made comparable to the variations observed on actual prototype structures.

4.5 TENSILE STRENGTH OF CONCRETE

The tensile strength of concrete is a fundamental and therefore an important property. The tensile stress or strain that can be developed in a prototype structure influences the behavior of the structure in many ways, including:

1. Strength in diagonal tension and resistance to shear
2. Bond strength of deformed bars
3. Cracking load levels and crack patterns
4. Effective stiffness of structure and degree of nonlinearity in response to load
5. Buckling behavior of thin shells

In addition, concrete tensile strength also influences the design process for prestressed concrete, pavements, and other structures where tensile strength is sometimes utilized in carrying load or the tensile stresses developed must be kept within certain limits.

It is evident that modeling of concrete tensile strength is an important consideration in the overall modeling process, and that the tensile properties of model concrete must be determined and understood. There are several tension tests available, and the final choice of the tests depends on the strain distribution existing in the member. For example, uniform tension across a member requires that the tensile strength be determined using direct tension tests that are difficult to perform and are not commonly used. In a direct tension test, a concrete cylinder is glued with epoxy to a set of coaxial platens that are in turn gripped in the jaws of the testing machine. Indirect tension test (split cylinder test) and torsion tests are used for sections with stress distributions in which the principal compressive and tensile stresses are of the same order. Compressive loads are applied to the cylindrical surface of a concrete cylinder at diametrically opposite ends; the cylinder is laid flat, and the load is distributed through two strips. The flexure test consists of testing a plain concrete beam with a square or a rectangular section under a central or third-point loading. This test is used in connection with reinforced and prestressed concrete flexure members, pavements, etc. Like the compressive strength, the indirect tensile strength and the flexural tensile strength of plain concrete specimens have been noted to be dependent on the following variables [Harris, Sabnis, and White (1966)]:

1. Type of test
2. Method and rate of loading
3. Specimen size
4. Maximum size of aggregate
5. Water-cement ratio, aggregate-cement ratio, and the method of casting
6. Effect of differential curing
7. Workmanship (quality of specimen)
8. Effects of strain gradient
9. Statistical volume effects
10. Differential temperature

Section 4.6 deals with tensile strength of prototype concrete and model concrete under strain gradient (flexural strength or modulus of rupture). Section 4.7 treats tensile strength under nearly uniform strain as well as the shear strength of model concretes. The various measured tensile strengths are compared in Section 4.7. The tensile properties of gypsum model concrete are presented in Section 4.10.

As is the case for the compressive strength of concrete, the tensile strength values obtained from any type of tension test are significantly influenced by the size of the test specimen. This effect of size on the tensile strength of concrete is discussed in Chapter 6.

4.6 FLEXURAL BEHAVIOR OF PROTOTYPE AND MODEL CONCRETE

The strength of nonreinforced concrete in flexure is usually obtained by testing prismatic beams under third-point loading in accordance with ASTM recommendations. The usual assumption of a linear distribution of strain across the depth together with a linear stress-strain relation is made. It is generally accepted that the tensile strength of concrete varies as $\sqrt{f'_c}$. The ACI Building Code 318-77 (1977) and the CSA Standard A23.3-M77 (1977) provide the following equations for the flexural tensile strength f_r of concrete

$$f_r = 7.5 \sqrt{f'_c} \text{ in imperial units} \tag{4.2a}$$

where f_r and f'_c are in pounds per square inch

$$f_r = 0.6 \sqrt{f'_c} \text{ in SI units} \tag{4.2b}$$

where f_r and f'_c are in megapascals.

The tensile strength of a model concrete is normally higher than that of a prototype concrete. It is therefore important to understand the flexural tensile properties of model concrete in the overall modeling process.

4.6.1 Specimen Dimensions and Properties

A series of model beams made of plain model concrete with mix proportions of 0.7 : 1 : 3.6 and sand passing the No. 8 sieve, and having a typical gradation curve shown on Fig. 4.17 were cast from the same batch and cured identically. The details of sizes and numbers of each are shown in Table 4.1. Each beam had a ratio of depth to width of 1.5 and a clear span between supports of four times the depth. Loading and support were accomplished with steel plates and rollers with dimensions as shown in Table 4.1. Third-point loads were applied to all the beams. Four beams from $1 \times 1\frac{1}{2} \times 6$ in. ($25 \times 38 \times 150$ mm) series (two cast vertically and two horizontally) and two from the $2 \times 3 \times 12$ in. ($50 \times 76 \times 300$ mm) series (one cast vertically and one horizontally) were instrumented with strain gages at midspan. Vertical casting meant that the larger dimension was vertical during casting. These instrumented beams were tested in a partially dry condition in order to ensure proper bonding of the strain gages, while the remaining beams were tested immediately after removal from the moist room at the age of 21 days.

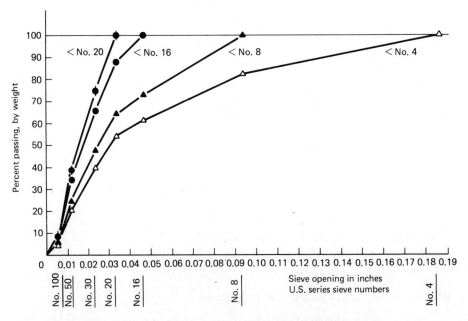

Figure 4.17 Gradation curves of Ithaca, N.Y. Glacial Deposits sand.

TABLE 4.1 Nonreinforced Beam Series

Section	Length		Number of Beams Cast		Rollers Diam.	Loading Details		
						Pad		
Breadth × Depth	Overall	Clear Span	Vertical	Horizontal		l	w	t
4 × 6	29	24	0	3	2	5	4	2
2 × 3	$14\frac{1}{2}$	12	3	3	1	$2\frac{1}{2}$	2	1
1 × $1\frac{1}{2}$	$7\frac{1}{4}$	6	6	6	$\frac{1}{2}$	$1\frac{1}{4}$	1	$\frac{1}{2}$
$\frac{1}{2}$ × $\frac{3}{4}$	$3\frac{5}{8}$	3	6	6	$\frac{1}{4}$	$\frac{5}{8}$	$\frac{1}{2}$	$\frac{1}{4}$
$\frac{1}{4}$ × $\frac{3}{8}$	$1\frac{13}{16}$	$1\frac{1}{4}$	6	6	$\frac{1}{8}$	$\frac{5}{16}$	$\frac{1}{4}$	$\frac{1}{8}$

Note: All dimensions are in inches.

4.6.2 Stress-Strain Curves

The experimental stress-strain curves were fairly linear except at higher loads, which agreed with Kaplan's findings (1963). Test results also showed that any extra drying of specimens resulted in additional shrinkage, which decreased the measured tensile strength. Hsu and Slate (1963) had made similar observations.

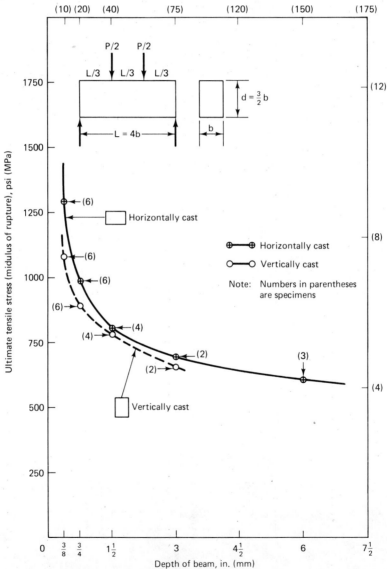

Figure 4.18   Variation of extreme fiber tensile stress with size of specimen.

4.6.3 Observed Variations in Modulus of Rupture with Changes in Dimensions

The results of the series of plain beams described in Table 4.1 were analyzed using the assumptions that the strain distribution is linear across the depth and that stress is directly proportional to strain. The extreme fiber tensile stresses were computed and averaged for each size group. The variation in the modulus of rupture with depth of beam is shown in Fig. 4.18. Horizontally and vertically cast beams show the same steep increase in strength with decreasing dimensions. The horizontally cast beams were on the whole 10 to 15% stronger because of their more homogeneous structure as cast. During casting it was observed that a gradual migration of water to the top layers takes place, resulting in partial segregation of the heavier sand particles to the bottom of the mold. The resulting structure of the vertically cast beams is thus less uniform than that of those cast horizontally.

4.6.4 Rate of Loading

Effects due to rate of loading of beams on the modulus of rupture have been investigated by Wright (1952) for concrete beams with an L/D ratio of 3 and an age of 28 days. Wright found that a linear relation exists between the modulus of rupture and the logarithm of the rate of increase of extreme fiber stress. Stress rates from 20 to 1140 psi/min (0.14 to 7.9 MPa/min) were investigated in his study, the results of which are shown in Fig. 4.19.

The rates of loading used in the model beam series were determined by the minimum rate of crosshead motion that could be read on the loading scale of the Tinius-Olsen machine. This minimum is 0.002 in./min. (0.05 mm/min) and was used for the smallest beams. The rates for the other sizes were increased in the ratio of the scale in order to keep the rate of stress increase constant. It was found, however, that the two larger-size beams could not be loaded at the scaled rates, so smaller rates of loading were used.

It can be seen from the result plotted in Fig. 4.19 that the rate of increase of the extreme fiber stress did not cause very large differences in the observed modulus of rupture values in these model beams.

4.6.5 Influence of Strain Gradient

Experimental data from studies of the effect of strain distribution in rectangular-section prismatic beams [Blackman (1958), Bresdorff and Hansen (1959), Harris et al. (1966)] shows that the ultimate strain at failure can vary considerably with size of beam. Blackman's data and its statistical analysis indicate that the effect of the strain gradient on ultimate strain for a particular shape of specimen can be approximated by a bilinear function.

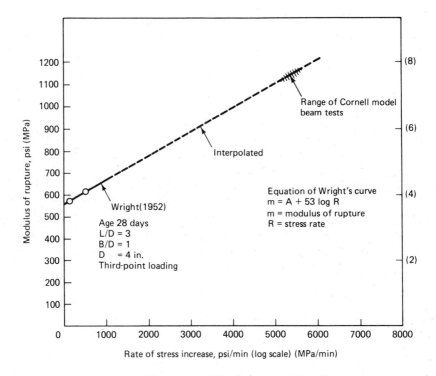

Figure 4.19   Effect of rate of loading on modulus of rupture.

4.7 BEHAVIOR IN TENSION AND SHEAR

Development of reliable and easily used methods for testing concrete in direct tension has been seriously hindered by the difficulties encountered in applying load to the specimen. Therefore, the flexure test was used to measure the tensile strength of concrete even though it does not measure the true tensile strength. However, the tensile split cylinder test and the ring test provide simple methods for determining the tensile strength under nearly uniform strain conditions. The split cylinder test has gained increasing acceptance as a tension test method during the last two decades; the modeling techniques for determining the splitting strength of model concrete and experimental data from some model tests are described in the following sections.

4.7.1 Tensile Splitting Strength

The split cylinder test measures the ultimate load needed to split a cylinder, lying on its side, with compressive line loads applied over a very small width on two opposite generators. This test, invented over 25 years ago by Carniero (1953) in Brazil and by Akazawa (1953) in Japan, apparently independently, has gained in popularity because of its simplicity and because it uses the same cylindrical specimen needed for compressive strength tests.

A stress state in the cylinder can be computed on the assumption of a linear elastic, isotropic, homogeneous material [Carniero and Barcellos (1953)]. The solution gives a uniformly distributed tensile stress over nearly the entire height of the diametral plane, as shown in Fig. 4.20, with high compression stresses developing in a small region near the loads. The exact distribution of

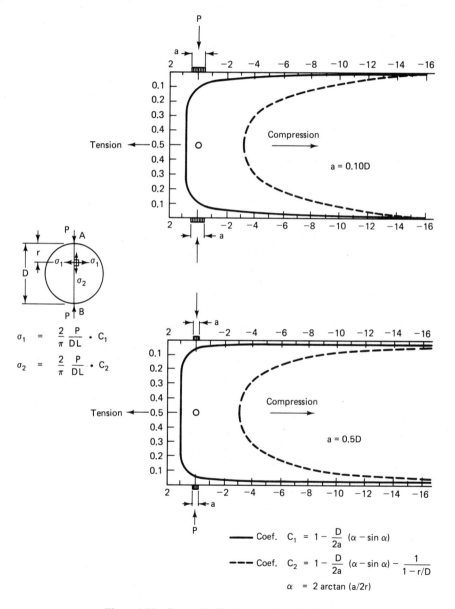

Figure 4.20   Stress distribution in split cylinder test.

stress is a function of the relative width and stiffness of the loading strip; Fig. 4.20 gives principal stresses σ_1 and σ_2 in terms of coefficients C_1 and C_2 for strip widths of $a/D = 0.1$ and $a/D = 0.05$.

When using the above relations to compute the stresses at failure, it is tacitly assumed that the stress distribution remains unchanged up to failure. This seems to be a questionable assumption for concrete inasmuch as the stress-strain relation deviates considerably from linearity as failure is approached.

Split cylinder specimens fail by splitting of the tensile zone, although the elastic analysis indicates very high compressive stresses near the load points. However, in the latter areas the two principal stresses are compressive; the apparent strength of concrete under such conditions is increased above its normal uniaxial strength sufficiently to cause failure to initiate in the region where the transverse principal stress σ_1 is tensile.

4.7.2 Results of Model Split Cylinder Tests

The results of split cylinder tensile tests on 1×2 in. (25×50 mm) model concrete cylinders are shown in Table 4.2. A comparison of the splitting tensile strength as a percentage of the compression strength is shown in Fig. 4.21 for the data of Table 4.2. A comparison is also made with prototype tests on 6×12 in. (150×300 mm) cylinders reported by various investigators. The model concrete cylinders have strengths of about $0.12f'_c$, while the concrete tensile strength range is slightly below this value.

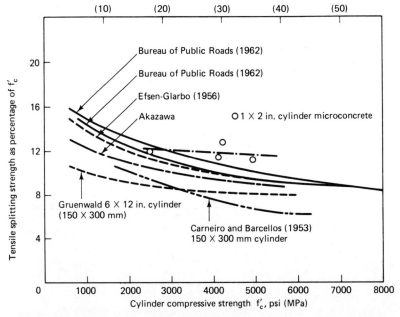

Figure 4.21 Relation between the tensile splitting strength and compressive strength of cylinders.

TABLE 4.2   Split-Cylinder Tensile Strengths of Microconcrete

| Size of Specimens, in. | Age at Test, days | Mix | Number of Specimens | | $f'_{t \, spl}$, psi | f'_c, psi | $\dfrac{f'_{t \, spl}}{f'_c}$ |
			Ten-sion	Com-pression			
1 × 2	28	0.8 : 1 : 3.6 < #16	8	8	300	2489	0.12
1 × 2	28	0.5 : 1 : 2 < #16	7	6	534	4175	0.128
1 × 2	31	0.6 : 1 : 3	6	5	594	4128	0.144
1 × 2	31	0.7 : 1 : 3.6	3	3	550	4940	0.112

4.7.3 Tensile Splitting Strength vs. Age

The strength-age relation for cylinders tested by splitting has been investigated by Carniero and Barcellos (1953) using 6 × 12 in. (150 × 300 mm) cylinders and three different concrete strengths. His results are shown in Fig. 4.22 and indicate that the tensile splitting strength has a similar dependence on age as the compressive strength.

Although the strength gain in tension is similar to compression strength gain as curing of the material progresses, there are indications that the proportionality between the two is not the same at all ages, and this affords at least a

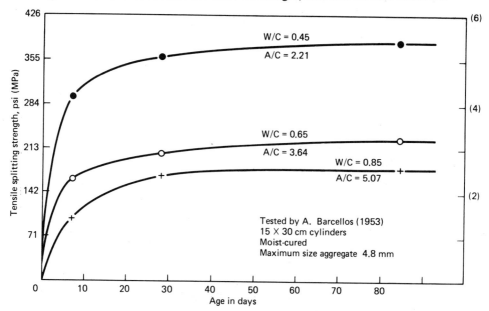

Figure 4.22   Splitting strength vs. age curves for mortars.

partial explanation of why there is so much scatter in combined stress test results performed at various ages. It has also been inferred that the bond strength between the paste and the aggregate deteriorates with age. If this is true, then the tensile strength will be affected more than the compressive strength and will start to decrease when the bond starts to deteriorate. However, up to ages of 3 months the tests of Carniero and Barcellos (Fig. 4.22) do not show this trend of reduced strength in tensile splitting.

4.7.4 Correlation of Tensile Splitting Strength to Flexural Strength

The correlation of tensile splitting and flexural strength of microconcrete has not been made. However, comparison of the individual strengths to proto-type concretes have been made and are presented in Fig. 4.23. It follows from the trends shown in these comparisons that the relations between the tensile strengths obtained from the three methods indicated will be approximately the same for model concrete as it is for prototype concretes.

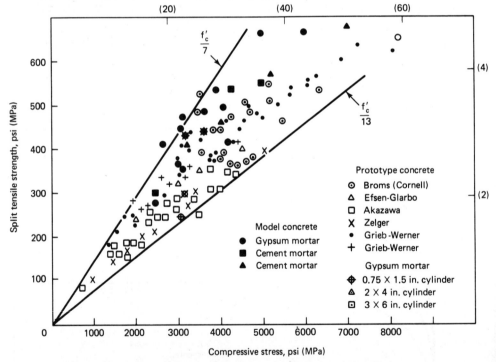

Figure 4.23 Relation between split tensile strength and compressive strength.

These relations for the prototype concrete are clearly indicated in Fig. 4.24. It is seen from this figure that the splitting tensile strength is closely related to the direct tensile strength over a wide range of compressive strength values,

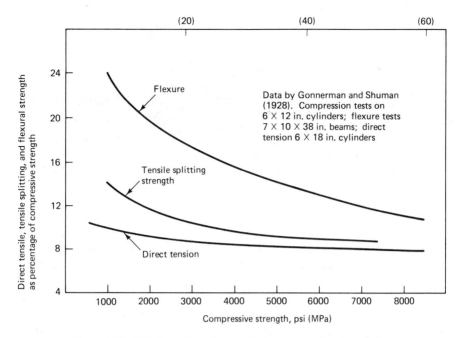

Figure 4.24 Relation of tensile strength to compressive strength.

while the flexural strength is considerably higher than direct tensile strength. At low values of compressive strength there is a greater discrepancy between the three tensile tests as shown in Fig. 4.24. The difference between splitting tensile strength and direct tensile strength decreases significantly as compressive strength increases.

4.8 DESIGN MIXES FOR MODEL CONCRETE

4.8.1 Introduction

The choice of a suitable model concrete mix is of considerable importance in direct modeling of concrete structures. After a model scale has been chosen by similitude and other considerations, the material scale is then well defined in terms of the relative size of the coarser particles of the heterophase. Thus, if a mean size of coarse aggregate exists in the prototype concrete, then a corresponding scaled-down mean size of (sand) particles exists in the model concrete. Then mix proportions must be chosen that will ensure mechanical properties similar to those of the prototype within specified limits. This method or design will generally achieve reproducibility within the confidence level common in engineering work and within the statistical variations of testing.

4.8.2 Choice of Model Material Scale

The cement is the same in model and prototype. The difference is in the aggregates. Aggregates used in structural concrete (with the exception of light-weight and high-density concretes) consist of particles whose size ranges from fine sand to coarse particles of a specified maximum average dimension. For the model material, ordinary well-graded concrete sand is used with scaling of the coarsest particles. It is implicitly assumed that the scale ratio between the maximum-size aggregates of model and prototype materials has the same ratio as the mean sizes of the two. In practice this is usually the case. The finer particles in the model mix are limited to less than 10% passing the U.S. No. 100 sieve (0.0059-in. or 0.149-mm mesh) and is done to prevent the necessity for very high water-cement ratios in order to obtain workable mixes. Using this method of modeling aggregate, it becomes apparent that the gradation curve of the sand becomes steeper and steeper as the size of the model decreases. Generally, the amount of aggregate and not the gradation has the greatest effect on the mechanical properties of model concrete.

In some localities some grades of very narrowly graded crushed sands are commercially available. The maximum aggregate size for these crushed sands normally varies from that passing U.S. No. 6 sieve to that passing U.S. No. 100 sieve. In such cases, a desired grading is first established by scaling of the coarsest particles and by limiting the amount of fines passing U.S. No. 100 sieve to less than 10%. Gradings of the various available grades of crushed sands are determined using the standard sieve analysis. Then by trial and error, four or five of the available grades of sands ranging from the maximum aggregate size selected for the model concrete down to the grade with a maximum sand particle size passing U.S. No. 70 sieve are combined to match the selected grading.

4.8.3 Properties of the Prototype To Be Modeled

It is generally required that a model material have the following properties under *short-time* load:

1. A specific ultimate compressive strength f'_c

2. A specified modulus of elasticity such as the secant modulus E_s at 0.45 or 0.5 of the ultimate strength f'_c

3. A specific ultimate compressive strain ϵ_u, or a strain at 95% ultimate, $\epsilon_{95\% \ ult}$

4. A specified ultimate tensile strength f'_c

In the case of prototype concrete all these properties are specified for a hypothetical 6×12 in. (150×300 mm) cylindrical specimen to be cast, cured, and tested under prescribed procedures set by the ASTM.

The values of the mechanical properties shown in Fig. 4.25 define bounds on the stress-strain relationship but do not define its *shape*.

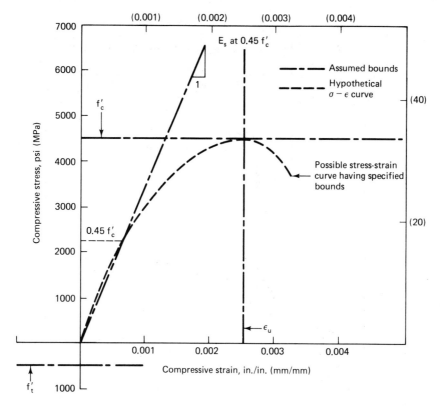

Figure 4.25   Specification of a hypothetical concrete stress-strain curve to be modeled.

4.8.4 Important Parameters Influencing the Mechanical Properties of Concrete

The mechanical properties of concrete (particularly those of interest to the models engineer) depend on a large number of parameters. The parameters that have the most influence on short-term behavior of concrete are: water-cement ratio, the percentage volume of aggregates in the system, aggregate-cement ratio, and the age of concrete at testing as shown in Fig. 4.26. For clarity, the dependence on time has been assumed to be linear in Fig. 4.26. These relations are plotted in Fig. 4.27 for an age of 28 days.

The choice of an appropriate mix is aided by the curves shown in Fig. 4.27, which are based on work done by Ruiz (1965); the extrapolated curves (shown dotted) have similar shapes to those obtained experimentally (full lines). It should be pointed out that in order for the choice to be valid, the model material should be tested at the same age, on the same size specimens, and under

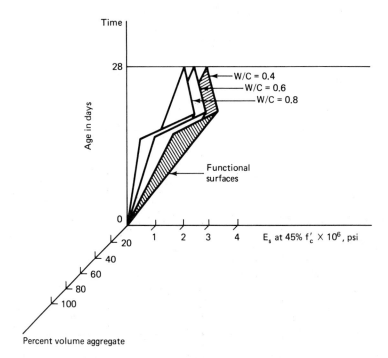

Figure 4.26   Schematic of functional relations existing between the property E_s and parameters W/C, age, and percent volume of aggregate.

similar mixing, casting, curing, and testing conditions as those used to derive such experimental curves as shown in Fig. 4.27.

With the accumulation of data and valid extrapolation, sets of curves can be established and used whenever a particular mix is to be designed. As an illustration, a model mix will be designed for a hypothetical prototype concrete with the following characteristics:

$$E_s = 3 \times 10^6 \text{ psi } (20.6 \times 10^3 \text{ MPa})$$

$$f'_c = 500 \text{ psi } (3.45 \text{ MPa})$$

$$\epsilon_{0.95f_{c'}} = 0.0025$$

In Fig. 4.27a, draw a horizontal line for $E_s = 3 \times 10^6$ psi (20.6×10^3 MPa) intersecting the various water-cement ratio curves at different percentages of aggregate volume. Similarly, by drawing horizontal lines for $f'_c = 500$ psi (3.45 MPa) in Fig. 4.27b and for compressive strain at $0.95f'_c$, $\epsilon_{0.95f_{c'}} = 0.0025$ in Fig. 4.27c, it is found that a water-cement ratio of 0.5 and a volume of aggregates of 65% (indicated by line 1) satisfy the specified values. The resulting mix is 0.5:1:4 by weight of water, cement, and sand; however, this mix will be too dry to use. A reexamination of Fig. 4.27 shows that a higher water-cement ratio of 0.70 and a 60% aggregate volume (indicated by line 2) can also result in a

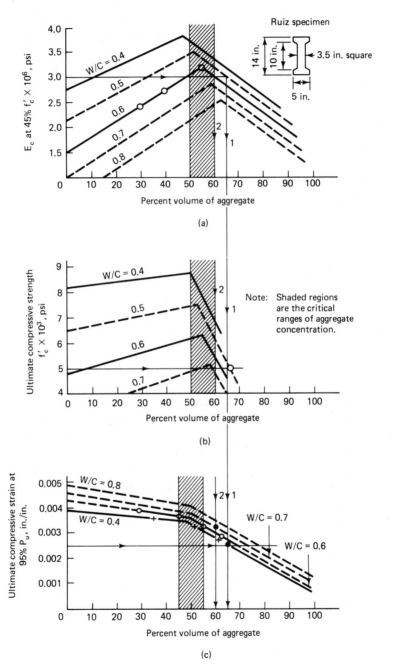

Figure 4.27 Relations between specified properties and most important parameters.

satisfactory mix. However, the ultimate strain will be about 28% higher than desired. Therefore the trial mix is 0.7 : 1 : 4, which is a workable mix.

In designing the above trial mix, we make the following observations:

1. Often, the specified values cannot be satisfied simultaneously, and compromises must be made.
2. To get low ultimate strains similar to those of prototype concrete, very high percentages of aggregate are needed, as shown in Fig. 4.27c. The resulting mixes are unworkable and have low modulus and strength since they are beyond the *critical volume* of 50 to 60%. This means that we must settle for somewhat higher strains in the model material.

Having designated a trial mix, one must cast sample cylinders and test them at the required age. If the mechanical properties are not satisfactory, some adjustments may be made.

4.9 SUMMARY OF MODEL CONCRETE MIXES USED BY VARIOUS INVESTIGATORS

Aldridge and Breen (1970) designed a model concrete mix based on an assumed aggregate gradation for the prototype 3000-psi (21-MPa) concrete mix (maximum aggregate size $\frac{3}{4}$ in., or 19 mm). The model concrete mix was a $\frac{1}{8}$-scale replica of the prototype mixture with the exception of the omission of the very fine particles. The maximum aggregate size used was No. 8 (0.0937 in., or 2.38 mm). Three other model concrete mixes were used for size-effect studies using maximum aggregate sizes of No. 4 (0.187 in., or 4 mm), $\frac{1}{4}$ in. (6 mm), and $\frac{3}{8}$ in. (10 mm) and cylinder sizes of 1×2 in. (25×50 mm), 2×4 in. (50×100 mm), and 3×6 in. (75×150 mm), respectively.

A number of gravels and sands, each with a limited distribution over only a few screen sizes, were blended to obtain the desired gradation. The grading of each basic aggregate was expressed by a linear equation indicating the portions retained on certain selected critical sieve sizes. If the basic aggregates are selected reasonably carefully, solution of a set of simultaneous linear gradation equations can indicate the appropriate proportions of each basic aggregate type for blending. The factors for each sieve size of a selected basic aggregate are multiplied by the same proportioning variable X_i, with the sum of all X_i values equal to 1.00. Moreover, the sum of the product X_i and the percentage of the basic aggregates retained on a given critical sieve must equal the desired percentage retained on the sieve in the model concrete mix. Any solution of these equations that results in positive values for all X_i will result in the formulation of a physically attainable model concrete mix. If negative values result for one or more X_i, the solution is not usable and one or more of the basic aggregates must be changed. The workability of the fresh model concrete was used as a visual measure of its consistency, since it was considered impractical to model any of the standard consistency devices. The compressive strengths of both the prototype and the model concrete, measured values of strains at $0.95f'_c$, and the values of the secant moduli at $0.45f'_c$ showed reasonable agreement. Stress-strain curves were obtained for the prototype and the model concrete from tests on 6×12 in. (150×300 mm) and 1×2 in. (25×50 mm) cylinders.

Mirza (1967) experimented with several trial mixes aimed at obtaining a compressive strength of 3000 psi using high early-strength cement (Type III) and local sand passing the U.S. No. 4 sieve (0.187-in., or 4-mm, mesh) and the U.S. No. 8 sieve (0.0936-in., or 2.38-mm, mesh), respectively. Large variations in strength were noted for the same water-cement ratio and aggregate-cement ratio; these were traced back to the inconsistent grading of the sand from batch to batch. A blended mixture of five grades of very narrowly graded crushed quartz sands, namely, No. 10, No. 16, No. 24, No. 35, and No. 70, was used as follows (percentage by weight):

No. 10 crushed quartz sand (0.0787-in., or 2.00-mm, mesh): 20%
No. 16 crushed quartz sand (0.0469-in., or 1.19-mm, mesh): 20%
No. 24 crushed quartz sand (0.0278-in., or 0.707-mm, mesh): 35%
No. 35 crushed quartz sand (0.0197-in., or 0.500-mm, mesh): 25%
No. 70 crushed quartz sand (0.0083-in., or 0.210-mm, mesh): 10%

The grading of the mix used is shown in Fig. 4.28 and is compared with the

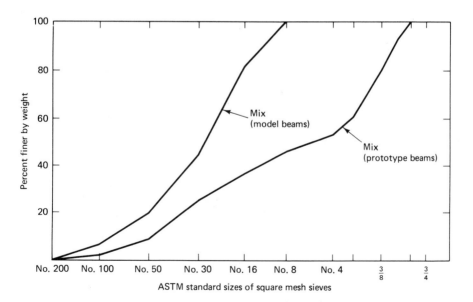

Figure 4.28   Mix details. [Mirza (1967).]

grading of the prototype mixes used in some McGill investigations. The water—cement—aggregate ratio used was 0.8 : 1 : 3.25.

A total of five hundred and eighty-eight 3 × 6 in. (75 × 150 mm) control cylinders were tested to determine the compressive strength of model concrete and resulted in an average compressive strength of 3135 psi (21.6 MPa) with a standard deviation of 137 psi (0.94 MPa). Similarly, one hundred and ninety-seven 3 × 6 in. (75 × 150 mm) control cylinders were tested in splitting tension; the average tensile strength was 395 psi (2.7 MPa) with a standard deviation of 14 psi (0.1 MPa). Thus, excellent quality control was achieved for model concretes used in this investigation. Compressive and splitting tension stress-strain curves for the model concrete were obtained from tests on 20 instrumented cylinders (14 in compression and 6 in splitting tension). Typical stress-strain curves for the model concrete in compression and indirect tension are shown in Figures 4.29a and 4.29b, respectively.

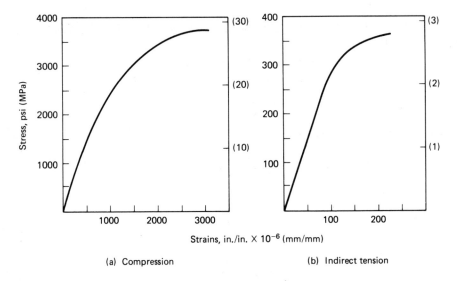

(a) Compression   (b) Indirect tension

Figure 4.29   Concrete stress-strain curves.

Tsui and Mirza (1969) developed model concrete mixes with design strength from 2500 to 6000 psi (17 to 41 MPa). High early-strength cement (Type III) was used throughout the investigation, and the aggregates consisted of two separate blended mixtures of fine, very narrowly graded crushed quartz sands passing sieve No. 10, No. 16, No. 24, No. 40, and No. 70. The combined gradings are shown in Fig. 4.30. The water-cement and the aggregate-cement ratios used for 14 trial mixes in these tests are detailed in Table 4.3. The statistical variation of these same mixes is shown in Table 4.4. Typical stress-strain curves are shown in Fig. 4.31. Variation of the direct compressive and the indirect tensile strengths with the water-cement ratio is shown in Figs. 4.32 and 4.33, respectively. Based

TABLE 4.3   Microconcrete Mix Details*

Mix No.	Water-Cement Ratio	Aggregate-Cement Ratio	Crading Curve Used	Remarks
A-1	0.8	3.50	1	Cement: Type III
A-2	0.775	3.50	1	Aggregate: Sand Nos. 10, 16, 24, 40, and 70
A-3	0.75	3.50	1	
A-4	0.725	3.50	1	
A-5	0.733	3.50	1	
A-6	0.675	3.50	1	
A-7	0.650	3.50	1	
A-8	0.63	3.50	1	
A-9	0.825	4.5	2	Cement: Type III
A-10	0.72	3.75	2	Aggregate: Sand Nos. 10, 16, 24, 40, and 70
A-11	0.635	2.5	2	
A-12	0.575	2.0	2	
A-13	0.535	1.5	2	
A-14	0.45	1.25	2	
B-1	0.72	3.75	2	Cement: Type III
B-2	0.635	2.50	2	Aggregate: Sand Nos. 10, 16, 24, 40, and 70
B-3	0.575	2.0	2	
B-4	0.535	1.5	2	
B-5	0.45	1.25	2	

*From Tsui and Mirza (1969).

Figure 4.30   Grading curves, Model concrete mixes. [Tsui and Mirza (1969).]

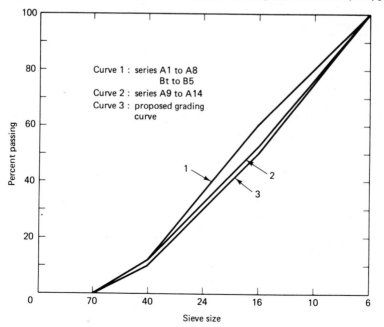

Curve 1 : series A1 to A8
           Bt to B5
Curve 2 : series A9 to A14
Curve 3 : proposed grading
           curve

TABLE 4.4   McGill Model Concrete Trial Mix Results*

Series No.	Direct Compressive Strengths			Indirect Tensile Strengths		
	Mean Strength, psi	Standard Deviation, psi	Coefficient of Variation	Mean Strength, psi	Standard Deviation, psi	Coefficient of Variation
Test Results, 3 × 6 in. (75 × 150 mm) Cylinders						
A1	2886	102	0.04	303	15.3	0.05
A2	2753	229	0.08	276	3.54	0.01
A3	2622	149	0.06	311	42.4	0.14
A4	3100	108	0.03	356	32	0.09
A5	2970	561	0.19	330	11.4	0.03
A6	3687	268	0.07	325	32.4	0.10
A7	3848	251	0.07	387	14.3	0.04
A8	3514	214	0.06	400	34.8	0.09
A9	2570	20.5	0.008	278	14.7	0.05
A10	3151	200	0.06	348	9.6	0.03
A11	3817	162	0.04	424	6.2	0.01
A12	4484	88.7	0.02	420	24.8	0.06
A13	6040	51	0.01	482	4.5	0.01
A14	5189	112	0.02	416	16	0.06
Test Results, 2 × 4 in. (25 × 50 mm) Cylinders						
B1	3533	177	0.05	488	12.1	0.02
B2	5013	242	0.05	631	19.3	0.03
B3	4795	66.3	0.01	597	17.2	0.03
B4	6573	512	0.08	656	13.8	0.02
B5	5942	247	0.04	574	13.6	0.02

*From Tsui and Mirza (1969).

TABLE 4.5   Recommended Water-Cement vs. Aggregate-Cement*

Water-Cement	Aggregate-Cement	f'_c expected
0.83	4.0	2500
0.72	3.75	3000
0.60	3.25	4000
0.55	2.75	5000
0.50	2.50	6000
0.40	2.25	7000

*From Mirza (1967).

on the workability obtained for different mixes, six mix designs have been recommended, as shown in Table 4.5.

From his model tests at Queen's University, Batchelor (1972) found that neither a model concrete with the properly scaled aggregates nor a mortar with

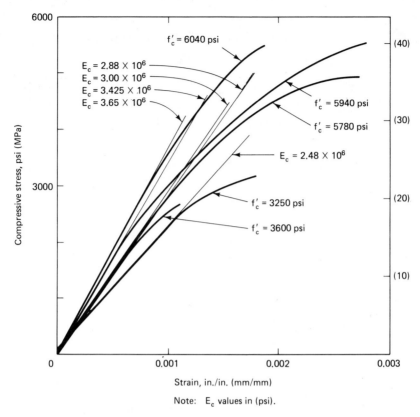

Note: E_c values in (psi).

Figure 4.31   Compression stress-strain curves, model concrete. [Tsui and Mirza (1969).]

Figure 4.32   Variation of compressive strength with water-cement ratio. [Tsui and Mirza (1969).]

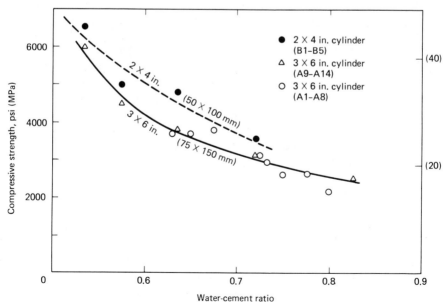

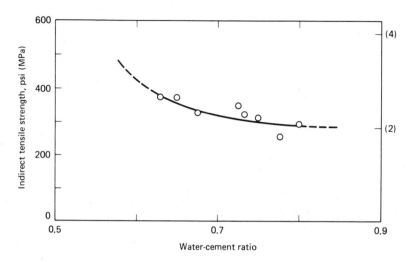

Figure 4.33 Variation of tensile strength with water-cement ratio. [Tsui and Mirza (1969).]

gypsum was suitable to model concrete. The fraction of a locally available fine limestone passing a $\frac{1}{4}$-in. (6-mm) sieve was used as the aggregate for a new model concrete. The maximum size of the aggregate was selected to match the minimum construction clearances rather than scaling them down, and the grading was then maintained within practical grading limits as shown in Fig. 4.34. Typical mix proportions of water, cement (Type III), and aggregate were 0.65 : 0.70 : 1 : 4 by weight. Typically, the concretes had slumps of 3 to 5 in. (75 to 125 mm), as determined by the standard slump cone, a 14-day compression strength of 5000 psi (35 MPa), and split cylinder tensile strength of about 400 psi (2.8 MPa), with a typical stress-strain curve as shown in Fig. 4.35. The ratios of the indirect tensile to compressive strengths were about 0.08 and thus were typical of the values to be expected for the prototype. The modulus of elasticity of the model concrete is in general somewhat lower than that for prototype concrete, and this factor has been recognized in research on models.

　　White (1976) experimented with 11 series of model concrete mixes detailed in Table 4.6. He used high early-strength cement (Type III) and varied the water-cement ratio from 0.40 to 0.63. The aggregate-cement ratio ranged from 2.5 to 5.6. River sand passing the U.S. No. 3 sieve (0.265-in., or 6.73-mm, mesh) and retained on the U.S. No. 100 sieve (0.0059-in., or 0.15-mm, mesh) and two different gradations of locally available Conrock aggregates, maximum aggregates sizes $\frac{3}{8}$ in. (10 mm) and $\frac{1}{4}$ in. (6 mm), were used, with the sand-gravel ratio being 1.0 for nine mixes and 3.0 for two mixes (Table 4.6). The Conrock aggregates were intended to be used as filter sands and were cleaned and kiln-dried by the manufacturer. The compressive strengths, the moduli of elasticity, and the indirect tensile (split cylinder) strengths at 14, 28, and 105 days were obtained

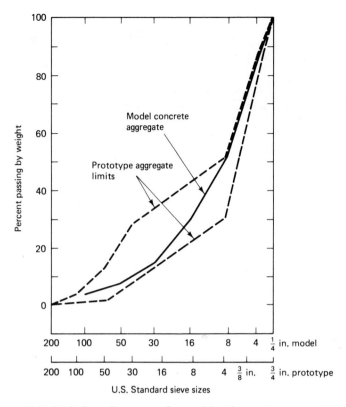

Figure 4.34 Typical grading curves for model and prototype aggregates.

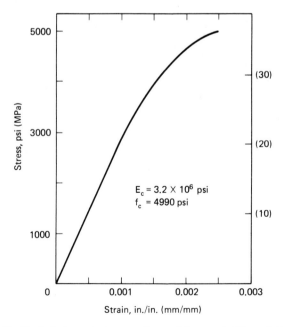

Figure 4.35 Typical stress-strain curve for model concrete (Queen's University).

TABLE 4.6 Mix Proportions and Summary of Properties of High-Strength Model Concretes*

Mix	1	2	3	4	5	6	1-A	2-A	3-A	4-A	5-A
W/C	0.455	0.481	0.50	0.50	0.459	0.40	0.547	0.60	0.547	0.60	0.63
A/C	3.0	3.41	3.8	3.8	3.41	2.5	4.5	5.2	4.5	5.2	5.6
S/G	3.0	1.0	3.0	1.0	1.0	1.0	1.0	1.0	1.0	1.0	1.0
Sand	S3-100	S3-100	S3-100	S3-100	S3-100	S3-100	S3-100	S3-100	S3-100	S3-100	S3-100
Gravel	G-$\frac{1}{4}$	G-$\frac{1}{4}$	G-$\frac{1}{4}$	G-$\frac{1}{4}$	Nelson $\frac{1}{4}$-#5	G-$\frac{1}{4}$	G-$\frac{1}{4}$	G-$\frac{1}{4}$	Nelson $\frac{1}{4}$-#5	Nelson $\frac{1}{4}$-#5	G-$\frac{1}{4}$
Workability	Low	Good+	Low	Good	Good+	Good−	Good	Good−	Good	Good−	Good−
14 day f'_c	9,577	9,037	8,440	8,703	8,587	10,755	8,156	7,003	7,623	7,085	6,121
E	3.92	4.44	4.32	3.79	4.19	4.38	3.79	3.90	3.57	3.81	5.37
f'_t	724	750	744	697	680	739	735	668	664	672	695
28 day f'_c	11,082	10,260	9,634	10,058	10,117	12,425	9,125	8,175	9,204	7,459	—
E	—	3.82	3.85	3.74	4.02	4.36	4.37	4.26	4.50	4.46	—
f'_t	—	893	892	790	807	890	732	704	740	725	—
105 day f'_c	10,816	10,122	9,114	10,090	10,398	12,159	9,634	8,497	9,458	7,676	7,395
E	5.40	4.54	4.23	4.41	6.64	4.32	4.38	4.66	4.04	4.45	4.20
f'_t	918	808	828	755	799	935	831	752	794	728	755

*After White.

using 2 × 4 in. (50 × 100 mm) cylinders. The average compressive strength at 28 days ranged from 7460 to 12,430 psi (51.4 to 85.7 MPa), while the average tensile strength ranged from about 700 to 890 psi (4.8 to 6.1 MPa). White's work shows that it is possible to design high-strength model concrete mixes; for example, a 12,000-psi (83-MPa) model concrete mix with reasonable workability can be obtained by using a water-cement ratio of about 0.4 and an aggregate-cement ratio of about 2.5 [with equal sand and gravel of maximum size $\frac{1}{4}$ in. (6 mm)]. The extremely high quality Conrock aggregate was the key to success.

4.10 GYPSUM MORTARS

Gypsum mortars, consisting of gypsum, sand, and water, are used as model concretes in some laboratories. Their main advantage is a fast curing time (1 day or less), while their main disadvantage is the strong influence of moisture content on mechanical properties. Gypsum gains strength and stiffness by drying and tends to become too brittle if allowed to air-dry for periods of several days or longer. Successful use of gypsum mortars in structural modeling involves preservation of a selected moisture content by sealing the surfaces of the models.

Several gypsum products have been used in model concretes, including Hydrocal 1-11, Ultracal 30, Ultracal 60, and Hydrostone, typically made in the North America. Most of the following discussion will concentrate on Ultracal mixes, as their properties have been studied most extensively [Sabnis and White (1967), (White and Sabnis (1968)].

The strength properties of Ultracal, as supplied by the manufacturer and for a water-gypsum ratio of 0.38, are:

Compressive strength (wet)	3000–5000 psi (21–35 MPa)
Compressive strength (dry)	6800–7800 psi (47–54 MPa)
Tensile strength (wet)	470–550 psi (3.2–3.8 MPa)
Tensile strength (dry)	670–720 psi (4.6–5.0 MPa)
Modulus of elasticity (dry)	2.5×10^6 psi (17×10^3 MPa)

The wet strengths are measured 3 hr after casting, and the dry-strength tests are conducted after specimens are dried to a constant weight at 110°F.

Ultracal 30 and Ultracal 60 differ only in the amount of set retarder added to the product at the mill; the numbers 30 and 60 refer to the time of set in minutes. It is recommended that Ultracal 60 always be used in order to maximize the available time for placing the model concrete in the forms and for making control specimens, although Ultracal 30 can be modified by the addition of a commercially available sodate set retarder (0.1 % by weight) to lengthen its setting time.

Ordinary mortar sands and those used for other concretes are satisfactory for gypsum mortars. No special gradations are required; normal practice at

Cornell University is to take out the coarser particles such that the ratio of the largest particle size to the smallest dimension of a model does not exceed $\frac{1}{5}$. The gypsum sand mixes used at North Carolina State University utilize sand that passes through a No. 16 mesh and is retained on a No. 40 mesh.

Suggested proportions for Ultracal mortars with compressive strengths in the range of 2500 to 4000 psi (17.2 MPa to 27.6 MPa) were developed by Sabnis and White (1967) and are given in Table 4.7. Higher-strength mixes are very difficult to control. Experience has shown that mixes that utilized Hydrostone and combinations of Ultracal 60 and Hydrostone are excessively inconsistent from batch to batch, and even within batches. The very early setting time of Hydrostone (about 10 min) may be a factor in these unsatisfactory results. The effect of the water-gypsum ratio and the aggregate-gypsum ratio on the compressive strength of gypsum mortar is shown in Figures 4.36 and 4.37, respectively.

TABLE 4.7   Properties of Ultracal Model Concrete Mixes

Mix	Sand	Ultracal	Water	Age, hr	f'_c, psi (MPa)
1	1.2	1.0	0.35	24	2400 (16.5)
2	1.0	1.0	0.35	24	2600 (18)
3	1.0	1.0	0.3	24	3000 (20.7)
4	0.8	1.0	0.3	24	3400 (23.4)
5	1.0	1.0	0.3	48	4000 (27.6)

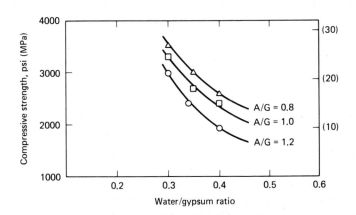

Figure 4.36   Effect of water-gypsum plaster ratio on compressive strength.

4.10.1 Curing and Sealing Procedures

The sealing process will be discussed before mechanical properties are presented inasmuch as the latter are a direct function of the moisture content. The cast material is allowed to air-dry for a specified time, after which all surfaces are sealed with a waterproof material. A seal consisting of two coats of shellac has proved to be better than most commercial sealing compounds.

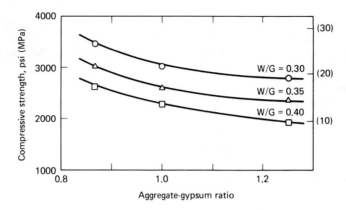

Figure 4.37 Effect of aggregate-gypsum plaster ratio on compressive strength.

One problem associated with the sealing process is the selection of the proper age for sealing. The drying rate of a specimen depends upon size and geometry. Data on this important matter is presented in Chapter 6 on size effects.

4.10.2 Mechanical Properties

No significant study of the failure criteria for gypsum mortars has been made; hence currently one must settle for data on those properties determined from the uniaxial compression tests and the various measures of tensile strength.

The shape of the uniaxial compressive stress-strain curve for several Ultracal gypsum mortars obtained by Sabnis and White (1967) is compared with curves for Portland cement model concrete and prototype concrete in Fig. 4.5; the gypsum mortars model this important property very well. Kandasamy (1969) arrived at the same conclusion from the results of his tests on eighty 3 × 6 in. (75 × 150 mm) gypsum mortar cylinders.

After some trials, Loh (1969) developed a casting procedure that resulted in uniform density of the test cylinders. Using this procedure, he cast some 100 compression cylinders and 100 splitting tensile cylinders from six separate batches. The cylinder sizes varied from 1 × 2 in. (25 × 50 mm) to 3 × 6 in. (75 × 150 mm) and were sealed with shellac to maintain uniform moisture content. All cylinders were capped with Hydrostone and were tested using accessories and loading rates that were scaled geometrically to match the specimen size. A summary of the test results follows:

Specimen Size, in. (mm)	Compression Strength, psi (MPa)	Splitting Tensile Strength, psi (MPa)
3 × 6 (75 × 150)	3015 (20.78)	312 (2.15)
2 × 4 (50 × 100)	3021 (20.83)	312 (2.15)
1.5 × 3 (38 × 75)	3061 (21.11)	306 (2.11)
1 × 2 (25 × 50)	3033 (20.91)	315 (2.17)

The consistency in strength within each size in a given batch was excellent; coefficients of variation ranged from 0.0005 to 0.064 (with an average of 0.033) for compressive strengths and from 0.004 to 0.102 (average of 0.048) for splitting tensile strengths.

Tests for modulus of rupture were conducted on two beam sizes, 1×1.5 in. (25×38 mm) deep and 2×3 in. (50×75 mm) in cross section. All beam dimensions and loading accessories were scaled. The compaction procedure and drying times were adjusted to achieve a uniform moisture content and density. Six small beams and three large beams were cast from each of six batches; the coefficients of variation for measured modulus of rupture averaged 0.046 for each of the 12 test series. A size effect was evident from the results:

Beam Cross Section, in. (mm)	Modulus of Rupture, psi (MPa)
2×3 (50×75)	531 (3.66)
1×1.5 (25×38)	574 (3.96)

It was concluded that there are no discernible size effects in the compressive and splitting tensile strength of Ultracal mortars when the density, moisture content, and loading conditions are identical for various size specimens (Loh (1969)]. A small size effect was measured for modulus of rupture; it is felt that this is produced by a combination of strain gradient effects and statistical variation in strength over the uniform moment region in the middle third of the beam. More about size effects is discussed later in Chapter 6.

It is evident that proper procedures can reduce size effects in gypsum mortars to negligible levels. These procedures are relatively easy to follow if the models are beams, columns, or frames, but they are difficult to implement in more complex or variable-thickness models.

4.11 MODELING OF MASONRY STRUCTURES

4.11.1 Introduction

Masonry structures represent one of the oldest forms of construction. Many different types of masonry components have been evolved over the years both in this country and elsewhere for constructing walls of low-rise and high-rise buildings. The large majority of masonry units fall into two main categories: burnt clay and concrete. Both of these types find extensive applications in civil engineering structures and have been successfully modeled at reduced geometric scales. A brief historical review of work using the modeling technique in masonry structural studies, together with similitude requirements, has been presented in Chapter 2.

4.11.2 Material Properties

4.11.2.1 Prototype Masonry Units

Concrete masonry units are molded of a mixture of sand, aggregate, cement, and water under pressure and/or vibration. Curing is by autoclave and steam curing. There are many types of aggregates used in concrete masonry; these include the normal weights, such as sand, gravel, crushed stone, air-cooled blast-furnace slag, and the light weights, such as expanded shale, clay, slate, expanded blast-furnace slag, sintered fly ash, coal cinders, pumice, and scoria [ACI Committee 531 (1970)].

The terms *normal weight* and *light weight* refer to the density of the aggregates used in the manufacturing process. Generally, local availability determines the use of any one type, and blocks made with any of the aforementioned aggregates are considered concrete blocks. However, the term *concrete block* is used in some locations to describe units made of sand and gravel or crushed stone [Randall and Panarese (1976)].

ASTM specifications classify concrete masonry units according to grade and type. The grade describes the intended use of the units, and the type of unit is either Type I, moisture-controlled, or Type II, non-moisture-controlled. From a design standpoint, dry shrinking due to loss of moisture could cause excessive stress buildup and cracking of walls. This specification is designed to eliminate moisture-less effects.

Concrete blocks are classified as *hollow* if the net cross-sectional area parallel to the bearing face is less than 75% of the cross-sectional area. *Solid* units have a net concrete cross-sectional area of 75% or greater. Most concrete masonry units have net cross-sectional areas ranging from 50 to 70% (30 to 50% core area) depending on unit width, face-shell and web thickness, and core configuration [Randall and Panarese (1976)].

ASTM C140 outlines the procedure for determining compressive strength of concrete masonry units. Attempts have been made to measure tensile strength, but no accepted technique has been adopted. The compressive strength of hollow concrete blocks is about 500 to 3000 psi (3.4 to 20.7 MPa) after 28 days, based on gross area. With different types of lightweight aggregates, values from 500 to 1140 psi (3.4 to 7.9 MPa) are reported [Sahlin (1971)]. Because curing plays an important role in concrete masonry strength, no conclusions of compressive strength can be drawn from the knowledge of unit weight, aggregate type, and water-cement ratio. The designer, therefore, must rely on test data of finished blocks.

4.11.2.2 Model Masonry Units

The configuration of the $\frac{1}{4}$-scale units resembles the double corner and regular stretcher type of $8 \times 8 \times 16$ in. nominal size concrete masonry blocks. Manufactured by the National Concrete Masonry Association, McLean,

TABLE 4.8 Dimensions and Physical Properties of Concrete Masonry Units

Masonry Unit	Number of Specimens Measured	Actual Dimensions			Minimum Face Shell Thickness, in.	Gross Area, in.²	Net Solid, %	Compressive Strength Gross Area, psi	Dry Weight Density, lb/ft³	Water Absorption, lb/ft³
		Width, in.	Length, in.	Height, in.						
Prototype: 8-in. (200 mm) hollow expanded slag block [Yokel et al. (1971)]	5	$7\frac{5}{8}$	$15\frac{5}{8}$	$7\frac{5}{8}$	$1\frac{1}{4}$	119.1	52.2	1100	103.0	14.3
¼-scale model: double corner units	12	1.91	3.91	1.89	0.323†	7.48	52.0	901*	114.4	13.9
Regular stretcher units	6	1.92	3.93	1.89	0.325†	7.54	50.9	1175	110.9	17.1

*Average face shell thickness.
†Combined average compressive strength of six saturated surface dry ($f' = 816$ psi) and six dry ($f' = 1086$ psi) specimens.

Virginia, these units measure $2 \times 2 \times 4$ in. nominally. Data on the physical properties of both model and prototype units are listed in Table 4.8. The prototype data was collected under the guidelines of the ASTM C140 by Yokel et al. (1971), and modeled as closely as possible. Figure 4.38 shows typical average sections of the units. Note that the outside dimension and face shells were scaled accurately while the webs were oversized by about 30%. The web-thickness dimensions shown in Fig. 4.38 are the average of 36 measurements taken at six locations on the webs of six regular stretcher units.

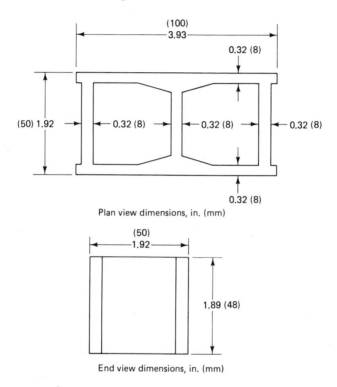

Plan view dimensions, in. (mm)

End view dimensions, in. (mm)

Figure 4.38   Model concrete masonry units at one-quarter scale.

Net solid area and adsorption for 12 units were calculated using the mean values of all dimensions and weights. Adsorption represents the difference of dry-weight density and wet-weight density after a 1-min immersion in water.

These model units were then capped on both bearing surfaces with Hydrostone* and tested in compression, with a loading rate that caused failure in 1 to 2 min. For the double corner units Table 4.8 shows a difference in strength between units tested wet and those tested dry, although this may be due only to statistical variation of such a small population.

*Hydrostone is a trade name of the U.S. Gypsum Corporation.

4.11.2.3 Mortars

ASTM Specification C-270 recognizes five mortar types, with acceptance based on either the proportion specification or property specification (Table 4.9). Compressive strength of mortars depends mainly on the type and quantity of cementitious material used [Sahlin (1971)] and the duration of curing [Isberner (1969)]. Compressive strength is measured in the laboratory by casting, curing, and testing 2-in. (50 mm) cubes in compression (as per ASTM C-270).

TABLE 4.9   Mortar Proportions by Volume (ASTM C-270)

Mortar Type	Minimum Compression Strength at 28 Days, psi (MPa) (2-in. cubes)	Portland Cement	Hydrated Lime Minimum	Hydrated Lime Maximum	Masonry Cement	Damp Loose Aggregate
M	2500 (17.2)	1	—	$\frac{1}{4}$	—	$2\frac{1}{2}$ to 3
		1	—	—	1	times sum
S	1800 (12.4)	1	$\frac{1}{4}$	$\frac{1}{2}$	—	of the
		$\frac{1}{2}$	—	—	1	cements
N	750 (5.2)	1	$\frac{1}{2}$	$1\frac{1}{4}$	—	or the
		—	—	—	1	sum of
O	350 (2.4)	1	$1\frac{1}{4}$	$2\frac{1}{2}$	—	cement
		—	—	—	1	plus
K	75 (0.52)	1	$2\frac{1}{2}$	4	—	lime

Because available test data on full-scale hollow-core masonry incorporates the use of ASTM Type N masonry mortar, it was necessary to develop a similar type model masonry mortar. Using the proportion specification as outlined in ASTM C-270-71 as a guide, three mixes were tested in an attempt to match the reported 28-day strength of 750 psi (5.2 MPa) on 2-in. (50-mm) cubes.

The proportion specification calls for a masonry cement-aggregate ratio of 1 : 3 by volume. After several trial mixes, which were in general too strong, a mix of 1 : 1 : 4 (water—cement—aggregate) proportion by volume was found adequate as a model mortar mix. Generally, the strength of this mix was about 750 psi. The workability and retentivity suffered to some degree because of the harshness of the mix, but steps were taken during casting of model masonry components to remedy these conditions. For these reasons, the 1 : 1 : 4 mix was chosen as the model of Type N mortar.

The aggregate used for model mortar was a commercially obtained natural masonry sand having the gradation shown in Fig. 4.39 and a fineness modulus of 0.8. To properly scale a nominal $\frac{3}{8}$-in. (10-mm) mortar joint as used in practice, a model joint thickness of $\frac{3}{32}$-in. (2.4-mm) was necessary. This required the removal of particle sizes greater than a U.S. No. 16 sieve. The result of this

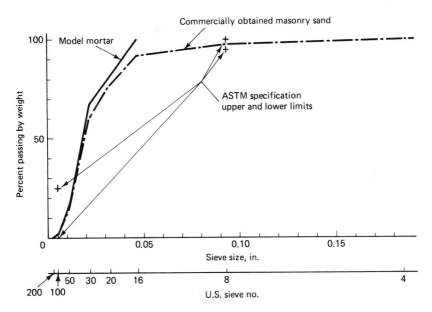

Figure 4.39   Aggregate gradation curves for masonry mortars.

reduction is shown in Fig. 4.39 and represents the aggregate gradation used to fabricate model masonry components. It has been shown [Harris et al. (1966)] that in testing concrete mortars it is the volume and not the gradation that most strongly influences model mortar behavior. As previously indicated, it was necessary to decrease the cement-aggregate ratio from the specified 1 : 3 by volume to 1 : 4 in order to achieve the required model mortar strength.

As is typical for cementitious materials, moisture plays a significant role in the strength development of prototype and model mortars. It is reported [Isberner (1969)] that mortar relative humidity must be a minimum of 85 % for hydration to continue and that as relative humidity decreases so does the rate of hydration. This can be of significance in small-scale modeling where instrumentation and testing over an extended period of time can cause changes in strength.

Figure 4.40 shows the results of strength-vs.-age tests on 1 × 2 in. (25 × 50 mm) compression cylinders of 1 : 1 : 4 mix. All specimens were moist-cured for 3 days, at which time sufficient specimens were removed from the wet room, shellacked, and left in room air until tested with their counterpart moist-cured cylinders. Each group of tests represents six specimens.

Note in Fig. 4.40 that there was an increase in strength of 14 % at 7 days for the dry-tested specimens. It has been observed [Harris et al. (1966)] that when wet-cured cylinders of normal-weight concrete are allowed to lose some of their moisture, strength increases of up to 25 % are obtained. Other studies of

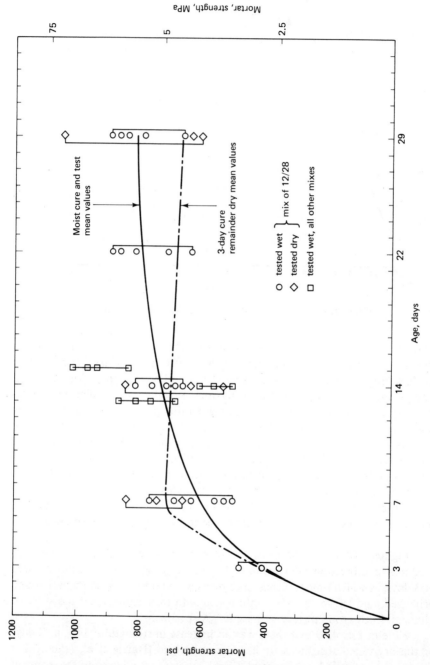

Figure 4.40 Strength-age curves for 1 × 2 in. cylinders of masonry mortar 1 : 1 : 4 mix.

curing effects on 6×12 in. (150×300 mm) cylinders show that in the first 28 days the moist-cured specimens showed considerable strength increase, and no increase in strength for the air-cured specimen was observed after 14 days. On the other hand, for small volumes of model material, this effect is accelerated to such a degree that stable strength are seen as early as 10 days for air-cured specimens.

Four 1×2 in. (25×50 mm) model mortar cylinders were instrumented with two 1-in. (25-mm) SR-4 strain gages on generators 180° apart and tested in compression, as shown in Fig. 4.41. Specimens No. 1, 2, and 3 are of the mix

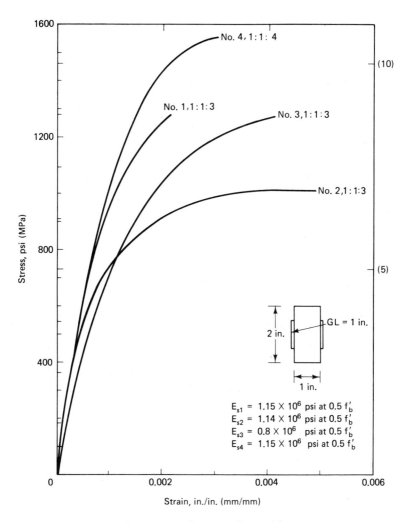

Figure 4.41 Compressive stress-strain curves for model masonry mortar.

1 : 1 : 3 and specimen No. 4 is of a mix 1 : 1 : 4 by volume. Specimen No. 4 is typical of a batch used in casting shear or diagonal compression specimens. Values for the secant modulus at 50% of the ultimate compressive strength are also shown in Fig. 4.41.

4.11.3 Fabrication Techniques

Concrete masonry models can be fabricated using procedures closely resembling prototype masonry construction. Horizontal casting of cored masonry work is very difficult to achieve because the mortar tends to run into the hollow cores. A vertical casting technique [Becica and Harris (1977)] has been developed for $\frac{1}{4}$-scale concrete masonry work. Unfortunately the accuracy of this method is greatly dependent on workmanship. To give vertical and horizontal alignment, a frame of the type shown in Fig. 4.42a can be used. The horizontal lines are set at a predetermined height providing for unit height plus a $\frac{3}{32}$-in. (2.4 mm) mortar joint. The plumb line establishes vertical reference. The entire frame is clamped and braced to a level table. Mortar is then troweled onto the face shells of presoaked units while adjoining units are placed accordingly with firm pressure, causing mortar to squeeze out of the joint. This movement allows for vertical and horizontal alignment and establishment of the joint thickness. As illustrated in Fig. 4.42b gentle tapping with the trowel is sometimes required to further consolidate joints. After the specimen is cast, it is removed from the casting frame, and all the joints are pointed using the tip of a small trowel, as shown in Fig. 4.42c. Care must be taken to ensure complete joint filling.

4.11.4 Basic Strength Properties

Modeling of structures in the inelastic range requires a choice of materials that have identical stress-strain characteristics (Chapter 2). In the case of concrete masonry structures, the validity of the direct modeling technique must be demonstrated for simple compression, tension, and shear behavior prior to the testing of structures with complex states of stress. Carefully documented prototype tests in compression, flexural bond, and shear were chosen and modeled at $\frac{1}{4}$-scale. The results are compared in the following sections.

4.11.4.1 Compressive Strength

At the present time there is no firm standard for determining the compressive strength of concrete masonry prisms. The Portland Cement Association (PCA) and the National Concrete Masonry Association (NCMA) presently recommend a prism not less than 16 in. (400 mm) high with a height-thickness ratio of 2. The ASTM E447-74 guidelines are more general, specifying prisms of

(a)

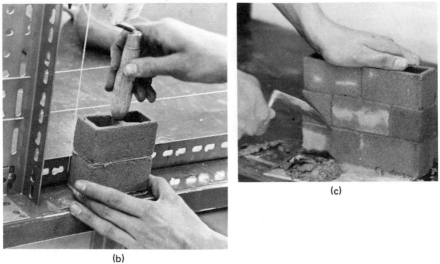

(b)

(c)

Figure 4.42 Fabrication techniques. (a) Casting frame. (b) Consolidating joints. (c) Pointing mortar joints.

height-thickness ratio from 2 to 5 with correction factors for slenderness effects. It is generally felt that end restraints have a large effect on the strength of two-block prisms of full scale. At a reduced scale, such effects are easily amplified and require careful consideration. The three-block prism was therefore chosen for comparison because prototype data using such specimens were more readily available.

Failure of the three-block prisms was by end splitting, as shown in Fig. 4.43a. This is also typical of prototype prisms. Both strain and deformation measurements were taken on some of the series tested in order to evaluate the consistency of the strain measurements. Two 4-in. (100 mm) SR-4 strain gages were bonded to the face shell of the prisms (Fig. 4.43b) at the center line of the specimen and across both the horizontal bond lines. The dial gages were attached to the face shells on stiff aluminum brackets with 5-min epoxy. These recorded the deformation over an identical 4-in. (100-mm) gage length on opposite faces of the three-block prisms.

(a)

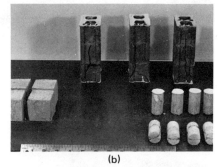

(b)

(c)

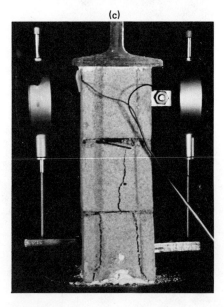

Figure 4.43 Compressive strength of quarter-scale masonry. (a) Specimen ready for testing. (b) Failed 3-block model prisms showing vertical end splitting. (c) SR-4 strain gages and dial gages to determine stress-strain relationships.

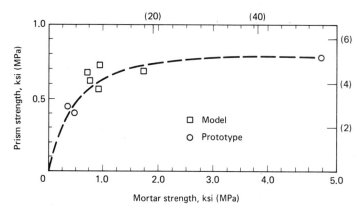

Figure 4.44 Comparison of model and prototype compressive strengths.

Since masonry strength is directly related to the mortar strength [Fishburn (1961), Sahlin (1971), Yokel et al. (1971)], the adjusted model strengths are compared with the prototype data in which the prism compressive strength on the gross area is related to the mortar strength. Comparison of six series of model compression prism data with prototype data [Yokel et al. (1971)] is shown in Fig. 4.44. Model data are obtained on hollow-core units that had a compressive strength of 1100 psi on the gross area. Some imperfections are noted in the scaling of the $\frac{1}{4}$-scale masonry units used. In particular, the oversized web structures (see Fig. 4.38) causes perhaps most of the deviation in the compressive strength of three-course prism test specimens. To compensate for the increased tensile strength that these units could sustain in splitting as a result of their increased web area, a correction to the apparent strength is made. In addition, the effect of volume of mortar in the joints on the compressive strength is taken into account by empirically determining the effect of size of specimen on the unconfined compressive strength (Section 6.7). The model data shown in Fig. 4.44 have been corrected for scale and geometry effects. Note the small deviation from the mean curve (shown dashed), which would indicate that the modeling technique developed for prisms in compression is indeed accurate.

4.11.4.2 Flexural Bond

The flexural bond strength of model masonry is determined by testing two-block-high prisms that are clamped in metal frames at both the top and bottom of the prism and loaded eccentrically 4 in. (100 mm) from the centroid of the prism, as in Fig. 4.45. This method is duplicated from prototype tests [Yokel et al. (1971)] and described in ASTM E149-66. The prisms are constructed of regular stretcher units with face-shell and end-web bonding. The prototype specimens in tests were constructed of similar units with only face-shell bonding. Because of the relatively weak bond developed between units, the mortar ulti-

TABLE 4.10 Summary of Flexural Tests on Masonry Prisms

(1)		Mortar Compressive Strength, 2 in. Cubes, psi (2)	No. of Specimens (3)	Age at Time of Test, days (4)	Masonry Flexural Strength		No. of Specimens (7)	Mode of Failure (8)	Test Conditions (9)
					Gross Area, psi (5)	Net Area, psi (6)			
Prototype: two-block high prisms, 8-in hollow block		345	3	180	6	9	3	Separation at mortar-unit interface	Dry
¼-scale model, two-block-high prisms	A	882	6	15	42	50	6	Separation at mortar-unit interface	S.S.D.*
	B	588	6	14	48	58	6	Separation at mortar-unit interface	S.S.D.*

*Saturated surface dry.

162

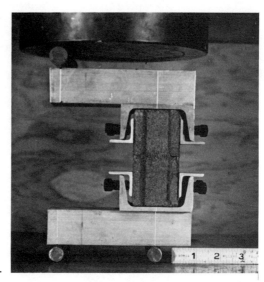

Figure 4.45 Flexural bond specimen.

mate loads are small, and thus failure occurs within 1 min of load application. The typical mode of failure is separation at the mortar-unit interface, with one unit remaining free of mortar. The test results of two model series consisting of six specimens each and test results of three prototype specimens are shown in Table 4.10. It should be noted that series A prisms were cast using dry blocks while series B were cast using saturated surface dry units. The effect of dewatering the mortar via the high block adsorption is a probable contributor to the reported strength difference within the model study. The effects of end-web bonding and high mortar strength have also contributed significantly to the differences between prototype and model results.

4.11.4.3 Shear Strength

Determination of masonry shear strength usually consists of testing square prisms by compression along one diagonal, with a resulting failure in diagonal tension [Fattal and Cattaneo (1974)]. Two model series of tests were designed after the work of Fishburn (1961) and the ASTM specification E519-74 (1976) for the determination of masonry shear (diagonal tension) strength. This procedure calls for the testing of small masonry walls with height-length ratio (H/L) of 1 in diagonal compression. The model specimen size was chosen to be two units long by four units high of the running bond pattern. After curing, the specimens were instrumented with SR-4 strain gages placed along the compressive and tension diagonals of both faces. These were wired in series to provide average strain readings. In lieu of the loading shoes recommended by the ASTM specification, a 3-in. structural tube with $\frac{1}{4}$-in. (6 mm) wall was cut into lengths that fit at opposite ends of the specimen diagonal such that the resulting bearing area encompassed one unit height (2 in. [50 mm]). These were then

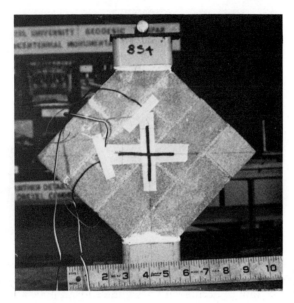

Figure 4.46 Shear specimen.

placed on a level surface and filled with a stiff mix of Hydrostone capping material. Figure 4.46 shows a typical shear specimen ready for testing. Figure 4.47 shows the plots of shear stress versus shear strain for model series B and prototype tests. There exist some differences in the methods used to measure the shear strains in model (strain gages) and prototype (deflection gages) tests, but the shear strains are considered essentially the same in the lower-load range prior to cracking.

The model shear specimens, which for reasons of economy had sides only one-third of the scaled prototype values, showed an average shear strength of 38.5 psi (265 kPa) based on six specimens. The prototype shear strength was found to be 18 psi (124 kPa) as an average of three specimens. In making a direct comparison of the model and prototype results, the assumption must be made that uniform tensile and compressive stresses exist all along the diagonal length of the specimen. Apart from the size difference mentioned above, however, the model mortar was 84% stronger than the prototype mortar based on 2-in. (50 mm) cubes. This increase in strength, amplified by the volume of stressed material differences that exist between the $\frac{1}{8}$-in. (3 mm) model joint and the 2-in. (50 mm) cube tested for strength determination, will result in a higher apparent shear strength in the model specimens. The influence of the higher mortar strengths of the model specimens are, however, more difficult to evaluate under the condition of combined stresses. Combined compressive and tensile stresses exist all along the compressed diagonal of the shear specimen, but the volume effects in such complex stress situations cannot be evaluated in the present comparison because no empirical data exist to relate the strength of different-size specimens under these stress states.

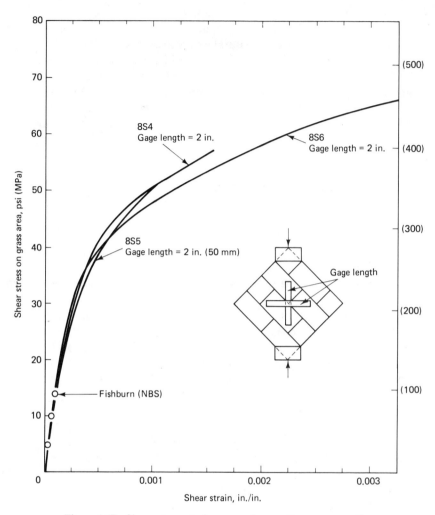

Figure 4.47   Shear stress-strain curves for small masonry walls.

4.12 SUMMARY

In summary, it must be pointed out that although it is not possible to completely simulate the constituents of concrete (aggregates, cement, water, and admixtures), it is adequate to have homologous stress-strain curves and to simulate the tensile strength–compressive strength ratio and the maximum aggregate size. Factors influencing the strength and the behavior of model mixes have been discussed in this chapter. Sufficient data and techniques are available to design model concrete mixes for any strength ranging from 2500 psi (17 MPa) to

12,000 psi (83 MPa) for use in model reinforced and prestressed concrete work. It is possible to use either local natural sand if it is suitably graded or a blended mixture of very narrowly graded natural or crushed sands. The sizes of the test cylinder or cube must be selected with due consideration of the minimum effective dimension of the model and the volumes of failure regions. The ACI Committee 444, Models for Concrete Structures (1979) has recommended that to enable a rational comparative analysis of size effects in concrete strength, 2×4 in. (50×100 mm) cylinders must be used along with the specimen size selected.

Suitably sealed gypsum plaster mixes can adequately model the behavior of practical concrete mixes in compression and tension. Gypsum mixes can be designed for a given strength and have the advantages of reducing the testing time to 24 hr or less.

The methodology for using small-scale direct models of concrete masonry structures has been presented. The basic strength evaluation tests for compressive, flexural bond, and shear strengths recommended for prototype structures have also been developed, with minor modifications for the evaluation of model masonry strength. A systematic analysis of the parameters that affect the strength and stiffness of masonry under compressive, flexural, and shear loadings has provided the means to compare model and prototype test results. Correlation of the model and prototype results ranged from excellent to good. For the case of masonry in compression, the testing of prisms has shown that the model masonry behavior is essentially the same as that reported for prototype tests. In the case of flexural bond and shear strength of masonry, proper considerations must be given to the tensile strength of the joint mortar, including the effects of the stressed volume. This implies that very small size control specimens (cubes or cylinders) must be tested in order to be sure that the small volume of the model joint bears the same relation to its control specimen as the volume in the prototype mortar joint bears to its respective control specimen. When proper consideration of the above effects are made, the correlation of model and prototype masonry strength tests for evaluating flexural bond and shear are shown to be satisfactory. Excellent correlation of the elastic modulus of concrete masonry in compression and shear was obtained from the model tests and the limited stress-strain data of prototype masonry reported in the literature.

Inelastic Models:
Structural Steel
and Reinforcing Bars

5

5.1 INTRODUCTION

In this chapter, we shall discuss the modeling of steel reinforcement and structural steel for small-scale direct models. The models are designed to study structural response to applied loads through both the elastic and the inelastic ranges of behavior until failure. As most structural concrete elements are underreinforced, that is, the reinforcing provided is not adequate to fully utilize the compressive strength of the concrete, the stress-strain characteristics of both the prototype and the model reinforcement are critical in determining the structural behavior in the inelastic range. A brief discussion of steel as a structural material follows.

5.2 STEEL

Most steels have a crystalline structure and consist of a basic iron-carbon system. Relatively small changes in the carbon content and/or other alloys result in significant changes in the mechanical behavior of the resultant steel. The mechanical properties of steel that are of interest to the design engineer are the stress-strain curve; the yield strength, if any; the percentage elongation at failure, or ductility; the strain hardening; and the ultimate tensile strength. While the mechanical behavior of a particular steel is significantly influenced by its carbon content, other factors that influence its properties are the chemical composition and the method used to shape the molten mass into its final form as steel. The mechanical properties of steel are affected by the following parameters:

1. Chemical Composition
   a. Carbon content
   b. Presence of alloying elements such as nickel, chromium, vanadium, and copper
   c. Presence of other elements such as sulfur, phosphorus, manganese, and silicon
2. Physical Conditions
   a. Slow cooling from the molten state or quenching
   b. Annealing
   c. Hardening characteristics
   d. Shaping operations (e.g., cold working)
   e. Weldability

5.2.1 Reinforcing Steel Bars

The ASTM Standard Specification A615-76a (1978) for reinforcing bar steel covers billet steel of Grades 40 and 60 with minimum yield strengths of 40,000 psi (276 MPa) and 60,000 psi (414 MPa), respectively, and ultimate tensile strengths of 70,000 psi (483 MPa) and 90,000 psi (621 MPa), respectively. The billet steel is newly made steel with a carefully controlled chemical composition to obtain the necessary ductility. Rail steel Grades 50 and 60 (ASTM A616-76, 1978) and axle steel Grades 40 and 60 (ASTM A617-76, 1978) are also used for manufacturing reinforcing steel bars. Both axle and rail steel bars are made from steel that is rerolled from old axles and rails and are in general less ductile than billet steel [Salmon and Johnson (1971)]. Low-alloy steel of Grade 60 is useful for applications of reinforcing steel bars that involve both welding and bending. The carbon content of these steels is approximately 0.25%. All bar sizes, No.3 through No. 11, No. 14, and No. 18 are available in Grade 60 billet and low-alloy steels; however, bars of sizes No. 14 and No. 18 are not available in Grade 40 billet steel nor in rail or axle steel.

5.2.2 Structural Steels

The 1970 AISC specification covering the *common* structural steels for buildings includes steels with specified minimum yield strength ranging from 36,000 psi (250 MPa) (ASTM A36-77a, 1978) to 50,000 psi (345 MPa) (ASTM A242-75, 1978 and ASTM A440-77, 1978). Since the chemical composition is constant (0.22% carbon for A242 steel and 0.28% for A440 steel), the increased hot working required to produce thin sections increases the yielding strength in thinner sections. The A36 steel has a maximum carbon content varying from 0.25 to 0.29% depending on the thickness. The structural steels show a marked yield point, as shown in Fig. 5.1. An increase in the carbon content raises the yield strength but decreases the ductility and causes problems with the welding operations if the carbon content is higher than 0.3%. Most of the structural steels have a low carbon content and therefore have good welding characteristics. Also, increased limitations have been placed on the carbon content in recent years as emphasis on good weldability has increased [Salmon and Johnson (1971)].

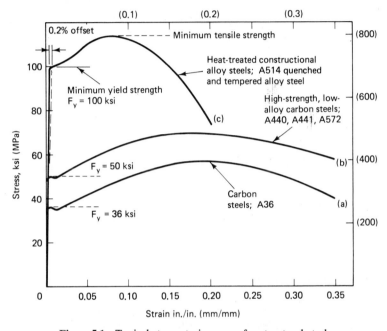

Figure 5.1 Typical stress-strain curves for structural steel.

5.2.3 Prestressing Steels

5.2.3.1 General

Prestressing steels are available in the form of single wires (ASTM A421-77, 1978), wire strands (ASTM A416-74, 1978) or high-strength bars (AISI 5160 and 9260, 1978). Wire strands are of the seven-wire type with six helically placed outer wires wound tightly around a central wire. Strand diameters vary from $\frac{1}{4}$ to $\frac{1}{2}$ in. (6 to 12 mm). Prestressing wires usually have diameters of 0.192 to 0.276 in. (5 to 7 mm). The high-strength steel bars range in diameter from $\frac{3}{4}$ to $1\frac{3}{8}$ in. (19 to 35 mm). The prestressing wires or bars are made by cold-drawing high-carbon (approximately 0.6% carbon content) steel bars. Typical stress-strain curves for prestressing strands and prestress bars are shown in Fig. 5.2.

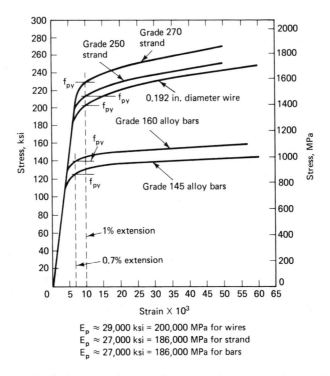

Figure 5.2 Typical stress-strain curves for prestressing strands, wires, and bars.

5.2.3.2 Stress-Strain Characteristics

An enlarged view of the stress-strain curve for an A36 steel and some other steels for the strain range 0 to 0.030 is shown in Fig. 5.3. The following points must be emphasized with respect to this stress-strain curve:

1. Even with the high quality control in the manufacture of steel, there will be property variations from structure to structure, and even within a structure. This suggests that the yield strength of steel that is used in design must be carefully determined.

2. Yield strength variations are not normally accompanied by similar changes in the slope of the initial portion of the stress-strain curve. The modulus of elasticity of the various steels varies between very narrow limits, seldom exceeding 30×10^6 psi (0.21×10^6 MPa). The values for prestressing steel wire strands is of the order of 27×10^6 psi (0.19×10^6 MPa) because of wrapping effects. For most structural steels and reinforcing steel bars, the modulus of elasticity has a typical value of 29×10^6 psi (0.2×10^6 MPa).

3. Steel is a structural material that exhibits the flat yield plateau shown in

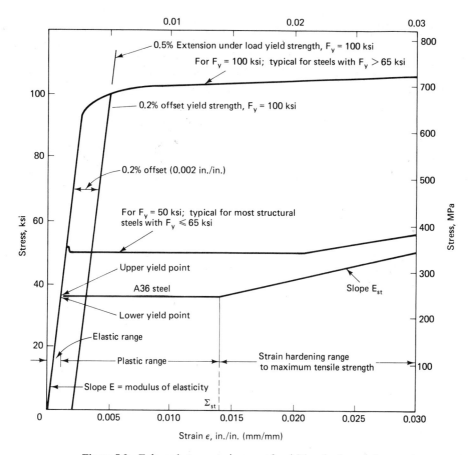

Figure 5.3 Enlarged stress-strain curve for A36 and other steels.

Fig. 5.3. There are many other metals that are ductile, but none possesses this property. Only phosphor-bronze has a definite yield point, as does mild steel; however, it strain-hardens at strains only slightly greater than the yield strain [Antebi et al. (1962) and Harris et al. (1962)]. In seeking a model material that will simulate steel in the elastoplastic range, consideration must be given to the yield plateau, including its extent.

4. It must be noted that the stress-strain curves that are ordinarily presented (e.g., Fig. 5.1) are normally obtained from small tensile coupons. If one were to test a complete wide-flange section (rolled or welded), the yield plateau might not be easily observed. The reason for the apparent difference in behavior would be due solely to the presence of initial or residual stresses in the wide-flange specimen that result from the differential cooling of hot-rolled or welded shapes. If such initial stresses influence the structural behavior, the initial stresses in the model must simulate those

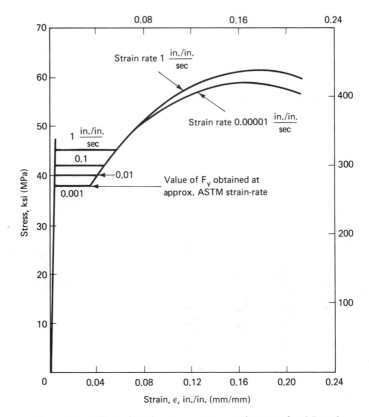

Figure 5.4   Effect of strain rate on stress-strain curve for A7 steel.

that would exist in the prototype. Finally, the load-deformation characteristics of steel materials are sensitive to the rate at which the strains
are induced. A typical example is shown in Fig. 5.4.

5.3 STRUCTURAL STEEL MODELS

Vastly improved computing facilities and sophisticated design techniques have
provided the design engineer with tools that can be used confidently in the
design of any conventional structure such as a building or a bridge. However,
it must be noted that most of the provisions in the existing codes are empirical
in nature and have been derived from interpretation of complex experimental
information on member behavior. A rigorous analytical approach may not
always be sufficient to obtain a design solution [Litle and Foster (1966)], and it
is in the cases of highly unconventional structures that model studies can be
useful.

Successful experimental research work on steel structures at Lehigh University and elsewhere has led to innovations and improvements in design techniques, e.g., plastic design, residual stresses, member design under combined loadings, etc. So far, full-scale tests have produced much useful information. However, they are generally restricted to simple members and very simple structures because of the space and laboratory facilities required and the expenditure involved. Use of small-scale models to study the complex behavior of whole structures or their component substructures can overcome these restrictions and can be useful as research and development tools.

Litle and Foster (1966) undertook a project to fabricate small-scale wide-flange sections, small-scale joints, and a small-scale building frame made of steel. This was accomplished through the testing of 44 tensile coupons, study of feasible fabrication techniques, testing of four milled wide-flange beams and eight fabricated joints, and construction of a $\frac{1}{4}$-scale space framework. For details on fabrication procedures, testing techniques, and test results see their work. Essential conclusions derived from various experiments include:

1. The chemical and mechanical properties of C1020 hot-rolled structural steel [Society of Automotive Engineers (1964)] are such that it may be used satisfactorily in the modeling of steel structures.

2. Milling wide-flange sections from hot-rolled bar stock is a reliable and accurate method for fabricating small-scale sections with element thickness as small as 0.025 in. (0.6 mm).

3. Tension and joint tests demonstrate that the heliarc process with industrial stainless 410 filler rods provides more than adequate strength and ductility for joining C1020 steel model sections.

4. High ultimate values obtained in the nonannealed joint tests are a result of the heating effect of the welding process and/or a change in the chemical properties of the steel that is caused by the filler rod.
   a. Preannealing of the sections before welding did not lower the ultimate moment values obtained to the postannealed value.
   b. Nonannealed welded tension samples failed outside the 1-in. (25-mm) gage length or just inside adjacent to the gage line.
   c. Beam tests demonstrated little difference between the annealed and nonannealed yield and ultimate moments.

5. Fabrication of a complete framework is possible, but it is necessary to fix elements during welding and to follow a predetermined sequence of assembly to reduce shrinkage deformations. This sequence may vary with each structure.

6. Until more refinements are made in the welding process, it is necessary to anneal the whole framework to obtain member behavior consistent with the stress-strain characteristics of the material.

5.4 REINFORCEMENT FOR SMALL-SCALE MODELS

Steel bars, rods, and cables are generally used for reinforcing structural concrete, with low- to medium-strength steels used for normal reinforcing and high-strength wires and rods for prestressing tendons. The two types of steel reinforcing commonly used in North America are deformed bars and welded wire fabric. Although the ACI Code 318-77 (1977) permits the use of steel with a yield strength of 80 ksi (550 MPa) and without a well-defined yield point, most conventional reinforced concrete members are reinforced with steels that have a well-defined yield point and sufficient ductility to fulfill the requirement of an underreinforced design. The properties of steel that must be considered in the modeling of reinforcement are:

1. Yield and ultimate strength in tension, plus yield strength in compression
2. Shape of stress-strain curve
3. Ductility
4. Bond characteristics at steel-concrete interface

5.4.1 Model Reinforcement Used by Various Investigators

Commercially available wires and rods of varying sizes and strengths have been used for model reinforcing steel at different research centers. However, a problem exists with the bond characteristics in conjunction with their use as reinforcing in model concretes, and generally deformed wires are necessary for model reinforcing to improve the nature and amount of bond strength and to achieve the best possible cracking similitude. Presently available wires and rods may be grouped as follows.

1. Round steel wire and rod in a variety of sizes and strengths.
2. Square steel rod. This form has a particular advantage of providing a flat surface for strain gages, but the use of square bars in reinforced concrete construction has been discontinued.
3. Cold-rolled threaded steel rods. The highly deformed surface provided by the threads leads to a high bond strength.
4. Commercially available deformed wires as used in manufacturing welded wire fabric.
5. Custom deformed wire.
6. No. 2, No. 3, and 6-mm deformed bars.

Careful choice of model reinforcing, combined with the proper annealing processes described later, will result in reinforcing of suitable properties for a given model study.

At present, a majority of North American investigators use No. 3 (10 mm) and No. 4 (12 mm) deformed steel bars for reinforcing large-scale models and deformed steel wires (ASTM A496, 1972) for reinforcing smaller models. However, various other techniques have been attempted by model investigators to simulate the prototype reinforcement. Plain wires with a rusted surface have been used with partial success at MIT, Cornell University, and McGill University. Brock (1959) and Lord (1965) have successfully used threaded wires in achieving both a high degree of bond and acceptable cracking similitude. Harris et al. (1970) developed a simple technique to cold-deform commercially available plain steel wires by passing them through a special device with two pairs of perpendicular knurling wheels. White and Clark (1978) used a similar technique at the Cement and Concrete Association, UK, to develop a $\frac{1}{6}$-scale model of the $\frac{1}{2}$-in. (12-mm) GK60 deformed bar used in the United Kingdom. Using an identical technique, Subedi and Garas (1978) deformed available 1.6-mm-diameter wires to produce "crimped" wires for use in their model work. They also used plain, threaded, and deformed bars in their investigations.

5.5 MODEL PRESTRESSING REINFORCEMENT AND TECHNIQUES

5.5.1 Model Prestressing Reinforcement and Anchorage Systems

There are several possible substitutes for prestressing tendons, which include:

1. Individual strand prestressing wire and strands, as used in making twisted-strand cables for prototype prestressing
2. Piano wire
3. Stainless steel twisted-strand cable
4. Bicycle spoke material, complete with threaded ends (particularly useful for posttensioning applications)
5. Stranded aircraft or marine cable

Typical stress-strain curves for some of these model prestressing wires are shown in Fig. 5.5. Prestressing wires and piano wires have been successfully used in model studies of prestressed concrete beams and slabs at McGill University [Pang (1965)]. Labonte (1971) used seven 7-mm wire strands to model the posttensioning system of the buttress of a prestressed concrete reactor vessel. The standard BBRV system was used to anchor the tendons in the $\frac{1}{6}$-scale model of the buttress. Typical details of the passive and active load cells are shown in Fig. 5.6. In a prestressed concrete model study at Cornell Uni-

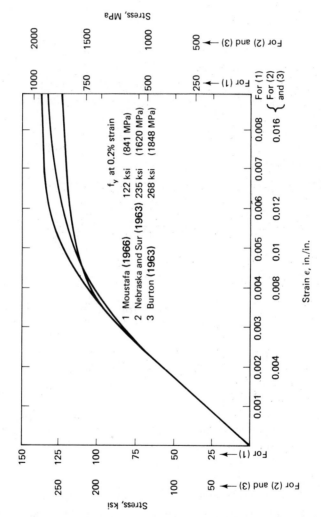

Figure 5.5 Stress-strain curves for prestressing wires.

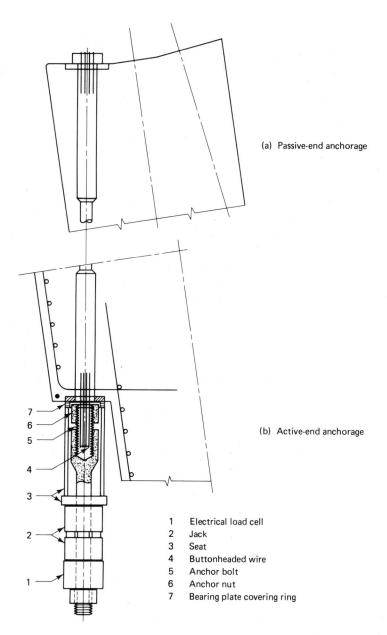

(a) Passive-end anchorage

(b) Active-end anchorage

1	Electrical load cell
2	Jack
3	Seat
4	Buttonheaded wire
5	Anchor bolt
6	Anchor nut
7	Bearing plate covering ring

Figure 5.6 Posttensioning system. (a) Passive-end anchorage. (b) Active-end anchorage.

versity, the surfaces of individual wires were mechanically roughened with an oversize thread-cutting die to improve their bond characteristics.

Harris and Muskivitch (1977, 1979) used stainless steel twisted-strand cables in a Drexel University model study of the nature and mechanism of progressive collapse in industrialized buildings. Kemp (1971) successfully used the smallest available plain wire (0.276 in., or 7 mm, diameter) prestressing system in posttensioned, nongrouted, prestressed slab models. In a Cornell University study Moustafa (1966) used 0.1032 in. (2.6 mm) diameter custom-length bicycle spokes in a prestressed concrete flat slab. Maisel (1978) used hot-dipped galvanized 1.55-mm-diameter prestressing cables in prestressed concrete model studies at Stuttgart. A typical stress-strain curve for this cable is shown in Fig. 5.7.

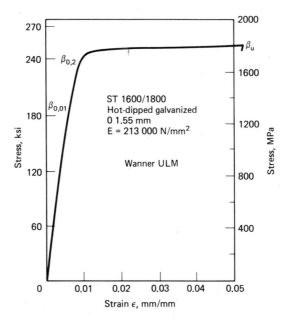

Figure 5.7 Stress-strain relation of prestressing cables for a 1:6 scale model. [Maisel (1978).]

5.5.2 Model Prestressing Techniques

5.5.2.1 General

In prestressed concrete models the reinforcing technique can be divided into two categories: (1) pretensioning and (2) posttensioning. As the nature of problems in the two cases is different, they are discussed separately in the following sections.

5.5.2.2 Pretensioning Technique—Single Wires in Beams

An easily used pretensioning technique was developed at Cornell University [Burton (1963), Chao (1964)] for single wires in beams. A prestressing bed was designed for pretensioning the steel wire as shown in Fig. 5.8. The

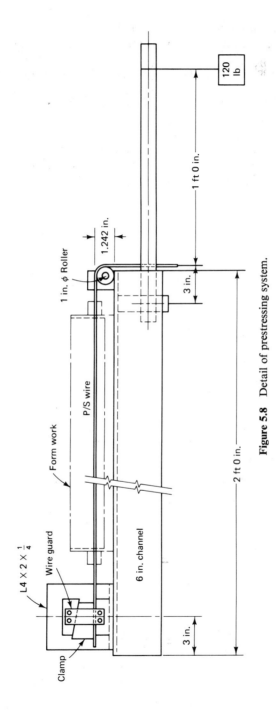

Figure 5.8 Detail of prestressing system.

tension is applied through a lever arm, with the other end of wire clamped. Slipping of wire at the tensioning end is prevented by means of a clamp, details of which are shown in Fig. 5.8. A detail of the end of the beam is shown in Fig. 5.9. The desired load is applied to the lever (with an arm ratio of 4), and the beam is cast. The loads are removed when the concrete has attained the required strength to take the prestress. The main shortcoming of this method is in measuring the prestress force accurately. This could be overcome by placing a dynamometer at the clamped end and measuring the true load in the wire.

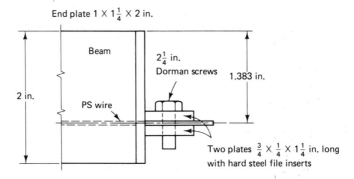

Figure 5.9   End detail of beam.

Another simple prestressing device consisting of three parts is shown in Fig. 5.10. It was developed at McGill University [Pang (1965), Mirza and McCutcheon (1974, 1978)] to pretension model prestressing wires. The jaw shown in Fig. 5.10a was designed so that the button (Fig. 5.10c) would just fit into the slot provided at one end. The jaw also contained a long, smooth rod welded to a rectangular steel block designed to slide in the frame shown in Fig. 5.10b. Two strain gages were installed 180° apart on this smooth rod and were calibrated to read the applied prestressing force.

The free end of the prestressing wire was flattened by passing it through the hole to develop the ultimate tensile strength of the wire. Other attempts to anchor the wire (without the flattened end) using techniques that involved welding, soldering, and cementing with an epoxy glue were not successful. It is suggested that an epoxy glue or low-temperature soldering be used in the $\frac{1}{4}$-in. (6-mm) diameter hole of the button to augment the anchorage capacity of the flattened end and to prevent any slip.

The button (with the wire) is slipped into the appropriate end of the jaw, which is then slid into the frame, and a nut is mounted at the free end of the threaded rod. The desired prestressing force can be applied by tightening this nut and reading the load from the calibrated strain gages on the jaw.

The model prestressing bed consisted of a W10 × 21 with four 7 × 4 × $\frac{7}{8}$ in. (175 × 100 × 22 mm) unequal leg angles as bulkheads bolted firmly to

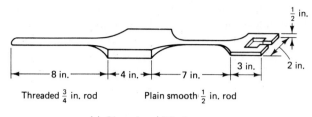

(a) Dimensions of the jaw

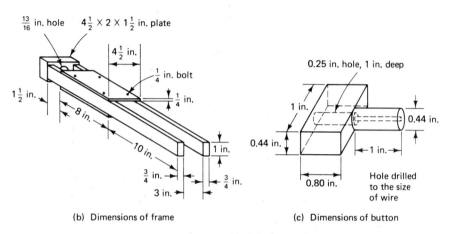

(b) Dimensions of frame   (c) Dimensions of button

Figure 5.10 Detail of the prestressing equipment. [Pang (1965).]

the top flanges (Fig. 5.11). The position of the $7 \times 4 \times \frac{7}{8}$ in. ($175 \times 100 \times 22$ mm) angles could be varied to suit the length of the model. Threaded rods, 3 in. (75 mm) long and 0.38 in. (10 mm) in diameter, were used to allow the wires to pass through a 0.12 in. (3 mm) diameter hole along the rod axis, and

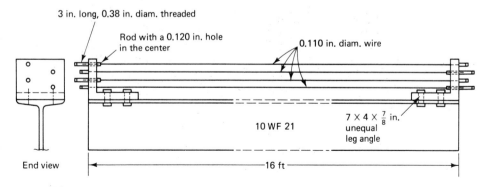

Figure 5.11 Prestressing end systems for the prestressed model. [from Pang (1965).]

the wires were tensioned to the desired load level, one at a time, using the prestressing device (Fig. 5.10). The appropriate threaded rod on the bulkhead was then adjusted to transfer the prestress to the abutment and release the prestressing device for the next tensioning operation.

5.5.2.3 Pretensioning Technique—Multiple Wires in Beams

A mechanical prestressing frame for model beams with multiple tendons has been used at the University of Illinois [Nebraska and Sur (1963)]. As shown in Fig. 5.12, the device utilized steel grips capable of accommodating up to ten $\frac{3}{64}$-in. (1.2-mm) stranded cables. Prestressing was applied by moving the jacking plate and monitoring the applied force with a dynamometer consisting of a short length of tubing instrumented with electrical resistance strain gages. The frame is sufficiently long to permit simultaneous casting of three beams.

5.5.2.4 Posttensioning Technique—Multiple Wires in Slab

As mentioned in Section 5.4.1, Moustafa (1966) used 0.10-in. (2.6-mm) custom-length bicycle spokes in a prestressed concrete flat slab model study at Cornell University. In order to prevent bond between the wire and the surrounding concrete, the wire was coated with grease just before casting of the slab. As there was free movement of the wire after the concrete was set, this condition of no bond was achieved. Prestressing was accomplished by torquing one nut against a steel bearing plate. The amount of torque required for the desired tension in the wire was determined by precalibration. The torque-wrench calibration was done in the testing machine, tensioning the wire and noting the corresponding torque shown by the torque-wrench. Thus, for any desired pretension the amount of torque required could be fixed prior to post-tensioning. Transfer of prestress was through steel bearing plates fixed to the slab. The prestress induced in the concrete was checked by electrical strain gages fixed on both surfaces of the slab along the length of prestressing wires.

In an investigation on simply supported, posttensioned, nongrouted slab models tested under uniformly distributed loads, Kemp (1971) used the Freyssinet monowire system with 0.28 in. (7 mm) diameter wires anchored by cone anchorages. Again, to prevent bond between the prestressing wire and the concrete, the wires were coated with grease before casting of the slab concrete. The monowire jack allowed the wire to pass through the center of the piston; each wire was stressed individually by the jack and then anchored by hammering in the radial friction-type wedges. Mild steel bearing plates of size $4 \times 2 \times 0.25$ in. ($100 \times 100 \times 6$ mm) were used at each anchorage. To reduce the tensile strains resulting from the jacking operation, two $\frac{1}{8}$-in. (3 mm) diameter mild steel deformed bars were placed behind the plates, along the entire slab edge, and spaced $\frac{1}{4}$ in. (6 mm) from each face of the slab. These prevented spalling of the concrete edges during handling and also, because of their small

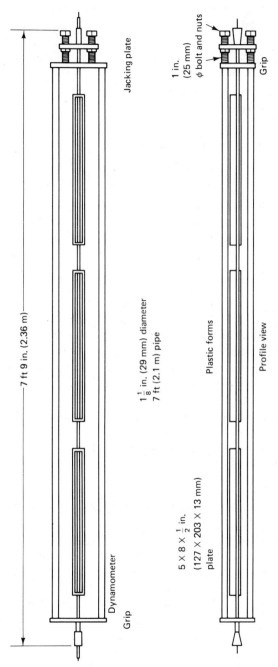

Figure 5.12 Prestressing frame. [Nebraska and Sur (1963).]

7 ft 9 in. (2.36 m)

Jacking plate

Dynamometer

Grip

$1\frac{1}{8}$ in. (29 mm) diameter
7 ft (2.1 m) pipe

$5 \times 8 \times \frac{1}{2}$ in.
(127 × 203 × 13 mm)
plate

Plastic forms

1 in.
(25 mm)
ϕ bolt and nuts

Grip

Profile view

sizes and locations, had no effect on the load-deformation characteristics of the simply supported slab.

5.6 REINFORCEMENT FOR REINFORCED CONCRETE MODELS

5.6.1 Wire Reinforcement for Small Models

The range of wire sizes generally used in small-scale models varies from SWG No. 11, diameter = 0.12 in. (3 mm), to SWG No. 21, diameter = 0.032 in. (0.8 mm). Although these wires are readily available in the form of rolls of annealed wire, straight lengths of annealed wires can also be specially ordered. Straight wires are generally desirable for ease of working in the laboratory and also for accurate placement of wires in the very small models.

Commercially deformed wires used in producing welded wire fabric are available in sizes above a diameter of 0.105 in. (2.7 mm). In addition, deformed wires as small as 0.062 in. (1.6 mm) are available on special order from some suppliers. Threaded wires in various sizes are readily obtained, but their cost is considerably higher than other types of reinforcement.

Some of the wires do not have a sharp yield plateau and have very limited ductility in tension because of the rather severe deforming process used in making them. The characteristic sharp yield of annealed wires may be significantly altered by the straightening of rolled wire and also by the deformation technique used in the laboratory. Thus proper strain relieving, or in certain cases annealing, is required before the wire is suitable for use as model reinforcing. The lack of a sharp yield point is sometimes desired when modeling high-strength reinforcement that does not exhibit sharp yielding; this type of model reinforcing may also be produced by careful use of selected heat treatments. During heat treatment, care must be exercised to check variation of properties along the entire length of wires.

A simple technique was developed in the Cornell Structural Models Laboratory to cold-deform plain wire, thus making it more suitable for use as scaled deformed bars. The desired deformations are obtained by passing the wire through two pairs of perpendicular knurls, as shown in Fig. 5.13. The extent of deformation, which is easily adjusted, is kept proportional to the various wire diameters. In order to get the optimum uniformity, the wire can be continuously deformed for the full length required for a particular model.

Five different sizes of wire deformed by this technique are shown in Fig. 5.14. In addition, three commercially deformed wires are shown. None of the deformation patterns is geometrically similar to the raised lugs used in prototype reinforcing; the threaded rod has a large continuous helical deformation, while the deformations on the laboratory-deformed and commercially deformed wires are rolled into the wire in the shape of internal valleys rather

Figure 5.13

(a)   (b)

Figure 5.14 (a) Typical model reinforcements (left to right): Fabribond, U.S. Steel, Cornell deformed, threaded rod, and plain bar. (b) Typical model reinforcements (all deformed in Cornell Lab).

than external protrusions. It is assumed that inward deformations of a given depth have the same mechanical bonding effect as a protrusion of the same height. Although the assumption is open to question, it serves a useful purpose in helping to arrive at a criterion for judging the bond characteristics of the various reinforcements, which are discussed in Section 5.8.

Figure 5.15 shows the effect of the laboratory deformation technique on the measured stress-strain curve of 0.078 in. (2 mm) diameter annealed wire.

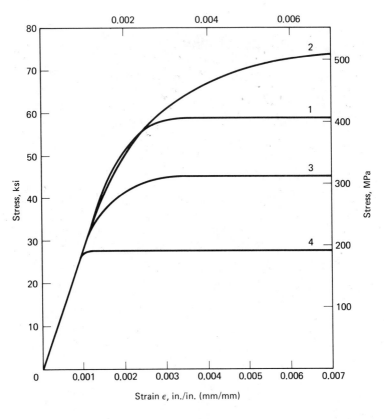

#	σ_y (ksi)	Condition
1	58	As delivered
2	70	$\frac{1}{3}$ turn* deformed — no heat
3	45	$\frac{1}{3}$ turn plus annealed 1000°F–45 min.
4	27	$\frac{1}{3}$ turn plus 1500°F–15 min.

*1 turn = $\frac{1}{28}$ in. motion of knurls developed in the laboratory

Figure 5.15 Effect of deformation and heat treatment on stress-strain behavior of reinforcing wire (size: 14 gage or 0.078 in. diam.).

The drop in yield strength with increasing annealing is shown in Fig. 5.16 for three different small-diameter wires. Detailed results of annealing effects are given by Harris et al. (1966). Suitable annealing temperatures ranged from 1000 to 1500°F (540 to 820°C), with the lower portion of the temperature range being most suitable. Experiments will quickly establish the appropriate annealing time for a given wire.

White and Clark (1978) modeled standard $\frac{1}{2}$-in. (12 mm) GK60 deformed bars with a laboratory-deformed 14-gage (2 mm diameter) mild steel wire. The deformations were produced using a knurling machine similar to that used at

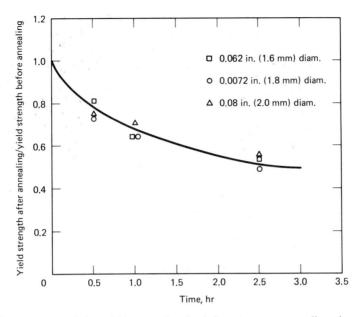

Figure 5.16 Relative yield strength of reinforcement vs. annealing time at 1000°F.

Cornell University. It was observed that the wire diameter reduced to 1.88 mm after deforming, thus resulting in a linear scale factor of 1/6.75. The prototype rib dimensions and the required model rib dimensions at a scale of 1/6.75 and the dimensions obtained after deforming are as follows:

		Model Dimension	
	Prototype Dimension	Required	Achieved
Bar diameter, mm	12.70	1.88	1.88
Rib spacing, mm	8.30	1.23	1.25
Rib width, mm	0.97	0.14	0.30
Rib height, mm	0.53	0.08	0.08
Rib angle, degrees	30	30	30

Thus excellent similitude was obtained for all dimensions except the rib width, which was about twice the required width. The model rib width is determined by the width of the transverse saw cuts made in the rolls of the knurling machine, and it is not considered feasible, by conventional workshop practice, to make these any narrower. In addition, if the cuts could be made narrower, the wires would then have to be hot-rolled in order to form the ribs.

The stress-strain curves of the prototype bar, the deformed model wire, the Cornell University knurled model wire, and the plain wire are compared

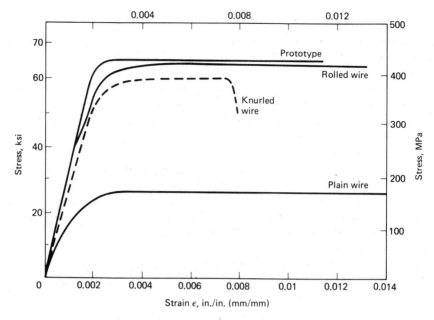

Figure 5.17   Reinforcement stress-strain curves. [White and Clark (1978).]

in Fig. 5.17. Good agreement was obtained between the prototype and models except in the vicinity of the yield point. White and Clark (1978) reasoned that this was because the prototype was hot-rolled whereas the model steels were cold-rolled.

5.6.2 Black Annealed Wire as Model Reinforcement

Microconcrete models at MIT have been reinforced with black annealed wire that is available commercially in the form of rolls. The wire is cold-drawn and annealed in the factory. Before it can be used in models, it must be straightened by pulling, and this process strains the wire sufficiently to destroy the yield point, as shown by curves 1, 2, and 3 in Fig. 5.18. These graphs have been obtained for a black annealed wire that was first straightened, then bent to form part of the reinforcement cage of an arch, and then straightened again before testing. If the cage is allowed to age for 2 months, the wire once more exhibits a definite yield point. The process can be speeded up by keeping the assembled reinforcement cage at 300°F (150°C) for 90 min. The stress-strain curves of the wire are then given by curves 4 and 5, which are in good agreement with the stress-strain curves of prototype reinforcing steel. Curve 4 demonstrates, however, that the yield strength of the wire can exceed that of mild steel considerably. The small-diameter wires that are commercially available frequently have too high a strength to be useful in model studies. It must be

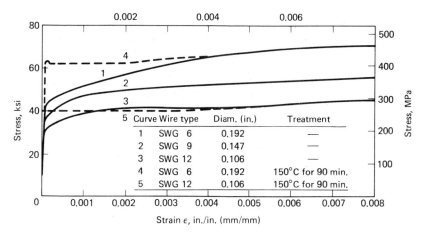

Figure 5.18   Stress-strain curves of wires.

noted that the yield strength of such wires can be reduced to the desired value by annealing.

Mirza (1967) used closed stirrups made from 11-, 13-, and 16-gage black annealed wires in a study of combined bending and torsion in $\frac{1}{4}$-scale model beams. Some of these stirrups were instrumented with strain gages to examine the postcracking and ultimate behavior of these model specimens.

5.6.3 Commercially Deformed Wire as Model Reinforcement

Commercially deformed wires (ASTM A-496, 1978) normally consist of cold-worked deformed steel wire intended for use in producing welded wire fabric and as reinforcement in concrete construction. These wires normally have a minimum yield strength of 70 ksi (480 MPa) and a minimum tensile strength of 80 ksi (550 MPa). These wires can be obtained by special order in sizes D1 through D10, with the wire size number indicating the nominal cross-sectional areas of the deformed wire section in one-hundredths of a square inch. Details of geometry of deformed steel wire are shown in Table 5.1, which is reproduced from the *Manual of Standard Practice of the Reinforcing Steel Institute* (1975).

McGill University and Queen's University have used 0.092 in. (2.3 mm) diameter wire (approximately 13 gage) in their reinforced concrete model work. This indented wire was supplied by the Steel Company of Canada and was specially annealed to conform to the ASTM specifications for intermediate-grade steel. A typical stress-strain curve for the wire is shown in Fig. 5.19. The wires used at Queen's University were initially cold-drawn to conform to ASTM Standard A496, and subsequently annealed to give the required stress-strain characteristics.

TABLE 5.1 Dimensional Requirements for Deformed Steel Wire for Concrete Reinforcement

Deformed Wire Size Number*	Nominal Dimensions				Deformation Requirements		
	Unit Weight, lb/ft	Nominal Diameter, in.	Cross-Sectional Area, in.²	Perimeter, in.	Spacing		Minimum Average Height of Deformations, in.†
					Maximum, in.	Minimum, in.	
D-1	0.034	0.113	0.01	0.355	0.285	0.182	0.0045
D-2	0.068	0.159	0.02	0.499	0.285	0.182	0.0063
D-3	0.102	0.195	0.03	0.612	0.285	0.182	0.0078
D-4	0.136	0.225	0.04	0.706	0.285	0.182	0.0101
D-5	0.170	0.252	0.05	0.791	0.285	0.182	0.0113
D-6	0.204	0.276	0.06	0.867	0.285	0.182	0.0124
D-7	0.238	0.298	0.07	0.936	0.285	0.182	0.0134
D-8	0.272	0.319	0.08	1.002	0.285	0.182	0.0143
D-9	0.306	0.338	0.09	1.061	0.285	0.182	0.0152
D-10	0.340	0.356	0.10	1.118	0.285	0.182	0.0160
D-11	0.374	0.374	0.11	1.174	0.285	0.182	0.0187
D-12	0.408	0.390	0.12	1.225	0.285	0.182	0.0195

D-13	0.442	0.406	0.13	1.275	0.285	0.182	0.0203
D-14	0.476	0.422	0.14	1.325	0.285	0.182	0.0211
D-15	0.510	0.437	0.15	1.372	0.285	0.182	0.0218
D-16	0.544	0.451	0.16	1.416	0.285	0.182	0.0225
D-17	0.578	0.465	0.17	1.460	0.285	0.182	0.0232
D-18	0.612	0.478	0.18	1.501	0.285	0.182	0.0239
D-19	0.646	0.491	0.19	1.542	0.285	0.182	0.0245
D-20	0.680	0.504	0.20	1.583	0.285	0.182	0.0252
D-21	0.714	0.517	0.21	1.624	0.285	0.182	0.0259
D-22	0.748	0.529	0.22	1.662	0.285	0.182	0.0265
D-23	0.782	0.541	0.23	1.700	0.285	0.182	0.0271
D-24	0.816	0.553	0.24	1.737	0.285	0.182	0.0277
D-25	0.850	0.564	0.25	1.772	0.285	0.182	0.0282
D-26	0.884	0.575	0.26	1.806	0.285	0.182	0.0288
D-27	0.918	0.586	0.27	1.841	0.285	0.182	0.0293
D-28	0.952	0.597	0.28	1.876	0.285	0.182	0.0299
D-29	0.986	0.608	0.29	1.910	0.285	0.182	0.0304
D-30	1.020	0.618	0.30	1.942	0.285	0.182	0.0309
D-31	1.054	0.628	0.31	1.973	0.285	0.182	0.0314

*The number following the prefix D identifies the nominal cross-sectional area of the deformed wire in hundredths of a square inch.

†The minimum average height of deformations shall be determined by measurements made on not less than two typical deformations from each line of deformations on the wire. Measurements shall be made at the center of indentions.

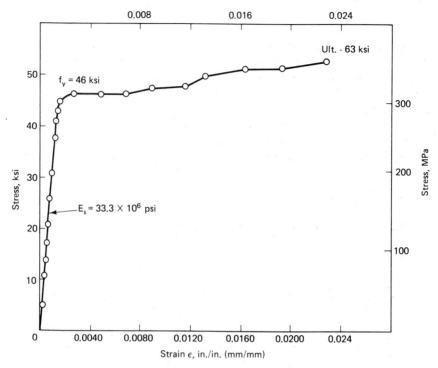

Figure 5.19 Stress-strain curve for wire reinforcement (Stelco—0.092 diam., indented).

Recent McGill work has used deformed steel wires obtained from Lundy Fence Company, Dunnville, Ontario, ranging in sizes from D2 to D10. The chemical composition of this wire steel is carbon, 0.13%; phosphorus, 0.007%; sulfur, 0.028%; manganese, 0.68%; and silicon 0.15%.

Labonte (1971) used D2 wires to model No. 8 (25 mm) bars in a model investigation of the behavior of anchorage zones in prestressed concrete containments. Figure 5.20 shows the stress-strain curve obtained using electrical resistance strain gages for the D2 wires as obtained from the supplier and after being annealed at 1700°F (925°C) for approximately 30 min, which lowered the yield point and increased ductility. Mirza (1972) also reported his work on heat-treatment of steel bars on the deformed wires obtained from the Lundy Fence Company. Sixteen sets, each set consisting of 20 samples (five bars for each size D2, D3, D4, and D8), were heated to temperatures ranging from 900 to 1600°F (480 to 870°C) for a period of 1 hr. Twelve of these sets were air-cooled, two groups were oil-quenched at temperatures of 1380 to 1600°F (750 to 870°C), and the remaining groups were water-quenched at 870°C. The stress-strain curves for the bars as obtained from the supplier are shown in Fig. 5.21. The effect of annealing followed by air cooling on the mechanical properties of the D2, D3, D4, and D8 wires is summarized in Fig. 5.22. It can be

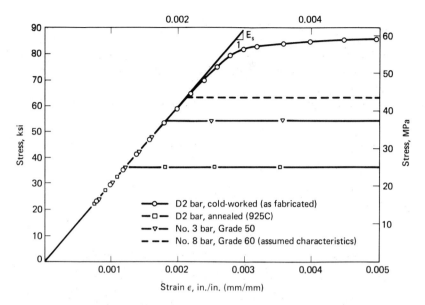

Figure 5.20 Reinforcement load-deformation characteristics.

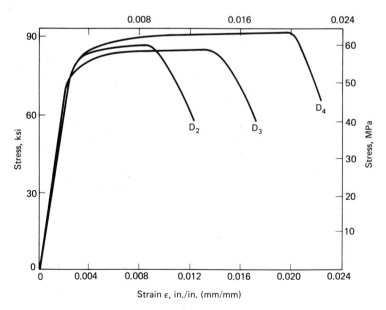

Figure 5.21 Stress-strain curves for D_2, D_3, and D_4 bars as obtained from the supplier.

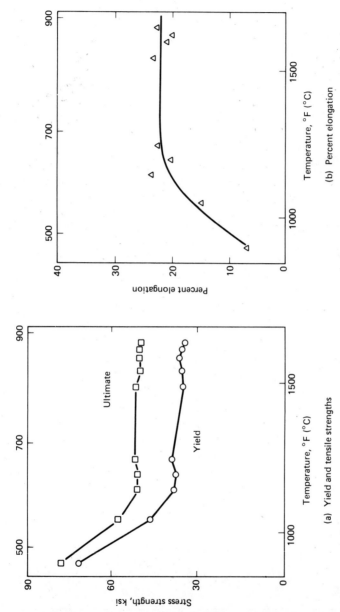

Figure 5.22 Effect of heat treatment on deformed wires. (a) Yield and tensile strengths. (b) Percent elongation.

noted that at temperatures between 1100 and 1200°F (600 and 650°C) the yield strength is lowered to a value between 35 and 40 ksi (240 and 275 MPa) and the ultimate tensile strength is lowered to approximately 50 ksi (345 MPa). Temperatures above 1200°F (650°C) do not have any additional influence on the yield and the ultimate strengths. For all temperatures above 1100°F (600°C) a percentage elongation of more than 20% was observed for all wires over a gage length of 5 in. (125 mm). It was noted that specimens that were oil-quenched or water-quenched had higher yield and ultimate strengths than the specimens that were air-cooled. The percentage elongation and therefore the overall ductility of oil- or water-quenched steel was much lower than the air-cooled steel.

An extensive study of the annealing characteristics of commercially deformed wires of diameter 0.11 in. (2.9 mm) and 0.16 in. (4 mm) was performed by Chowdhury and White (1971). The wires had four lines of rectangular patterns embossed into the surface at $\frac{1}{4}$-in. (6-mm) spacing. The stress-strain curves for the wires as delivered are shown at the top of Fig. 5.23. Because of the cold working involved in the embossing process, the wires did not have a well-defined yield point. The effects of various heat treatments on the wire properties are shown in Figs. 5.23 and 5.24. One hundred and twenty-seven specimens were tested to failure in tension. Full annealing at about 1600°F (870°C), with slow cooling through the critical range, produced very low yield points. The cooling rate is crucial in full annealing. Normalizing, which consists of heating about 100°F (38°C) above the critical temperature (1675°F or 910°C in this case) and then cooling in air, gives higher yield strengths and lower ductilities than full annealing. Process annealing was used to achieve the proper model steel yield points in this study. Process annealing is done at 900 to 1200°F (480 to 650°C) and does not involve the pearlite-austenite transformations of full annealing. It produces recrystallization of deformed crystals into undeformed or less deformed ones without changing their nature because the temperature is held below the critical point. A more complete description of these heat-treatment processes is given by Chowdhury and White (1971).

The variation of properties of individual lengths of model steel is an important quantity, particularly in sections with relatively few reinforcing wires. For wires annealed in a single batch, the maximum coefficient of variation was 0.09. In 81% of the batches the coefficient of variation was below 0.05. Such differences may be produced among others by (1) differences in wire quality and (2) nonidentical surface deformations, which lead to differences in net area. When comparing one batch with another, the differences in heat-treatment temperature and in cooling rates also must be considered.

It is recommended that heat treatment of reinforcements be done under closely controlled conditions using a single furnace with temperature control of ±5°F (3°C). This precision is necessary to reduce the typical variation of steel strength from the desired value to within ±5%. It is necessary to clean

Symbol	Diam., in.	Temp., °F	Annealing time, hr	Type of cooling
●	0.159	As delivered	—	—
△	0.123	As delivered	—	—
○	0.113	1010	$1\frac{1}{2}$	FC (16 hr)
▽	0.159	1075	$1\frac{1}{2}$	FC (36 hr)
▲	0.113	1100	1, 2	FC (16 hr)

Symbol	Diam., in.	Temp., °F	Annealing time, hr	Type of cooling
+	0.159	1675	$\frac{1}{2}$	AC (78°F)
□	0.113	1675	$\frac{1}{2}$	AC (78°F)
▼	0.159	1600	$\frac{1}{2}$	FC (16 hr)
■	0.113	1600	$\frac{1}{2}$	FC (16 hr)
AC - air cooling, FC furnace cooling				

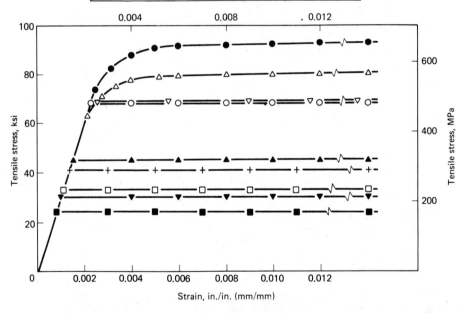

Figure 5.23 Stress-strain curves for commercially deformed wires under various kinds of conditions.

the wires with diluted hydrochloric acid to remove grease and improve the bond to model concrete.

Model stirrups and ties can be made of plain annealed wire, 0.041 in. (1.04 mm) and 0.54 in. (1.4 mm) in diameter. Reinforcing steel may be fabricated by tying with 26 SWG wire.

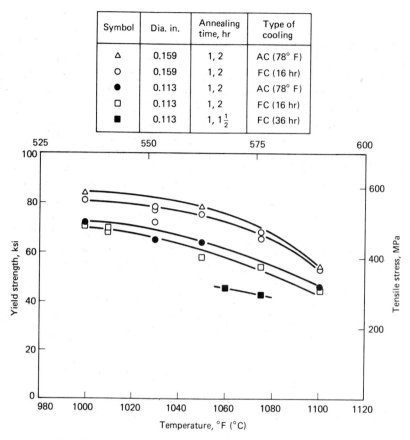

Figure 5.24   Relation between yield strength of commercially deformed bars and temperature of annealing.

5.7 FABRICATION OF MODEL REINFORCEMENT

5.7.1 General

Fabrication of reinforcing in small-scale models forms an important part of the model process. One might think of bending the wires as is done in prototype bars and tying them with very thin wires as in the case of prototype reinforcement cages. Although this can be done in certain cases, it proves to be extremely difficult for very small scale structures. Methods used for positioning the wires during fabrication include welding, the use of epoxies, and soldering. The following sections describe the methods and techniques used at Cornell, McGill, and Drexel Universities.

5.7.2 Fabrication Methods

5.7.2.1 Bending of Reinforcement

Bending of reinforcement must be done carefully so that all the wires are accurately located and placed. Main reinforcement for beams can be easily bent over a template after marking the bending positions accurately. In order to avoid very small hooks at the ends, extra lengths covering the length of hook need to be provided. In bending stirrups a template of Plexiglas or steel is made and all the stirrups are bent to give the desired uniform shape. Commercially available bar benders have also been used to bend bars and stirrups.

5.7.2.2 Use of Epoxies

There are various epoxies that are useful in reinforcement fabrication. One can select a good metal-to-metal epoxy, with a curing period that varies from a few minutes to a few hours. In beams, main reinforcing wires may be held by end blocks and stirrups positioned as desired. The epoxy is then placed at the stirrup-bar joints using small steel wire or nails.

5.7.2.3 Welding of Reinforcement

The welding of model reinforcing enables easy fabrication of complicated reinforcing cages in a short time. It also helps to keep better control on forming joints.

The Raytheon spot welder (Weldpower-Model: QB) was used for all welded fabrication discussed here (Raytheon Company, 1962). Straight wires and small stirrups were welded using the main unit. For curved reinforcing, as in shells, a hand unit (Weldplier-Model: 2-205) can be conveniently used. A complete assembly is shown in Fig. 5.25.

While using a welder, however, proper precaution should be taken to

Figure 5.25

avoid excess of heat input at the joint. This becomes very important for welded cages with closely spaced wires. Tension tests were conducted on wires with cross welds of different spacing to study this effect. The main wire was 0.062 in. (1.6 mm) in diameter, and the cross wire pieces of 0.041-in. (1.4 mm) diameter were welded at various spacings of $\frac{1}{4}$, $\frac{1}{2}$, $\frac{3}{4}$, and 1 in. (6, 12, 20, and 25 mm), respectively. These were compared with a specimen with no welds. A reduction of about 15, 8, 5 and 0% in the yield strength of wires was found for the above spacings.

Mirza (1967) tested several coupons of 5 and 8 SWG steel wires with one, two, or three 11, 13, or 16 SWG wire pieces spot-welded across the coupon using a low heat cycle. He compared their yield strength with those of the corresponding unwelded wires and noted that there was no reduction in the strengths of 6 SWG wires. However there was a strength decrease between 7 and 10% in the cross-welded 8 SWG wire coupons.

5.7.3 Accuracy of Placing of Reinforcement

It is necessary to ensure that the reinforcement is placed in the proper position in the model. Accurate placement is one of the more difficult problems met in fabricating small-scale models. The wires should be straight; they should be held in place with miniature bar chairs (fabricated from wire) along the length of the member. Exact positioning of main reinforcement in beams is achieved by passing it (reinforcement) through the end blocks of the forms. Accurate placement is easier to achieve in prestressed beams because the initial tension in the wire holds it in place during casting. The load in the wire also removes any kinks or curvature along its length, and accurate end positioning is sufficient to produce adequate models.

5.8 BOND CHARACTERISTICS OF MODEL STEEL

The requirements of bond similitude are discussed in Section 2.5.1. Some of the highlights of the above discussion are summarized below:

1. The use of standard deformed bars does not necessarily ensure true bond modeling because of the difference in bond strength of small- and large-size bars.

2. Modeling of bond is seriously complicated by our limited knowledge of the bond mechanism in prototype members.

3. The ultimate bond strength per unit length of the bar is proportional to $\sqrt{f'_c}$. It would be grossly misleading to model the dimensionally non-homogeneous relationship for bond with small-size reinforcement and wires.

4. "True" bond-stress distribution bears little resemblance to the computed average unit bond stress, thereby making it difficult to model such a poorly defined quantity.

5. The effect of concrete cover upon bond strength and the increased bond strength afforded by stirrups and other steel, which tends to prevent splitting from developing adjacent to bars under high bond stress, are difficult to predict and to model.

Several different types of specimens and test procedures may be employed in studying the bond strength of reinforcement, including the concentric and the eccentric pullout tests (Fig. 5.26), the embedded bar tensile test, and various types of flexural specimens, e.g., the University of Texas beam (Fig. 5.27), the National Bureau of Standards beam, and the McGill doubly symmetrical bond beam (Fig. 5.28). The flexural tests are preferable because flexural bond is of prime concern and these tests do simulate the practical situation more closely. The concentric pullout test is economic, simple, and less time-consuming; however, the main disadvantage of this type of test is that the concrete at the loaded end is in compression and eliminates or significantly decreases the

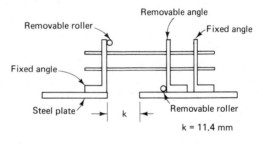

Support equipment

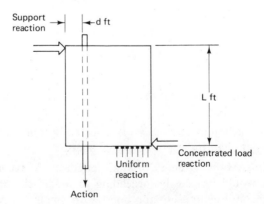

Figure 5.26 Eccentric pullout test details.

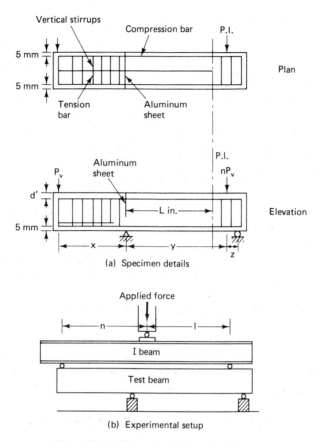

Figure 5.27   University of Texas bond beam.

transverse cracking. This also results in an increased relative slip between the concrete and the steel.

Earlier investigations of bond between plain wire and model concrete [Harris et al. (1963), Taher (1963), Aldridge (1966), Lim et al. (1968)] showed that for black annealed wires, an embedment length wire diameter ratio of about 20–25 was adequate to cause the wire to yield. As expected, the smaller diameter wire (16 and 18 SWG) exhibited better bond characteristics than larger diameter wires (12 SWG and longer); this trend is similar to that observed in prototype bars. Using smooth wires in pullout and flexure specimens, Taher (1963) found that rusting increased the bond resistance, so that for 12 SWG wires and smaller, a yield point of 50,000 psi (345 MPa) could be developed with length over 20 diameters, corresponding to a bond strength of 625 psi (4.3 MPa). When rusted wire is used, caution must be taken to prevent deterioration of the wire as a result of excessive rusting. Harris et al. (1963)· also noted that rusting the wires for 7 days improved their bond characteristics.

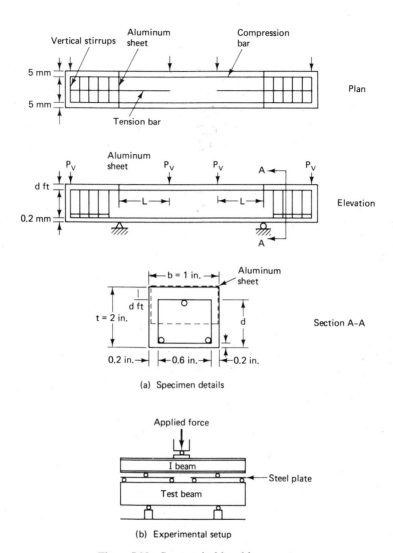

Figure 5.28 Symmetrical bond beam test.

Unfortunately, smooth or rusted wires will not be adequate if the purpose of the model study is to investigate bond failure; number, size and distribution of cracks; postcracking deflections; effects of reversed or repeated loads; or redistribution of internal stresses. To accurately simulate bond characteristics, the model reinforcement should have surface deformation similar to those of prototype reinforcement. In a comprehensive study, Harris, Sabnis, and White (1970) investigated the bond characteristics of plain wire, commercially available deformed wire, threaded rod, and laboratory-deformed steel wires with cement- and gypsum-based model concretes using concentric pullout, tension, and flexure tests, and compared the results with similar tests conducted on proto-type specimens. It was possible to develop the yield strength of a 0.039 in. (1 mm) diameter deformed steel wire with length-diameter (L/D) ratio of 15, and to develop yielding of a 0.059 in. (1.5 mm) diameter deformed wire with an L/D ratio of 8.

The pullout test used by Harris et al. (1970) consisted of a model bar embedded in a 1 × 2 in. (25 × 50 mm) cylindrical specimen, and this test was used to study the bond characteristics of the various model reinforcing wires used in model work. Results of a number of tests are plotted in Fig. 5.29, along with other model and prototype data. Plain wires showed a marked decrease in average ultimate bond stress with increasing L/D ratio. Deformed model wires had an ultimate bond stress comparable to large prototype bars, with best agreement at the lower L/D ratios. The comparison of ultimate bond stresses as presented in Fig. 5.29 indicates that suitably deformed wires will have pullout bond strengths reasonably close to those measured for prototype bars.

Hsu (1969) investigated the influence of concrete strength, clear cover, end anchorage, vertical stirrups, and rust on bond between plain and deformed bars and model concrete. The average ultimate bond stress values for the eccentric and doubly symmetrical bond beam tests were in good agreement with each other for both plain and deformed bars. This behavior was anticipated since the free end of the test bar was subjected to concentrated loads in both cases. Average ultimate bond stress values calculated for the University of Texas beams were generally higher than the results for the eccentric pullout and the doubly symmetrical bond beam tests. The University of Texas speci-mens provided for a point of contraflexure at the free end of the test bar, thus eliminating the disturbance caused by the concentrated loads in other types of bond specimens and resulted in higher value of the average ultimate bond stress.

As expected, the bond between steel and concrete improved with an increase in the concrete compressive strength, and the average ultimate bond stress was approximately proportional to $\sqrt{f'_c}$. The average ultimate bond stress increased with the embedment length–bar diameter (L/D) ratio up to 15, after which the average ultimate bond resistance decreased gradually. It was noted that for D2- and D4-size wires, and a 3000-psi (21 MPa) concrete, an embedment length of approximately 15 diameters was adequate to cause these wires to yield at steel stresses between 64 and 76 ksi (439 and 528 MPa) and an L/D ratio between 25 and 30 necessary to cause the steel wire to fracture. It was also observed that as the concrete compressive strength decreased from 5000 to 3000 psi (35 to 21 MPa), the L/D ratio required to cause the steel wire to fracture increased from approximately 12 to about 25 (Fig. 5.30). It was concluded from these various experiments that deformed bars exhibit bond characteristics that are comparable to those of the prototype reinforcing bars, as was observed in the Cornell University tests.

White and Clark (1978) conducted concentric pullout tests on specimens

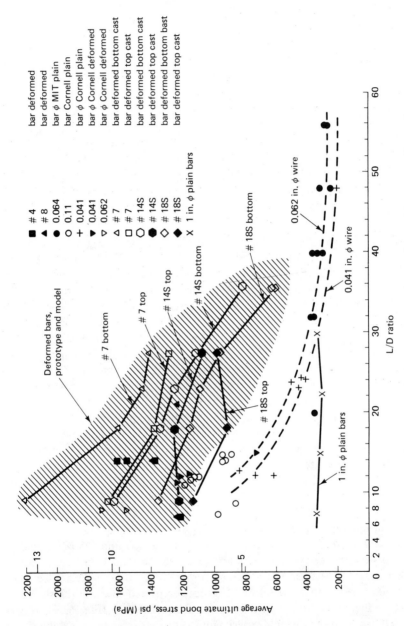

Figure 5.29 Average ultimate bond stress vs. L/D ratio from pullout tests. Comparison of prototype and model results.

204

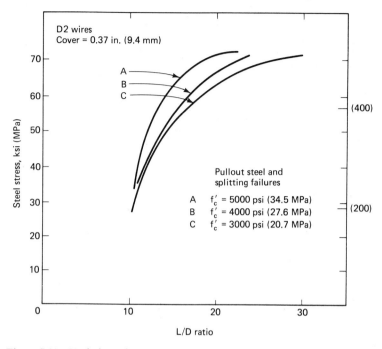

Figure 5.30   Variation of steel stress developed with L/D ratio. [Hsu (1969).]

reinforced with knurled and rolled wires and also, for comparison purposes on plain wire, plain rusted wire, and threaded rod. By instrumenting the pullout specimen with a transducer and a load cell, a direct plot of bond force–slip relationship was obtained. The resulting bond stress–slip curves are shown in Fig. 5.31. The principal conclusions can be summarized as follows:

1. Both types of deformed wires gave higher average ultimate bond strengths with the knurled-type reinforcement showing a stiffer bond strength–slip relationship. The failure mechanisms were significantly different from those obtained in prototype tests.

2. As a result of tests with mixes of different single aggregate sizes, it was observed that for all types of wires, "coarse" aggregates gave higher ultimate bond strengths and stiffer bond behavior than "fine" aggregate mixes. When compared with plain wires, the aggregate size had a greater influence on bond with deformed wires. However, experiments with two different aggregate shapes showed that the aggregate shape did not appear to have a significant influence on bond characteristics.

3. Posttest examination of the model concrete-reinforcement interface showed that in many of the specimens the concrete on the underside of the reinforcement (as cast) was badly compacted. This was due to air

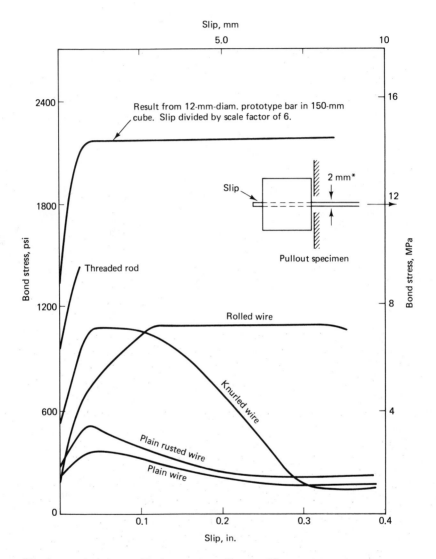

Figure 5.31 Bond stress distributions.

*Bond stress calculations are based on an average diameter of 2 mm.

becoming trapped under the reinforcement during casting. In an attempt
to solve this problem, the specimens were cast in a vertical position. The
effect of direction of casting was investigated by Rehm (1961), who showed
that bond stresses obtained for horizontally cast specimens were as low
as 25% of the value obtained for the vertically cast specimens. The model
tests at the Cement and Concrete Association (C & CA), UK, have not

agreed with Rehm's results. Little difference was observed in the bond strength and behavior of the horizontally and vertically cast specimens.

According to White and Clark (1978), compaction is one of the main factors influencing the bond characteristics. Results of the C & CA tests indicated an improvement in bond by using less workable mixes that require long vibration times and by using molds of greater mass.

5.9 BOND SIMILITUDE

Although some investigators have studied the bond characteristics of plain and deformed steel wires, experimental data on the similitude of bond is almost nonexistent. Harris, Sabnis, and White (1966) modeled two prototype specimens (scale $1: 8 \cdot 33$) to study the effect of wire deformation on bond and cracking in reinforced concrete models. The crack patterns in the prototype and its models reinforced with deformed wires were very similar with respect to secondary and tertiary cracks. The primary cracks in these models compared excellently with the primary cracks observed in the prototypes and predicted using Brom's theoretical work (1965). However, the number of cracks in the models reinforced with plain wires was between 25 and 50% of the number of cracks in the prototype.

Stafiej (1970) tested 23 direct models of the prototypes tested by Gergely (1969) to study the interaction between the applied pullout and dowel forces in causing splitting cracks along the main reinforcing bars in the beam end zones. Three scale factors (1/5.95, 1/4.83, and 1/4.03) were used to simulate the No. 6, No. 8, and No. 10 prototype deformed bars with locally available deformed steel wires that conformed to ASTM A492-78 specifications.

The test procedure consisted of leveling the specimen on a loading frame (Fig. 5.32) and applying the dowel and pullout forces by using two center-hole jacks bearing against plates as shown. The wires were gripped with standard prestressing wire grips. The pullout force, the horizontal reaction in the compression zone, the vertical reaction, and the dowel force were monitored with a strain indicator and four load cells.

The specimens failed in one of the following five modes and showed excellent correlation with the cracking modes observed in the prototypes.

1. Steel yielding and fracture
2. Concrete splitting at the bottom only
3. Concrete splitting at the sides only
4. Concrete splitting at both the bottom and the sides
5. Complete bar pullout with no splitting

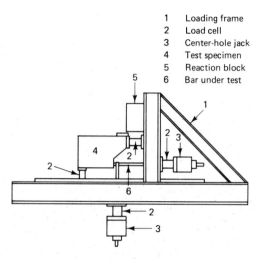

1	Loading frame
2	Load cell
3	Center-hole jack
4	Test specimen
5	Reaction block
6	Bar under test

Figure 5.32   Bond similitude arrangement.

Similitude relations were developed from dimensional analysis, and the resulting prediction equations were used to compare the pullout force values in the prototypes and their models. Most of the predicted values were within a range of ±15% of the experimental results. A statistical analysis of the experimental data showed that the mean and the standard deviation of the experimental strength–predicted strength ratio was 1.04 and 0.128, respectively.

5.10 CRACKING SIMILITUDE AND GENERAL DEFORMATION SIMILITUDE IN REINFORCED CONCRETE ELEMENTS

The inelastic load-deflection response of a reinforced concrete structure is often strongly dependent upon the degree and manner of cracking [Zia et al. (1970)]. Cracking modes can also influence behavior under reversed or repeated loading, moment and force redistribution in indeterminate systems, and occasionally even the final mode of failure. The modeling of cracking is just as difficult as modeling of bond; the two are intimately related phenomena. Moreover, an incomplete understanding of the cracking mechanism contributes greatly to the difficulties. It must also be noted that the cracks that first become visible in a prototype structure are not visible to the naked eye at the cracking load level scaled from the prototype because the width of these cracks gets scaled down, thus rendering these cracks "invisible." Normally, these can be detected by using a magnifying glass with an appropriate magnification factor and by the sudden change in the slope of the load-deformation curve.

Borges and Lima (1960) analyzed the similitude conditions for cracking and deformation in reinforced concrete both analytically and experimentally.

They reported good success in modeling cracking in beams; however, their smallest model was a $\frac{1}{4}$-scale model of a beam 40 in. (1 m) deep.

Kaar (1966) presented a study on similitude of flexure cracking in T-beam flanges in which $\frac{1}{2}$- and $\frac{1}{4}$-scale models of prototype T beam 40 in. (1016 mm) deep with 7×90 in. (178×2286 mm) flanges were tested to failure. Deformed bars were used in all specimens, with No. 8 bars in the prototype and No. 2 bars in the $\frac{1}{4}$-scale model. Crack patterns at identical steel stress levels in the three specimens revealed that the total number of cracks decreased with decreasing beam size. However, overall cracking patterns were similar, and the scaled load-deflection curves for the prototype beam and its two models were practically identical. Thus it appears that overall deformational similitude was achieved in Kaar's tests despite the differences in number of cracks.

Sabnis (1967) modeled two beams C2 and C5 from Mattock's tests (1964) at the Portland Cement Association Laboratories. The geometrical scale factor used was $\frac{1}{10}$, and all beams were 0.6×1.1 in. (15×28 mm) in cross section. Of the four beams, two had a steel ratio of about 0.0145 and two had a steel ratio of 0.025. All main reinforcement was from the same batch of 0.078 in. (2 mm) diameter laboratory-deformed wires. The mix properties were $1:1:0.35$ by weight of Ultracal, sand, and water, respectively, and gave a strength of approximately 3000 psi (21 MPa) at 24 hr. The crack patterns obtained in the model beams showed a striking resemblance to those of the prototype beams. A high consistency of cracking for identical models was also most encouraging.

Further work at Cornell on $\frac{1}{10}$-scale gypsum plaster models of eccentrically compressed columns and $\frac{1}{6}$-scale gypsum plaster model (Fig. 5.33) of a

Figure 5.33   Details of portal frame.

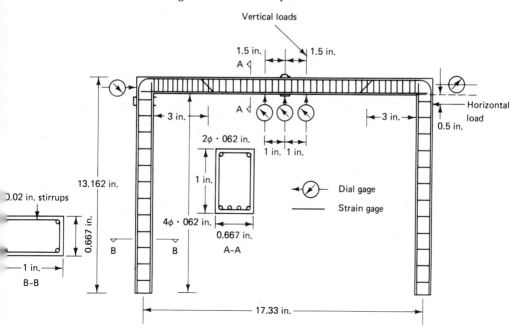

portal frame indicated good agreement between the experimental load-deflection and the predicted values. It was inferred from the moment-deflection relations that the overall behavior of the models was adequate and the crack patterns obtained were quite satisfactory. It must be noted that the model frame with minimum cross-sectional dimensions of 0.667 × 1 in. (16.9 × 25.4 mm) reinforced in each corner with one 0.062 in. (1.6 mm) diameter annealed laboratory-deformed wire and 0.020 in. (0.5 mm) diameter stirrups is a $\frac{1}{15}$-scale model of a full-scale structure since the full-size frame tested was, in itself, a reduced-scale structure. Although there were fewer cracks in the models (Fig. 5.34), a linear scaling of the load-deflection curves for the models gave results nearly identical to those of the prototype (Fig. 5.35). Therefore, there must be many very small cracks, invisible to the naked eye, that give proper stiffness scaling in the model load-deformation relationships. A reasonable degree of confidence can, therefore, be placed in the use of very small scale models in limit design studies [Sabnis (1967)].

Figure 5.34 Crack pattern in the model frame.

Mirza (1967) tested three prototype beams and their quarter-scale models under combined bending, torsion, and shear and obtained excellent similitude between cracking patterns, failure mechanisms, and strengths at initial cracking and failure. Pang (1965) also observed satisfactory similitude of cracking and mechanism of failure between posttensioned prestressed concrete prototype beams and their $\frac{1}{2}$- and $\frac{1}{4}$-scale models subjected to combined bending and shear [see also Mirza and McCutcheon (1974)].

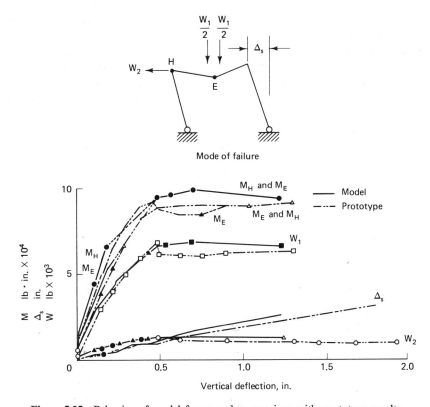

Figure 5.35 Behavior of model frame and comparison with prototype results.

Syamal (1969) and Mirza (1978) reported results from tests on 56 reinforced concrete specimens that included 11 prototypes (built at a local precasting company) and their 45 direct models, at different geometrical scale factors ($\frac{1}{2}$, $\frac{1}{4}$, and $\frac{1}{6}$), constructed at McGill laboratories using a model concrete mix with a nominal strength of 3000 psi (21 MPa) and locally available deformed and plain steel wires. The tests can be grouped in the following three categories.

1. Prototype and model specimens under concentric and eccentric loads (combined axial force and bending)
2. Prototype and model specimens under combined bending and shear
3. Prototype and model specimens under pure torsion and combined bending and torsion

For all loading conditions, it was observed that the crack patterns were reasonably similar except that more cracks appeared in the prototype and the $\frac{1}{2}$-scale models than in the $\frac{1}{4}$-scale and the $\frac{1}{6}$-scale models. Comparison of the crack patterns between two identical specimens, one reinforced with plain wires and the other with deformed wires of mild steel, showed that the beam re-inforced with the plain galvanized wires was poor in reproducing the crack patterns; however, the beam reinforced with deformed wires reproduced crack patterns of the prototype reasonably well, although the number of cracks was generally less than in the prototype. Excellent similitude was obtained for deflections in short and normal beams, but there was a variation of $\pm 30\%$ in deflections for the deep beams at cracking. However, this scatter decreased to between 10 and 15% at 80% of ultimate load. The higher tensile strength of model concrete delayed the formation of cracks and led to discrepancies in deflections at cracking load for deep beams. For the specimens tested under pure torsion and combined bending and torsion, excellent deformation simili-tude was noted up to cracking load values of 0.4 to 0.45 of the ultimate load. However, postcracking torque-twist similitude was not as good, and the differ-ence between the measured and the predicted values of angles of twist ranged between 10 and 40%. Some nondimensional moment-deflection and torque-twist curves are shown in Figs. 5.36 and 5.37, respectively. It was concluded that it is possible to obtain reasonably good deformation similitude (deflections, twists, etc.) for the entire loading range between the prototype and its small-scale microconcrete models (minimum dimension 2 in.) reinforced with deformed steel wires.

Work on models of reinforced concrete slabs showed that size effects in

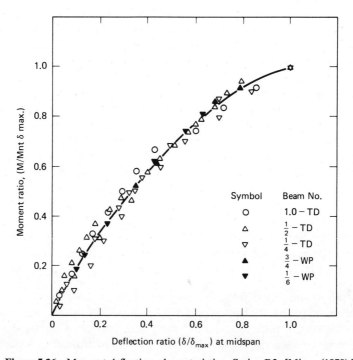

Figure 5.36 Moment-deflection characteristics, Series B5. [Mirza (1978).]

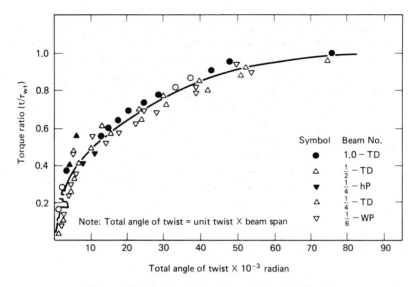

Figure 5.37   Torque-twist characteristics (M/T = 0).

the model slabs are reflected not only in ultimate load but also in the load-deflection relationships. Mastrodicasa (1970) noted that discounting the fact that exact similitude was not obtained in model material properties, the relative model deflections tended to be larger as the size of the model slab decreased from 55.5 × 55.5 × 1.85 in. (1410 × 1410 × 47 mm) to 13.88 × 13.88 × 0.463 in. (353 × 353 × 11.8 mm). However, the crack pattern and crack spacing in the models were similar to those in the prototype slabs. Also, the modes of failure in the prototype and the model slabs were identical. Clark (1971) observed from the results of a series of tests on eight 1/3.7-scale model one-bay slabs that cracking did not scale. The main reason for this was that the model specimens, as compared with the prototypes, exhibited greater cracking strains and more ductility in tension. This scale effect was shown to be a function of the absolute size of the model, in particular its total depth, rather than a function of the model scale. He concluded that prototype strains and crack widths can be predicted accurately from a model test only if the material properties of both the model and the prototype are known accurately.

Clark (1971) suggested that more research is needed on the fundamental properties of materials used in modeling reinforced concrete. The property that has the greatest influence on the correlation of model and prototype behavior at the service load level is the tensile stress-strain relationship for concrete, for which very few data are available for either normal-size concrete or model concrete. It is clear that an investigation is needed to establish a tensile stress-strain relationship for concrete that takes into account variations in maximum aggregate size, aggregate grading and type, and mix proportions.

Finally, it must be pointed out that most of the investigators have observed good to excellent similitude between load-deformation characteristics of the prototype structures and their small-scale models. Each investigator observed a smaller number of cracks in the model than in the prototype. However, the overall pattern of cracking was faithfully reproduced in each case.

5.11 SUMMARY

Materials suitable for modeling structural steel and for modeling reinforcement in reinforced and prestressed concrete structures are presented in this chapter. Properties of commercially available plain and deformed wires and the laboratory-deformed wires and the techniques for annealing these wires to simulate prototype steel yield and ultimate strengths, ductility, etc., are discussed.

Bond characteristics of available steel wires used to model reinforcing steel are presented. The tests used by various investigators and a comparison of model test results with prototype test data are also presented. Influence of different parameters on bond characteristics of steel wires are given. An exploratory investigation on bond similitude showed that certain phenomena, including bond, can be modeled with reasonable reliability if care is exercised in selecting model materials and in constructing and testing the model.

Experimental results show that the total number and width of cracks decrease as the model size is decreased. Existing information shows that in spite of the difficulty of simulating the number and the width of cracks, the overall load-deformation characteristics (load-deflection, moment-rotation, torque-twist, etc.) can be reproduced with reasonable accuracy in small-scale models built from microconcrete or gypsum plaster mixes.

PROBLEMS

(Note: Some of these problems assume familiarity with strain measurement techniques covered in Chapter 8.)

5.1. A W24 × 100 beam, with a depth of 24.00 in. and an I of 2990 in.[4] supports a concentrated load of 20 kips at the center of a 30-ft span.

    (a) Design a distorted direct model for direct analysis in accordance with the following data: $S_L = 10$; aluminum alloy plates to be 0.100 in. thick; width of flange plates to be 1 in. The overall depth of the model is arbitrarily selected as 2 in. Strains are to be measured with SR-4 strain gages. $E_s = 29,000$ ksi, and $E_a = 10,000$ ksi.

    (b) Determine the value of the concentrated load that should be applied to the model.

(c) What is the value of the prediction factor by which the observed strains should be multiplied to correct for the fact that the requirements for similitude have not been satisfied relative to the depth of the model?

(d) Check your model design by computing and comparing midspan deflections, stresses, and strains of the model and prototype.

(e) Discuss the suitability of the model for predicting lateral buckling resistance of the prototype.

(f) Choose suitable SR-4 gages for the model, and describe thoroughly the instrumentation you would use, including the disposition of the temperature-compensating gage(s). (Hint: See Chapter 8.)

5.2. Determine load-deflection and load–bending stress curves for a steel bar 24 in. (60 mm) long used as a beam with central loading on a simple span of 20 in. (50 mm.) Use a minimum number of strain gages. Suggested load increments: Estimate yield moment by assuming $F_y = 40$ ksi (270 MPa) for the steel. No less than four increments of load should be used from $P = 0$ to $P = P_{yield}$.

5.3. Given: Aluminum beam section, strain gages and strain indicator equipment, dial gages, loading device, load cell, and other accessories. For loading in bending to some stage substantially beyond yield, determine:

(a) Load-deflection behavior (deflection measured at center of span).

(b) Load-strain behavior (strain measured at or near center of span; at least one strain gage).

(c) E from both the measured deflection and the measured strain. How do these differ?

5.4. In order to determine the buckling behavior of a steel-tied arch ($E = 30 \times 10^6$ psi), a true-scale model is made from aluminum ($E = 10 \times 10^6$ psi). The span length of the model arch is 10 ft, and of the prototype arch 200 ft. All dimensions are to the same geometric scale. A sketch of the model is given below.

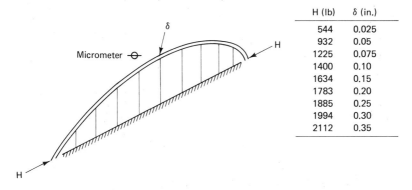

H (lb)	δ (in.)
544	0.025
932	0.05
1225	0.075
1400	0.10
1634	0.15
1783	0.20
1885	0.25
1994	0.30
2112	0.35

(Courtesy: W. G. Godden of U.C. Berkeley)

The load is applied to the model in the form of a horizontal force H at the boundaries, and the response of the model arch is measured by a transverse micrometer that measures δ, the *transverse* deflection movement of the arch at midspan.

The experimental results of this test are given in the table.

(a) What can you deduce about the buckling characteristics of the arch model from this data? In particular what can you say about its behavior at small and large values of loading?

(b) What is the small displacement elastic buckling load of the model?

(c) What is the limit of small displacement response of the model, for acceptable accuracy?

(d) What is the prediction for the elastic buckling load of the prototype?

(e) What is the prediction for the limit of small displacement response of the prototype?

H, lb	δ, in.
544	0.025
932	0.05
1225	0.075
1400	0.10
1634	0.15
1783	0.20
1885	0.25
1994	0.30
2112	0.35

(Courtesy: W. G. Godden, University of California, Berkeley.)

Size Effects in Materials
Systems and Models

〜〜〜〜〜〜〜〜〜〜〜〜〜〜

6

6.1 GENERAL

There are a number of factors that influence the strength of materials and the structures built from these materials. In prototype structures we usually regard the basic material strength properties as measured from control specimens, and as they occur in the members, to be independent of the absolute size of the structural members. For scaled models, in particular, the size of the structure becomes more important since the eventual result of a model testing is to enable the experimenter to predict the strength of the prototype. In the case of concrete, which is highly heterogenous, the reduction in the size of a specimen may change its properties significantly. The presence of reinforcing or prestressing bars acting in conjunction with concrete poses additional problems in interpreting the results because of the difference in size of the model and prototype.

Size effect, a phenomenon observed by many researchers, is related to the change, usually an increase, in strength that occurs when the specimen size is decreased. Sabnis (1980) has reviewed a large number of theoretical studies available in the literature during the last several decades; however, the size-effect phenomenon is not yet fully understood. Many theories for explaining size effects have been put forth, including those of Weibull (1939), Tucker (1941), Nielsen (1954), Pahl and Soosaar (1963), Glücklich and Cohen (1968), and other investigators, who have concentrated on obtaining extensive experimental evidence in testing different size specimens. Various experimental investigations into size effects have been summarized by Sabnis and Aroni (1971) and Sabnis and Mirza (1979).

6.2 FACTORS INFLUENCING SIZE EFFECTS

A number of factors influence the strength properties and hence the behavior of material systems. The strength properties include compressive and tensile strengths, bond and fatigue strengths, and creep and various dimensional changes. Along with these properties, the nature of the material and the geometrical configuration of specimens are also important. The materials range from the naturally occurring timber and rocks to the manufactured materials, such as concrete and steel.

Some of the above properties and materials are affected more by changes in size than by other variables. Some properties may not influence the final interpretation of model investigation to the same degree as the size effect because of their minor influence on the behavior of a structure; e.g., in the case of reinforced concrete, the change in compressive strength is not as important as the yield strength of the reinforcement in an underreinforced *or* lightly reinforced beam. On the other hand, in the investigation of shear strength of a slab or heavily reinforced beam, the compressive strength (as related to its tensile strength) plays a direct and important role. If not considered properly, the observed difference in the strength of specimens of two scales might be attributed incorrectly to the size effect.

Generally, theoretical studies treat the behavior of material systems and their physical results on a statistical basis, the basic philosophy being that the failure in heterogenous materials is a statistical phenomenon. Thus, the larger the volume, the greater the chances for failure which will result in lower strength.

In general, variations in strength of similar shape but different size specimens of concrete are produced by the following factors:

1. Differential curing rates of the various size specimens
2. Differences in the quality (density) of the material cast into the various size molds
3. Change of quality of the cast material as a result of the water gain of the top layers and water leakage through the forms
4. Differential drying of the various size specimens during testing
5. Difference in induced stress conditions because of variation of quality of end capping of different size compressive specimens
6. Statistical variations in strength as a result of volume effects
7. Loading rate and method
8. Strain gradient effects in flexural specimens

6.3 THEORETICAL STUDIES OF SIZE EFFECTS

Generally specimens of smaller size are observed to have higher strength. Also, the scatter in strength is generally greater in the smaller specimens. These factors are shown in Figs. 6.1a and b. This phenomenon of size effect and

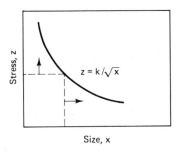

(a) strength vs. size
[adapted from Freudenthal (1968)]

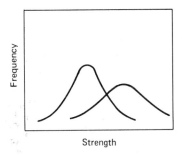

Figure 6.1 Size effects on strength properties (schematic). [Adopted from K.V.S. Rao (1972).]

Strength

(b) Scatter vs. strength

scatter led to theories for explaining this behavior. The basic approach in statistical theories of strength is to evolve a statistical distribution function that adequately characterizes the random heterogeneity of materials and the variation (*scatter*) in their strength. In effect, it identifies admissible forms of distribution functions with suitable parameters that reflect accurately the true material behavior and a realistic mechanism of failure for which the distribution function is applied.

Among the theoretical studies of size effects are those of Weibull (1939), Tucker (1941), Wright and Garwood (1952), Nielsen (1954), and Glücklich and Cohen (1968). Basically, two distinct approaches were used to study the statistical aspects of strength of materials and size effect. These approaches are based on the classical bundle concept, as presented by Freudenthal (1968), and the weakest-link concept. In the classical bundle concept the strength is not only determined by that of the weakest element but is dependent on the strength of the elements in the neighborhood. In this model the specimen is assumed to be made up of parallel fibers (elements), and in such a situation the gross strength at failure is influenced by the strength of all constituent fibers. In the weakest-link concept, the presence of a single severe defect in any of the constituent elements is adequate to cause failure of the total material. Consequently, overall strength of a specimen subjected to uniform stress is determined by the strength of the weakest element present. These two approaches are, in a way, idealizations used to make the problem tractable. However, in reality the actual characteristics of materials fall in between the two theories. A summary of these theories is presented to give the reader a basic idea of the size-effect phenomenon from the theoretical point of view.

6.3.1 Classical Theory of "Bundled Strength"

The concept of this theory lies in the assumption that the bulk specimen consists of parallel elements and that the instability or failure of one element will not lead to total fracture, rather it will be stopped before it propagates from local to bulk scale. This means that the weakest volume element containing the failure is surrounded by elements of such high local strength that the stress carried by the weakest element prior to its failure can indeed be transferred to the elements. The statistical model according to Freudenthal (1968) replaces the bulk specimen by a classical bundle and this discussion is based on his work. This bundle consists of a large number of parallel filaments of identical length L and cross section A, all of which come from a common source, so that the statistical distribution of local filament strength is homogenous and constant; a random filament is the weakest link of the specimen. The strength of this classical bundle model, as well as of the bulk specimen that it is designed to represent, is represented by the forces under which a *chain reaction* process of consecutive filament failures resulting from the successive overload carried by the surviving filaments leads to final failure of all filaments. The fracture process starts at the weakest point in the bundle, but, contrary to the *weakest-link* model, it does not necessarily propagate unless, in a bundle of total filaments (n) of (weakest-link) strength $\sigma_n, \sigma_{n-1}, \ldots, \sigma_2, \sigma_1$, arranged in the order of their consecutive failure, the following conditions are satisfied:

$$0 \leq \sigma_n \leq \frac{S}{nA} = s_n$$

$$\sigma_n \leq \sigma_{n-1} \leq \frac{S}{(n-1)A} = s_{n-1}$$

$$\sigma_3 \leq \sigma_2 \leq \frac{S}{2A} = s_2 \qquad (6.1)$$

$$\sigma_2 \leq \sigma_1 \leq \frac{S}{A} = s_1$$

where

$$S = \text{total applied force}$$

$$s_i = \text{stress in the individual element}$$

Introducing the probability density of the filament strength $p(\sigma)$, the probability of the event defined by Eqs. (6.1) can be expressed in the form

$$p_n(s) = n! \int_0^{s_n} p(\sigma)\, d\sigma \int_{s_n}^{s_{n-1}} p(\sigma)\, d\sigma \ldots \int_{s_3}^{s_2} p(\sigma)\, d\sigma \int_{s_2}^{s_1} p(\sigma)\, d\sigma \qquad (6.2)$$

where the factor $n!$ provides for all possible ways of arranging the filaments.

Developing an asymptotic form of $p_n(S)$, Daniels (1945) has further shown that the specific strength $s = S/nA$ of a bundle of n filaments has a probability distribution which, for large n, tends to a normal distribution with expectation

$$E(s) = \sigma_r[1 - P(\sigma_r)] \qquad (6.3)$$

and variance,

$$\mathrm{Var}\,(s) = \sigma_r^2\{P(\sigma_r)[1 - P(\sigma_r)]\}n^{-1} \qquad (6.4)$$

where σ_r is the value of σ which makes the expression $\sigma[1 - P(\sigma)]$ a maximum:

$$\frac{d}{d\sigma}\{\sigma[1 - P(\sigma)]\}_\sigma = \sigma_r = 0 \qquad (6.5)$$

Introducing $P(\sigma) = 1 - \exp[-(\sigma/v)^{2\alpha}]$, with the modal value

$$v = k\sqrt{u} = \sigma_u$$

the solution is obtained

$$\sigma_r = v(1/2\alpha)^{1/2\,\mathrm{a}} \qquad (6.6)$$

and therefore,

$$E(s) = v(1/2\alpha)^{1/2\alpha}\, e^{-1/2\alpha} \qquad (6.7)$$

The ratio, γ, between the mean filament strength in the bundle and the mean individual filament strength $(\bar{\sigma})$ is given by

$$\gamma = (1/2\alpha)^{1/2\alpha}\, e^{-1/2\alpha}[\Gamma(1 + 1/2\alpha)]^{-1} < 1 \qquad (6.8)$$

the mean strength $\bar{\sigma}$ is given by:

$$\bar{\sigma} = v\Gamma(1 + 1/2\alpha)$$

In Eq. (6.8), $(1/2\alpha)$ can, as a first approximation, be replaced by the coefficient of variation of the filament strength; e.g., the effective mean strength of a filament with coefficient of variation of 0.25, i.e., $(\alpha = 2)$ is reduced in a large (theoretically infinite) bundle to about 65% of the mean strength of the component individual filament using Eq. (6.8).

According to Eq. (6.4), the variance of s is an inverse function of n and therefore tends to zero for very large n. The dispersion of bundle strength is, therefore, much narrower than the dispersion of individual filament strength, with the result that, in a bulk specimen with a fracture process represented by a bundle model, no pronounced effect of volume on either mean strength or variance is anticipated, except that the variance might show a tendency to decrease with increasing specimen size. In particular, it is this last trend (not usually found in tests leading to brittle fracture) which suggests that the bundle model is applicable only to the description of fracture processes in materials (1) in which bundles such as groups of long-chain molecular filaments physically exist or (2) in which fracture takes place following large strain.

6.3.2 The Weakest-Link Theory

The weakest-link concept has been used widely in developing various statistical strength theories, which differ from each other only in the way in which the use of the concept is justified or in which the form of distribution function of local strength is assumed. Early attempts were made by many to formulate a theory for the strength of cotton yarn and later for strengths of solid volumes. The development put forth by Weibull (1939) is a culmination of these earlier efforts.

Weibull pointed out the inadequacy of specifying material strength by a single quantity as is usually done in deterministic approaches. In developing his theory Weibull used additional parameters to characterize the strength of a material. He regarded a specimen as an ensemble of a very large number of primary elements. He considered the failure of total material the same as that of any one of the primary elements or of the weakest link.

If the probability of failure of the primary element for a stress between 0 and σ is S_0, then the probability of survival of the element is given by $(1 - S_0)$. Also, if S denotes the cumulative probability of failure of a specimen of total volume V, then the probability of survival of the total specimen is given by

$$(1 - S) = (1 - S_0)^V \tag{6.9}$$

or

$$\log (1 - S) = V \log (1 - S_0) = -B \tag{6.10}$$

where B is the risk of failure. The risk of failure of the primary element is then given by

$$dB = -\log (1 - S_0)\, dV \tag{6.11}$$

If we assume S_0 as some function $n_0(\sigma)$ of stress level σ, then

$$dB = n_0(\sigma)\, dV \tag{6.12}$$

or

$$B = \int n_0(\sigma)\, dV \tag{6.13}$$

and

$$(1 - S) = e^{-\int n_0(\sigma)\, dV} \tag{6.14}$$

so that

$$S = 1 - e^{-\int n_0(\sigma)\, dV} \tag{6.15}$$

Using the last expression for the probability of failure S, one can evaluate S knowing $n_0(\sigma)$.

For the case of a simple uniaxial stress field, Weibull intuitively suggested the use of the form of function $n_0(\sigma)$ given by

$$n_0(\sigma) = \left(\frac{\sigma}{\sigma_0}\right)^m \tag{6.16}$$

where

$$m = \text{flaw density parameter}$$
$$\sigma_0 = \text{scale effect parameter}$$

The parameters m and σ_0 thus characterize the strength distribution of the material.

With Eq. (6-16), S takes the form,

$$S = 1 - \exp\left[-V\left(\frac{\sigma}{\sigma_0}\right)^m\right] \qquad (6.17)$$

It can be seen from the above equation that the specimen of volume V has a finite probability of failure unless the stress level is zero. This implies that the material can have a zero strength. In order to account for a lower bound on failure stress, Weibull subsequently generalized the distribution function as follows:

$$S = 1 - \exp\left[-V\left(\frac{\sigma - \sigma_u}{\sigma_0}\right)^m\right] \qquad (6.18)$$

in which he defined $\sigma_u = $ as the *zero probability fracture stress* or *minimum strength*.

Corresponding to the above form of distribution Eq. (6.18), the relation between the size and the mean strength is given by

$$\bar{\sigma} = \int_{\sigma_u}^{\infty} \sigma \frac{dS}{d\sigma}\, d\sigma$$

or

$$\bar{\sigma} = \sigma_u + \frac{\sigma_0}{V^{1/m}}\Gamma\left(1 + \frac{1}{m}\right) \qquad (6.19)$$

where $\bar{\sigma}$ is the mean strength and $\Gamma(\ )=$ the gamma function defined by

$$\Gamma(x) = \int_0^{\infty} z^{x-1}e^{-z}\, dz \qquad (6.20)$$

Also, the variance of strength S_d^2 is given by

$$S_d^2 = \sigma_0^2 V^{-2/m}\left[\Gamma\left(1 + \frac{2}{m}\right) - \Gamma^2\left(1 + \frac{1}{m}\right)\right] \qquad (6.21)$$

where the complete gamma function is defined as before.

The above expressions for $\bar{\sigma}$ and S_d^2 indicate that as specimen size increases, the mean strength and variance decrease, which is consistent with the experimental observations of size effect on some materials. Thus Weibull's theory is simple to apply. Eq. (6.18) can be rewritten in another form as

$$\frac{1}{1-S} = \exp\left[V\left(\frac{\sigma - \sigma_u}{\sigma_0}\right)^m\right] \qquad (6.22)$$

or

$$\underbrace{\log\log\frac{1}{1-S}}_{y} = \underbrace{m.\log(\sigma - \sigma_u)}_{x} - m.\log\sigma_0 + \log V \qquad (6.23)$$

Equation 6.23 indicates a linear relationship between $\log\log 1/(1-S)$ and $\log(\sigma - \sigma_u)$; this is shown in Fig. 6.2.

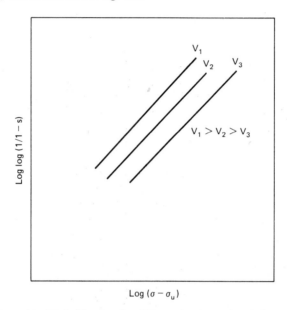

Figure 6.2 Weibull's plots for different specimen sizes (schematic).

Weibull applied his theory to a number of cases, including the strength of glass rods in tension, the bending strength of porcelain, and the tensile strengths of Portland cement, wood, plaster of Paris, malleable iron castings, and other materials. In all cases, he verified the applicability of the linear fit of the distribution function with test data. The *weakest-link concept* has since been used by several researchers.

6.3.3 Other Theoretical Studies

Tucker's (1941) *strength summation theory* is based on the assumption that the strength of a specimen is equal to the sum of the strengths contributed by the component parts, or elements. In this theory the specimen is assumed to have a parallel link assembly. Thus, it is implied, for example, that the flexural strength of a beam specimen decreases with increasing length and increasing depth, as governed by the weakest-link theory, but is independent of the width.

Nielsen (1954) assumed in his *surface theory* that failure of a beam is determined solely by the thin surface layer of the tension zone, and if this surface layer is assumed to be brittle, Weibull's theory may be applied to the tensioned surface. This implies that the flexural strength would decrease with increasing length and width but should be independent of depth. Since the strain distribution in a beam is triangular, and since only the extreme fibers govern the flexural strength, Johnson (1962) showed the influence of depth to be very small. Nielsen's tests on concrete beams cured in water until testing show no influence of beam depth on the tensile flexural strength. This process essentially prevented any drying and thus caused the modulus of rupture to be independent of depth.

Pahl and Soosaar (1964) indicate in their work that concrete and mortar are fairly brittle materials, so that failure at a few points in the mass is soon followed by overall collapse. A small specimen has fewer points than a large specimen at which failure can be initiated by local weaknesses, so that the strength of the small specimens on the average might be larger than that of the larger specimens. They suggest that this size effect can be represented by a best-fitting equation:

$$f = a + bV^{-c} \qquad (6.24)$$

where

f = strength of concrete

V = volume of specimen

a, b, c = positive constants depending on the concrete mix

Glücklich and Cohen (1968) drew attention to the influence of size, in certain materials, on the brittle-ductile transition and strength, with both ductility and strength decreasing with increasing size. In most metals, ductility is mainly due to plastic deformation prior to nucleation of cracks, which is coincident with fracture. These are Griffith-type materials, and Weibull's statistical theory describes the size effect of crack nucleation. However, materials such as concrete exhibit no permanent deformation prior to crack nucleation but appreciable permanent deformation during slow crack growth. Thus, crack nucleation and fracture are not coincident, and Weibull's weakest-link theory is not sufficient to explain the behavior during crack growth. Crack propagation depends on equilibrium between the respective rates of strain energy release due to cracking, and the energy demand for continued propagation of cracking. Any sudden drop in the energy demand, for example, as a result of the propagating crack encountering preexisting cracks in its plane of advance, will create an excess of released energy. This energy, converted to kinetic energy, can work against the remaining uncracked material and bring about reduced ductility and premature fracture. If there is a large amount of stored energy, the rate of energy adsorption is low and a premature fracture will occur, as described above. Specimen size thus enters into the picture not as a size effect per se, but insofar as it governs the amount of stored energy.

6.3.4 Evaluation of Theoretical Studies

It becomes apparent that the theoretical investigations discussed above are based solely on the statistical treatment of heterogeneity of the material. The weakest-link concept focuses attention only on the most critical or potential flaw, but disregards the interaction between the flaws that will exist in continuous systems. The experimental results, however, have been used extensively to explain size effects, and therefore should also be examined.

Thus, the theoretical investigations presented in this chapter are all valid from the *theoretical* point of view, but should be considered with caution. For concretes, in particular, the weakest-link theory applies extremely well. However, such a conclusion should be confirmed by very carefully executed experiments, with controlled curing, and considering the factors mentioned earlier (Section 6.2).

6.4 SIZE EFFECTS IN PLAIN CONCRETE—EXPERIMENTAL WORK

The heterogeneity of any material leads to the theoretical arguments presented in the last section that predict the existence of size effects. When investigating the behavior of concrete, it is important to recognize factors that contribute directly to the observed changes in properties with size. An understanding of these factors can help minimize the effect of size, before the intrinsic variability is inaptly used to explain the observed data.

6.4.1 Experimental Factors Influencing Size Effects

6.4.1.1 Compaction, Density, and Loss of Water

Compaction is an important variable influencing concrete strength, yet cannot be scaled, and therefore, smaller specimens will tend to achieve better compaction, higher density, and thus higher strength. This is particularly true when standard compaction procedures are followed, involving a given time of vibration or specific number of tampings. Larger specimens will undoubtedly have more internal voids and entrapped air. When uniform compaction is achieved, the size effect due to this factor will be minimized. This has been confirmed for gypsum mortars by Loh (1969).

Water loss from specimens during casting can vary with size and cause different quality in the cast material. To minimize this source of variability and size effects, a controlled humidity room during casting and watertight molds such as Plexiglas are needed.

6.4.1.2 Curing and Drying

Curing is another important factor. Curing of two specimens of different size will take place at different rates, because the surface-volume ratio increases with decrease in specimen in size, and the length of moisture migration paths will differ. The strength of the material will vary from the surface of the specimen to its center, depending on its size, since hydration may not be uniform throughout the specimen at the time of testing. Studies summarized by Sabnis and Aroni (1971) indicate that if curing is controlled, for example, by sealing (discussed later in detail) the surfaces of the test specimens, the increase in strength due to the reduction in size can be minimized. Tests on cores drilled from a massive concrete dam after a period of 5 years showed an insignificant difference in strengths between 10 in. (250 mm) and 22 in. (560 mm) diameter cylinders. This is probably due to the more uniform curing conditions inside the dam as well as the advanced age at which hydration was almost complete. The density of the cores would be identical.

Drying of the specimen also results in higher strength and will depend on the surface to volume ratio, which varies inversely with the specimen size. Slower surface drying of larger specimens will result in smaller flow gradients and greater resistance to drying due to the longer distance to the surface. This also influences size effects in sustained loading (creep).

6.4.1.3 Strain Rate

High rates of loading lead to higher strengths. In a given testing machine, with the rate of crosshead movement kept constant, smaller specimens will experience higher strain rates. As the specimen size is decreased, the crosshead movement rate should be reduced accordingly. However, with very small specimens it is not always possible to achieve the required low rates of crosshead movement with existing testing equipment, which contributes to their apparent increased strength. Of course, in normal testing the influence of increased strain rate is not critical, except in cases of dynamic loading (see Chapter 11).

6.4.1.4 The State of Stress

The stress state, such as compression, tension, and flexure, influences the strength of the specimen. The strength of compressive specimens depend on the accuracy of the loaded ends, and on parallelism, if rotating heads are not used. It is possible to achieve a higher level of capping accuracy in smaller cylinders, which result in higher strength. Tests by Wright and Garwood (1952) showed that flexural stress increased with a decrease in the specimen size. The effect of strain gradient was discussed in detail in Chapter 4.

6.4.1.5 Testing Machine and Loading Platens

In addition, the properties of the testing machine and, in particular, the stiffness of the loading platens at the ends of the test cylinders have a significant effect on test results. Stiff end platens tend to apply *uniform strain* conditions to the specimen under test and result in higher strength than thinner platens, which tend to lead to a state of *uniform stress*. Also, end platens restrain lateral movements of the specimen and induce lateral stresses at both ends. The higher the lateral restraint, the higher the compressive strength will be. The mode of failure may also change with end conditions.

6.4.2 Experimental Research on Size Effects

A recent paper by Sabnis and Mirza (1979) reviews the experimental work done to date and discusses its effects on the size effect phenomenon. Some notable contributions are reviewed briefly here.

Gonnerman (1925) conducted the earliest study on size effects in concrete with an extensive investigation into the compressive strength of cylinders with height/diameter ratio of 2. He varied the cylinder diameter from 4 to 10 in. (100 to 250 mm) and examined the influence of age, cement-aggregate ratio, relative consistency and aggregate fineness as shown in Fig. 6.3. Note that the aggregate size is less than 40% of cylinder diameter. Each point in Fig. 6.3 is an average of 5 to 30 tests.

Johnson (1962) investigated the influence of "scaling" of aggregate with cylindrical and cube specimens. He used four different aggregates scaled in a linear ratio and high early-strength cement and natural sand. He tested $1\frac{1}{2} \times 3$ in. (38×75 mm) and 6×12 in. (150×300 mm) cylinders and 3-in. (75-mm) cubes for each series. Variation of the concrete compressive strength with water-cement ratio and the specimen size for the $\frac{1}{4}$- and $\frac{1}{8}$-scale mixes is shown in Fig. 6.4. It was found that the 6 in. (150 mm) diameter cylinders had approximately 75 to 85% of the compressive strength of $1\frac{1}{2}$ in. diameter cylinders for both $\frac{1}{4}$- and $\frac{1}{8}$-scale mixes.

Harris et al. (1963) tested several series of model cylinders to investigate the effect of size on the compressive strength at various ages. The results in Fig. 6.5 show the variation of compressive strength with age and volume of specimen. Figure 6.5 indicates that at the earlier ages of testing, the smallest specimens, $\frac{1}{4} \times \frac{1}{2}$ in. (6×12 mm), reached almost their full strength because they cured faster. The nature of the f_c'-vs.-volume curves with respect to curing time indicates that part of the size effect is due to differential curing of the specimens. The second series of tests was designed to show the influence of aggregate surface area on the size effect by using different gradations of sand with a larger percentage of fines. There was no significant reduction of the size effect by increasing the amount of fines. The surface area of the aggregate

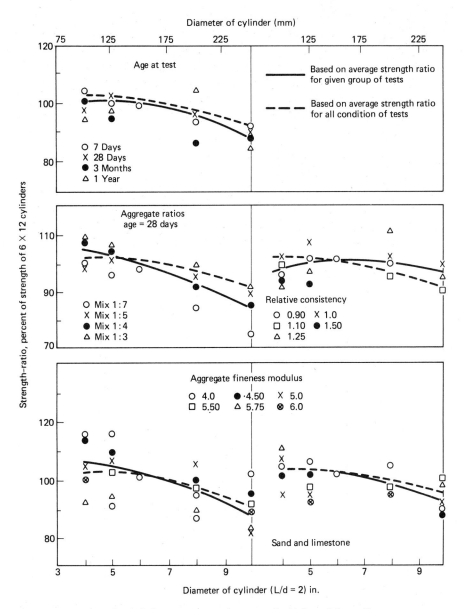

Figure 6.3 Effect of size on compressive strength. [Adopted from Gonnerman (1925).]

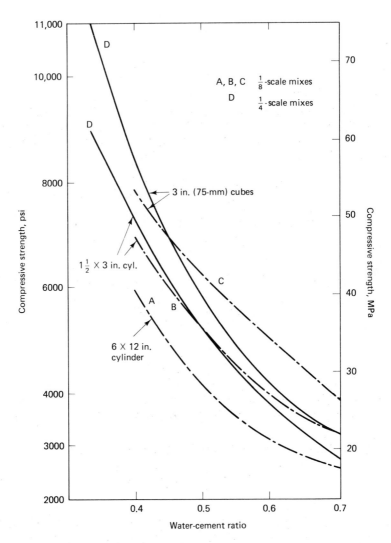

Figure 6.4   Compression tests on $\frac{1}{4}$- and $\frac{1}{8}$-scale mixes at 7 days. [From Johnson (1962).]

did not seem to influence the distribution of flaws. Part of the size effect was again attributed to the faster curing of the smallest specimens.

Neville (1966) analyzed the experimental data from 12 different investigators on many types of concretes, cured in various ways and tested at a number of ages. He considered the compressive strength of concrete P to be a function of three variables: V, the volume of the specimen; d, its maximum lateral dimension, and h/d, its height–lateral dimension ratio. He ignored the fact that the strength of concrete may also be influenced by other factors, such as

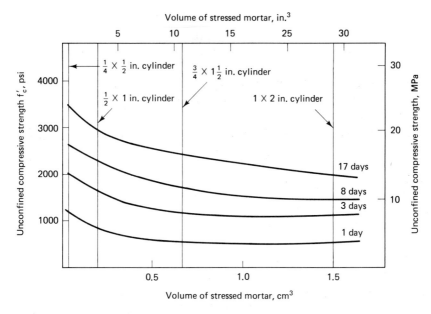

Figure 6.5 Unconfined compressive strength vs. stressed volume at various ages. [Adopted from Harris et al (1963).]

the modulus of elasticity of the aggregate, its Poisson's ratio, and the aggregate-cement ratio (since no experimental data was available on these factors). His regression analysis yielded the following dimensional relation:

$$\frac{P}{P_6} = 0.56 + 0.697 \frac{d}{(V/6h) + h} \qquad (6.25)$$

as shown in Fig. 6.6. Note that d and h are in inches and V is in cubic inches and the subscript 6 refers to a 6-in. (150-mm) cube of the same concrete chosen as a standard specimen for comparison.

Fuss (1968) investigated both solid and hollow cylinders varying in size from $\frac{1}{4} \times \frac{1}{2}$ in. to 1×2 in. (6×12 mm to 25×50 mm) and attempted to maintain constant moisture in the specimens by surface sealing as Sabnis and White (1967) had done. He concluded that when properly sealed the size of cylinder had no significant effect on its strength.

Meininger (1968) cored 2, 4, and 6 in. (50, 100 and 150 mm) diameter specimens from a slab, and from a concrete wall each 16 in. (400 mm) thick and moist-cured for 3 months. Prior to testing, half of the cores were soaked in water 40 to 44 hours while the others were immersed for 28 days before testing. The test results indicated no significant effect of core diameter on strength for 2, 4, and 6 in. diameter cores all having a length-diameter ratio of 2.

Mirza, Labonte, and McCutcheon (1972) investigated size effects with a major emphasis on method of curing and method of compaction. They tested

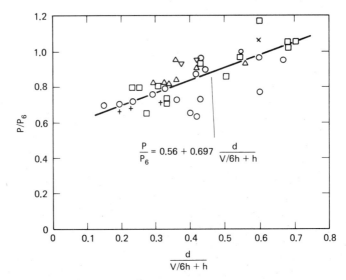

Figure 6.6   Relation between P/P_6 and $d/[(V/6h) + h]$. [From Neville (1966).]

cylinders ranging from 1×2 (25×50 mm) to 6×12 in. (150×300 mm) at ages of 3, 7, and 14 days. Over 500 cylinders were tested to study these variables, and their conclusions about curing methods, stated below, are particularly important.

Curing procedures in which cylinders were kept in a continuously moist environment were equivalent for all sizes and were more effective than coating procedures and air drying. The effectiveness of moist curing compared with air drying is indicated by a gradual decrease in the compressive strength with a decrease in cylinder size. This is due to the comparatively more rapid moisture migration from smaller cylinders. Curing and sealing with chemical coatings allowed some exchange of moisture with the environment and resulted in strengths intermediate between fully cured and air-dried conditions. Spray paint (lacquer or polyurethane) was ineffective in maintaining internal moisture and also caused deterioration of the concrete. It was shown for their mixes that 3 and 4 in. (75 and 100 mm) diameter cylinders exhibited strength increase of approximately 5 to 15%, and for 2 in. (50 mm) diameter cylinders strength increased up to 40% over the strength of 6 in. (150 mm) diameter cylinders.

6.4.3 Evaluation of Experimental Research

Size effect is usually not an important factor in the selection of prototype test cylinder size, since all cylinders above 2×4 in. (50×100 mm) fall on the flat section of the strength curve. However, for small-scale models, the selection of smaller-size control cylinders can have a considerable variation in the compressive strength. Based on the available data to date, the relative strengths of various cylinder sizes that have typical curing and drying histories with no surface sealing are as shown in Fig. 6.7.

In general, the scale of cylinder to measure the compressive strength of

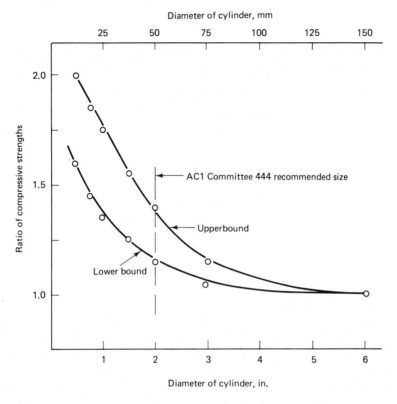

Figure 6.7 Relative strengths of different size cylinders with 6-in. cylinder as a unit. Note: All cylinders have height = twice the diameter.

concrete in the structure should be consistent with the scale factor of the model structure itself. ACI Committee 444 (1979) recommends that 2×4 in. (50×100 mm) cylinders be accepted as a standard for comparing model concrete mixes. In addition, ACI Committee 444 (1979) suggests that for model elements of very small minimum characteristic dimension [less than $\frac{1}{2}$ in. (12.5 mm)], the apparent strength of the control cylinders may not be representative of the strength of the model material. In such situations, additional tests should be conducted on model cylinders with a length diameter ratio of 2 to 1 and diameter equal to the characteristic dimension of the model. Pahl and Soosaar (1964) suggest the following empirical rules:

1. The strength of the model material should be equal to that of a test cylinder with diameter equal to the minimum dimension of the structure in the region of failure, e.g., the shell thickness or the width of the beam.
2. The size of the largest sand particle used in the mix may not be larger than one-fifth of the cylinder diameter, nor larger than 80% of the clear distance between reinforcing bars in the model.

6.4.4 Tensile and Flexural Strength of Cement Mortar

The tensile strength of concrete is a fundamental property, and it has a significant influence on several important phenomena in reinforced concrete elements, such as shear strength, bond strength of deformed bars, cracking load and crack patterns, effective moment of inertia, and nonlinear response. Several tension tests are available, and the best choice depends on the strain distribution existing in the member; for example, uniform tension requires direct tension tests. These are difficult to perform and are seldom used. The indirect tension test (split cylinder test) and the torsion test are used for sections in which the principal compressive and tensile stresses are of the same order. The flexure test (modulus of rupture) would be used in connection with reinforced and prestressed concrete flexural members, pavements, etc.

The strain distribution was studied by Blackman et al. (1958) on three identical specimens: one loaded axially, one in pure flexure, and one in between these two conditions (Fig. 6.8). Although each of the stress distributions caused a failure at the tensile strength of the material, the influence of a strain gradient became obvious in that the ultimate tensile strain increased with the applied gradient. The situation is exaggerated in smaller beams, which have higher strain gradients, when subjected to a similar flexural loading. Beam tests by Wright and Garwood (1952) indicated increase in the flexural tensile strength with an increased strain gradient, i.e., with a reduction in depth of the beam.

With regard to the method of casting and testing, experimental results in flexural strength and size effects have been shown for both gypsum and cement mortars, for beams cast on their sides (horizontal position), and in the usual manner, on their bottom faces (vertical position). In Fig. 6.9 data clearly indicates that the strength of the horizontally cast beams is consistently higher than that of those cast in the vertical position. In horizontally cast beams, the tensile face of the beam, as tested, contains material cast at various depths and is thus more heterogeneous, causing the observed larger scale effect.

Abrams (1922) was probably the first to report extensive flexural tests on concrete; his tests considered a number of variables associated with strength of concrete. Although his objective was not to investigate the effect of size on flexural strength, his results on beams with depth varying from 4 to 10 in. (10 to 250 mm) indicated that deeper beams had lower strength, the variation being approximately 10% from shallower to deeper beams.

Blackman, Smith, and Young (1958) conducted axial tensile and flexural tests on 1×1 m (25×25 mm) specimens to investigate the effect of strain distribution on the ultimate tensile strength. They used a parameter β, which was defined as the ratio of strain to the depth of the specimen (see Fig. 6.8). Since β can be varied by changing the loading system or depth, they concluded from tests that flexural specimens with larger depths would have relatively lower values of β and, therefore, lower ultimate stresses or strains.

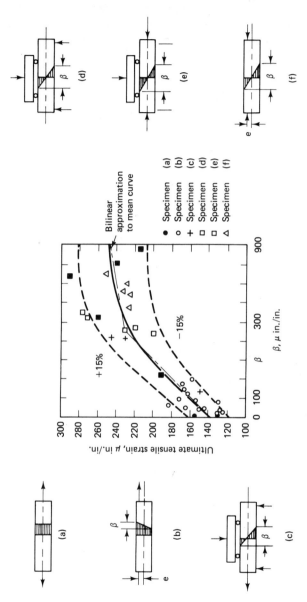

Figure 6.8 Effect of strain distribution and gradient on ultimate tensile strain. [After Blackman et al. (1958).] Figures (a) through (f) represent various strain gradient conditions. β represents the total non-linear strain between the end faces.

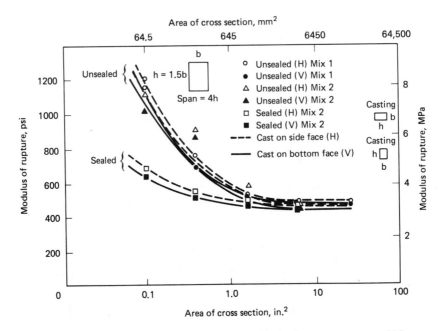

Figure 6.9 Variation of modulus of rupture with size for gypsum mortar. [After White and Sabnis (1968).]

Harris et al. (1963) tested beams of sizes ranging between $\frac{1}{4} \times \frac{1}{2}$ in. (6 × 12 mm) to 1×2 in. (25 × 50 mm) using microconcrete with 2×4 (50 × 100 mm) cylinders as control cylinders. These beams were tested under third-point loading over spans of 3, 6, 9, and 12 in. (75, 150, 225, and 300 mm). The results showed that the flexural tensile strength increased as the beam size decreased from 1×2 in. to $\frac{1}{4} \times \frac{1}{2}$ in. The variation of the flexural tensile strength with the beam size is shown in Fig. 6.10 in terms of depth, cross section, and volume of the beam.

Harris, Sabnis, and White (1966) conducted tests on similar beams ranging in size from $\frac{1}{4} \times \frac{3}{8}$ in. (6 × 9 mm) to 4×6 in. (100 × 150 mm) to investigate the effect of strain gradient on strength. The smaller flexural specimens have a larger strain gradient to attain the same fiber stress. Each beam had a depth-width ratio of 1.5 and a clear span of 4 times the depth. The loading and support details were modeled to the scale as the test beams. These beams were tested under third-point loading, and the elastic bending theory was used to calculate the extreme fiber stress. They carried out analyses similar to Blackman, Smith, and Young (1958) and concluded that the relation between ultimate tensile strain and strength may be represented by a bilinear relation and that the relation between ultimate tensile strain and tensile stress gradient is similar for all model beams (see Chapter 4).

Mirza's (1967) tests were based on series of cylinders cast from the same

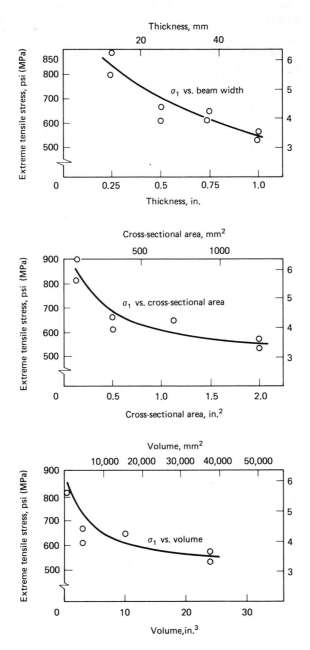

Figure 6.10 Variation of extreme fiber tensile stress with size of specimen. [After Harris et al. (1963).]

batch of microconcrete for tensile splitting and square beams for flexural
tests: they ranged in size from 1 × 2 in. (25 × 50 mm) to 6 × 12 in. (150 ×
300 mm) cylinders and 1 × 1 × 4 in. (25 × 25 × 100 mm) to 4 × 4 × 6 in.
(100 × 100 × 400 mm) beams respectively (see Fig. 6.11). The mean strength
and the standard deviation decreased as the size increased. The principal tensile
and compressive strains from split cylinders showed that the strains in all size
specimens are of the same order at half ultimate load and a load stage just before
failure for all sizes. The beam tests with a third-point loading over spans ranging
from 3 to 12 in. (75 to 300 mm) showed that both the mean flexural tensile
strength and the standard deviation decrease with an increase in size.

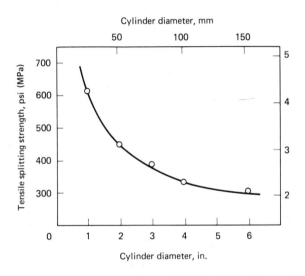

Figure 6.11 Variation of split tensile strength with cylinder diameter (l/d = 2).
[After Mirza (1967).]

Kadlecek and Spetla (1967), using both cylinder and prism tests in direct
tension to investigate size effects, developed the following relation:

$$f_t = AV^{-B} \tag{6.26}$$

where

f_t = tensile strength in kilogram per square centimeter

V = test volume of specimen in cubic centimeter × 10^{-3}

A, B = constants for best fit of data.

The values of A and B ranged between 23.32 to 29.56 and 0.021 to 0.041,
respectively. This relation implies that tensile strength decreases indefinitely as
the size of the specimen increases. A correction made by Rao (1972) improved
the relation with a finite limitation to the form:

$$f_t = f_{\text{limit}}(1 + CV^{-D}) \tag{6.27}$$

where

f_t = minimum value of the tensile strength

f_{limit} = the strength of standard specimen in the investigation

C, D = experimental constants

Using Eqs. (6.26) and (6.27), the variation of strengths is shown in Fig. 6.12.

Malhotra (1969) carried out extensive tests on 276 specimens made from 21 different mixes of concrete to investigate the size effect on tensile strength

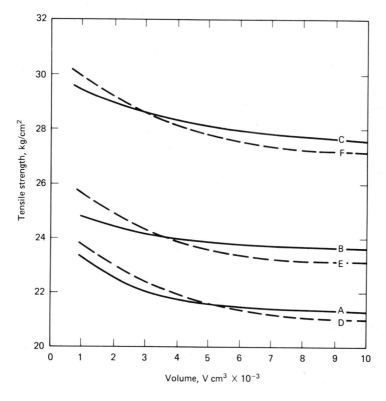

Cylinders	Kadlecek and Spetla (1967)		C.V.S.K. Rao (1972)	
H/D = 2	A	23.32 V$^{-0.041}$	D	21(1 + 0.16 V$^{-1.08}$)
H/D = 3	B	24.78 V$^{-0.021}$	E	23(1 + 0.085 V$^{-0.5}$)
Prisms				
H/D = 3	C	29.56 V$^{-0.03}$	F	27(1 + 0.118 V$^{-0.72}$)

Figure 6.12 Relation between tensile strength and size of specimen. [Based on C.V.S.K. Rao (1972).]

using direct tension, ring tension, and splitting tension tests. Statistical analysis of his tests showed that, irrespective of the test method used, the tensile strength increased with decrease in size of specimens; this increase was more apparent in the case of the ring tension test than in the other two types of tests.

Mirza, Labonte, and McCutcheon (1972) observed that the indirect tensile strength f_{sp} was observed to be 10% of the compressive strength f_c'; f_{ct}, the split tensile strength, was related to f_c' by the following empirical equation:

$$f_{ct} = 6.45 \sqrt{f_c'} \qquad (6.28)$$

where f_{ct} and f_c' are expressed in psi.

6.4.5 Evaluation of Experimental Work on Tensile Strength

Results of various tensile strength tests indicate that the three different types of tensile specimens give different values which also vary with the compressive strength (Table 6.1). For model concretes the variation with compressive strength is not as pronounced. In addition, the relative tensile strength is found to be a higher percentage of the compressive strength.

TABLE 6.1   Relative Tensile Strength of Prototype and Model Concrete

Concrete Compressive Strength, psi (MPa) (1)	Type of Test (2)	Tensile Strength as Per Cent of Compressive Strength		
		6 × 12 in. (150 × 300 mm) cyl. (3)	2 × 4 in. (50 × 100 mm) cyl. (4)	1 × 2 in. (25 × 50 mm) cyl. (5)
2000 (14)	Direct tension	9.2	11	12.2
	Split cylinder	11.8	12	12
	Modulus of rupture	19.7	20	—
4000 (28)	Direct tension	8.5	10	12
	Split cylinder	9.8	11	13
	Modulus of rupture	15.7	17	20*
6000 (42)	Direct tension	8.3	9	11
	Split cylinder	9	10	11
	Modulus of rupture	13	17	30*

*Gypsum mortar, the others being cement mortar.

Direct modeling is hard to achieve in modeling the prototype tensile strength. This affects both the strength and deflections in the model structure because the tensile strength controls cracking; further research is needed in this area of direct modeling.

6.4.6 Size Effects in Long-Term Properties of Concrete

The long-term properties of concrete involve creep and shrinkage. They affect structural behavior in many ways; creep is especially important in long-term deflections. An understanding of these properties in model materials will undoubtedly enhance the use of the direct model approach for studying the sustained load behavior of structures. At the present time, however, models are not used to investigate these properties. The influence of size effects, which was discussed earlier for short-term loading, should also be investigated for creep and shrinkage. There has been only exploratory research done in this field, and the discussion presented here merely indicates some potential uses, as well as problems, in the application of direct models to study long-term effects.

6.4.6.1 Shrinkage

The size and shape of a concrete member will influence the rate at which moisture moves to or from the concrete, and will therefore affect the rate of shrinkage. Carlson (1937) showed that at a relative humidity of 50%, drying would only be felt to about 3 in. (75 mm) from the surface of a large concrete member during the first month of exposure. After 10 years, the drying will be felt about 24 in. (0.6 m) from the surface. This means that a small or slenderly proportioned test specimen will shrink much more rapidly and uniformly than a large and bulky member.

Ross (1944) suggested that the most suitable parameter in comparing the shrinkage of members of different sizes and shapes is the ratio of exposed surface area S to the volume V of the member. He showed that the ratio S/V can be used to correlate the shrinkage at three different ages of a series of small mortar specimens having rectangular, circular, triangular, and annular cross sections.

Hansen and Mattock (1966) tested three different-size specimens with two different aggregates to investigate the influence of specimen size on shrinkage and creep. They observed that both the rate and the amount of shrinkage at a given age decreased as the size of the specimen increased. The sealed specimens stored at 100% relative humidity had a negligible amount of shrinkage (about 5% or less than that of the exposed unsealed specimens of corresponding size). This is an important conclusion to consider if one were to undertake tests on small-scale models under sustained load.

6.4.6.2 Creep

While shrinkage is caused by the loss of water from cement gel and takes place under no load, creep of concrete under a sustained load can occur in fully saturated concrete and in concrete sealed to prevent loss of moisture. The rate of creep is observed to increase if there is simultaneous moisture movement, either into or out of concrete. Creep that occurs without the exchange of moisture between the concrete and its surrounding environment has been called basic creep. Thus, the basic creep rate independent of moisture movement will not be subject to the size of specimen. It may, however, be a function of applied stress and a function of the strength of concrete, which is influenced by the size of the specimen.

No tests are available on creep effects in specimens smaller than 3 in. (75 mm) thick, which approximates the upper limit in terms of practical small-scale model dimensions. A general conclusion from the available tests indicates that even in larger sizes, creep decreases with the increase in specimen size; the largest difference occurs during early days of curing. Surface drying is also important because it involves moisture transfer. Tests on sealed specimens or cores from large dams show much smaller variation in creep as a result of change in the size of speciment. Tests by Hansen and Mattock (1966), who were investigating the size effect in creep, showed that for large specimens, creep reduces to the value of basic creep corresponding to a sealed specimen.

6.4.7 Size Effects in Gypsum Mortar

Gypsum mortar has been used successfully as a substitute for cement mortar to model prototype concrete. Investigators have considered mortars made of gypsum because of two main disadvantages of Portland cement mortars from the modeling point of view: First, cement-based model concrete tends to have excessive tensile strength compared with prototype concretes, and this becomes even higher as the size of specimen is reduced; second, cement-based model concretes require a longer curing period to develop the required stable-strength level. The very short curing time (less than 1 day) and better control of tensile strength of gypsum mortars make them an attractive substitute for model concretes.

The major variables on which the size effect of gypsum mortar should be evaluated are the same as those for cement mortar. The earliest reported work on this type of model material was done at Cornell University, where gypsum mortars were used extensively for the first time for direct modeling purposes [White and Sabnis (1968), Loh (1969)]. They presented experimental evidence and the related techniques that could be used to minimize the so-called size effect phenomena.

Although the major factors influencing size effects were discussed earlier for cement mortars, some of these are repeated with particular reference to gypsum mortar.

1. Differential drying of specimens, both before and during testing (if the latter is extended over a period of many hours). The time required to gain

a certain strength level will increase with increasing specimen size because the lower ratio of surface area to volume retards the rate of drying and strength. Different-size specimens will have a different drying history if they are to have the same strengths at a given age.

2. Difference in quality of material as cast in various size molds. It is not possible to scale precisely the compaction process; also water gain in the upper portion of the specimens as well as its loss from imperfect molds will be different for different sizes. This is crucial because gypsum gains strength only as the specimen water content decreases.

3. Statistical variations in strength because of differences in volume. It should be pointed out that one must be very careful in attempting to explain size effects with statistical strength theories in the case of gypsum mortar because of the strong possibility of including other effects [as in (1) and (2) above] that have all the appearances of a statistical strength variation.

An attempt was made to separate the variables affecting size effects on the behavior in compression and flexure [Loh (1969)]. Based on the results presented in section 4.10.2 the following conclusions were drawn:

1. *Compression and split cylinder tests.* It was concluded that when the density, moisture content, and loading conditions are identical for various size specimens, size effects both in uniaxial compressive and split cylinder tensile strength are negligible.

2. *Modulus of rupture tests.* It was found that there exists some size effect in flexural strength; this was mainly attributed to the strain gradient (see Fig. 6.8).

6.5 SIZE EFFECTS IN REINFORCED AND PRESTRESSED CONCRETE

Overall size effects in reinforced and prestressed concrete models are important since the behavior of the model is to be extrapolated for predicting prototype behavior. With this in mind, three types of behavior are important in comparing model and prototype structures made of reinforced and prestressed concrete.

1. Bond characteristics
2. Cracking similitude (service conditions)
3. Ultimate strength and deformation

Relatively few tests have been reported with the specific objectives of studying size effects per se in reinforced concrete. Details of tests on reinforced concrete elements are presented in Chapter 4. The general discussion given here is to make the reader aware of the importance of size effects in the behavior of model structures.

6.5.1 Bond Characteristics

Investigation into the bond characteristics is complicated by our limited knowledge of the bond phenomenon in the prototype concrete. The bond strength of prototype deformed bars is mainly due to the mechanical wedge action and eventual cracking of concrete beams against the deformations. This action reduces size effect on bond to a certain degree, if bar deformations are reproduced in the model reinforcement.

Limited number of pullout tests by Aroni (1959) on smooth and square-twisted bars, indicated the existence of scale effects in bond strength of different size bars tested with the bars in their usual condition; however when the surface was polished, this scale effect disappeared, suggesting that it was related to the surface condition associated with a given size, rather than the size itself. Alami and Ferguson (1963) concluded from beam tests that models fail to predict the behavior of reinforced concrete prototypes as the result of due to inadequate bond when it is the primary reason of failure, thus casting doubt on the use of models in the cases where the bond may be the expected cause of failure. For investigations concerned with precracking behavior, or if the flexural or shear resistance is required, it is not necessary to satisfy all the requirements of bond similitude. It is sufficient to ensure that there is sufficient bond resistance so that premature cracking or bond failure does not occur. This can be achieved by providing sufficient embedment length to develop the yield strength of the bar.

The results of tests by Harris, Sabnis, and White (1966), Mirza (1967), and many others indicate that certain phenomena involving the bond as the primary cause for failure can be modeled with reasonable reliability if these models are constructed carefully to eliminate any variation and also if the variables such as the concrete strength, steel yield strength, and mechanical deformations [see, Harris, Sabnis, and White (1970)] are controlled accurately. Clark (1971) showed that the bond between concrete and reinforcement has a significant effect on service load behavior; this is so particularly in small-scale models where the reinforcement may take a variety of forms, such as threaded rods or deformed wires, which would exhibit entirely different bond characteristics.

From these various experiments, it may be concluded that certain deformed model bars exhibit bond characteristics that are comparable to those of the prototype reinforcing bars (see Chapters 4 and 5).

6.5.2 Cracking Similitude (Service Conditions)

The inelastic load-deflection response of a reinforced or (partially) pre-stressed concrete structure is often strongly dependent upon the degree and manner of cracking. Cracking modes can also influence behavior under reversed or repeated loading, moment and force redistribution in indeterminate systems, and service load conditions. The existence of size effects in cracking is defined as follows: crack width should vary with the size of model, and the number of cracks will be reduced with decreased model size. Initiation of cracking is a function of the tensile strength of concrete. As seen earlier, tensile strength increases as the size of specimen is reduced. Therefore, it may be stated that on reducing the size of a structure the load level at which the first crack forms will be somewhat higher. Crack spacing and width will both be dependent on the bond between the two materials. In the case of deformed bars or wires in models, the crack spacing will be evenly distributed depending on the distribution of mechanical lugs or deformations. If the bond properties are inadequate, there will be a reduced number of cracks with relatively fewer and wider cracks. Variation of strain-gradient can also affect cracking, but very little research has been done on this parameter.

Many tests conducted on small-scale reinforced concrete beam specimens reveal that the total number of major visible cracks decreases with decreasing beam size; however, the overall cracking patterns are found to be similar, and load-deflection behavior is properly modeled. These tests also indicate that only a small size effect is associated with cracking in scaled models, provided that the other conditions of similitude (mainly the properties of materials and bond strength) are satisfied.

6.5.3 Ultimate Strength (Load-Deflection Behavior)

In many model studies, the objective is to obtain the ultimate strength as well as the load-deflection behavior of a scaled model.

The different types of behavior in reinforced concrete include:

1. Underreinforced beams (both simple spans and two spans) to study the entire behavior due to redistribution of stresses, and failure
2. Overreinforced beams
3. Underreinforced slabs
4. Punching shear behavior of slabs
5. Behavior under combined axial force and bending, and bending and torsion
6. Seismic (or reversed) loading tests on ductile frames

Although a large number of tests are given in the literature, only items (2), (4), (5), and (6) are discussed here because of their greater dependence on the *concrete* properties.

Tests of *overreinforced beams* have been reported by Sabnis (1969) with models at scale factors of $\frac{1}{10}$ and $\frac{1}{6}$ to compare the prediction of ultimate loads, moment-rotation behavior, and effectiveness of helical binders in the compression zone. The cross section of specimens were 0.6×1.1 in. $(15 \times 28$ mm) and 1×1.833 in. $(25 \times 47$ mm) to match the scale of the available reinforcement. Comparable size cylinders were tested to determine the various properties of the model concrete. It was found from moment-rotation curves that for prototype and two different sizes of models there was no size effect in these beams and that the predictions of other related behavior was within $\pm 10\%$ for models of both scales.

In punching shear investigations of slabs, Sabnis and Roll (1971) used a scale factor of 2.5 for model slabs as well as for the control cylinders. Since this type of failure is directly related to the tensile strength of concrete, great care in determining tensile strength is required in such tests. Predictions of the parameter $P_u/bd\sqrt{f_c'}$ and the punching shear strength of slabs were excellent, with no scale effect observed.

From tests on specimens tested under pure torsion and combined bending and torsion, Syamal (1969) reported excellent deformation similitude to cracking load values of about 40 to 45% of the ultimate load. However, postcracking torque-twist and the predicted values of angles of twist showed considerable variation. He concluded that it is possible to obtain reasonably good deformation similitude (deflections, twists, etc.), for the entire loading range, between the prototype and its small-scale models [minimum dimensions $= 2$ in. (50 mm)] reinforced with deformed steel wires without excessive size effects.

Chowdhury's (1974) tests on model beam-column joints are discussed in Chapter 10. At the scale factor of 10, his tests were successful in predicting the complete behavior of reinforced concrete joints, subjected to fully reversed loads; this included flexural and shear behavior and load-deflection response. Cracking patterns obtained were also very similar to those in the prototype.

6.6 SIZE EFFECTS IN METALS AND REINFORCEMENTS

Compared with the detailed discussion given above on size effects in brittle or semibrittle materials, such as concrete, relatively little experimental evidence exists for the investigation of size effect in metals. One reason for this is the homogeneity of ductile materials (metals). This section considers the basic size effects in metal specimens.

Morrison (1940) carried out tests on small steel beams to investigate the influence of specimen size on load-carrying capacity. He concluded that the magnitude of the upper yield-point stress at which a beam yielded was increased with a decrease in the beam size.

Davidenkov et al. (1947) investigated the effect of size on the embrittlement of steel in a liquid-air environment. They observed that the strength and the standard deviation increased with a decrease in the specimen size. The variation of strength and dispersion in the results was explained by Weibull's theory (1939).

Sidebottom and Clark (1954) reported test data using steel beams with rectangular cross section. In this study a total of 18 beams with depths of 3, 1, 0.5, and 0.25 in. (75, 25, 12, and 6 mm, respectively) were used. The experimental moment was 11, 8, and 10% below the theoretical plastic moment for the first three depths, respectively, and for the last one it was 1.5% above the theoretical. They concluded that there was a definite increase in the load-carrying capacity as a result of the decrease in depth. The increase was attributed to the higher stress gradient at ultimate moment stages in the beam as the depth was decreased.

Richards (1954) investigated size effect on the tensile strength of mild steel; he concentrated on both upper and lower yield points. He differentiated between the two by considering the mild steel model to consist of two components, one brittle and the other ductile. Under the increasing rate of loading, the system behaves elastically up to the upper yield point, at which time the brittle component fails suddenly and the load drops to the lower yield point, which is then taken by the ductile portion. With this proposition, he concentrated on upper yield (the brittle fracture), which would depend on nonhomogeneity, microcracks and so on. He tested bars $\frac{1}{8}$, $\frac{1}{2}$, and $1\frac{1}{2}$ in. (3, 12, and 37 mm) in diameter with a corresponding volume ratio of 1:64:1000 and demonstrated that the upper yield strength was an inverse function of the stressed volume. In his later work, Richards (1958) tested beams of mild steel to investigate size effect on yielding in flexure. These beams were dimensionally similar, in five different sizes and the ratio between the largest and smallest dimensions was 6.3. The results indicated that the upper yield point of mild steel in flexure was influenced by size effect.

6.7 SIZE EFFECTS IN MASONRY MORTARS

Harris and his associates at Drexel University observed in modeling of masonry structures that 2-in. (50-mm) cubes and 1×2 in. (25 × 50 mm) cylinders of model masonry mortar were not representative of model joint strength in compression. Tests on the prototype compression specimens 2×4 in. (50 × 100 mm.) cylinders or 2-in. (50-mm) cubes are also poor representatives of actual joint strength. However, the goal is usually to verify modeling techniques by correlating model data with prototype data. To examine the effect of specimen size on mortar strength for both companion specimens and mortar joints, a study [Becica and Harris (1977)] was designed incorporating a range of cube and cylinder sizes. Cylinders having the dimensions of 2×4, 1×2, $\frac{1}{2} \times 1$, and $\frac{1}{4} \times \frac{1}{2}$ in. and cubes of 2, 1, and $\frac{1}{2}$ in. were cast of mortar mix $1:1:4$. These were tested at 7 days after 4 days of continuous moist curing. Figure 6.13 shows

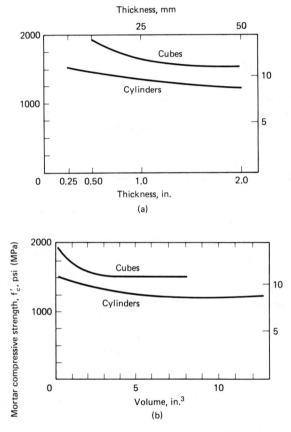

Figure 6.13 Strength-volume relations for model mortar.

the population size and relative dimensions of the specimens. Strength vs. size effects from this test are presented in Fig. 6.14. Note that the volume of the $\frac{1}{4} \times \frac{1}{2}$ in. (6×12 mm) cylinder most closely approaches the unit volume of an ideal model mortar joint. The results of these tests tend to support the previous findings [Harris et al. (1963)]: that is, with decreasing stressed volume, the strength is increased. From the modeling standpoint, this effect greatly influences the behavior of masonry assemblages [Khoo and Hendry (1973)].

In reducing the data of Fig. 6.14 it was noted that quality of specimen (workmanship) plays an important role in the strength of smaller specimens. For the $\frac{1}{4}$ in. (6 mm) diameter cylinders, an increase in strength of 14% was observed

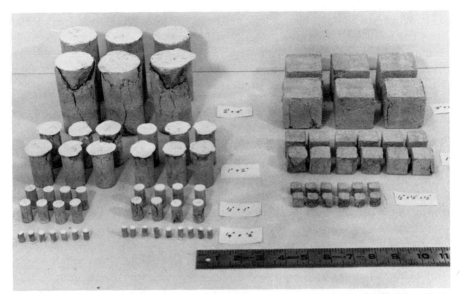

Figure 6.14 Model companion specimens to study scale effects of model masonry mortars. [After Becica and Harris (1977).]

between two experienced workers. For the $\frac{1}{2}$-in. diameter cylinders, however, only a 3 % increase was observed. The difficulty of accurately casting the $\frac{1}{4} \times \frac{1}{2}$ in. cylinders is the main cause of a wide scatter in the strength values obtained.

6.8 SUMMARY

Theoretical studies have been described, based primarily on the heterogeneity of materials. Experimental evidence has been presented that indicates the higher strength of smaller specimens of various types of materials. With particular reference to concrete, various physical factors have been identified that produce differences in quality of cast material and of testing, and hence contribute to the size effect. Procedures have been indicated which can lead to a reduction of this effect.

Why are size effects important? One reason is the use of various size specimens, in addition to different shapes, in various countries and at various times for the characterization of material properties. However, a much more important reason is in connection with model studies. Models tested to failure to determine ultimate load capacity may be subject to size effects that would give unconservative predictions of prototype strengths. Thus, in addition to efforts in minimizing size effects, there is a need to be able to predict them. A somewhat empirical (but very sound) practice is to use a suitably small specimen to measure model material properties. This tends to compensate for the size effect of the model itself.

What are the future research needs in this area? On the experimental side, much work still remains to be done. The emphasis should be on carefully planned and executed tests, with suitable statistical analysis of the results so that proper conclusions can be reached on the significance of the strength variation with size. Special care should be given to provide full details of the test to enable a thorough analysis.

On the theoretical side, the problem of predicting size effects needs to be further examined. In concrete, this is connected with the understanding of the failure mechanisms involved, which is still incomplete. The strength of concrete, for example, can be regarded as a function of the strength of aggregates, the cementitious matrix, and the aggregate-matrix bond. Each of these elements has its own size effect characteristics. Other concrete characteristics, for example, the influence of compaction or the type of mold for specimens, affect its behavior under load.

In conclusion, a substantial amount of experimental data and theoretical studies have been accumulated on the subject of size effects. There are, however, still many questions to be answered and much research to be done in this important area.

Loading Systems
and Laboratory Techniques

〜〜〜〜〜〜〜〜〜〜〜

7

7.1 INTRODUCTION

Prototype structures are normally designed for either concentrated forces or uniformly distributed loads. Using direct models, the concentrated loads must be scaled down to small concentrated loads on the model, and the uniform loading may be represented on the model by either a series of discrete loads or by suitably scaled pressure. Various loading methods will be covered in this chapter. Errors introduced by discretization of distributed loads will be examined. Particular attention will be given to modeling of loading for shell structures and for buckling studies. Brief comments will be made on thermal models and on special techniques for simulation of self-weight effects. The careful reader will recognize that development of adequate yet simple loading techniques requires a great deal of common sense as well as a certain degree of ingenuity. Ideas developed for one type of loading usually can be refined and adapted to situations that appear to be totally different.

In this chapter we shall concentrate on static loadings, with some discussion of quasistatic representation of earthquake effects. Dynamic loads will be treated in more depth in Chapter 11.

Any model loading system should:

1. Accurately represent the prototype loads (both magnitude and direction)
2. Be easy to apply to the structure, and to remove and reapply
3. Present no undue safety hazards

4. Offer no restraint to the model, particularly for buckling studies
5. Be capable of being "caught" if the model should fail suddenly and catastrophically

7.2 TYPES OF LOADS AND LOAD SYSTEMS

Loading systems for models must be very carefully designed and constructed if they are to function properly throughout the model test. Prior to considering detailed provisions for several categories of loading systems, we shall review some fundamental ideas and concepts of laboratory-applied loadings.

7.2.1 Load Reaction Systems

Discrete or point loads may be applied mechanically or hydraulically, or with dead weights. Mechanical and hydraulic loads must react against another structure that is substantially stronger and stiffer than the model structure, and the provision of such reaction structures is often the most expensive portion of a new loading system. Many modern structural testing laboratories are built on a strong floor that serves as the reaction system for loading. The system in the Portland Cement Association Laboratory in Skokie, Illinois, is illustrated in Fig. 7.1, which shows a $\frac{1}{2}$-scale prestressed concrete bridge girder being loaded with a series of hydraulically applied loads through the transverse beams that are spaced along the girder. The beams are loaded by tensioning the steel tie

(a)

Figure 7.1 Strong floor loading system [Hognestad et al. (1959)]. (a) (Courtesy of Portland Cement Association.)

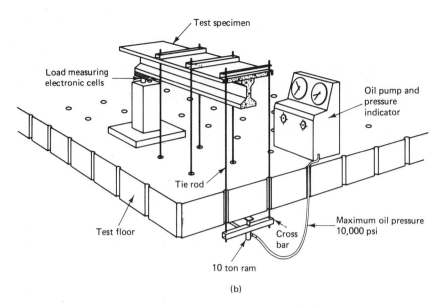

Figure 7.1 (*cont.*) (b) Application of vertical load.

bars that extend through the test floor to another cross beam, which in turn is jacked away from the underside of the floor with a hydraulic ram. This system and similar systems in other laboratories have been used to load many large models.

The same concept is used in the Cornell University Structural Models Laboratory for loading small models (Fig. 7.2). Each testing table is made from

Figure 7.2 Stiff testing table system for small models. (Courtesy of Cornell University.)

steel bridge deck and is supported 36 in. (0.9 m) off the floor by tubular steel legs. Aluminum reaction frames are bolted to the table, along with support devices for the model, and the applied loads react against the frame and transmit all forces back into the table. The distinguishing feature of both the strong floor laboratory and the stiff models testing table is that the force system is self-equilibrated by reacting against the structure and integral parts of the testing facility. New structures are tested simply by changing the locations of the loading devices.

7.2.2 Loading Devices—Discrete Loads

Hydraulic and electrically driven universal testing machines are perhaps the most common method of applying a single discrete load to a structure or test specimen. With suitable load-distribution devices, these machines can also produce a series of concentrated loads on a beam span.

Individual hydraulically actuated rams (or jacks) are familiar pieces of loading equipment in any laboratory, with load capacities ranging from 1 kip to hundreds of kips. They can be attached to loading frames and used to apply loads in any direction. Quick-release hose connectors are used with manual or electric pressure systems operating at pressures from 3000 to 10,000 psi (21 to 69 MPa). Approximate load intensities may be obtained from a pressure gage and the known area of the piston in the hydraulic ram; load cells placed between the ram and either the structure or the reaction system must be used to obtain more accurate values of load.

Mechanical loading devices such as those being used in Fig. 7.2 are very convenient for many model tests. They are operated by turning the crank, and feature zero backlash (no lost motion in either loading or unloading). The applied load must be measured with a load cell such as the small compression load cell attached to the mechanical loading devices in Fig. 7.2. This type of load cell uses electrical resistance strain gages as the sensing elements (see Chapter 8).

Suspended dead weights are often used for both augmenting the scaled dead load and for live loads. For small suspended loads [say up to 5 lb (1 Newton) per load point] one can use either lengths of steel bars with hooks at both ends or tin cans filled with lead shot or steel punchings. Both types of load are shown in Fig. 7.3a. When heavier loads are needed, such as in the modeling of dead weight during construction of a prestressed concrete bridge model, suspended weights made from household bricks or blocks of concrete are normally the least expensive to use.

Often a single concentrated load is needed to load bridge models or similar structures at many points. A convenient dead load for these applications is shown in Fig. 7.3b. Mechanical loading systems can also be built for these models, but at considerably more cost. The convenience and cost of the mechanical system must be balanced against the large amount of physical effort

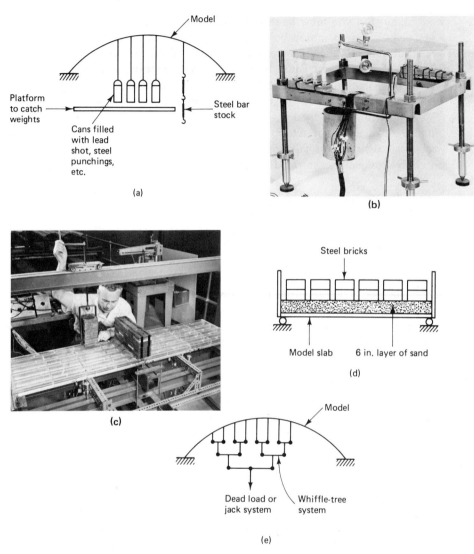

Figure 7.3 (a) Suspended dead weights. (b) Single point load. (Courtesy of LNEC.) (c) (d) Dead loads applied on top of model. (Courtesy of Portland Cement Association.) (e) Whiffle-tree system.

involved in using the suspended dead weights. A mechanical system for moving stacks of lead sheets to different locations on another bridge model is shown in Fig. 7.3c.

Dead weights may also be applied to the surface of the model. Figure 7.3d illustrates the application of dead loads to a slab through a layer of sand, which helps produce a uniform distribution of load. However, care must be taken to prevent any arching action. Sand, prepacked into plastic bags, is often convenient. But you must beware of changes in moisture content, which affect the weight of any common building material such as sand, concrete, or brick.

Substantial difficulties are met with suspended dead weights when a large number of load points must be loaded simultaneously, particularly in controlling the simultaneous application of loads, in unloading the model quickly, and even in reaching into the system to apply incremental loadings.

The whiffle-tree load system of Fig. 7.3e is ideal for applying a large number of identical discrete loads. This system, which involves an articulated set of load distribution bars and either one large dead load or a jacking system to apply a single concentrated load, is further described in Section 7.4.2.

7.2.3 Loading Systems—Pressure and Vacuum

Either air pressure or vacuum loading systems are usually used to apply uniformly distributed loads normal to the surfaces of a model. Again the provision of a strong reaction structure is the most difficult part of design and construction of loading equipment.

The use of a pressurized gas directly against the model (Fig. 7.4a) is usually not recommended in static tests because of construction difficulties in making a container that does not restrain the model at the connection between the model and the container. The use of an air bag between a reaction device and the model (Fig. 7.4b) is a better adaptation of using air pressure to distribute a single applied force over a structure.

Vacuum loadings are more widely used because they are more safe, the edge-sealing problem is easier to deal with, and it is easy to generate the small

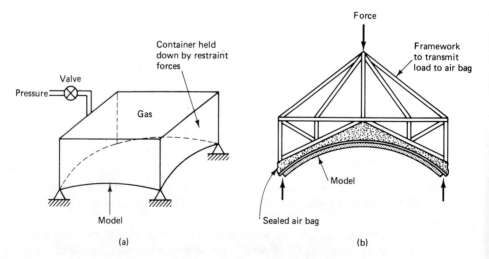

Figure 7.4 Pressure loading system for uniform load application. (a) Use of gas pressure as loading. (b) Air bag loading.

vacuum load needed using only a commercial-size shop vacuum. The following procedure is recommended for vacuum loading of a model structure:

1. Any commercial-size vacuum cleaner will suffice for small-scale work in which load requirements are less than 300 or 400 psf (14 to 19 kN/m²).
2. The space between the model and vacuum chamber is left at about $\frac{1}{8}$ to $\frac{1}{4}$ in. (3 to 6 mm) and is closed off with polyethylene film and petroleum jelly as shown in Fig. 7.5.

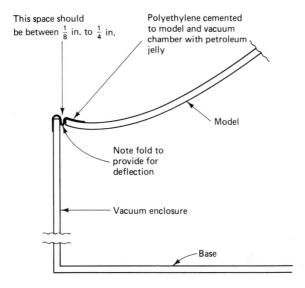

Figure 7.5   Vacuum loading system.

3. A simple open water manometer is the best method for measuring the negative pressure. The addition of colored water makes it easier to read the instrument. An electric pressure gage should be used if automatic recording is needed.
4. An adjustable opening in the vacuum chamber wall is necessary to control the load.

Applications of the vacuum loading method will be seen later in this chapter and in the case studies of model structures presented later in the text.

Air pressure may also be utilized for modeling blast effects on structures. An ingenious design for blast effects on model slabs, perfected at MIT, is shown in Fig. 7.6. This device permits control of both rise time and pressure dropoff on the slab. Both chambers are initially pressurized to the same level. Then one of the two-element diaphragms covering the port of one chamber is ruptured by pressurizing the diaphragm with a separate pressure line. This permits the gas to escape from one chamber, at a rate controlled by the initial pressure and

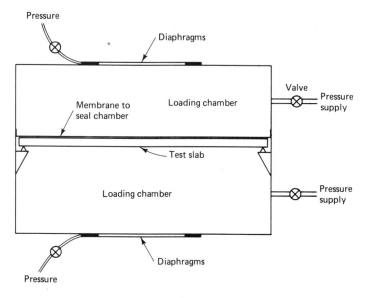

Figure 7.6   Pressure system for dynamic loading of slabs.

port size, and the slab "feels" the pressure from the other loading chamber. Unloading is controlled by rupturing the diaphragm in the other loading chamber.

7.3 DISCRETE VS. DISTRIBUTED LOADS

A recurring question facing the models engineer is: How many discrete loads are needed to adequately represent a uniformly distributed load? Two points are important here: (1) accuracy of bending moments and other stress resultants induced by the loading, and (2) local effects, such as local bending in thin shells. The latter point is particularly important in shell stability studies, where the local deformations induced by discrete loads may influence buckling behavior very strongly.

One can often profit by analytical study of effects of load spacing on internal forces in the model. By choosing simple structures that are representative of the actual model, it is possible to perform relatively quick analyses to give insight into the degree of inaccuracies produced with discrete loads. Litle et al. (1970) studied several cases of discrete load spacing effects on beams and arches. The first, shown in Fig. 7.7, gives the bending moment at the center of a simple span beam as a function of the number of equal-spaced concentrated loads used to represent the uniformly distributed load w. It is seen that any even number of loads produces $M = wL^2/8$ at midspan, while the use of only three loads leads to an error of $+11\%$ in bending moment.

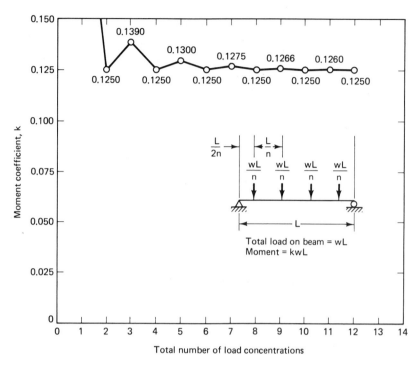

Figure 7.7 Effect of load discretization on simple beam bending.

A more revealing study is summarized in Fig. 7.8, where a parabolic two-hinged arch with a rise of $0.3L$ is analyzed for extreme fiber stresses produced by three different load cases: uniform load on the horizontal, 15 concentrated loads, and 5 concentrated loads. The highly variable stresses produced by 5 loads bear no resemblance to the uniform state of compression from a uniform loading. Even with 15 loads the stress variation remains quite severe. The question of how severe such variations are depends upon what one is expecting from the model. Clearly an experimental stress analysis done on such a structure with discrete loading would be quite misleading if the prototype did have a truly uniform loading. Some ideas of potential problems to be met in interpreting strains measured on the surface of a discretely loaded shell structure model should be gained from this simple example.

The third study, which was done to determine the influence of discrete loading on buckling of a flat, parabolic fixed-base arch with rise = $0.1L$, is given in Fig. 7.9. The buckling load reduces as fewer loads are used because of the more severe departure from the desired state of uniform compression. Using 10 loads gives a predicted buckling capacity of 90% of the true value. Even with 20 loads the predicted capacity is about 5% too low.

These studies on the arch structures are not meant to be extrapolated to other cases, but instead are intended to illustrate the desirability of performing

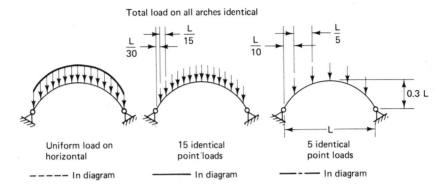

Total load on all arches identical

Uniform load on horizontal

— — — — In diagram

15 identical point loads

——————— In diagram

5 identical point loads

—— - —— In diagram

Geometry and loading of parabolic arches

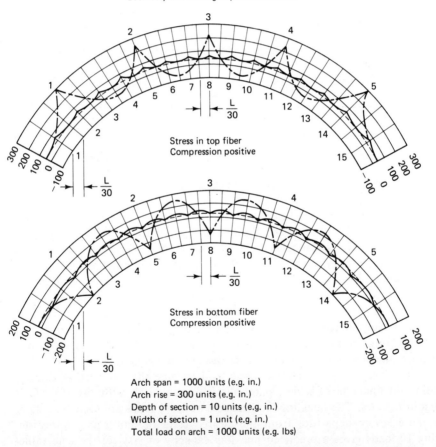

Stress in top fiber
Compression positive

Stress in bottom fiber
Compression positive

Arch span = 1000 units (e.g. in.)
Arch rise = 300 units (e.g. in.)
Depth of section = 10 units (e.g. in.)
Width of section = 1 unit (e.g. in.)
Total load on arch = 1000 units (e.g. lbs)

Figure 7.8 Effect of load discretization on stress distribution in a parabolic arch.

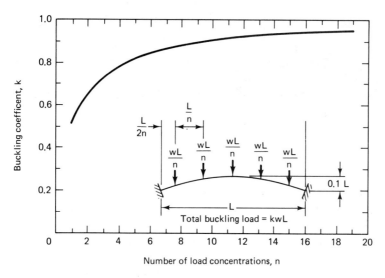

Figure 7.9 Effect of load discretization on buckling of a parabolic arch.

analytical studies to help assess the inaccuracies met in using discrete loads for modeling uniformly distributed loads. Additional discussion of local load effects in shells is given in Section 7.4.2.

The other main problems encountered in using distributed loads is that of modeling gravity loads (such as snow) on inclined roof surfaces with vacuum or pressure loads that act normal to the surface rather than in a desired gravity mode. This problem is discussed further in the next section.

7.4 LOADINGS FOR SHELL AND OTHER MODELS

Prototype shells transmit distributed surface loading (dead weight, snow, and wind) to discrete support locations. Proper modeling involves not only the correct magnitude of loads but also the correct distribution. Loading can be interpreted as also including reactions to applied loads, or the particular support and boundary conditions for a given shell. It is easy (but dangerous) to overlook proper modeling of the boundary conditions as well as the load system. Litle (1964) and others have investigated the effect of varying boundary conditions on shell behavior; some conclusions are presented in Section 7.5.

As described earlier, loading systems for surface loads on shells are usually one of three types: discrete, pressure, or vacuum. The choice of one system over the other is often based on the personal preference of the models engineer and on the type of loading equipment available in the particular models laboratory. The vacuum system is certainly one of the best ways to load shells. Load can be applied and released quickly. The sole disadvantage of the vacuum loading technique for shell models is that it exerts a uniform pressure normal to the surface instead of a gravity-type loading. Thus its use on steep shells may be questioned, and it cannot be used where a partial snow load or variable wind pressure is to be considered. The errors associated with not having the loading vertical are not appreciable for most practical shells; however, unless buckling is a crucial factor, and this "idealization" of the loading certainly seems reasonable when compared with the accuracy with which we know the assumed loadings or with the types of assumptions we make whenever an analytical approach is used.

7.4.1 Vacuum and Pressure Loadings

Two vacuum-loaded shell modeling studies are illustrated in Fig. 7.10. In the first model, the shell was supported directly on a metal base plate and the load was applied by applying a vacuum to the enclosed space defined by the shell and the base plate. Lubricating oil around the base of the shell formed the seal. In the second model, where nonrotational edge member conditions were desired, two identical models were cemented together and the vacuum was applied in the defined cavity of the two models. This loading method also illustrated another important concept—that of using symmetry to dictate what is happening at a support condition. Since each half of the double model tends to have the same rotational tendencies at the support, the effects cancel out and a clamped edge condition is achieved.

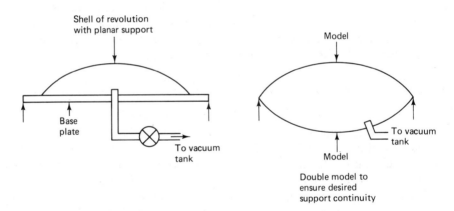

Figure 7.10 Vacuum loading of shell models.

An air bag pressure loading system is illustrated in Fig. 7.11 for a hyperbolic paraboloid umbrella shell [White (1975)]. The air bag was built to conform to the shape of the hypar shell and then pressurized slightly to maintain its shape. The upper surface of the bag was formed from a sheet of plywood to serve as the loading surface. Dead weights placed on it transmitted a uniform pressure through the air bag to the shell surface. No instrumentation is needed to measure this type of loading since it is precisely controlled by the amount of dead weight applied to the top of the air bag. Precise balancing of load is

Figure 7.11   Air bag loading on hyperbolic paraboloid shell models. (Courtesy of Cornell University.)

essential, however, to prevent "drooping" of one side of the bag, and membrane effects at the edge of the model must be minimized.

A unique ring loading device for loading shell models was developed at McGill University. Harris (1964) used two simple techniques to load spun aluminum spherical shells 28 in. (711 mm) in diameter and fixed at the ends (shell thickness = 0.080 in. or 2.03 mm). To apply a load of 40 lb (179 N) on a circular area 1.466 in. (37.24 mm) in diameter and centered on the shell crown, he used 5-lb (22-N) weights on a loading platform (Fig. 7.12a); this was designed to eliminate any horizontal force between the platform and the loading device and to ensure that the load applied to the shell was truly vertical. Spring balances were used to keep the platform stable under full load.

The two ring loads [517 lb (2.32 KN) on a ring 0.481 in. (12.22 mm) wide and with a mean diameter of 9.724 in. (247.0 mm); 917 lb (4.11 KN) on a ring 0.460 in. (11.7 mm) wide and with a mean diameter of 19.152 in. (486.5 mm)] were applied using loading rings (Fig. 7.12b) and a 150-gal (682-liter) domestic fuel oil tank that could be filled with water. The tank was calibrated on a scale to an accuracy of 50 lb (224 N) in the lower load range and 100 lb (448 N) in the upper range, and the loads were read using a glass manometer. The conical points on the loading rings were heat-treated to make them very hard and to prevent blunting by the loads. This also eliminates any appreciable bending moments from being transmitted from the loading platform or tank to the loading rings. The loading rings were fitted with an $\frac{1}{8}$-in.-thick (3 mm) neoprene gasket to distribute the load evenly. The oil tank was carefully bal-

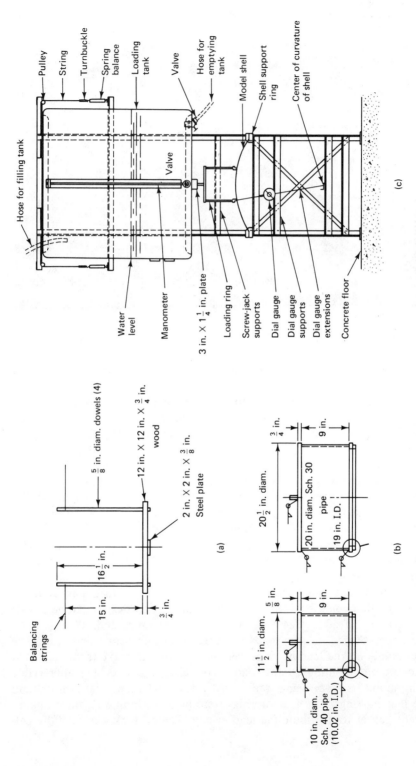

Figure 7.12 General arrangement of shell test apparatus. (a) Loading platform. (b) Loading rings. (c) Test setup. [After Harris (1964)].

anced with four horizontal strings and spring balances at the top. The entire assembly was supported by a framework made of perforated Dexion 225 standard angles and Dexion punched strap (Fig. 7.12c).

Mufti (1969) successfully used a similar ring loading device to apply an intermittent ring load to a similar spherical cap with a hole at the crown.

7.4.2 Discrete Load Systems

Suspended dead weights are often utilized in discrete loading. Any dead-weight system presents the problems of how to quickly unload the model and how to catch the loading system if the model fails suddenly. A typical dead-weight system for shell loading used at MIT is illustrated in Fig. 7.13, where a supporting mold fits the underside of the shell, with the weight strings passing through the mold. Before loading, or to unload, the mold is jacked up against the shell, transmitting the load to the mold and its supports. Loading occurs when the supporting mold is lowered away from the model. The mold is always kept close to the shell lower surface (but not in contact with it) so as to catch the model as buckling or other large displacements occur.

Figure 7.13 Suspended dead loads on shell model, with a catching mold for shell and weights. (Courtesy of MIT.)

Figure 7.14a and b illustrate other shell modeling studies that utilized suspended discrete loads. In the second case the loads are not meant to be removed quickly, nor can the structure be caught if it begins to fail. In another variation of suspended loading, cans are suspended from the model and are temporarily supported by a movable table; the table is lowered to apply the loads to the shell. This system was employed for a buckling study described in Section 7.5.

Three different types of load-distribution pads are shown in Fig. 7.14c. In each type the underlying principle is to minimize the effect of the concentrated load by spreading it out to several loading feet or legs, or to make it even more uniform by transmitting the load through a foam rubber pad.

A 64-point whiffle-tree load system for ultimate strength modeling of hyperbolic paraboloid shells [White (1975)] is shown in Fig. 7.15a. A similar

(a)

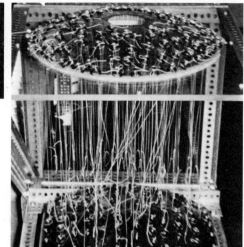

(b)

Plastic disk
with 3 rubber
legs (Type 1)

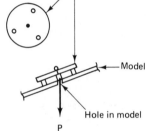

Model

Hole in model

P

Masonite (hard-board)

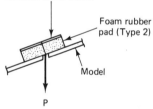

Foam rubber
pad (Type 2)

Model

P

Pinned joint

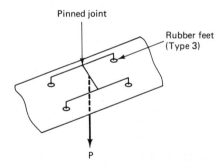

Rubber feet
(Type 3)

P

(c)

Figure 7.14 Shell models—discrete loading methods. (a) Point loads on shell. (Courtesy of MIT.) (b) Buckling studies on shells. (Courtesy of MIT.) (c) Load pads.

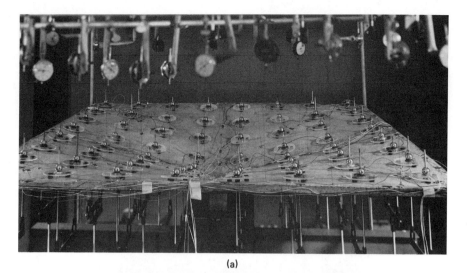

(a)

(b)

(c)

Figure 7.15 Whiffle-tree systems for plates and shells. (a) 64-point whiffle-tree system on hypar shell. (Courtesy of Cornell University.) (b) Whiffle-tree system on medium length cylindrical shell. (Courtesy of Cornell University.) (c) Whiffle-tree system on short cylindrical shell. (Courtesy of Cornell University.)

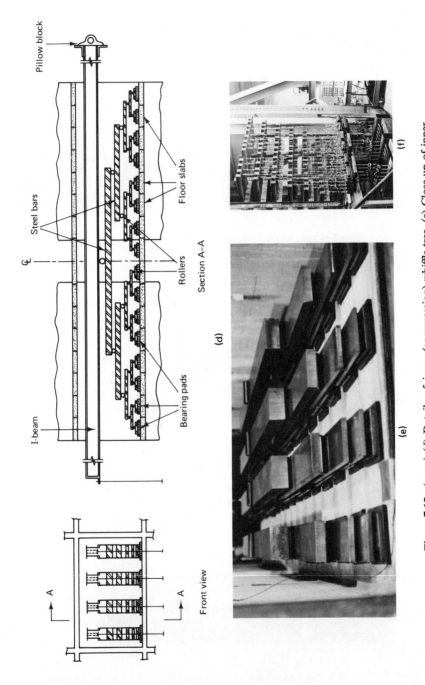

Figure 7.15 (*cont.*) (d) Details of inner (compression) whiffle-tree. (e) Close-up of inner whiffle-tree. (Courtesy of Drexel University.) (f) Front view of loading system. (Courtesy of Drexel University.)

Pillow block

Steel bars

Floor slabs

Rollers

Bearing pads

I-beam

Ç

Section A-A

(d)

Front view

A

A

(e)

(f)

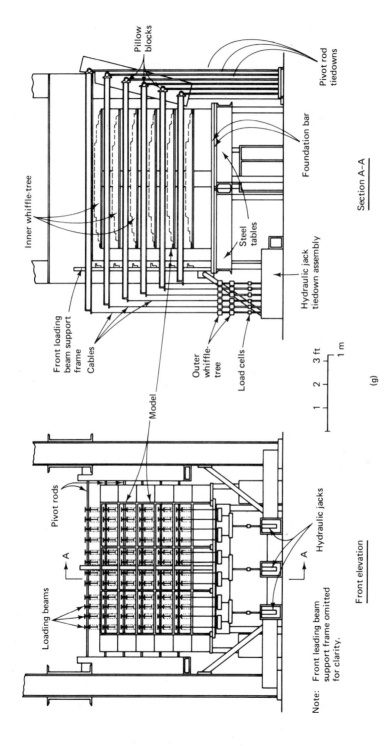

Figure 7.15 (*cont.*) (g) Loading system for three dimensional model.

system for cylindrical shells is shown in Figs. 7.15b and c [Harris and White (1972)]. One potential disadvantage of the whiffle-tree system is its relatively high dead weight, which cannot be removed from the shell. On the other hand, a well-constructed whiffle tree converts a single load into a large number of very accurate equal loads that will remain almost equal even under severe deformations of the shell (Fig. 7.15c).

An interesting whiffle-tree loading system was developed at Drexel University to load a three-dimensional model of a precast, large-panel concrete building [Muskivitch and Harris (1979), Harris and Muskivitch (1980)]. It consists of an interior whiffle tree in compression spreading a point load supplied by a pivoted I beam to 16 bearing pads (Fig. 7.15d). There are four of these compression arrangements per bay of the three-bay six-story model (Fig. 7.15e). Each of the four compression whiffle trees are collected by a tension whiffle-tree arrangement to a single tension jack. The resulting 18 jacks attached to a common hydraulic system are shown in Fig. 7.15f. An overall view of the discrete point mechanical-hydraulic loading system for applying gravity loads to the $\frac{3}{32}$-scale model building is shown in Fig. 7.15g. Thus whiffle-tree load systems are most useful for ultimate strength models where applied loads are high and the dead weight of the whiffle tree is not an important factor. Whiffle trees of the size shown in Fig. 7.15 would be useless for thin elastic models made of plastic: the shell would most likely be overloaded from the weight of the whiffle tree alone.

7.4.3 Effects of Load Spacing

The effects of load spacing were examined at MIT by Soosar (1963) and later summarized by Litle (1964). He showed that replacement of a continuous loading with discrete loads may lead to a very large systematic error in shell stress distribution. However, discrete loading does not affect buckling pressure if the grid spacing is sufficiently small. It should not be necessary to go to less than a 1-in. (25 mm) grid spacing for shell stability studies, and the spacing can usually be higher.

The ultimate strength hypar models of Fig. 7.15a had discrete loads at a 6-in. (150 mm) grid spacing, which is one-eighth of the shell-span dimension. Each load in turn was distributed to three points on the shell surface (Fig. 7.14c, Type 1). While this spacing certainly influenced the local shell stresses, it did not seem to affect the development of major inelastic action and the failure mode. The precise definition of satisfactory discrete load spacings needs additional exploration. It is felt that the examples given here represent acceptable practice for the particular model being studied. One cannot overemphasize the danger of looking at someone else's loading system, which might have been designed for heavy loads on ultimate strength models, and then using a similar setup for light loads on elastic models. A final point on load systems is that they must not exert significant restraint on the model. At the same time, the supports for a model must not move if they are intended to remain fixed in position. Proper planning of supporting fixtures for shell models requires considerable thought, common sense, and perhaps a little luck.

7.5 LOADING TECHNIQUES FOR BUCKLING STUDIES AND FOR STRUCTURES SUBJECT TO SWAY

The major requirement for buckling- and sway-prone structures is that the loading must not restrain the movement of the structure because such restraint will tend to increase the load capacity. Plane trusses, frames, space trusses, and similar structures that are being modeled for instability are usually loaded with gravity loads if the applied load is not excessively high. If it appears unfeasible to use dead load, one may have to resort to a gravity load simulator of the type shown in Fig. 7.16. This simulator is merely a mechanism that permits transverse motion without any vertical motion; thus it is capable of maintaining a constant vertical load while moving laterally. Six such devices were used in a frame test to be described subsequently.

Figure 7.16   Frame fastened to models testing table. (Courtesy of Cornell University.)

7.5.1 Shell Instability

Shell structures have been known to suffer various forms of instability. A large dome covering a market in Bucharest reportedly collapsed from creep-buckling action. The failure mode of the Ferrybridge hyperbolic cooling towers in England included a snap-through type of behavior that had been demonstrated in earlier model studies by Der and Fidler (1968). A similar type of buckling action occurred after extensive inelastic deformation in some of the intermediate-length cylindrical shell tests reported by Harris and White (1972).

The very thin shell surfaces utilized in metal skin construction are almost always subject to buckling failure modes, and light-gage steel hypars and folded plates often are limited in strength by the buckling capacity of the corrugated decking. Instability can also be a serious problem for certain folded-plate geometries.

Loading for a buckling model must be very carefully planned in order to prevent any restraint against motion that might occur when buckling takes place. Swartz et al. (1969) utilized a whiffle-tree loading system fastened to pull-type hydraulic rams that produced vertical concentrated loads on a 3.38 × 3.38 in. (84 × 84 mm) square grid (Fig. 7.17) over the surface of a folded-plate model. Horizontal movement of the hydraulic rams during buckling was accomplished by having the rams anchored to thrust bearings that permitted the necessary movement.

Figure 7.17   Stability model for folded plate structure. (Courtesy of IIT.)

The shell roof of the Providence, Rhode Island, post office building [described by Litle and Hansen (1963)] was designed for suitable stiffness against instability failures by using plastic models loaded with dead loads (Fig. 7.18). The concept of using a "catching device" under the model was also employed here. This not only prevents the destruction of the model but also permits its retesting if the instability mode does not produce stresses beyond the linear range of behavior for the plastic. The plastic models shown in Figs. 7.13 and 7.14a were also intended primarily as instability models.

In his definitive work on the reliability of models for predicting shell stability, Litle (1964) presents considerable practical information on loading techniques and on other laboratory techniques needed in shell modeling. The reader is urged to study this important reference before embarking on any shell modeling problem. Litle emphasizes that buckling behavior is extremely sensitive to changes in boundary restraint. If the model loading rig induces

Figure 7.18   Shell roof model. (Courtesy of MIT.)

initial edge bending, the buckling behavior will almost certainly be affected. Any buckling model must reproduce prototype boundary conditions as closely as possible. When the prototype boundary conditions are not well defined, it is recommended that alternate model tests be conducted in which the upper and lower bounds on edge restraints are both modeled to give bounds on the actual prototype conditions.

Litle also recommends that closely spaced hanging loads be used instead of pressure (or vacuum) loads for shells that have any appreciable slope, since buckling can be quite dependent upon load direction as well as magnitude.

7.5.2 Structures Undergoing Sway

A model study for the Australia Square Tower in Sydney [Gero and Cowan (1970)] is shown in Fig. 7.19 for a design wind condition that produces substantial sway. A whiffle tree is used in an ingenious fashion for this uniform loading, and a hydraulic jack is used to produce the single horizontal force. Note that a mechanical lever is used to amplify the jack force and that the entire weight of the whiffle tree is suspended from a support. Each line of the tree is adjustable to align the applied forces in the horizontal direction. Although the gravity loads are not applied simultaneously with wind effects in this model, a gravity load system could be superimposed and the two systems could act independently without restraining each other.

The three-story, two-bay frame model [Chowdhury and White (1977)] of Fig. 7.20b illustrates several important features of loading, including:

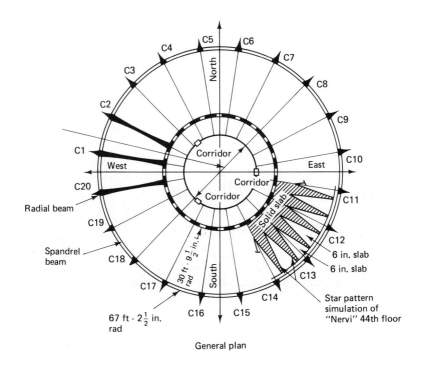

General plan

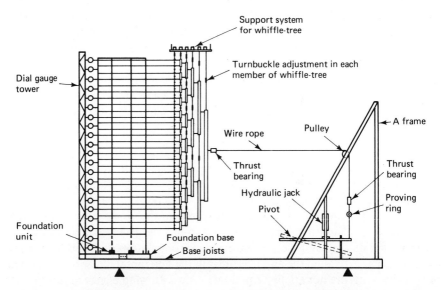

Figure 7.19 Lateral load system for high rise building frame.

1. Testing a vertical structure in a horizontal plane
2. Combined gravity and simulated static lateral seismic loads
3. Combined axial force and reversing bending loads on a component of the frame (Fig. 7.20a)

The frame has fixed bases that are modeled by bolting the heavy base beam to the stiff steel testing table. The frame was tested in the horizontal plane because it was relatively easy to provide the needed out-of-plane restraint with the testing table. Also, the table was a convenient attachment surface for the many gages and loading devices employed in the experiments.

The gravity loads are applied at the quarter points of each of the six beams with six gravity load simulators. These mechanisms are fixed to the table and fastened to the beams as shown. They are activated by small hydraulic rams connected on each level to a common hydraulic pressure source to give a fixed gravity load on each floor. This gravity load, which is measured with tensile load cells, is maintained constant during the test, and permits sidesway of ± 3 in. (75 mm)

The reversing simulated seismic loads are applied at each floor level with hand-cranked mechanical loading devices. The major difficulty met with this type of load system is in achieving the proper lateral loads at each floor level. The load induced at each level is dependent on loads at other levels, and considerable adjustments are needed, particularly at high loads when the frame is approaching failure.

Prior to the frame test, component testing was done on the beam-column joint specimen of Fig. 7.20a. This specimen represents an exterior connection with its half columns and half beam that would extend to the normally assumed inflection points when the frame is under lateral load. The loading on this specimen was constant axial force on the column and fully reversing bending moment on the cantilever beam section. This rather complex load condition was achieved quite simply by some minor adaptations to an existing 30,000-lb (133 KN) capacity universal testing machine. Column axial load was applied vertically with the testing machine. The force to create moment was applied vertically through a double-acting hydraulic ram that was anchored to a steel pipe column spanning between the floor and ceiling, immediately in front of the machine. The horizontal reactions at the column ends induced by the bending force were taken back to a building column immediately behind the machine, through steel angles fastened to the end reaction devices on the model column. This loading system functioned extremely well and was inexpensive. It is another example of how ingenuity must be used to solve rather complex loading designs.

Gravity loads can also be applied to frames that are subject to sway by using long, flexible tensile elements that span between the load point and the floor (base of the frame). These elements offer very little resistance to lateral load as long as the sway is small, but one set of tensile elements may interfere with others when the frame is more than a single story high.

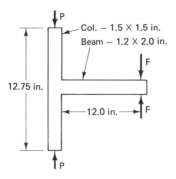

(a) Beam-column joint

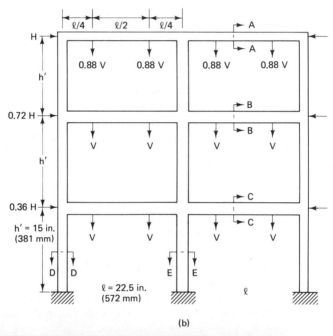

(b)

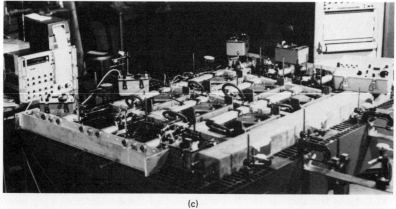

(c)

Figure 7.20 Simulated seismic and gravity loads on 1/10 scale model frame. (a) Beam column joint. (b) Model frame. (c) Test set-up for frame. (Courtesy of Cornell University.)

7.6 MISCELLANEOUS LOADING DEVICES

There are many special situations met in devising loading schemes for structural models, including thermal effects in dams, nuclear reactors, and other structures; self-weight effects where additional surface loadings do not produce the necessary state of dead-weight stresses; internal pressure loadings for nuclear reactor vessels and other pressure vessels; and dynamic loads for earthquake, wind, and oscillating machinery effects. We shall discuss several of these topics briefly here and return to others in case studies and special modeling applications later in the book.

7.6.1 Thermal Loads

Thermal loads require a heat source and a medium to properly distribute the heat to the model. Electric heating elements are the most common heat source because they are compact and can be controlled with a high degree of accuracy. Typical heating coils for modeling thermal effects in dams are shown in Fig. 7.21. As described by Rocha (1961), the dam is first covered with a waterproof membrane, and the shaped coil assemblies are then inserted into upstream and downstream cavities that contain water. The coils are used to heat the water to the proper temperatures.

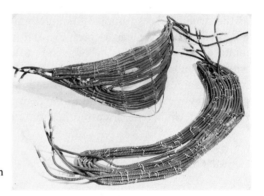

Figure 7.21   Upstream and downstream heating coils. (Courtesy of LNEC.)

It should be noted that the normal critical factor in thermal modeling is the presence of a thermal gradient across the model thickness. The absolute temperatures are normally not important, which means that cooling can also be used to induce the proper gradients. Thus room temperature on one face of a model and a refrigerated condition on the other face could produce the desired gradient.

Thermal effects in reactor structures are also usually generated by electric heating elements and either an oil or water bath to properly disperse the heat. Local hotspots might be modeled with heating elements alone, with the thermal load being distributed by the material itself.

7.6.2 Self-Weight Effects

In using small-scale models for structural engineering studies, the dead-weight stresses are not properly modeled in most situations. The only exception is where the stress scale can be chosen in such a way as to satisfy the dead-load similitude conditions, but this is uncommon in most models for reinforced concrete structures, and even other types of structures. Several methods have been used to satisfy, at least approximately, dead-weight stresses in models. In dams, for instance, where the final stress state is a strong function of dead weight, models are sometimes built in many lifts (layers), and prestressing is applied to each lift to approximate the state of stress that is desired during the construction phase. This is not entirely satisfactory because of the stress concentrations and disturbances created by the prestressing elements and anchorages.

Models have been tested in a centrifuge to induce a higher acceleration of the mass of the model, thus satisfying the dead-weight similitude condition. The complexities inherent in such a test are rather obvious. Also, the dangers involved in unbalancing the centrifuge when the model fails are quite serious. Soils modeling in the Soviet Union (U.S.S.R.) has been done with centrifuges. This approach may be the only feasible method of getting proper dead-weight effects in soils models.

Durelli, Parks, and Ferrer (1970) studied gravity and surface load stresses in hollow spheres by using the immersion technique. In this method the model of the structure being studied for gravity dead-weight stresses is placed upside-down in a fluid that has a specific gravity that is substantially higher than the model itself. The model, which tends to float, is held down at those points where it would normally rest in its upright position. The resultant load on the model becomes the weight of the model times the factor $(k - 1)$, where k is the ratio of the density of the fluid to the density of the model material. This rather ingenious technique is discussed in more detail in the reference cited above.

7.7 SUMMARY

Loading systems for structural model tests take on many forms and require great care and substantial ingenuity in their design and construction. They range from the very simple, single-hanging concentrated dead-weight system to automated, computer-controlled systems. The reader is urged to study carefully the many experimental studies reported in the journals of ASCE, ACI, SESA, and other technical societies, and in special models publications and conference proceedings. Most of the references on models in the bibliography include descriptions of loading systems. The development of an ability to design efficient and inexpensive loading systems for complex models can be extended directly to tests for large structures and thus represents a highly valuable skill.

Instrumentation—
Principles and Applications

ᏉᏉᏉᏉᏉᏉᏉᏉᏉᏉᏉᏉᏉ

8

8.1 GENERAL

In earlier chapters we dealt with similitude analysis, material properties, and loading techniques, which essentially enable us to conceive, fabricate, and test a model. However, meaningful interpretation of such model tests is not possible unless proper instrumentation is used for measuring the many quantities related to the behavior of the structure.

The instrumentation process includes careful identification of the quantities to be measured; selection of the appropriate sensors and the necessary auxiliary equipment; installation of the sensors on the completed model; calibration of sensors and checkout of equipment prior to the model test; acquisition of data; and reduction of data into meaningful stresses, forces, and force-deformation relationships. This process can be quite demanding, particularly for the more sophisticated models involving dynamic response, thermal loading, high internal pressure, and other complex loadings. Instrumentation can be the most time-consuming and expensive part of some modeling studies, and the serious models engineer must know not only how to measure everything in a given experiment, but also how to strike the rather delicate balance of "getting enough data" vs. overinstrumenting and running up the costs unnecessarily.

This chapter deals with the major aspects of instrumentation, including quantities to be measured, means and techniques of measurements, and the underlying theory. Applications to actual case studies are only referenced here and are treated in more detail in Chapters 10 and 11.

8.2 QUANTITIES TO BE MEASURED

As mentioned earlier, the behavior of a structure is reflected in the forces and deformations that result from subjecting it to the different loading conditions. These are measured through the instrumentation on the structure, on the surface, at the boundaries and loading points, and sometimes inside the model. In general, in a reinforced or prestressed concrete structure or its model, the following quantities need measurement:

1. *Strain*: its distribution across the section under consideration. Strain may be measured in concrete, either by instrumenting the surface *or* by suitably embedding gages inside, or on the steel reinforcement and the prestressing strands. Knowing the stress-strain characteristics, the stresses associated with these strains in a structure can then be determined.

2. *Deflection*: its distribution along the structure and its variation with the applied load and magnitude in a structure or a constituent element. Deflection measurements are needed to define the load-deformation characteristics and can be helpful in determining the limits of elastic behavior, curvature, and changes in curvature.

3. *Cracks*: their locations, patterns, and widths related to the loading. This information is used to determine satisfactory service load conditions and also to obtain the ultimate or limit load stress conditions.

4. *Forces*: their magnitudes and nature in the concrete or the steel reinforcement, at the boundary supports, and sometimes at loading points. Knowledge of these internal forces, which are in equilibrium with the applied forces, is especially useful in the study of indeterminate structures.

5. *Temperature*: its distribution within the mass of concrete, where the structure is subjected to differential temperature conditions.

6. *Creep and shrinkage*: their measurements in a structure subjected to sustained loading. These are similar to item (1) above, but care must be exercised to ensure that the instrumentation is stable over the entire period of measurement.

7. *Properties of materials*: they must be determined in order to translate other measurements (such as strains) into overall structural behavior, and to correlate test results with theory. Measurement of properties of concrete are particularly important since they are subject to variations from environmental conditions, such as relative humidity and temperature.

8. *Dynamic response*: various types of responses of a structure when subjected to dynamic loads, e.g., seismic, fatigue, and repeated loadings. Accelerations, velocities, and displacements are measured.

The equipment to measure the above quantities varies from simple hand instruments to the more sophisticated electronic devices. The former are used manually from point to point, and the latter, although they work much faster, require elaborate setups for monitoring. The readout instruments accompanying these measuring devices also vary from hand-operated to continuous scanning, recording, and monitoring systems. The choice of using one or the other will depend on the type of quantities to be measured, loading, reliability of measurements, and economics.

The outcome of any experimental program depends significantly on the accuracy and reliability of measurements. In the case of small-scale models, the quantities to be measured are much smaller in magnitude, based on the principles of similitude, thus magnifying the error possibility and the associated need for accuracy. To achieve the same level of accuracy as in the prototype tests, accuracy in measurements of model quantities must be achieved, theoretically, to the order of at least the scale factor. For example, if in the case of a prototype beam, the deflection is read to within ± 0.01 in., then the deflection in its $\frac{1}{10}$-scale model should be read within ± 0.001 in. to achieve the same order of accuracy and the corresponding reliability of the model test results. More about the accuracy of modeling will be discussed in Chapter 9.

The foremost measurement in all model testing is that of the strain; the basic reason behind this is that stress and strain are related to each other by a fundamental relation, the modulus of elasticity, for linearly elastic materials. Although strain is a fundamental physical quantity and stress is a derived quantity, more use is made of the word *stress* to express the ability of a material to support applied loads or forces because it is easier for the engineer to visualize in terms of stress rather than in terms of strain. Because of its basic importance, strain measurement should be carried out with the utmost care and accuracy.

8.3 STRAIN MEASUREMENTS

Strain is measured with strain gage, which in essence is a means of magnifying the change in length over a given length; strain gages are therefore classified according to the type of magnification system they use. To achieve accurate strain measurements, all strain gages must fulfill the following basic requirements, which define the "perfect universal strain gage" according to Perry and Lissner (1962).

1. It must be extremely small and of insignificant mass (inherent need for this exists because strain should be measured at a "point," and dynamic response of the structure may be affected by any additional mass attached to it).

2. It must have a high sensitivity to strain and attach easily to the structure.

3. It must not be susceptible to variation in ambient conditions such as temperature, vibration, or humidity that will be encountered in the test.

4. It must be capable of indicating both static as well as dynamic strains by remote indication and recording.

5. It must be inexpensive and convenient to use.

Unfortunately, if one were to develop the "perfect" strain gage with all of the above characteristics, the cost of such a gage would be astronomical. As a compromise, over the last few decades efforts have been made to achieve many of these properties with emphasis on economy and convenience in use.

The major types of strain gages are:

1. Mechanical

2. Optical

3. Electrical

The optical strain gage has become obsolete because of advancements in electrical gages and will not be discussed here. Some techniques using optics have been developed recently; the details may be obtained from Photoelastic Inc.

8.3.1 Mechanical Strain Gages

The mechanical strain gage in its basic form uses mechanical systems such as levers, gears, or similar means for magnification of strain. Although this appears simple, the magnification from one gear and/or lever to another causes mechanical interaction, such as friction, lost motion, inertia, and flexibility of the parts, and if not overcome, some of these shortcomings cause reduction in accuracy. In most instances, mechanical strain gages are limited to static measurements of strain, since their size and inertia rule out any reasonable frequency response, which is required in the dynamic applications. In spite of some of the disadvantages of mechanical gages, they are used often, primarily because they are self-contained. The strains to be measured are shown on scales or dials, and no additional equipment is required for readouts. They are reusable, which makes them economical.

The Whittemore gage, manufactured by Baldwin Locomotive Works, has been in use for many years. As shown in Fig. 8.1, it is self-contained and consists essentially of two frame members connected together by two elastic hinges, which provides a parallel frictionless motion. Conical points are attached to the frame legs, which are inserted into the attachment holes on the structure; these points define the gage length. The strains are measured with an integral dial indicator.

The main disadvantage of this gage is the potential error induced when the gage is repositioned on the structure for each strain reading. This error is minimized by having the operator that reads the gage develop a consistent technique and also by having the same operator read the same gage throughout

Whittemore strain gage

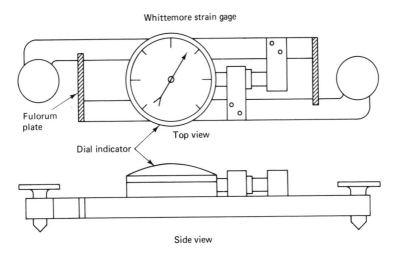

Figure 8.1   Whittemore gage showing its various components.

a given test. In spite of this disadvantage, the gage is extremely useful for long-term (creep) measurements on concrete members, for measuring distortion in shear panels, and in other similar applications where measurement over a relatively long gage length is permissible.

8.3.2 Electrical Strain Gages

Electrical strain gages use the principle of change in some electrical characteristic of the gage material caused by strain in the structure. The electrical variables commonly used are resistance, capacitance, and inductance. The important advantage of electrical strain gages is the relative ease with which the output can be amplified, recorded, and displayed. Of the various kinds of electrical strain gages, the resistance type is the most commonly used because of its many advantages, and this type will be discussed separately.

8.3.2.1 Electrical Resistance Strain Gages

The resistance type of strain gage functions as a resistance element in an electric circuit. Strain produces a change in the magnitude of the resistance associated with the gage. This change is recorded and related to strain during calibration, after the gage is bonded to the structure.

Although the principle of metallic resistance was observed by Kelvin in 1856, the first use of this principle related to strain measurement was by Carlson and Eaton in 1930. This first gage developed was unbonded, of the metallic type, with a single wire wound over two pins that were embedded in the structure. This gage had appreciable mass, size, and gage length, and consequently it was not used very much as a practical strain gage. Simmons and Ruge in 1938 independently conceived the idea of bonding the wire either directly to the test specimen or to a thin paper backing which was in turn bonded to the specimen.

Since that time, a considerable amount of development has taken place in bonded electrical resistance strain gages, using etched foil circuits as well as wire circuits (Fig. 8.2). Today we have strain gages in a large variety of shapes and sizes, and gage lengths as small as $\frac{1}{64}$ in. These resistance-type gages, such as the SR-4* strain gage (so designated to embody the initials of both inventors, Simmons and Ruge), are the best tools for strain measurements in all types of structures and structural elements.

Because of their relatively low cost and lightness, a large number of these gages can be used to investigate the behavior of a structure. Electric resistance strain gages have achieved the widest applications for strain measurements, including use in airplane parts, boat hulls, structures, buildings, bridges, and machine parts. They have a number of advantages that make them extremely versatile, as shown in the table below.

Property	Comment
Size	Very small
Weight	Insignificant
Ease of attachment	Relatively simple
Sensitivity to strain	Good; higher output very advantageous
Static and dynamic strain measurement	Equal ease
Remote indication and recording	Easily accomplished
Expense	Relatively inexpensive
Gage length	As small as $\frac{1}{64}$ in.
Sensitivity to ambient variables	Slightly affected, but gages can be usually protected or variables involved can be properly compensated for
Linearity of output	Good

The main disadvantages of electrical resistance gages are the drift that can occur over a long period of time because of instability of the mounting cement, and some sensitivity to ambient environmental conditions. This gage also needs appropriate signal conditioning and readout equipment (either a manual or automatic strain indicator system). These disadvantages are rather minor as compared with their many advantages, however. The sensitivity of this type of gage can be as high as 0.000001 in./in. (commonly called one microstrain).

Recently, weldable resistance strain gages have been devised for use in testing jet aircraft, nuclear containment vessels, rockets and missile surfaces, and heavy construction projects in severe environments, such as shock, vibra-

---

*SR-4 is a registered trademark of the Baldwin-Lima-Hamilton Corp., Electronics Division, Waltham, Mass.

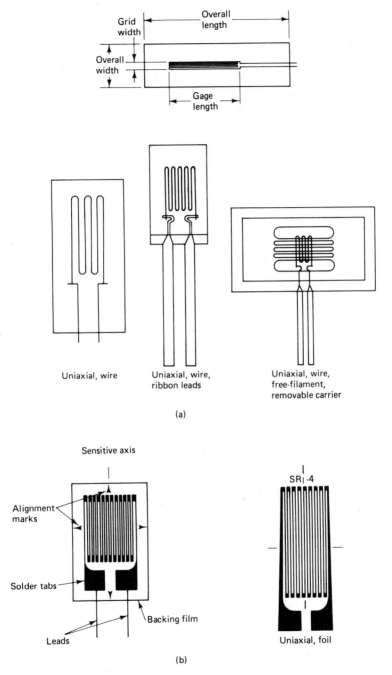

Figure 8.2 Typical electrical resistance strain gages. (a) Wire. (b) Foil.

tion, steam, and saltwater. These testing applications call for rugged and stable gages capable of withstanding temperature variations up to as high as 1200°F (649 °C) over a period of time. The weldable wire resistance strain gage can be installed in a matter of minutes. This unique technique, utilizing capacitive discharge spot-welding equipment, eliminated the need for all bonding materials and overcame the weaknesses of the bonding processes used for conventional electrical resistance strain gages. The gage consists of a filament configuration in which the strain-sensing filament and lead-out wire are of unitized construction. The lead-out section is electroformed (using gold) to a diameter of 0.007 in. A controlled taper between this diameter and that of the strain wire is provided. Such a construction technique eliminates joint fatigue problems or instabilities due to erratic electrical connections and contact resistance problems. Strain-sensitive alloys generally used are nickel-chrome and platinum-tungsten, both of which are excellent for strain measurements. The former is used for temperatures up to 650°F and the latter at higher temperatures up to 1200°F.

Two geometric configurations are used for the active and dummy filaments. The active element is a V-shaped simple element, while the dummy is an identical filament wound in a helix. The gage factor is adjusted to zero by selecting a proper pitch angle of the helix. The Poisson's ratio is matched so that no net dimensional change results when the gage is strained. Thus, a temperature compensation is provided by the dummy element, which is in the same environment, but without the applied strain. The strain filament is encased in a strain tube that is made by welding a tubular metallic shell to a flat flange stock. The filament is mechanically coupled to the strain tube but electrically isolated from it by highly compacted (metallic) high-purity magnesium oxide powder. This compaction is obtained by using a programmed high-speed centrifuge and a swaging operation.

When the gage is welded to a specimen and the test specimen put into a stress field, the stress is transmitted through the welds to the mounting flange, into the strain tube, and through the magnesium oxide powder. Because of the high compaction of the powder, the strain is transmitted to the sensing element all along its length, with no slippage. The property of stable, reversible, strain transfer is due to the large value of the ratio of surface area to the cross-sectional area. Weldable gages (Fig. 8.3) are equipped with a thin flange welded to the strain tube; the flange is spot-welded to the test structure, thus providing the bond required for strain transfer. A guide for the selection of weldable gages for different applications is given in Table 8.1.

8.3.2.2 Other Electrical Strain Gages

Other types of electrical gages make use of properties affected by straining, namely capacitance, inductance, and the piezoelectric effect. In the capacitance gage, change in capacitance of the condenser is observed as strain varies. The condenser consists basically of two plates separated by a short distance and

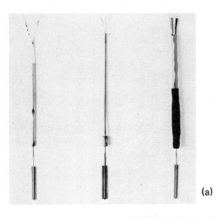

(a)

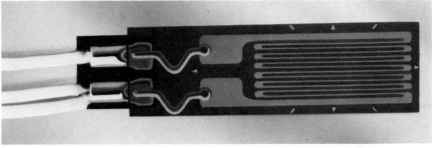

(b)

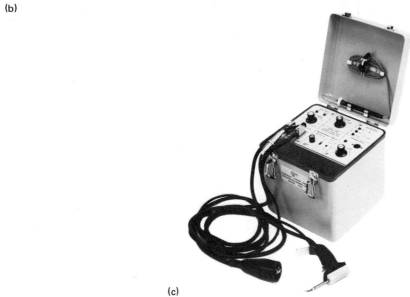

(c)

Figure 8.3 Weldable gages. (a) Gages. (Courtesy of Ailtech.) (b) Weldable gage type LWK. (Courtesy of Measurements Group.) (c) Strain gage welding unit. (Courtesy of Measurements Group.)

TABLE 8.1 Strain Gage Selection Chart*

Application	Preferred Gage Type[1]	Filament Material	Description
Strain Measurements in the Field Bridges, buildings, pilings—whenever humidity, rain, direct immersion, or a simple, rugged installation procedure is required	SG 189	Nickel Chrome Alloy	Flexlead Strain Gage[2,3] $\frac{1}{4}$ bridge, 120 ohm, self-compensated gages with pre-waterproofed, vinyl insulated; 3 conductor, shielded cable.
Extended Temperature Ranges to 650°F static or 1500°F dynamic even when water, steam or corrosive media surrounds gages	SG 125 SG 128	Nickel Chrome Alloy	Integral Lead Strain Gage[2,3] $\frac{1}{4}$ bridge, 120 ohm, self-compensated gages with hermetically sealed, stainless steel jacketed cable. Leads insulated with fiberglass (SG 128) or magnesium oxide (SG 125) for the most severe applications.
Cryogenic Temperatures up to 650°F—coupled with water, steam, corrosive media	SG 325 SG 328	Nickel Chrome Alloy	Integral Lead Strain Gages[2] $\frac{1}{2}$ bridge versions of SG 125 or SG 128 gage provide optimum thermal compensation in the cryogenic or sub-zero ranges. Also available for minimizing drift at 650°F.
Ultra-Temperature Environments to 950°F static (1200°F for short times) with protection from corrosive media	SG 425	Platinum Tungsten Alloy	Integral Lead Strain Gages $\frac{1}{2}$ bridge Pt-W gage for static strain measurements to 1200°F, equipped with stainless steel jacketed, magnesium oxide insulated integral leads.
Embedment in Concrete for models or full size structures	CG 129	Nickel Chrome Alloy	Flexlead Strain Gage designed for direct embedment in concrete, 0-180°F. Other temperature ranges available.

—Hydrostatic testing, special temperature ranges, etc. are available.

[1] Basic gage types (without integral leads) are available. They provide features of ruggedness and simplicity of installation when integral lead, hermetically sealed cables are not required. Some basic gages are useful to 1800°F for dynamic measurements.
[2] 350 ohm gages are available.
[3] Short length gages are available.
*Courtesy of Ailtech, Inc.

insulated from each other. The strain gage is attached to the structure at the point where strain will alter either the spacing or the arc of the above two plates. This change in turn causes the change in the impedance of the ac circuit and is a function of the capacitance of the condenser. Because of the separation of the two plates, it has inherent disadvantages, which include oversensitivity to vibration and difficulty in attaching the gages. This has considerably restricted the commercial development of these gages.

The inductance strain gage has an iron-core coil, whose inductance can vary with the applied strain. The variation in the inductance is achieved by changing the position of the armature with respect to the coil or by altering the space of the air gap between the two. This type of strain gage has disadvantages similar to capacitance gage and it is heavier. However, it has a very high electrical output compared with a resistance strain gage, which helps simplify the readout circuitry. Because of its bulk and relatively large gage length, it is used more as a displacement measuring device in the form of a variable inductance transducer rather than as a strain gage.

The piezoelectric principle, related to the ability of a material to generate an electric charge when strained, was observed first by Curie. Some crystalline materials are naturally piezoelectric; other polycrystalline ceramics can be given this property artificially. In the last 20 years, this principle has been used in various kinds of transducers for pressure and acceleration measurements. Recently, Mark and Goldsmith (1973) have reported a successful use of barium titanate as a piezoelectric material in strain gages. The main advantage of this type of gage lies in the very high signal produced per unit strain and also in the fact that it is self-generating, i.e., no external power supply is needed. Their main disadvantages are: (1) They are not suitable for static strains, hence they are difficult to precalibrate; (2) they cannot be precalibrated because of oversensitivity while mounting; and (3) they produce considerable reinforcement at the point of attachment.

The development of these various gages has remained somewhat specialized and limited because of the success and versatility achieved by the resistance-type strain gages.

In view of the short space, only the basic principles have been discussed here. A number of leading manufacturers of the various types of strain gages maintain up-to-date literature and the specialized circuitry for specific types of gages; this information is readily available. An extensive list of these manufacturers is provided in Table 8.2.

8.3.2.3 Choosing Electrical Resistance Strain Gages and Cements

Various types of resistance strain gages with different properties and configurations have been developed for a variety of environmental conditions. Gage length is determined by the particular application; in general, the longest

TABLE 8.2   Partial List of Manufacturers of Different Types of Instrumentation

Manufacturer	Address	Types of Accessories
Acurex Corporation	485 Clyde Avenue Mountain View, Calif. 94042 Tel: (415) 964-3200	Data acquisition system for strains
AILTECH Corporation	19535 East Walnut Drive City of Industry, Calif. 91748 Tel: (213) 965-4911	Weldable strain gages
BLH Electronics	42 Fourth Avenue Waltham, Mass. 02154 Tel: (617) 890-6700	All types of strain gages, and related accessories for strain reading
Blue M Electric	Blue Island, Ill. 60646	Ovens with range of about 700°C
Comtel Systems Corp.	688B Alpha Drive Highland Heights, Ohio 44143 Tel: (216) 461-0171	Data acquisition system
Magnaflux Corp.	7300 W. Lawrence Ave. Chicago, Ill. 60656	Stress coats of different types
Micromeasurements Div. of Vishay Inter- Technology, Inc.	P.O. Box 27777 Raleigh, N.C. 27611 Tel: (919) 365-3800	All types of strain gages, and related accessories for strain reading
PCB Piezotronics, Inc.	P.O. Box 33 Buffalo, N.Y. 14225 Tel: (716) 684-0001	Quartz transducers for dynamic measurements
SATEC Systems, Inc.	Grove City, Pa. 16127 Tel: (412) 458-9610	Electro mechanical testing systems
Strainsert Company	Union Hill Industrial Park West Conshohocken, Pa. 19428	Flat load cells of various capacities

feasible gage is used to make installation easier, but in cases where the strain gradient is high, very short gages may be necessary. Gages on the surface of concrete should have a gage length of at least several times the maximum aggregate dimension to avoid stress concentration effects. The range of strain to be measured is an important factor in selecting type of gage, as is the type of test (static vs. dynamic). The user is advised to keep up-to-date copies of manufacturers' catalogs on hand and to consult them carefully before making a choice of gage type and length. Additional discussion on gage configuration is given in Section 8.3.3.6.

Strain Sensing Filaments.   Copper-nickel alloy (constantan alloy 400) is primarily used in static strain measurements because of its low and controllable temperature coefficient. This alloy may be operated safely between 0 and 400°F.

Nickel-chrome alloy (Nichrome alloy 200) can be used for high-temperature static and dynamic strain measurement of up to 1800°F with the use of a proper ceramic cement. Nickel-iron alloy (Danaloy alloy 600) is recommended for dynamic tests where its larger temperature coefficient minimizes the temperature compensation problem. The higher gage factor (see Section 8.3.3.2), which leads to higher electrical output, is an advantage for measuring dynamic strains of small magnitude. Platinum alloy (alloy 1200) is used for its excellent stability and fatigue resistance at elevated temperatures. Its only disadvantage is that this alloy cannot be adjusted for self-temperature compensation and therefore, as will be shown in Section 8.3.3.3, an additional (dummy) gage must always be used along with each active (measuring) gage. A typical strain gage selector chart from BLH, Inc., is given in Table 8.3.

Cements (Adhesives).   A large number of cements are available for use in various types of measurements. More commonly used cements are: Duco,* Eastman 910,† and various epoxies.

Duco (single-component, nitrocellulose, room-temperature curing cement) is used particularly with the paperbacked strain gage and has a maximum operating temperature of 150°F. Eastman 910 or its equivalent, single-component contact cement, fast-curing at room temperature, is especially used with polyimide-backed strain gages and is excellent for short-term testing. Fast-curing, two-part epoxy resins, available at local hardware stores, are also adequate for bonding strain gages and work particularly well on concrete surfaces where the Eastman 910 type cements are less satisfactory because of the porous nature of concrete. Other special types of cements include epoxy and ceramic base cements that have better long-term stability characteristics; their use is recommended on some types of transducers, for long-term strain measurement, and under thermal environments.

It cannot be overemphasized that the cement is the only medium available to transfer the strain in the structure into the gage sensing element. The cement layer should be as thin as possible. A poor cementing operation means inaccurate strain measurement. As in selecting gages, up-to-date literature from the manufacturer should be studied carefully before choosing a cement. A typical cement-selection chart from BLH literature is shown in Table 8.4.

8.3.3 Resistance Strain Gage Circuitry and Applications

The theory of resistance strain gages is based on the Wheatstone bridge principle with four resistances, R_1, R_2, R_3, and R_4, as shown in Fig. 8.4. A summary of the following is presented in this section.

1. Circuit analysis

---

*Registered trademark of E. I. duPont de Nemours and Co., Wilmington, Delaware.
†Registered trademark of the Eastman Kodak Co., Rochester, New York.

TABLE 8.3 Strain-Gage Selector Chart*

Gage Type, Prefix	Carrier Material	Operational Temperature Range °F					Test Application Primary/Sec.
		Lowest Extended†	Lowest Recommended	Temperature Compensated	Highest Recommended	Highest Extended†	
CONSTANTAN GRID—GENERAL PURPOSE							
FAE	Polyimide	−320	−100	+50 to +150	+400	+600	Static/Dynamic
FAP	Thin Paper	−320	−100	+50 to +150	+150	+180	Static/Dynamic
FAB	Phenolic-Glass	−320	−100	+50 to +250	+450	+600	Static/Dynamic
NICHROME V GRID—CRYOGENICS							
FNB	Phenolic-Glass	−452	−452	None	+600	+600	Static/Dynamic
NICHROME V GRID—TRANSFERABLE FROM TEMPORARY CARRIER							
FNH	Strippable Vinyl	−452	−452	None	+600	+700/1800	Static/Dynamic
FNO	Free-Filament	−452	−452	None	+600	+700/1800	Static/Dynamic
NICHROME V GRID—PLATINUM ELEMENT—UNIVERSAL TEMPERATURE COMPENSATION							
FNB-E	Phenolic-Glass	−452	−452	−452 to +600	+500	+600	Static
FNH-E	Strippable Vinyl	−452	−452	−452 to +700	+700	+850	Static
SR-4 CONSTANTAN WIRE GAGES—GENERAL PURPOSE							
A	Paper-Wrap	−320	−100	None	+100	+150	Static/Dynamic
A, AF	Paper-Flat	−320	−100	+50 to +150	+150	+180	Static/Dynamic
AB	Bakelite-Wrap	−320	−100	None	+250	+350	Static/Dynamic
ABD	Bakelite-Wrap	−320	−100	None	+250	+350/400	Static/Dynamic
AB, ABF	Bakelite-Flat	−320	−100	+50 to +250	+300	+400	Static/Dynamic
SR-4 DUAL ELEMENT WIRE GAGES—CLOSE COMPENSATION PARTICULAR MATERIALS							
EBF (+)	Bakelite-Flat	−320	−100	+50 to +250	+300	+400	Static
EBF (−)	Bakelite-Flat	−320	−100	−50 to +300	+300	+400	Static

SR-4 ISO-ELASTIC WIRE GAGES—GENERAL PURPOSE

Gage	Type						Application
C, CD	Paper	−320	−100	None	+150	+180	Dynamic
CB, CBD	Bakelite	−320	−100	None	+250	+350	Dynamic
DYNALASTIC WIRE GAGES—WIDEST RANGE OF APPLICATION							
DLB-A	Phenolic-Glass	−320	−100	+50 to +250	+450	+450/600	Static/Dynamic
DLB-MK	(Welded Etched	−452	−452	+75 to +600	+600	+600/600	Static/Dynamic
DLB-PT	Foil Leads)	−452	−320	None	+600	+600	Dynamic
HT FREE-FILAMENT WIRE GAGES—CERAMIC AND FLAME SPRAY BONDING							
HT-200		−452	−452	None	+600	+1800/700	Dynamic/Static
HT-400	Removeable	−320	−100	+50 to +250	+450	+450/600	Static/Dynamic
HT-800	Teflon-Glass	−452	−452	+75 to +600	+600	+600/600	Static/Dynamic
HT-1200		−452	−320	None	+1200	+1500/1200	Dynamic/Static
FSM GAGES—HIGH TEMPERATURE—STABILOY FOIL/POLYIMIDE GLASS BACKING							
FSM	Polyimide Glass	−452	−452	+75 to 600	+600	+750	Static/Dynamic
SPECIAL PURPOSE STRAIN GAGES							
PA	Post Yield-Wire	−50	+50	None	+150	+180	High Elongation
AS	Brass Envelope Valore Type	−100	−50	+50 to +150	+100	+150	Concrete Imbedment
FABW		−320	−100	+50 to +250	+500	+600	Static
FNW	Weldable	−452	−452	None	+700	+1200	Static
FNWFB		−452	−452	−452 to +1000	+1000	+1200	Static
FAES, FAES-4	Polyimide	−320	−100	+50 to +150	+400	+500	Diaphragm Strain
FAB-28, -33	Phenolic-Glass	−320	−100	+50 to +250	+400	+450	Stress Gage

*Adopted from BLH Corp. literature.

†Operation in extended ranges requires special precautions in compensation or correction of temperature-induced errors.

TABLE 8.4 Strain-Gage Cement Selector Chart

Cements	SR-4	Duco	Eastman 910	EPY-150
		Room-temperature Curing		
Base of Cement	Nitro-Cellulose		Acrylic	Epoxy
Strain Gage Compatability	Use with quick drying thin paper backed gages	All std. paper backed gages	All except paper wrap-around construction	All
BLH Designation	FAP and A (as specified)	A (as specified)	FAE	FAE, FAB
Maximum Operating Temperature °F	180	150	150	150
Cure Temperature °F	Room to 150	Room to 150	Room	Room to 150
Cure Time	2–10 hours	12–48 hours	1–5 minutes	1–70 hours
Cure Pressure psi	Contact 1–5 for 1 minute	1–5	Contact 1–15	5–15
Specimen Material Compatability	All except plastics soluble in MEK and Acetone and unbondable plastics	All except plastics soluble in MEK and Acetone and unbondable plastics	All except some plastics	All except some plastics
Strain Limit	>10% at room	>10%	>10% at room	>10% at room >$\frac{1}{2}$% at −320
Electrical Properties	Excellent over operating temperature range	Excellent over operating temperature range	Excellent	Excellent
Humidity Resistance	Fair, absorbs up to 2% water	Fair, absorbs up to 2% water	Fair, absorbs up to 0.3% water	Good, absorbs up to 0.1% water
Moisture Proofing Recommendations	Low RH, short test-none. Cerese and Dijell Waxes; Barrier A, B, C and E for all others	Low RH, short test-none. Cerese and Dijell Waxes; Barrier A, B, C and E for all others	Low RH, short test-none. Cerese and Dijell Waxes; Barrier A, B, C and E for all others	Normal RH, except paper gages-none. Cerese and Dijell Waxes; Barrier A, B, C, D and E for all others
General Application Remarks	Easiest, most reliable for large-scale testing	Good, general purpose cement with adequate drying	Fastest for short tests, poor thermal mechanical shock resistance	Easy and durable for long-term tests, normal ambients

*Adopted from BLH Corp. literature.

TABLE 8.4 (*cont.*)

EPY-350 EPY-400 EPY-500 EPY-550 EPY-600	PLD-700	Bakelite	AL-PBX	Rokide-BLH
	Heat Curing			Flame Spray
Epoxy	Polyimide	Phenolic	Phosphate	Refractory Oxide
All that will stand cure temperature Exception: All paper backed	Phenolic and polyimide backed	Phenolic backed only	Strippable, transferable free-filament	Transferable, free-filament wire only
FAE, FAB, DLB	FSM	FNB, AB	FNH, FNO, HT	HT
400 normal, 500 after proper cure to 600 short time	>750	300 continuous to 500 for time	>1000	>1500
150–600	500	250–350	600	None
1–30 hours	$2\frac{1}{2}$ hours	5–6 hours	1–6 hours	None
15–30	35–45	50–100	None	None
All except some plastics and reactive metals	All except some plastics and reactive metals	All except some plastics and reactive metals	All except some plastics and reactive metals	All except some plastics
>5% at room >1% at −320	>2% at room	>2% at room >$\frac{1}{2}$% at −320	$\frac{1}{2}$%	1%
Excellent	Excellent	Excellent to 300°F. Poor in 400–500°F range	Deteriorates with increase in temperature above 1200	Excellent but deteriorates at high temperature above 1500
Good, absorbs up to 0.1% water	Fair, absorbs up to 1% water	Fair	Poor, is hygroscopic	Poor, is porous and hygroscopic
Normal RH, none. Cerese and Dijell Waxes; Barrier A, B, C D and E for all others	Conventional protection to 600°F; Barrier H	Normal RH, none. Cerese and Dijell Waxes; Barrier A, B, C D and E for all others	Conventional protection to 600°F, replace upon return room temp; or Barrier H	Conventional protection to 600°F, replace upon return room temp; or Barrier H
Tranducers and best long-term stability, varying ambient conditions	Fast $2\frac{1}{2}$ hour cure for high temp. organic backed gages	Some preference for transducers but generally replaced by epoxies	Easier application with free-filament wire and foil gages	Fastest application, no appreciable heat on test specimen, rough surface recommended

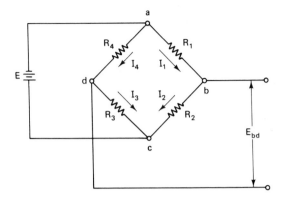

Figure 8.4 Wheatstone Bridge circuit.

2. Gage factor and transverse sensitivity
3. Temperature compensation (external and internal)
4. Typical applications
5. Types of bridges (full, half, and quarter)
6. Various configurations
7. Reduction of stresses from strain measurements

Details of some of the above items may be found in books by Perry and Lissner (1962) and Dove and Adams (1964).

8.3.3.1 Circuit Analysis

The circuit most commonly used with resistance strain gages is a four-arm bridge with a constant voltage excitation E (Fig. 8.4). In this circuit, the condition of balance in the bridge is first established; this is followed by determining the imbalance due to change in resistance (i.e., change in strain) in one or more gages mounted at positions R_1, R_2, R_3, and R_4.

For a balanced condition, the potential E_{bd} across the diagonal *bd* must be equal to zero, or the voltage at *d* equals the voltage at *b*. Therefore, the voltage drop from *a* to *d* $(I_4 R_4)$ must equal the drop from *a* to *b* $(I_1 R_1)$. Similarly, the voltage drops in the two legs in the bottom half of the bridge must be equal. These conditions can be summarized in

$$I_3 R_3 = I_2 R_2 \quad \text{and} \quad I_4 R_4 = I_1 R_1 \tag{8.1a}$$

In addition, with the bridge balanced, $E_{bd} = 0$, and thus the current I_1 must be the same as I_2, and I_3 must equal I_4, or

$$I_1 = I_2 \quad \text{and} \quad I_3 = I_4 \tag{8.1b}$$

Substituting Eq. (8.1b) into (8.1a) and forming the ratio of the two new equations to eliminate the values of current I, we obtain

$$\frac{R_4}{R_3} = \frac{R_1}{R_2} \tag{8.2}$$

which is the condition for a balanced bridge.

Let us establish the expression for a change in E_{bd} due to change in resistance, say R_1, with all other parameters held constant. From Fig. 8.4:

$$E = I_1(R_1 + R_2) \quad \text{or} \quad I_1 = \frac{E}{R_1 + R_2} \tag{8.3}$$

which gives

$$E_{ab} = R_1 I_1 = \frac{ER_1}{R_1 + R_2}$$

$$\frac{dE_{ab}}{dR_1} = \frac{(R_1 + R_2)E - ER_1(1)}{(R_1 + R_2)^2} = \frac{R_2 E}{(R_1 + R_2)^2}$$

or

$$dE_{ab} = \frac{R_2\, dR_1}{(R_1 + R_2)^2} E$$

But

$$E_{bd} = E_{ad} - (E_{ad} + dE_{ab}) = -dE_{ab}$$

which results in

$$E_{bd} = -\frac{R_2\, dR_1}{(R_1 + R_2)^2} E \tag{8.4}$$

Similar expressions can be obtained for changing each of the other resistance R_2, R_3, and R_4. If all these resistances are changed simultaneously, then

$$E_{bd} = \sum_{i=1}^{4} E_{ibd}$$

Using $F = dR/R/\epsilon$ (where F is defined as a gage factor* for each resistance R and the strain ϵ), we get

$$E_{bd} = E\left[\frac{-R_2 R_1 F_1 \epsilon_1}{(R_1 + R_2)^2} + \frac{R_1 R_2 F_2 \epsilon_2}{(R_1 + R_2)^2} - \frac{R_4 R_3 F_3 \epsilon_3}{(R_3 + R_4)^2} + \frac{R_3 R_4 F_4 \epsilon_4}{(R_3 + R_4)^2}\right] \tag{8.5}$$

In the case when all resistances and gage factors are equal, we obtain

$$E_{bd} = \frac{FE}{4}(-\epsilon_1 + \epsilon_2 - \epsilon_3 + \epsilon_4) \tag{8.6}$$

This means that the imbalance of the bridge is proportional to the sum of the strains in opposite arms and to the difference of strains in adjacent arms of the bridge.

*Defined in next section.

8.3.3.2 Gage Factor

The term *gage factor* is used to describe the sensitivity of output characteristic of the bonded resistance strain gage. It is defined as

$$F = \frac{\dfrac{\Delta R}{R}}{\dfrac{\Delta L}{L}} = \frac{\dfrac{\Delta R}{R}}{\epsilon} \tag{8.7}$$

where

F = gage factor

R = electrical resistance

L = length of strain sensitive element

ϵ = normal or axial strain

Thus, gage factor is the ratio of the change in resistance per unit of original resistance to the applied strain. The gage factor, which has a value close to 2 for gages commonly used in static strain analysis, is quite important since the higher the gage factor, the higher will be the electrical output and the higher the sensitivity. Successful attempts have been made to achieve strain gages with relatively high gage factors. The limiting factor in determining the gage factor is the conductor material. Materials that provide for high gage factors have other undesirable characteristics to render them less suitable as strain gage material.

In wire gages, when the filament (or wire) of the strain gage is wound back and forth to form a grid, there are a large number of "turnarounds" or bonded ends, which results in a significant decrease in ΔR, thereby causing variation in the value of the gage factor. To account for this, proper calibration of the gage factor is performed in the commercially available gages by installing several identical gages on a specimen bar and applying a known mechanical strain. Statistical averages of all the gages are then proportioned to the value of the specimen strain to obtain the gage factor value for that type of gage.

8.3.3.3 Temperature Compensation in Strain Gage Circuits

Throughout the above discussion, we have considered that the temperature is constant during the test in which strains are measured. However, temperature influences the electrical properties of the metals used in constructing the gages, and we must account for this effect as well as for the thermal strains that will result in the structure itself if the temperature of the testing area changes during the test. The separation of mechanical strains (induced by the physical loading of the structure) from unwanted thermally-induced strains is an important part of the experimental stress analysis process.

Temperature compensation can be achieved by forcing adjacent arms of the Wheatstone bridge to have identical thermal strain components. Since the bridge output is proportional to the algebraic difference of strains in adjacent arms, the thermal strain increments will cancel each other out. The objective is accomplished by mounting a gage identical to the active strain gage on an unstressed piece of the same type of material as in the structure. If the active gage is R_1 in the bridge, then the so-called "dummy" gage is placed in either position R_2 or R_4. Writing Eq. (8.2) for the case with R_2 being the dummy gage and R_1 the active gage,

$$\frac{R_4}{R_3} = \frac{R_1 + \Delta R_1}{R_2 + \Delta R_2} \tag{8.8}$$

where the changes in R_1 and R_2 are produced by thermal effects. It is obvious that the bridge is still balanced and that the output is zero. The reader should verify that making R_4 the dummy gage will also lead to a fully balanced bridge and hence inherent temperature compensation for the active gage R_1.

Thermal problems produced in plastic models by local heating from the bridge current are discussed in Section 3.5.4.

8.3.3.4 Self-Temperature Compensating Gages

As described in the last section, it may not always be possible to find a suitable location for installing a compensating gage *or* the temperature variation in both active and compensating gages may not be identical. In such situations, self-temperature-compensated (STC) gages should be used. The term *STC* is applied to strain gages in which the resistance change of the gage due to temperature effects is held to a very low value. The available STC gages can be classified into three basic types: (1) the two-element, (2) the single-element, and (3) the universally compensated gage.

The two-element gage is made with two wire element grids in series. One element has a negative temperature coefficient of resistance, while the other has a positive coefficient of resistance. The lengths of two grids are adjusted such that the net temperature coefficient of resistance will compensate for this change in resistance as the result of a differential coefficient of expansion when the gage is mounted on a particular material. The limitation of this gage is that a particular design is restricted to the material with the corresponding coefficient of expansion, and a variety of gages must be kept to accomodate the common construction materials.

The single-element gage is made by producing a grid where the change in resistance due to a change in specific resistance will be just equal in magnitude but opposite in sign to the change due to the difference in the coefficients of linear expansion of the gage and the material on which it is mounted.

The universal-type STC gage is also a two-element gage, but the relative effect of the two elements can be adjusted by making changes in the external circuitry. Figure 8.5 shows a gage of this type and the circuitry. In this circuit,

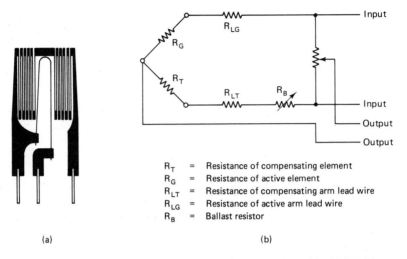

R_T = Resistance of compensating element
R_G = Resistance of active element
R_{LT} = Resistance of compensating arm lead wire
R_{LG} = Resistance of active arm lead wire
R_B = Ballast resistor

(a)   (b)

Figure 8.5   Universal temperature compensating gage (Courtesy of BLH Electronics, Inc.). [From Dove and Adams (1966).] (a) 2-element STC gage. (b) Circuitry for installation.

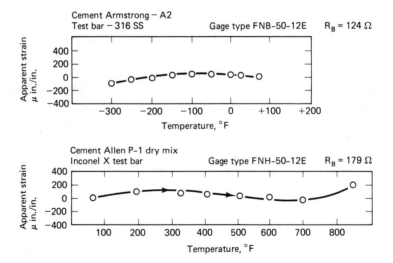

Figure 8.6   Apparent strain vs. temperature characteristic. (Courtesy of BLH Electronics, Inc.) [From Dove and Adams (1966).]

R_G and R_T are the active compensating elements, while R_{LG} and R_{LT} are the respective lead wire resistances. R_B is the variable resistor, which is adjusted to give least apparent strain over the range of interest. Figure 8.6 shows a typical plot between the apparent strain and temperature for a gage circuit of this type.

For further information on commercially available STC gages the reader is referred to the manufacturer's literature in Table 8.2.

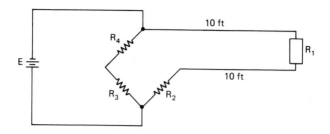

Figure 8.7 Temperature effect on lead wire resistance.

Another problem that arises is due to temperature variation in lead wires of different lengths. An example is shown in Fig. 8.7, where the single active gage R_1 is a STC gage with #18 copper lead wires totaling 20 ft in length, and the other three gages are internal resistances in the strain indicator system. The lead wire resistance R_w is about 0.12 ohms. If the wires are subjected to a temperature increase of 30°F, the change in wire resistance is

$$\Delta R = (\text{Temperature coefficient of resistance})(\Delta T)(R)$$

$$= (0.0022/°F)(30)(.12) = 0.00792 \text{ ohms}$$

With a gage factor $F = 2.1$, and the gage resistance $R_1 = 120$ ohms, the apparent strain is

$$\epsilon = \frac{\Delta R}{(R/F)} = \frac{0.00792}{(120.79/2.1)} = 0.000031$$

If the gage is mounted on a steel structure, with $E = 30,000,000$ psi, then the apparent strain produced by the temperature change is $E\epsilon = 937$ psi.

It should be apparent to the reader that the same problem exists in circuits with active gages having long lead wires and dummy gages having short lead wires.

8.3.3.5 Practical Circuits and Their Applications

Although the basic Wheatstone bridge has four arms, as discussed earlier, it is often necessary to use only one or two arms with active gages to measure individual strains. Furthermore, Eq. (8.6), which established the relation between the strains ϵ_1, ϵ_2, ϵ_3, and ϵ_4 and the change of potential E_{bd}, enables us to make use of three practical circuitries as follows:

1. Full bridge (four active gages)
2. Half bridge (two active gages)
3. Quarter bridge (one active gage)

Full bridge is one in which all four arms of the bridge are active, as shown in Fig. 8.8a, and the bridge output E_{bd} is given by $E_{bd} = (FE/4)(-\epsilon_1 + \epsilon_2 - \epsilon_3 +$

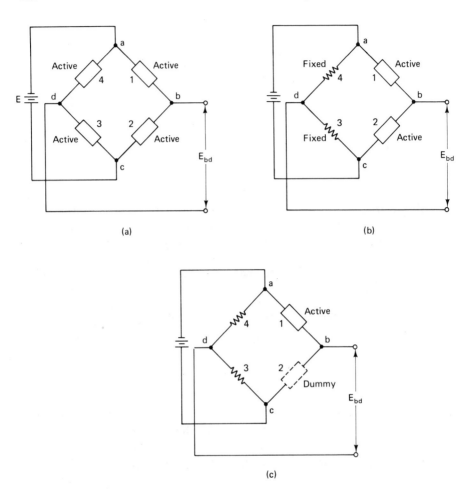

Figure 8.8 Typical practical circuits. (a) Full bridge. (b) Half bridge. (c) Quarter bridge.

ϵ_4). The inherent differences in sign for the strain terms in this expression may be used to good advantage in a load cell. The compressive cylindrical load cell in Fig. 8.9a is instrumented with gages R_1 and R_3 in the longitudinal direction and with gages R_2 and R_4 around the circumference. The latter gages will go into tension because of the Poisson effect. Hence E_{bd} becomes $(FE/4)[-\epsilon_1 + v(-\epsilon_1) - \epsilon_3 + v(-\epsilon_3)] = 2(1 + v)(FE/4)$(strain in either gage 1 or 3), and the output of the bridge circuit is magnified by a factor of $2(1 + v)$ over that of a single gage. Temperature compensation is automatically achieved because of the relation

$$\frac{R_4 + \Delta R}{R_3 + \Delta R} = \frac{R_1 + \Delta R}{R_2 + \Delta R} \tag{8.9}$$

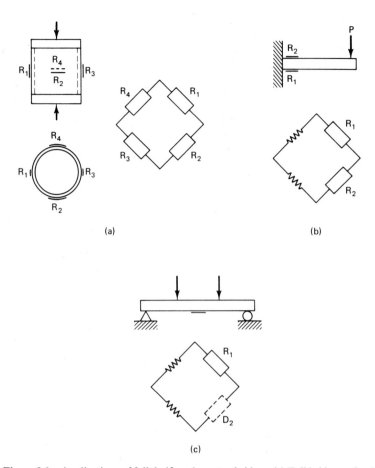

Figure 8.9 Applications of full, half, and quarter bridges. (a) Full bridge on load cell. (b) half bridge. (c) quarter bridge.

Half bridge is one with only two active gages R_1 and R_2 (or any two adjacent arms of the bridge) and two other fixed resistances to complete the bridge. The bridge output is $E_{bd} = (FE/4)(-\epsilon_1 + \epsilon_2)$. For example, a test on a cantilever beam to determine bending strains as shown in Fig. 8.9b may be instrumented with R_1 and R_2 so that E_{bd} is proportional to 2ϵ. The sensitivity of the circuitry is thus improved, and temperature compensation is simultaneously provided for.

Quarter bridge has only one active gage and is used when single strains in a stress field are to be measured. It can immediately be realized that proper precaution for temperature compensation must be exercised by using a dummy gage and lead wires, of proper length, as discussed earlier.

8.3.3.6 Gage Configurations

Various configurations of strain gages are shown in Figs. 8.10 and 8.11. Gage geometries can vary from general-purpose, single-grids to 2-element and 3-element rosettes that measure strains in different directions. Single grids

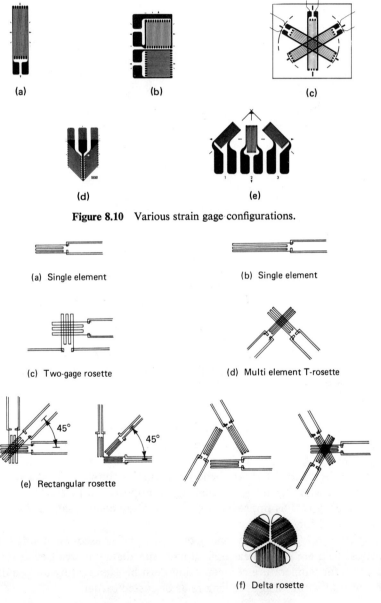

(a)   (b)   (c)

(d)   (e)

Figure 8.10   Various strain gage configurations.

(a) Single element   (b) Single element

(c) Two-gage rosette   (d) Multi element T-rosette

45°   45°

(e) Rectangular rosette

(f) Delta rosette

Figure 8.11   Various strain gage configurations.

(Figs. 8.11a and b) are used for measuring strain in one direction only; the length and width needed for a particular application depends on the strain gradient, the specimen size, the material used, etc. The usual range of lengths is from $\frac{1}{64}$ in. to 4 in., and the width varies from 0.03 in. to $\frac{1}{4}$ in.

The 2-element rosettes in Figs. 8.11c and d are used to measure strains in a general state of strain where the direction of principal strains is known. Two different types of 3-element rosettes are shown in Figs. 8.10e and f; these gages are used when nothing is known about the direction or magnitudes of strain in a general strain field, and three independent values of strain are needed to define the two principal strains and their direction (as will be discussed in the next Section 8.3.3.7). *Delta* rosettes with angles between gages of 120° (or 60°) are shown in Fig. 8.11f. *Rectangular* rosettes with two gages at 90° to each other and the third bisecting this angle are illustrated in Fig. 8.11e. In the second figure the gages are located adjacent to one another while in the first figure they are placed on top of each other to facilitate measurement of all three strain components at a single point.

8.3.3.7 Calculation of Stresses from Measured Strains

Frequently, the objective of the model study is to determine the stress distribution in a structure from strain measurements, using the techniques described earlier. The development of the basic relations between stresses and strains may be found in any textbook on the theory of elasticity [e.g., Timoshenko and Goodier (1965)]. In this section, these relations will be presented directly without proof and with the various procedures for their rapid solution. These can be classified as (1) direct reduction (algebraic), (2) graphic (geometric or nomographic) reduction, and (3) computer reduction. The first two are suitable only for a few point readings; the last one can be used for large numbers of reductions, either by using data on computer tape directly or using a programmable hand calculator.

Before discussing these methods, it must be pointed out that the strain readings are made on a finite area whose dimensions are governed by the size of the gage. However, the measurements are assumed to be at the center of the gage, and it is normally assumed that variations in the strains are linear in the area. Special care is therefore necessary to plan the number and size of gages in the areas of stress concentration. Furthermore, the number of strain measurements necessary for the stress determinations at the point of strain measurement depends on the knowledge (or lack of knowledge) of the directions of principal stresses (or strains). If the directions are known, only *two* measurements are necessary. However, if they are unknown, which may generally be the case, then *three* strain measurements are necessary to determine both the magnitudes and direction of the principal strains and stresses. The discussion thus relates to the various rosette arrangements of strain measurements as noted in the previous section.

According to the theory of elasticity, the direct stress on a surface in any direction x is given by

$$\sigma_x = \frac{E}{(1 - v^2)}(\epsilon_x + v\epsilon_y) \tag{8.10a}$$

where

$E =$ modulus of elasticity

$v =$ Poisson's ratio

$\epsilon_x =$ unit strain in x direction

$\epsilon_y =$ unit strain in y direction perpendicular to x direction

and the shearing stress is given by

$$\tau_{xy} = \gamma_{xy} G \tag{8.10b}$$

where $G =$ shearing modulus $= \dfrac{E}{2(1 + v)}$

In these expressions it is assumed that effects from pressure applied normal to the surface may be neglected.

If the x and y directions are known principal stress axes (known from symmetry considerations, for example), then two measurements of ϵ_x and ϵ_y are sufficient to determine the principal stresses σ_1 and σ_2 from the following expressions given in Eq. 8.11:

$$\begin{aligned} \sigma_1 &= \frac{E}{(1 - v^2)}(\epsilon_1 + v\epsilon_2) \\ \sigma_2 &= \frac{E}{(1 - v^2)}(\epsilon_2 + v\epsilon_1) \end{aligned} \tag{8.11}$$

and

$$\tau_{\max} = \frac{E}{2(1 + v)}(\epsilon_1 - \epsilon_2) = \frac{\sigma_1 - \sigma_2}{2} \tag{8.12}$$

On the other hand, without a prior knowledge of direction of the principal stresses, it is necessary to measure strains in three arbitrary directions and then solve for ϵ_x, ϵ_y, and γ_{xy} for use in the stress Eqs. (8.10). The strain ϵ_θ at any direction θ from the x axis (Fig. 8.12) is related to the orthogonal normal strains ϵ_x and ϵ_y and the shearing strain γ_{xy} by the expression

$$\epsilon_\theta = \epsilon_x \cos^2 \theta + \epsilon_y \sin^2 \theta + \gamma_{xy} \sin \theta \cos \theta$$

i.e.,

$$\epsilon_\theta = \frac{\epsilon_x + \epsilon_y}{2} + \frac{\epsilon_x - \epsilon_y}{2} \cos 2\theta + \frac{\gamma_{xy}}{2} \sin 2\theta \tag{8.13a}$$

From Eq. (8.13a) it is obvious that three unique values of ϵ_θ (three separate strain gage readings) are needed to solve for ϵ_x, ϵ_y, and γ_{xy}.

If three strain gages a, b, and c were applied at a particular point with directions θ_a, θ_b, and θ_c from an arbitrarily established x axis, forming a rosette,

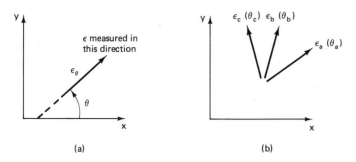

Figure 8.12 Measurements of strains and reference axes. (a) Location of strain. (b) Measured from x-axis.

as shown in Fig. 8.12b, Eq. (8.13a) can be written for each gage as follows:

$$\epsilon_{a,b,c} = \frac{\epsilon_x + \epsilon_y}{2} + \frac{\epsilon_x - \epsilon_y}{2}\cos 2\theta_{a,b,c} + \frac{\gamma_{xy}}{2}\sin 2\theta_{a,b,c} \qquad (8.13b)$$

These equations can be solved for ϵ_x, ϵ_y, and γ_{xy}. If ϵ_x, ϵ_y, and γ_{xy} are known, the principal strains ϵ_1 and ϵ_2 can be obtained from the following equations:

$$\epsilon_{1,2} = \frac{\epsilon_x + \epsilon_y}{2} \pm \left[\left(\frac{\epsilon_x - \epsilon_y}{2}\right)\right]^2 + \left(\frac{\gamma_{xy}}{2}\right)^2\right]^{1/2} \qquad (8.14a)$$

$$\tan 2\theta_p = \frac{\gamma_{xy}}{\epsilon_x - \epsilon_y} \qquad (8.14b)$$

Equation (8.14) can be derived either by maximizing ϵ in Eq. (8.13a) or using Mohr's circle for strain as shown in Fig. 8.13. These strains can be converted into stresses using the stress-strain relations of Eq. (8.11).

The general procedure is now established. One places three strain gages

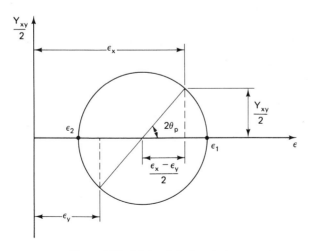

Figure 8.13 Mohr circle for strains.

in some arbitrary (but known, measurable, and different) directions. Corresponding strain measurements (ϵ_a, ϵ_b, ϵ_c) in these three directions are made and used in Eq. (8.13), which is then solved for any orthogonal strains ϵ_x, ϵ_y, and γ_{xy}. The values of ϵ_x, ϵ_y, and γ_{xy} are substituted into Eqs. (8.14) to determine ϵ_1, ϵ_2 and the direction of principal planes given by angle θ. These are then used in Eq. (8.11) to obtain the magnitude of the two principal stresses along the principal planes given by the angles θ_p and ($\theta_p + 90°$) from the x and y axes.

For practical purposes, the preferred orientations for gages a, b, and c are the rectangular or 45° rosette and the delta or 60° rosette discussed earlier. Fig. 8.14 illustrates typical commercial rosette gages.

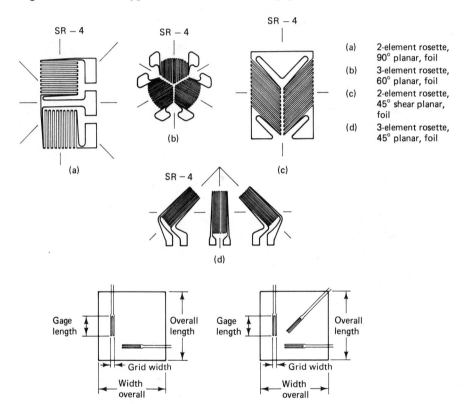

(a)   2-element rosette, 90° planar, foil

(b)   3-element rosette, 60° planar, foil

(c)   2-element rosette, 45° shear planar, foil

(d)   3-element rosette, 45° planar, foil

Figure 8.14   Various types of rosettes.

Fig. 8.15 shows the rectangular rosette geometry and the corresponding sine and cosine functions. Thus from Eq. (8.13), one obtains

$$\epsilon_a = \epsilon_x$$
$$\epsilon_b = \frac{\epsilon_x + \epsilon_y}{2} + \frac{\gamma_{xy}}{2} \qquad (8.15)$$
$$\epsilon_c = \epsilon_y$$

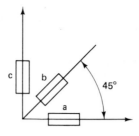

Double angle functions:

Figure 8.15 Geometry of rectangular rosette.

$2\theta_{ab}$ = $90° \rightarrow \cos = 0$ and sin = 1

$2\theta_{ac}$ = $180° \rightarrow \cos = -1$ and sin = 0

Solving,

$$\epsilon_x = \epsilon_a$$

$$\epsilon_y = \epsilon_c \qquad\qquad\qquad (8.16)$$

$$\gamma_{xy} = -\epsilon_a + 2\epsilon_b - \epsilon_c = 2\epsilon_b - (\epsilon_a + \epsilon_c)$$

Substituting into Eqs. (8.14),

$$\epsilon_{1,2} = \frac{\epsilon_a + \epsilon_c}{2} \pm \frac{1}{2}\sqrt{(\epsilon_a - \epsilon_c)^2 + [2\epsilon_b - (\epsilon_a + \epsilon_c)]^2} \qquad (8.17)$$

$$\tan 2\theta_p = \frac{2\epsilon_b - (\epsilon_a + \epsilon_c)}{\epsilon_a - \epsilon_c}$$

These results can in turn be substituted into Eqs. (8.11) and (8.12) to obtain:

$$\sigma_{1,2} = E\left\{\frac{\epsilon_a + \epsilon_c}{2(1 - v)} \pm \frac{\sqrt{(\epsilon_a - \epsilon_c)^2 + [2\epsilon_b - (\epsilon_a + \epsilon_c)]^2}}{2(1 + v)}\right\}$$

$$\theta_p = \frac{1}{2}\tan^{-1}\frac{2\epsilon_b - (\epsilon_a + \epsilon_c)}{\epsilon_a - \epsilon_c} \qquad (8.18)$$

$$\tau_{max} = \frac{E}{2(1 + v)}\sqrt{(\epsilon_a - \epsilon_c)^2 + [2\epsilon_b - (\epsilon_a + \epsilon_c)]^2}$$

Relations between the principal stresses and strains for delta and T rosettes can be derived in a similar fashion; results for these three major types of rosettes are summarized in Table 8.5. In addition to these algebraic solutions, which are easily programmed for a hand calculator, there are a number of graphical and nomograph-type solutions as discussed by several authors, such as Murphy (1945).

Example 8.1

A delta rosette on a model of a steel bin structure has the following strain values: $\epsilon_{a,b,c} = -78, +135,$ and $+173$ microstrain, respectively. Determine principal stresses and the direction of the maximum principal strain from an axis coincident with gage a. $E = 30 \times 10^6$ psi and $v = 0.3$.

TABLE 8.5 Strain Rosette and Principal Stresses

	Rectangular	Delta	T Delta
Pattern			
Strains	Three strains measured 45° apart	Three strains measured 60° apart	Three strains measured 60° apart and a fourth one 90° to one of them

$$\sigma_{max} = K_1 A + K_2\sqrt{B^2 + C^2}$$

$$\sigma_{min} = K_1 A - K_2\sqrt{B^2 + C^2} \qquad \text{in which}$$

Relations	$\theta_p = \frac{1}{2}\tan^{-1}\left(\dfrac{C}{B}\right)$		$K_1 = \dfrac{E}{1 - v}$

$$\tau_{max} = K_2\sqrt{B^2 + C^2} \qquad\qquad K_2 = \dfrac{E}{1 + v}$$

	Rectangular	Delta	T Delta
A	$\dfrac{\epsilon_a + \epsilon_c}{2}$	$\dfrac{\epsilon_a + \epsilon_b + \epsilon_c}{3}$	$\dfrac{\epsilon_a + \epsilon_d}{2}$
B	$\dfrac{\epsilon_a - \epsilon_c}{2}$	$\epsilon_a - \dfrac{\epsilon_a + \epsilon_b + \epsilon_c}{3}$	$\dfrac{\epsilon_a - \epsilon_d}{2}$
C	$\dfrac{2\epsilon_b - (\epsilon_a + \epsilon_c)}{2}$	$\dfrac{\epsilon_c - \epsilon_b}{\sqrt{3}}$	$\dfrac{\epsilon_c - \epsilon_b}{\sqrt{3}}$

From Table 8.5,

$$K_1 = \frac{30 \times 10^6}{1 - 0.3} = 42.9 \times 10^6 \text{ psi}$$

$$K_2 = \frac{30 \times 10^6}{1 + 0.3} = 23.1 \times 10^6 \text{ psi}$$

$$A = \frac{-78 + 135 + 173}{3} = +77 \text{ microstrain}$$

$$B = -78 - 77 = -155 \text{ microstrain}$$

$$C = \frac{+173 - 135}{\sqrt{3}} = +22 \text{ microstrain}$$

and

$$\sigma_{max, min} = K_1 A \pm K_2\sqrt{B^2 + C^2} = +6920, -310 \text{ psi}$$

$$\theta_p = \frac{1}{2}\tan^{-1}(22/-155) = -4.04° \text{ (clockwise)}$$

A key question is whether the angle θ_p is from the x axis to the minor principal strain or to the major principal strain. In this case it is to the minor principal strain (or stress), as shown in Fig. 8.16a. It can be shown rather easily that the direction of the maximum principal strain will always be within $\pm 30°$ from the direction of the gage with largest (in a positive sense) strain value; this handy rule applies only to delta rosettes. If the delta gage configuration is redrawn with the largest positive strain value being on the intermediate gage, with the other gages each 60° from this direction, then the maximum principal strain (or stress) direction will always be within the 30° sector that lies on the side closest to the second highest gage (the $+135$ gage in this case), as shown in Fig. 8.16b. Users of rosette gages are encouraged to use these simple relationships to check calculated results.

Mohr's circle for stress is plotted in Fig. 8.16c.

Murphy (1945) demonstrated a graphical solution by constructing a strain circle to convert measured strains into principal strains. Equation (8.13) can be put into the following alternate form to indicate that a Mohr's circle for strains is possible.

$$\epsilon_{a,b,c} = \frac{\epsilon_1 + \epsilon_2}{2} + \frac{\epsilon_1 - \epsilon_2}{2} \cos 2\theta_{a,b,c} \tag{8.19}$$

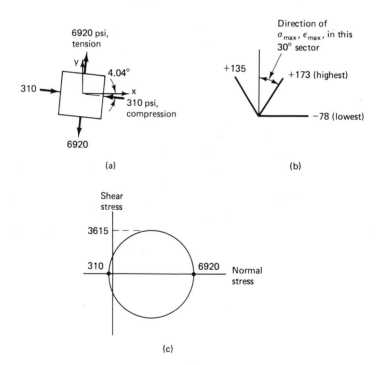

(a)

(b)

(c)

Figure 8.16 Delta rosette stress calculations. (a) Principal stresses. (b) Rosette with highest strain on intermediate gage position. (c) Mohr's circle for stress.

Since all the points ϵ_a, ϵ_b, ϵ_c must lie on a circle, they must be separated by angle 2θ (i.e., twice the angle in the rosette) and have the same orientation. Thus,

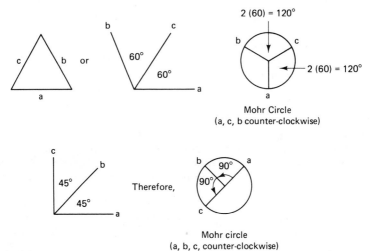

Mohr Circle
(a, c, b counter-clockwise)

Mohr circle
(a, b, c, counter-clockwise)

Example 8.2

Construct Mohr's circle for strain for the rosette with strain values of $\epsilon_{a,b,c} = +100$, $+150$, and $+300$ microstrain, respectively. The solution proceeds as follows:

1. Plot the values of strain ϵ_a, ϵ_b, and ϵ_c on a strain ($\epsilon - \gamma/2$) axis system as shown in Fig. 8.17a. We know that point A on Mohr's circle for strain must lie on line ϵ_a, point B on line ϵ_b, and point C on line ϵ_c. The problem of finding the center of Mohr's circle is addressed in the next four steps.

2. Redraw the rosette configuration to have the gage with intermediate strain value (gage b in this case) between the other two gages (a and c), and with the total included angle of the new configuration not greater than 180°. This configuration is shown in Fig. 8.17b.

3. Choose an arbitrary point P on the intermediate line (ϵ_b for this combination of strain results) and draw a replica of the gage configuration from Step 2 on the intermediate line. The gage direction must always be vertical. Extend the gage configuration lines such that the gage a extension intersects line ϵ_a at point A and gage c intersects line ϵ_c at point C. (Note that this step often involves plotting the rosette configuration pointing downward in order to have the extensions of the gage configuration lines intersect their corresponding strain lines.)

4. Construct perpendicular bisectors to the lines PA and PC as shown in Fig. 8.17d; their intersection defines the center of Mohr's circle for strain, and the construction is essentially completed.

Several points may be made about the completed Mohr's circle in Fig. 8.17d. The angular orientation of points A, B, and C is clockwise, just as in the gage configuration in Fig. 8.17b. The angles between A and B, B and C, and C and A can be shown

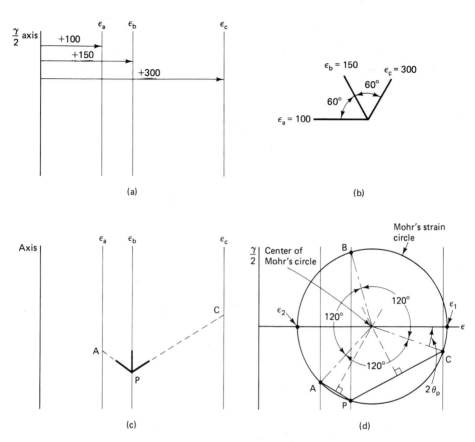

Figure 8.17 Murphy's method for plotting strain rosettes. (a) Plot strain lines. (b) Rosette drawn with intermediate strain value in intermediate position. (c) Plot rosette configuration on intermediate strain line and define points A and C. (d) Complete Mohr's circle.

to be 120°, thus maintaining the double angle relationship that always exists between Mohr's circle plots and real directions of stress and strain.

The angle θ_p is defined by the double angle $2\theta_p$ in Fig. 8.17e. Maximum principal stress direction is θ_p (about 5°) counterclockwise from the direction of gage c.

Stresses are calculated by measuring the maximum and minimum values of strain, ϵ_1 and ϵ_2, and then using Eqs. (8.11 and 8.12).

This plotting procedure is equally useful for any type of rosette and is particularly convenient to use when a "home-made" rosette is used that does not have the usual angular orientations (the standard algebraic equations in Table 8.5 will not work in such cases). The other advantage of the approach is that it forces the user to think about actual strain values and directions. Hence it is particularly useful for the new user of rosette gages who wants to get a better feel for what is going on. Quick free-hand sketching of Mohr's circle with Murphy's method will give surprisingly accurate results and will also fully clarify any questions about directions of principal strain (stress).

8.4 DISPLACEMENT MEASUREMENTS

Deflection measurements on a scaled model need more care than on a full-scale structure because the model displacements are reduced by the geometrical scale factor in comparison to the full-scale displacements. The problem becomes particularly acute with very small-scale models where displacements on the order of a few thousandths of an inch must be measured accurately.

Deflections may be measured using mechanical, electrical, or optical techniques, and recording methods vary from an individual reading by eye to a sophisticated continuous recording system. Mechanical dial gages are popular because of their low initial cost, ease of attachment, and manual recording, but they tend to be less accurate than some of the electrical methods and cannot be recorded continuously. Optical methods are relatively specialized and will not be discussed here.

8.4.1 Mechanical Dial Gages

Deflections of external surface points of a structure can be measured economically using mechanical dial gages. These are quite compact, easy to apply, self-sufficient (for readouts), and accurate. Because of these advantages they can also be used for strain measurement over a fairly large gage length, as discussed in Section 8.3.1.

In general, a dial gage consists of an encased gear train, which is actuated by the rack cut out in a spindle and follows the motion to be measured. A spring attached to this spindle maintains a small force to maintain a positive contact with the structure (Fig. 8.18). The gear train is connected to a pointer

Figure 8.18 A typical dial gage.

that indicates spindle travel on a graduated dial, generally with 0.001 in. or 0.0001 in. divisions, and with ranges of travel from 0.25 to 6.0 in.

Sometimes more than one gage may be selected to achieve the acceptable accuracy; e.g., if the range is large, say 1 in., this accuracy might only be 0.001 in. In this case, if 1-in. travel includes postyield deflections, it would be preferable to use two gages, first one with a range of 0.2 in. and an accuracy of 0.0001 in., then another one of 1-in. range to measure larger deflections. Precaution should be taken to ensure that at no stage of the test does the dial gage itself exert a significant force on the structure, particularly the very small scale models made of light gage material. By the nature of this measurement technique, it is not possible to measure dynamic deflections with mechanical dial gages.

8.4.2   Linear Variable Differential Transformer (LVDT)

The LVDT is a reasonably compact electrical device that can be used for precision measurement of displacements. The LVDT is a transformer with outer coils and a movable central core. As the core is moved from a structural displacement, it produces an electrical signal that can be recorded remotely and continuously, for both static and dynamic response. Three LVDTs being used to determine displacements of a model frame are shown in Fig. 8.19, with the outer casing of each LVDT attached to an auxiliary supporting frame and

Figure 8.19 LVDT's   for   deflection measurements.

(a)

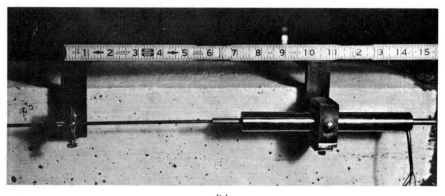

(b)

Figure 8.20 Use of LVDT as a strain gage. (a) LVDT used as strain gage in model frame. (b) LVDT used as a strain gage in prototype frame.

the movable cores attached in a spring-loaded fashion to the model. The LVDT can also be adopted to act as a strain gage to monitor strains over a moderate gage length; two applications are shown in Fig. 8.20.

Hanson and Curvitts (1965) discuss the details of LVDT instrumentation and Herceg (1976) gives considerable information on the theory and application of the LVDT. This instrument does require substantial electronics plus an appreciable amount of time for proper calibration before use, which means that in some applications the dial gage is still to be preferred. A similar device, the DCDT (direct current differential transformer) also has wide usage in modern experimental work, and special rotation meters for measuring the rotation on a beam or column are also available. These inclinometers operate on the same principles as the linear displacement LVDTs and DCDTs.

8.4.3 Linear Resistance Potentiometers

This typical transducer element is principally useful for measuring relatively large linear displacements because most potentiometers use wire-wound resistors that limit resolution; this can be increased if a film resistor is used. The potentiometer circuit, Fig. 8.21, has a constant voltage input E_i applied to the ends of a fixed resistance that has a movable slider contacting the resistance element. The output voltage E_o is taken from the slider and from one of the other terminals into a high input resistance indicator.

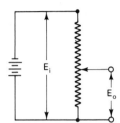

Figure 8.21 Circuit diagram for a q resistance potentiomenter. [After Herceg (1976).]

A drawback of the potentiometer is the variation of output impedance with slider position. The signal conditioning equipment must have high input impedance. A change in the applied voltage or the shunting effects of a low resistance in the output circuit changes the current through the resistance element, becoming nonlinear. Consequently, linearity may be difficult to attain when a simple meter is used to indicate slider position.

The inertia and friction present in the mechanical portion of the linear resistance potentiometer elements represent significant restrictions to the use of these devices for dynamic measurements. Furthermore, the sliding contact wears rapidly in continuous service, severely reducing reliability.

8.5 FULL-FIELD STRAIN MEASUREMENTS AND CRACK DETECTION METHODS

The electrical resistance strain gage is unequalled for measuring strain at a designated point, but in many situations the location of peak strain is not known. This situation calls for techniques that give a picture of strain distribution over a complete region of a structure or model. Such techniques are called *full-field strain measurement methods*, and include the following:

1. Whitewash (lime and water) brushed on the specimen before testing.
2. Brittle lacquers that crack at certain tensile strain levels.
3. Photoelastic coatings cemented on the surface of the structure or model.
4. Optical methods (Moire fringe and speckle holography) in which strain

fields may be interpreted from optical patterns produced by the distorted structure. These methods will not be discussed here.

Most of these techniques are also useful in detecting cracking in concrete models and structures. Two levels of accuracy are needed for crack studies in concrete—detection and location of cracks, and definition of actual crack widths.

8.5.1 Brittle Coatings

A simple solution of lime and water brushed on the surface of any structure prior to testing will help interpret the behavior. Cracks in concrete will be accentuated and yield lines in steel structures will become visible as the white-wash coating cracks and peels under high strains. A particularly useful technique for studying cracking in wall-type specimens is to apply the whitewash and then to superimpose a gridwork of black lines to help establish a background reference system for studying motion of the cracked segments of the wall under high strain fields.

A more sophisticated brittle coating is a spray-on commercially-available lacquer that will crack at a specified level of tensile strain, such as Stress-Coat [see literature from BLH (1975)]. The coating cracks normal to the direction of principal strain, thereby defining the directions of the principal stresses in the specimen or model. This makes the brittle lacquer a useful preliminary testing method prior to application of electrical resistance strain gages; with known directions it is possible to use 2-element rosettes rather than 3-element rosettes.

The brittle lacquer method has a number of advantages. Its effective gage length is essentially zero. It can be used on any surface of the structure, regardless of material, shape, or type of loading. When properly used in an environment with controlled humidity and temperature, cracking can be quite closely correlated with a certain tensile strain level, thus providing quantitative as well as qualitative information on the development of strain (and stress) with loading. Since the coating material relaxes under load, it is possible to detect high compressive strains by loading the coated structure, leaving the load in place for a number of hours, and then unloading very suddenly. The relaxed coating sees the unloading from heavy compression as a tensile load and cracks.

Brittle coatings are useful for crack detection in concrete only if the concrete cracks at a strain level below the natural cracking strain threshold of the coating.

8.5.2 Photoelastic Coatings

Birefringent materials may be bonded to the surface of structures and examined with a polariscope after the structure is loaded. The resulting optical patterns show directions of principal stresses and magnitude of shear stress (difference in principal stresses). With proper use, accurate stress analysis may be done with this method.

Several manufacturers can provide information on this technique (see list in Table 8.2). In particular, Photoelastic, Inc. provides such information in their Bulletin SFC-200. The coating used is in the form of a very thin, birefringent plastic sheet, which is attached to the surface with an epoxy adhesive. It is recommended that preliminary tests be conducted on a simple flexure specimen to get some experience with the technique before actually using it on a complicated structure.

Photoelastic coatings have been used quite successfully for detecting cracking in concrete structures before the cracks are visible to the naked eye [Abeles (1966) and Corum and Smith (1970)]. The basic mechanism of failure of concrete (or model concrete) under load is precipitated by the initiation and propagation of small cracks that extend and interconnect to form major structural cracks. As these microcracks form, the resulting strain concentrations over a very short gage length produce extremely high local stresses in the photoelastic coatings bonded to the concrete surface, and a literal explosion of optical effects (fringes) results when a crack forms.

8.5.3 Other Crack Detection Methods

Recent research has led to the development of a crack detection coating; the basic information is published by BLH (1975). The coating will crack whenever the underlying structure cracks; this is evidenced both visually and electrically. The crack detection coating is so designed that, when a crack occurs, an electrically conductive liquid is released and the crack is immediately visible. This conductive medium presents a low resistance path, so that with an electrical connection at the surface of the coating, a change in resistance or current flow between the metal base and the electrical connection can be detected by a simple ohmmeter. Its advantage is in its capability for remote observation, which can be followed (after electrical response indicates failure) by visual observation of the precise location of the crack. The technique is particularly useful for coated areas that are not readily accessible or that consist of a large area, such as a long, welded joint.

The simplest method of crack detection in concrete structures is by visual inspection, either by the naked eye or with a magnifying glass or a hand-held microscope. A graduated hand-held microscope is commercially available that enables one not only to detect cracks but also to measure their widths.

The individual strain gages of single-element type can be used for crack detection by mounting on a crack detection strip. They are connected to a continuous strip chart recorder and monitored remotely. Such a setup is suitable for crack detection of the internal surface of a vessel when it is subjected to pressure test and is not accessible during testing. It may be noted that these gages are also convenient for cracking strain measurement in concrete because the nonlinearity due to the cracking will be indicated on the recorder.

Similar to the electrical resistance gages, short-length LVDTs, described earlier, can be used for crack detection. Because of its more elaborate setup and initial adjustments, LVDT is more suitable for mounting on the external surface of the structure.

The last crack detection method to be mentioned here relies on sensing and measurement of the noise generated by the cracking process. The *acoustic emission* technique uses rather sophisticated electronic equipment. It has been used quite extensively in fracture mechanics testing and in mechanical engineering applications. It has great potential for further applications in structural testing and structural model analysis.

8.6 STRESS AND FORCE MEASUREMENT

For complete analysis in model testing, measurements of both forces and stresses are also needed. Forces are either measured directly or obtained using equilibrium principles, whereas the stresses in a structure are derived from the measured strains at a number of locations.

Various types of instrumentation are available for directly measuring different forces, such as compression or tension. They include load cells (for measuring reactions and external forces), embedded stress plugs or meters (for measuring stresses and strains inside a concrete structure), and stress-sensitive paints (between washers to measure forces by the electrical resistance of these paints). Although most of these are available commercially, often their use is precluded because of economic factors and the nature of the experiment; e.g., a load cell required for measuring reactions in a small-scale beam test may not be available in that small size or else may not fit in the available space for the measurement. In such a case, available laboratory equipment can be easily used to fabricate the required load cell. The following discussion will be focused on both commercially available and laboratory-made load cells to show that a proper combination may be made to achieve the best and most economical results.

8.6.1 Load Cells—Types and Sizes

Load cells are used for measuring loads and reactions and other forces and can be classified into three major categories, depending on the type of loading. Accordingly, they are called either *compression*, *tension* or *universal* (to measure either tension or compression) load cells. Tension cells are the simplest because of their innate stability in a state of tension and can vary from simple tension bolts, for rough estimation of forces, to more accurate and expensive load cells.

A basic load cell of any of these kinds consists of a complete strain gage bridge shown in Fig. 8.9a. The strain gages 1, 2, 3, and 4 are arranged so as to eliminate the effect of the undesired stress components. From bridge theory, the output of the bridge may be expressed as

$$\epsilon_0 = -\epsilon_1 + \epsilon_2 - \epsilon_3 + \epsilon_4 \qquad (8.20)$$

where ϵ_1 through ϵ_4 are strains of the different arms shown in Fig. 8.9.

If the strain gages reading strains ϵ_2 and ϵ_4 are placed in the load cell so as to read strains opposite in sign to strains ϵ_1 and ϵ_3, the sensitivity and accuracy of the load cell improves. This is accomplished by placing R_1 and R_3 in the axial direction of the applied force, and R_2 and R_4 in the transverse direction, as shown earlier in Fig. 8.9a. Thus

$$\epsilon_0 = K\epsilon_1 \qquad (8.21)$$

where K = bridge multiplication factor. $K = 2.6$ if Poisson's ratio is 0.3. It may further be noticed that any torsion in this cell and any bending during the applied loading is automatically eliminated and thus will not influence the output.

An ideal load cell should be compact in shape and size for a given load, be inexpensive and easy to use, have high sensitivity, have elastic behavior (for repeated use) at the upper end of the loading range, have inherent stability for long-term loading, and be relatively rigid. Several of these qualifications are generally found in the available load cells, and force measurement does not pose a problem, at least for static model testing. However, the above characteristics are discussed to help the reader plan his or her own load cell fabrication when required.

Sensitivity of a load cell may be expressed in units of strain per unit load. Thus it is directly proportional to the maximum stress used in the design of the cell and inversely proportional to its maximum load capacity. Rewriting Eq. (8.21), we have

$$\epsilon_0 = K\epsilon_1 = K \frac{\text{design stress}}{E} \qquad (8.22)$$

Then

$$\text{Sensitivity} = \frac{\epsilon_0}{\text{design load}}$$

$$= \frac{\text{design stress}}{\text{design load}} \cdot \frac{K}{E} \qquad (8.23)$$

This means that for a given design stress and design load the optimum sensitivity will result from a maximum value of K and a minimum value of E. Although the value of K can be as high as 4, in a load cell of usual design and with the advantages mentioned earlier it has a value of 2.6 with a Poisson's ratio value of 0.3. Various metals, such as aluminum and steel, are used for fabricating load cells. An aluminum load cell is three times as sensitive as a steel load cell because of its lower modulus of elasticity. But, it has a disadvantage in that aluminum has a nonlinear stress-strain relationship at a smaller design stress level than does steel.

A typical load cell, commercially available, is shown in Fig. 8.22. It consists of an enclosed tube that is instrumented with strain gages as explained earlier. The cross section of the tube depends on the highest permissible level of stress in the cell material and on the yield properties of the material. The outside cover essentially protects the core, which may be damaged during testing or as a result of environment. At both ends of the universal gage, threaded ends are provided in the core tube to provide suitable attachments for either tension or compression force measurements. A wide range of load capacities is available, ranging from loads as small as 20 lb to as high as 1,500,000 lb.

Figure 8.22   Load cells of various capacities. (Courtesy of BLH Electronics, Inc.)

For very small scale models, the commercially available load cell may pose problems with regard to size or perhaps cost, and one may construct his or her own cell. The load cell should be cycled through its design range after construction and prior to use in an experiment in order to eliminate any hysteresis in its behavior. Careful consideration should be given to the material used and to the stability of the strain gage cement as well as to the instrumentation needed over the designated loading period. Heat-cured epoxy resin cements are often used in load cell strain-gaging because of their superior stability. All load cells should be recalibrated periodically to check linearity and the calibration constant. A typical load cell made at the Cornell Models Laboratory to measure reaction is shown in Fig. 8.23. It was made using a

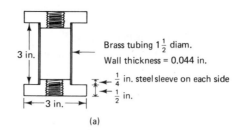

(a)

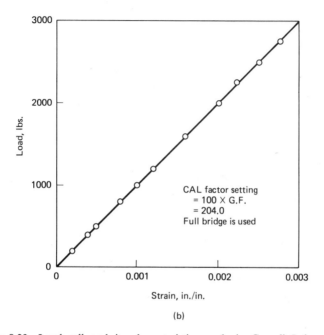

(b)

Figure 8.23 Load cell and its characteristics made in Cornell Laboratory. (a) Details of load cell. (b) Calibration chart for 3000 lbs load cell.

machined brass tube $1\frac{1}{2}$ in. in diameter and instrumented with four SR-4 strain gages to form a complete bridge. A calibration chart that gives the different ranges of accuracy for a particular set of measurements is also shown. Commercially made load cells typically specify a variety of basic characteristics. For load cell U3GI from BLH Electronics, Inc., for example, these data are:

Full-scale range	20 to 10,000 lb
Calibration accuracy	0.25 % of rated output
Repeatability	0.1 % of rated output
Creep	0.1 % of rated output
Safe operating temperature	-30 to $+175°$F

8.6.2 Embedded Stress Meters and Plugs

In order to obtain information in the core of a concrete structure (such as in dams or thick pressure vessels), one has to resort to embedded instrumentation. This is particularly important because the triaxial stress condition that exists is hard to investigate from surface strains, especially if material-related effects such as creep and shrinkage must also be studied. The embedded gage gives strain and/or stress readings at interior points in the structure to enable one to correlate the experimental work with theoretical predictions.

The problem with this kind of instrumentation is basically related to the heterogeneity that is introduced into the system because of these embedments. In addition to changing the properties of the structure in the immediate area, the instrumentation is susceptible to damage during the construction stage and requires a good deal of protection. Because of the time needed for curing of the structure, time-dependent stability of this instrumentation must be carefully considered. Reports by Corum and Smith (1970) and Geymeyer (1967) treat embedded gages.

In the *stress plug method* [Brownie and McCurich (1967)], a small volume of concrete is replaced by a plug, also made of concrete (or mortar), which is instrumented with strain gages. The plug shape and characteristics of its material are determined so that the strains can be converted into stresses. It is interesting to determine the error that may be introduced by the different properties of plug material and the surrounding concrete. Figure 8.24 shows a plug embedded in concrete. Meter and concrete properties are designated by subscripts m and c, respectively. By means of Hooke's law:

$$\sigma_c = E_c \epsilon_c, \qquad \sigma_m = E_m \epsilon_m \tag{8.24}$$

from which it can be shown that

$$\sigma_m = \sigma_c (1 + C_s) \tag{8.25}$$

and

$$\epsilon_m = \epsilon_c (1 + C_e) \tag{8.26}$$

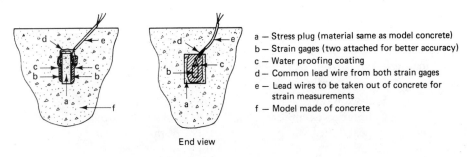

End view

a — Stress plug (material same as model concrete)
b — Strain gages (two attached for better accuracy)
c — Water proofing coating
d — Common lead wire from both strain gages
e — Lead wires to be taken out of concrete for strain measurements
f — Model made of concrete

Figure 8.24 Stress plug details (typical).

where

$$C_s = \text{stress concentration factor}$$

$$C_e = \text{strain increment factor}$$

Constants C_s and C_e are essentially the errors introduced in these measurements as a result of imperfect matching of the physical properties of the concrete and the meter. Thus, the error will be magnified because of the difference in moduli, Poisson's ratio, and thermal expansion. Although the error can be significant, by proper calibration and by keeping the mismatch to a low value, it can be reduced to a minimum. Tests have indicated reasonable accuracy— less than 0.5% variation for stresses up to 7000 psi, with a negligible drift over a 1-year period.

The *Carlson meter*, which is commercially available, has been used in the United States over a long period of time for strain and stress measurements in mass concrete. Although the original meters are somewhat too large for use in small-scale model structures, recent advances have reduced their size for successful use in large-scale models. The description as presented by Geymeyer (1967) applies to both small and large meters. The strain meter, in general, consists of a long cylinder with anchors that engage in the surrounding concrete. Inside, there are two equal coils of very fine steel music wire, 0.0025 in. in diameter, wound under a tensile stress of 100,000 psi. When the displacement at the ends of the tube (or meter) takes place, these coils undergo equal but opposite changes in resistance. When they are connected in a suitable position to a Wheatstone bridge, strain can be measured within 5 μ in./in. It also measures temperature variation up to 0.1°F. A cross section and some of the details of this meter are shown in Fig. 8.25.

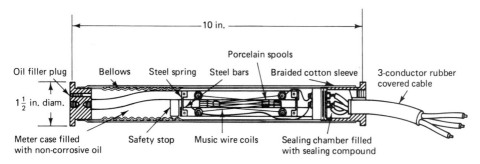

Figure 8.25 Carlson meter.

The Carlson stress meter is very similar to the strain meter, except that the pressure acting on a circular plate (Fig. 8.26) is measured. This plate is a mercury-filled diaphragm, designed so that the pressure in the mercury is always substantially equal to the pressure in the concrete normal to the plate. The center portion of the plate is made slightly flexible by cutting away part

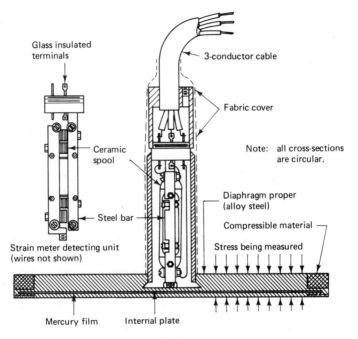

Figure 8.26   Carlson stress meter.

of its thickness. Since the mercury is in contact with the more flexible central part, it deflects in direct proportion to the intensity of the applied stress. The sensitivity and stability of the meter are excellent over a period of time.

Embedded strain gages as used by Corum and Smith (1970) are generally encapsulated either in plastic, resin, or an epoxy protection cover and may be used as single gages or rosettes depending upon the application. A typical rosette assembly is shown in Fig. 8.27. Thus foil gages are used as sensing elements and attached on each side of the epoxy strip so as to average the strains. Additional epoxy coating is given to the gage for further protection. The sensing and performance of these gages is similar to the electrical resistance gages described earlier. The embedded gage shown in Fig. 8.27 is laboratory-made since commercially available gages for small-scale models are lacking.

Another economical solution to the use of embedded gages is the application of a waterproof plastic cover to provide adequate protection to the regular strain gage. Many researchers have used these gages, however, with contradicting opinions and results. The problems inherent in using this type of gage are threefold: (1), the plastic material is not fully waterproof and might cause some drift in strain readings; (2) the mismatch of thermal expansion between the plastic and the concrete and also the self-heating effect from the gage itself produces problems; and (3) creep and relaxation of the plastic relative to the surrounding concrete could result in an apparent strain indication. Plastic-encapsulated gages, however, have a fairly good short-term stability.

Figure 8.27 Embedded strain gage. [From Corum and Smith (1970).] (Courtesy of the Oak Ridge National Laboratory, operated by Union Carbide Corporation under contract with the U.S. Department of Energy.)

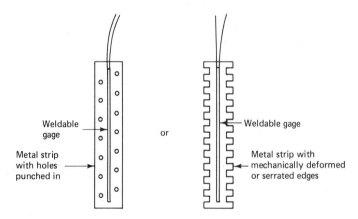

Figure 8.28 Weldable gage for embedment in concrete.

Two experimentally produced embedded gages are shown in Fig. 8.28. In each gage the sensing element is a weldable resistance gage, welded to a metal strip. The roughness of the strip permits a strong bond with the surrounding concrete and a transmission of the concrete strain field into the gage.

The *vibrating wire gage* shown in Fig. 8.29 operates on the principle of the natural frequency of a tensioned wire. The steel wire enclosed in the middle of the metal tubular body of the gage can be "plucked" with the actuation device at its mid-length, and the resulting vibrations are sensed by the same device and transmitted to a recorder. With the circular end of the gage plates cast into concrete, any strain in the concrete will result in a relative motion of the end plates and hence a change in the initial tension in the wire. The change in vibration frequency of the wire is then translated into an equivalent change in length of wire and then into effective strain.

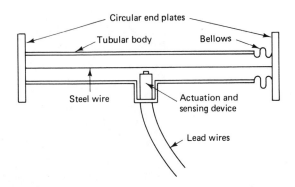

Figure 8.29 Vibrating wire strain gage for embedment in concrete.

This type of gage has seen extensive use in thick-walled prestressed concrete reactor vessels, particularly in Great Britain and on the continent. A welded stainless steel vibrating wire gage has been fabricated and tested to have the following characteristics: resolution of 0.1 microstrain with the best available recording equipment, drift of one microstrain per year at room temperature, and drift of about 10 microstrain per year at 200°F. These are rather remarkable specifications, particularly at low temperatures, and the chief disadvantage of the gage is its relatively high cost.

8.6.3 Other Measuring Devices

The proving ring, a device resembling a short length of metal pipe, is used to measure loads by sensing its change in diameter when compressed or extended. The diameter change is measured with a dial gage. Provided that applied loads are kept within the elastic limit, this type of device is very accurate and repeatable.

A newer stress measuring device is based on stress-sensitive paints that generates an electrical output when loaded in compression. The paint is applied to either washers or wafers made of steel or aluminum. A typical wafer-type device (called *Micro-ducer*) is shown in Fig. 8.30, also shown is its calibration chart between the current measured in milliamperes and the applied load. The circuit is similar to a balanced Wheatstone bridge. This is basically a miniature load cell and can be used successfully for measuring very small loads. Recently, washer-type load cells have been developed to measure loads up to 250 lb. The washers (or miniature load cells) have a diameter of 1 in. and a thickness of only 0.15 in.

The load-indicating washer (shown in Fig. 8.31) is very similar to the regular washer. They have three small protrusions that are compressed elastically when the load is applied; hence they can be calibrated and used repeatedly. They are commercially available for full-scale load measurements; however,

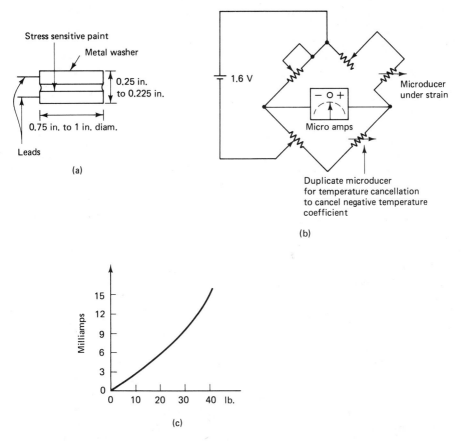

Figure 8.30 Microducer used as a load washer. (a) Microducer. (b) Circuitry. (c) Relation between load and current.

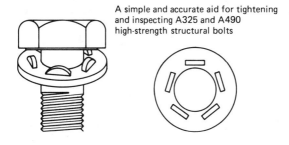

A simple and accurate aid for tightening and inspecting A325 and A490 high-strength structural bolts

Figure 8.31 Load indicator washer. (Courtesy Bethlehem Steel Co.)

this principle can be extended to smaller washers for measuring loads in small-scale models.

8.7 TEMPERATURE MEASUREMENTS

Temperature is an important quantity, particularly in the case of prestressed concrete reactor vessels or containments. In these structures, the temperature variations in the concrete is of interest to the researcher. Since it is difficult to calculate temperature inside the mass, it is determined by actual measurement.

Thermocouples and thermistors are generally used for measuring internal temperatures in concrete. The basic principle of the thermocouple is that an electric current is maintained in a circuit of two dissimilar metals when their junctions are held at different temperatures. The substantial variation in the electromotive force with temperature gives the thermocouple practical importance in that it provides a device with an accurately measurable electrical potential over a wide range of temperatures.

A thermistor, in principle, is a thermally sensitive electric resistor based on the semiconductor effect. The resistance of a thermistor decreases with an increase in the temperature. Advantages of the thermistor are the reduced degree of amplification required, reduced size, and the thermal inertia of the sensing unit compared with conventional thermal resistors.

A more recent development in this area is the thermistor-type thermometer and the method developed for embedding it in concrete. The thermistor is enclosed in a brass tube surrounded by a length of rubber welding hose, which protects the leads against mechanical damage and moisture movement. The whole assembly is moistureproofed and has moistureproof leads. Moisture-proofing is accomplished by sealing the thermistor and the PVC insulated leads in a thermoplastic material and applying a layer of rubber cement between the brass tube and the moisture tight rubber welding hose (Fig. 8.32). The whole assembly is prepared, calibrated, and ready-mounted in the required sections. All thermometer cables belonging to each section are enclosed in a $\frac{3}{4}$-in. compressed-air rubber tube, connected from the section to the readout unit. A separate thermometer is connected to an individual cable via the distribution box at the other end of the compressed-air tube. These are mounted on each section during concreting. It is obvious that the thermistors are not recoverable after the test.

Sometimes it is necessary to create differential temperature conditions across the section of concrete under test to reproduce operating or extreme accident conditions. Since concrete has a small thermal conductivity, the desired gradient cannot be obtained by merely artificially heating one surface, and cooling has to be done internally by artificial means. In several tests by Labonte (1971) on models of the anchorage zone of a prestressed concrete

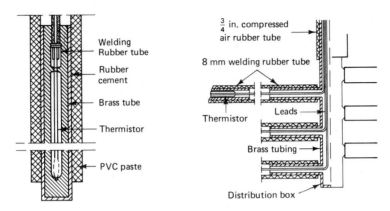

Figure 8.32   Installation of thermistors in concrete.

containment, the desired gradient was successfully created by embedding thin copper tubing across the wall and running boiling water through the tubes until the operating temperature was reached.

8.8 CREEP AND SHRINKAGE CHARACTERISTICS AND MOISTURE MEASUREMENTS

Under long-time loading on a concrete structure, in addition to strains from the applied load, there are additional strains due to creep and shrinkage. Simultaneous observations should, therefore, be taken to isolate the effect of these quantities. As described in the earlier sections, the necessary strains can be measured without too much difficulty. Very often, only mechanical gages are used since the electrical ones are usually not reliable for measuring strains over long periods.

Information on the progress of drying in massive concrete structures is of considerable value, not only in assessing the consequent changes in concrete properties (e.g., effectiveness as a neutron shield, thermal expansion, etc.) but also in determining its shrinkage characteristics. The latter may lead to stress development, cracking or dimensional changes, and loss of prestress.

Electrical resistance moisture gages have been used to measure the changing moisture distribution within concrete structures. The basic principle in this method is the relation between the resistance of the concrete and the concrete moisture content. In general, the contact resistance between the electrodes and the concrete may exceed the actual resistance of the moist concrete by a considerable magnitude. To overcome this difficulty, the absorbent type of gage shown in Fig. 8.33 has been used. The electrodes, instead of being placed

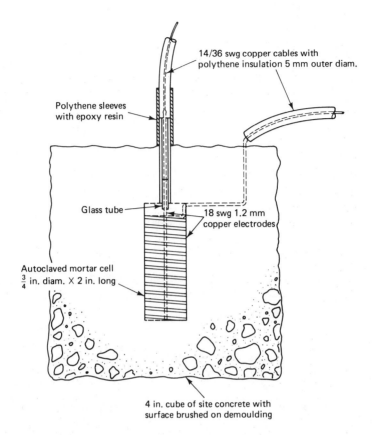

14/36 swg copper cables with
polythene insulation 5 mm outer diam.

Polythene sleeves
with epoxy resin

Glass tube

18 swg 1.2 mm
copper electrodes

Autoclaved mortar cell
$\frac{3}{4}$ in. diam. $\times$ 2 in. long

4 in. cube of site concrete with
surface brushed on demoulding

Figure 8.33   Details of moisture gage embedded in concrete.

directly in the concrete, are enclosed in an absorbent material, which in turn is
buried in the concrete. To eliminate the contact resistance between the absorbent
material and the concrete, a concentric electrode system is used. With such an
arrangement, the relationship between the gage resistance and the concrete
moisture content is no longer a direct one. Any change in the moisture content
of the concrete is normally accomplished by a change in partial pressure. A
pressure gradient between the gage and the concrete is thereby created, which
in turn causes a change of moisture content within the absorbent material and
a corresponding change in the electrical resistance.

8.9 DATA ACQUISITION AND REDUCTION

Recording of data is one of the crucial parts of any experiment, and it must be
carefully planned and checked out well in advance of the experiment. Modern
data acquisition equipment ranges from a simple, manually operated strain

indicator box to automatic, sophisticated systems that record data continuously on magnetic tape or on paper charts. The range of satisfactory equipment for a static test is very broad because recording time is usually not crucial except when failure is approached and strains and displacements are changing rapidly. On the other hand, dynamic tests require a data acquisition system that is capable of monitoring and recording many channels in a fraction of a second. The rather specialized topic of large-scale dynamic instrumentation systems is beyond the scope of this book, not only because of the complexities of the topic, but also because state-of-the-art systems change so rapidly that standard book treatments are quickly obsolete. Information from equipment manufacturers is of crucial importance in selecting a new data acquisition system.

Reduction of data must be considered in selecting data acquisition systems. Numbers (strains, displacement, etc.) that are written down by hand must later be reduced by hand or after putting the data manually into computer storage; both processes are prone to human error and should be avoided whenever possible. Fortunately, the electronics revolution has made it possible for even a modest-budget laboratory to have a data acquisition system tied to a desktop calculator, to a microcomputer, or to a small minicomputer. These systems not only make data acquisition itself much easier, but perhaps even more importantly, the data is stored permanently on magnetic tape, ready for reduction and conversion into stresses, stress resultants, and plots and tables that can be used directly in reports on the experiments.

Data acquisition systems may be classified as follows:

1. Intermittent

2. Semicontinuous

3. Continuous

Intermittent recording indicators are usually manual; i.e., each strain is read manually and recorded separately, one at a time. The strain indicator is usually connected to switch and balance units to permit multiple channels to be read by a single indicator. The indicator contains the Wheatstone bridge circuitry, internal resistances for completing the bridge when half or quarter bridges are used without external dummy gages, and other electronics and mechanical gear needed to convert the electrical signals into numerical values of strain. Each switch and balance unit accomodates a number of strain gages (usually 10, 20, or 30). It also has bridge completion resistances and terminals that permit the use of external dummy gages when needed. The balancing capability permits zero values to be set for each channel to facilitate later reduction of data. Commercially available indicator units are very compact and easy to use, and are employed extensively when a small number of readings is to be taken. A typical indicator is shown in Fig. 8.34. The reader is referred to commercial literature for more details.

Figure 8.34 Digital strain indicator. (Courtesy of Measurements Group.)

Semicontinuous recording systems permit a large number of measurements from strain gages or other sensing devices to be made rapidly. Commercially available units have up to hundreds of channels and consist of a power supply, signal conditioning circuitry such as bridge completion units for strain gages, a multiplexer or scanner, a digital voltmeter with a display unit, and a data storage capability such as a magnetic tape. The scanner sequentially scans many channels per second (typically 10 or 20) and controls the power supply and voltmeter for each channel. Sensing devices that may be used include strain gages, LVDTs, DCDTs, potentiometers, load cells, thermocouples, and any other device that generates a voltage. A printer and at least one plotter are the usual accessory devices for this type of system. While it may be tempting to let this automatic system acquire all data during a test without spending any time looking at the data, this is the wrong approach. Critical quantities (pressure during a pressure test; maximum displacement during a beam, frame, or slab test; load during a prestressing operation; etc.) should be output on recorders so that the progress of the test can be continuously monitored while it is running.

A continuous recording system records measurements continuously on magnetic tape or on a paper chart for later reduction and analysis. This type of system also accepts essentially any electrical device that generates a voltage. Important applications in structural testing include dynamic testing (where the analog-to-digital conversions are then done later), recording of complete moment-rotation relationships for beams and other structural components subjected to reversing loads, and monitoring of instruments as failure occurs.

Before considering a specific application in data acquisition systems, it is worth pointing out that tape recorders and television cameras are often useful in helping document experiments. The tape recorder may be used for storing visually read data and for recording remarks by the test engineer on the progress of the test. A video recorder, or a movie camera, is useful in capturing the behavior of a test specimen, particularly at the moment of failure. Stop-frame analysis after the experiment is over may well lead to a greatly improved understanding of the precise behavior modes during the failure process.

8.9.1 Various Data Acquisition Systems

Many different commercially available data acquisition systems are in use in structural testing laboratories. Some are used in the as-furnished condition, but most have at least some custom electronic components and interfaces to suit the needs of the individual users. The services of a good electronics technician are usually indispensable in getting a modern data system operating properly.

A general block diagram for an automatic data acquisition system is given in Fig. 8.35 [adapted from Orr and Breen (1972)]. The front end circuits are the part of the system that are most often custom made. In a typical unit that accommodates 100 separate data sensors, 75 would be bridge completion circuits for strain gages and 25 would be designed to take other transducers such as LVDTs. The scanner, which is a high speed switching device, controls the analog electrical signals going into the integrating digital voltmeter (IDVM in the block diagram). The IDVM is a crucial component in the system in that it must maintain high accuracy while operating at high scan rates (many channels per second). If it has adequate accuracy and resolution (i.e., down to 1 microvolt), then amplification of the data sensor electrical signal is not necessary. The controller is a microcomputer or minicomputer; it may or may not be linked to a mainframe computer in addition to data storage and display devices.

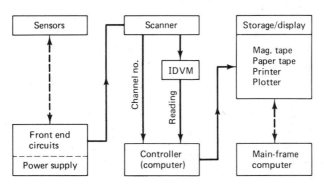

Figure 8.35   Block diagram for automatic data acquisition system.

A more specific block diagram for the Hewlett Packard 3052A automatic data acquisition system used at Cornell University is shown in Figs. 8.36 and 8.37. Various types of analog electrical inputs are shown flowing from sensors into scanner and front end circuits. The scanner units and front end circuit boxes are in modules of 40 channels each with 30 channels for strain gages and the other 10 for other transducers; once again the front end circuit boxes are custom made. 160 channels may be used for any given test and the system is

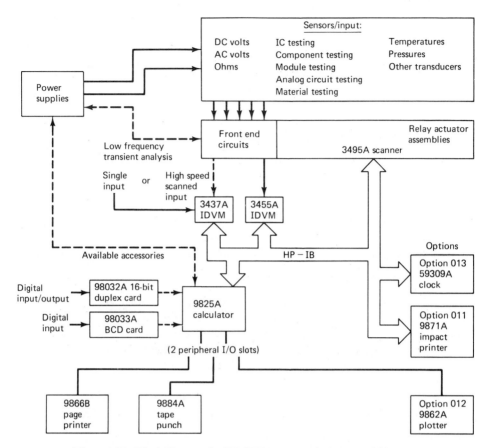

Figure 8.36 Block diagram for HP 3052A automatic data acquisition system.

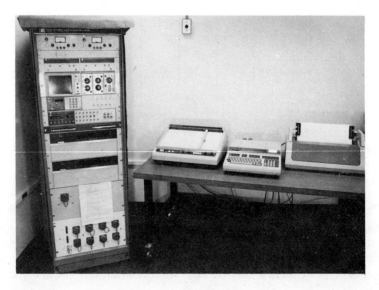

Figure 8.37 80-channel automatic data acquisition system with printer and plotter.

expandable as needed. The system incorporates two integrating digital volt-meters—one with high accuracy and a scan rate of 19 channels per second (3455A), and a second system 3437A voltmeter that has lower accuracy but a higher scan rate of 4900 readings per second from a single channel or 125 channels/second when scanned.

Two power supplies are used in the Cornell system—one that operates at a nominal two volts for strain gages, and a second operating at higher voltages for LVDTs and other transducers. The power supplies, scanners, and IDVMs are all controlled by a desktop calculator (either 9825A or 9845T) connected to the components through the HP-IB interface bus. Either calculator can be programmed to read the various channels in any desired sequence. Most experiments are custom programmed to permit interactive communications between the test engineer and the data acquisition equipment.

The digital data output is stored in the calculator memory and then buffered into an integral magnetic tape storage unit. It is also printed with a 9871A high-speed impact printer. Selected channels may be plotted on XY recorders during the experiment. The digital clock option shown on the right of the block diagram can be used as a 24-hour clock to record times during an experiment or to pace the system when timing to the nearest second is sufficient.

Such a system has several key characteristics that are indispensable in performing experiments with many data sensors: (1) high scan rate to capture a complete set of data in a minimum elapsed time; (2) data storage and display capabilities; (3) completely flexible control by an easily programmed computer; (4) sufficient calculation capabilities to permit subsequent reduction, tabulation, and plotting of results; and (5) portability and no need for air-conditioning (although the environment should be as clean as possible).

The HP calculators used in the system described above are also capable of being interfaced with the loading equipment to completely control the application of load. This is particularly important on experiments that involve repeated or reversed loading cycles. The MTS servo-controlled loading system at Cornell is now controlled by either the 9825A or 9845T calculators. This capability affords an important new dimension in testing capability, freeing the test engineer from the many routine (and energy-sapping) tasks involved in loading and measuring the response of a structure or model.

The last system to be discussed here is a high-speed data acquisition system developed at the University of California at Berkeley and described by Stephen and Bouwkamp (1975). Quoting from their paper,

> In case of the *high-speed system*, a relatively high rate of recording is desirable for simulating the earthquake response in the structure. In addition, concurrent readings of all channels are required during inelastic responses in order to avoid the observational difference due to structural creep. The high speed system is designed to read a maximum of 128 transducers at a sample rate of 20,000 samples per second. The data acquisition and control center is located in a con-trolled environment room adjacent to the main structural test bay, and has large

windows overlooking the laboratory to facilitate coordination with the test site. All of the sensitive electronic equipment is located in this dust-free room.

A unique feature of this data processing system is a multi-channel analog graphical display system. The most meaningful presentation of transducer data often is a plot of one or more transducer readings against another reading, for example, one or more displacements versus load. Often it is necessary to observe structural behavior during a test program. The data to be displayed may be a function of several transducer readings. With the fast computer processor and multi-channel display system, up to 8 XY plots may be created and updated almost simultaneously. The system is also adaptable to any type of recording device, including XY recorders, CRT (cathode ray tube) displays and film recording equipment.

8.10 SUMMARY

A brief introduction to the instrumentation needed for a successful structural model study has been given. Techniques for the measurement of strains, deflections, and temperatures and some examples of commercially available equipment have been presented. The measurement of strains and their interpretation to obtain stresses in the model have been discussed. It has been emphasized that the correct measurement and interpretation of relevant physical quantities is a crucial step in the modeling process. Several large-scale automated instrumentation systems have been described, which help interpretation and the data reduction process.

PROBLEMS

8.1. A cantilever beam is loaded at its end by a load of unknown magnitude and angle of inclination.

Describe strain gage location and type of circuitry to measure:
(a) Axial load component
(b) Shear force
(c) Bending moment at A
All circuits should have inherent temperature compensation.

8.2. A tubular steel section (round pipe) is to be used for a transducer to measure applied torque. Design the layout of strain gages (number and location) and the circuitry, and express the voltage output of the bridge as a function of the strain change seen by the gage. Is the system you have selected sensitive to axial force? Is it sensitive to bending moment? The circuit should have inherent temperature compensation.

8.3. Establish a circuit using a single Wheatstone bridge to measure the value of P on the cantilever beam below, where P is a load applied somewhere in the outer half of the beam length. *Hints:* Shear may be expressed as differences in moment, that is, $V = dM/dx$. Provide temperature compensation.

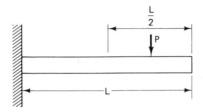

8.4. Derive equations for principal stresses σ_1 and σ_2, maximum shear stress τ_{max}, and the direction of principal stress θ_p for a rectangular strain gage rosette with gages a, b, and c as shown:

8.5. A delta rosette has the strains shown below. Sketch freehand an approximate solution to the principal strains (using Murphy's graphical method), and indicate the approximate principal strains and their directions on a sketch of a free body element properly oriented in direction. (You need not take the time to scale values and make an accurate plot of Mohr's circle for strain—a quick sketch is fully adequate.)

strains in micro in./in.

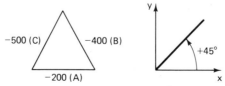

8.6. A delta rosette has the strains as shown below. Compute principal stresses and θ_p from the expressions derived in text. Then apply Murphy's method and compare results. Use $E = 10,000$ ksi and Poisson's ratio $= 0.32$. What are the stresses at an angle of $+45°$ from the x axis? Give answers on an element sketch.

8.7. A rosette made of three separate gages has the angular orientation and strain values as shown below. For $E = 29 \times 10^6$ psi and $v = 0.3$, determine the values and directions of principal stress and the maximum shearing stress. Give answers on elements at the proper orientation.

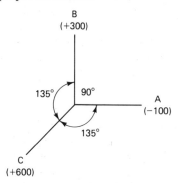

8.8. A deep, thin-webbed plate girder is instrumented with strain gages aligned in the x direction to measure the strain distribution at high load levels when it may become nonlinear. Gage 1 is located on the top flange. The beam is subjected to combined bending and shear.

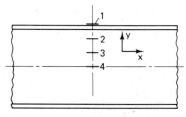

Discuss the effects of combined bending and shear on the strains measured by each gage. *If bending strains only are needed,* is the gage layout satisfactory, or does shearing strain have an effect on the strain output of the gages? Assume a vertical loading was applied to the beam in the vicinity of the gages. How does this affect the bending strains? Be specific.

Your assistant suggests that gages be installed at $\pm 45°$ to measure shear effects in the beam. Would this arrangement be satisfactory, or would the readings be influenced by the bending strains present in the beam?

8.9. For the rosettes shown below:

(a) Apply the equations for principal stresses and θ_p.

(b) Check your results by constructing Mohr's circle for strain.

$$E = 29{,}000{,}000 \text{ psi}, \qquad \text{Poisson's ratio} = 0.30$$

Show your results on sketches properly oriented with respect to stress direction. Also show the stress state that produces maximum shearing stress.

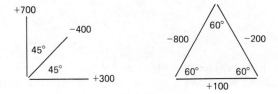

8.10. The rosettes below were used to measure strains on a Plexiglas model with $E = 450,000$ psi and $v = 0.35$.

(a) Delta rosette

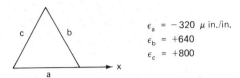

$$\epsilon_a = -320 \ \mu \ \text{in./in.}$$
$$\epsilon_b = +640$$
$$\epsilon_c = +800$$

Solve for $\epsilon_1, \epsilon_2, \sigma_1, \sigma_2, \tau_{max}$, and θ_{1-x} by (1) derived expressions and (2) graphical method.

(b) Rectangular rosette

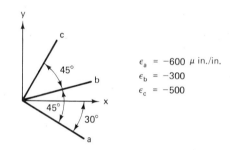

$$\epsilon_a = -600 \ \mu \ \text{in./in.}$$
$$\epsilon_b = -300$$
$$\epsilon_c = -500$$

(a) Using the graphical method, solve for $\epsilon_1, \epsilon_2, \epsilon_x, \sigma_x, \tau_{xy}$, and θ_{1-x}.

(b) What are the stresses at an angle of $45°$ counterclockwise from the x axis? Give your answers on a sketch of an element.

8.11. Using graphical construction, *sketch* a solution to the strain field as measured by the rosette shown below. You can do this *freehand*, without accurate scaling or use of a compass. Estimate the principal strains and show them in their proper orientation and directions on a sketch of an element.

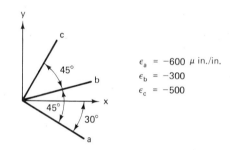

$$\epsilon_a = +1000 \ \mu \ \text{in./in.}$$
$$\epsilon_b = -700$$
$$\epsilon_c = +700$$

8.12. One of the difficult problems faced in strain gage instrumentation is the extremely small change in resistance produced by strain in the gage. Using the basic definition of gage factor F in Section 8.3.3.2, determine the change in resistance produced in a 120 ohm resistance gage with a gage factor $F = 2.1$ when the gage is applied to a steel member and is stressed by 100 psi.

8.13. A load cell is made from a 2 in. outside diameter steel pipe with 0.08 in thick walls and is instrumented with a full bridge (2 gages in the axial direction and 2 gages in the Poisson configuration). If the bridge is powered with 2 volts, gages are 120 ohms, and gage factor is 2.05, what is the output of the bridge in volts? Use modulus = 30,000,000 psi.

8.14. The blade of a metal screwdriver is to be instrumented with electrical resistance strain gages to measure the ratio of torque to axial force needed to drive screws into various species of wood. Using no more than two separate circuits with no more than four active strain gages, show the location of gages and the circuits needed to measure the axial force and the torque, and prescribe the Wheatstone bridge output in terms of the strain values ϵ_1, ϵ_2, ϵ_3, and ϵ_4. Each circuit should have inherent temperature compensation. You may assume that any accidental bending strains may be neglected.

8.15. The modulus of elasticity of thin rigid planished vinyl plastic sheeting was measured two ways—by using foil electrical resistance strain gages, and with an extensometer mounted on the tensile specimen to measure the increase in length with increasing tensile load. The results were as follows:

Thickness	Modulus E
0.010 in.	728,000 psi (with strain gages)
	425,000 psi (with extensometer)
0.015 in.	544,000 psi (with strain gages)
0.020 in.	493,000 psi (with strain gages)
	425,000 psi (with extensometer)

The differences are produced by the stiffening effect of the gage on the plastic. Both the metallic strain-sensing material (foil) and the epoxy backing of the gage have elastic modulus values higher than that of the vinyl model material. The problem is accentuated with extremely thin material, such as that listed above (which was used in building a model of a steel bin structure).

What value of E would you use in reducing strain gage readings to stresses?

Accuracy and Reliability of Structural Models

~~~~~~~~~~~~~~~~~~~~~~~~~~~~~~~~~~~~~~

9

9.1 GENERAL

We have so far presented basic information on the laws of similitude, various materials for model fabrication, instrumentation, etc., needed by the models engineer to understand and to undertake a research or design project that will enable him or her to predict the behavior of a prototype structure. However, many users of structural models may raise questions as to the reliability of such a program. Once the basic question "Why a model?" is answered, other questions relating to the reliability and accuracy are automatically posed. These include:

> How accurate must the results of a model study be to satisfy the objectives of the program?
>
> What can one do to optimize the chances that the results from any particular model study will be accurate?
>
> What are the various factors that affect an acceptable accuracy?

Furthermore, the value of the results, either from a research or a design model study, is related to the confidence that the engineer can place in the model study. This confidence level will in turn depend on the errors that are introduced at various stages of the modeling process. These errors are summarized in Fig. 9.1, which is taken from Pahl (1963). It is seen that certain types of errors and their causes occur in the total study. This chapter deals with

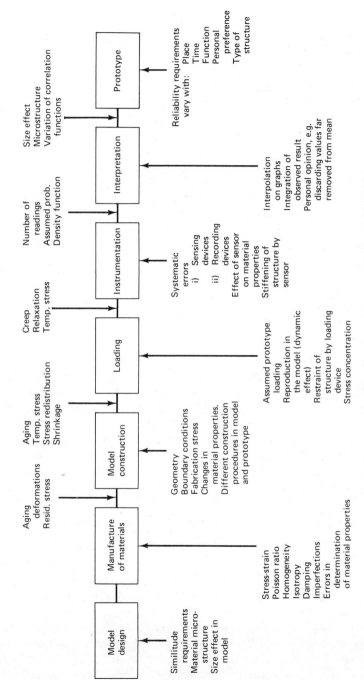

Figure 9.1 Parameters affecting model confidence level. [After Pahl (1963).]

various types of errors and their causes and focuses on the questions posed above.

9.2 ERRORS IN STRUCTURAL MODEL STUDIES

In earlier chapters, it was noted that structural model analysis can be considered as a process incorporating a sequence of five steps, namely: (1) planning, (2) fabrication, (3) loading, (4) data recording, and (5) interpretation and/or extrapolation to the prototype. Errors may enter in each of these five steps, and the following is a list of possible error sources. It is not exhaustive but merely illustrative.

1. Planning
   (a) Mistake in dimensional analysis
   (b) Failure to recognize a relevant variable
   (c) Mistake in proportioning model
   (d) Choice of inadequate material

Planning is very important, since all variables involved in the model project should be accounted for as adequately as possible at this stage. Some variables may be eliminated, however, after preliminary testing, dimensional analysis, or other relevant considerations are made. In the planning phase, the models engineer must define the scope and, thus, the acceptable accuracy of the rest of the project. *Errors* in this phase are of major concern and should be avoided.

2. Fabrication
   (a) Geometry: thickness, length, etc.
   (b) Material properties
       (i)   Poisson's ratio, for example, $\nu_{\text{plastics}} = 0.3$ to 0.5 as compared with $\nu_{\text{concrete}} = 0.15$ to 0.2
       (ii) Modulus of elasticity
       (iii) Complete stress-strain-time characteristics
       (iv) Coefficient of thermal expansion
       (v)   Microscopic and macroscopic structure
       (vi) Creep characteristics
       (vii) Initial stresses

In the *fabrication* process, material properties should be very carefully modeled. If the experiment is short-term and loads do not exceed service load conditions, the behavior of the structure is approximately elastic, therefore, material properties become somewhat secondary. The geometry should be reproduced with minimal error to give maximum accuracy. In ultimate strength models, the correctness of geometry or dimensions become paramount, especially with respect to the placement of steel reinforcement in concrete structures.

3. Loading
   (a) Boundary conditions
   (b) Magnitude of load
   (c) Direction of load
   (d) Distribution of load
   (e) Time history of load
   (f) Effect of gravity loading

The *loading phase* is equally important, because all loading conditions of interest for the prototype structure must be reproduced as faithfully as possible. Any errors in applying the loading will be reflected in the results, and therefore in the prediction of prototype behavior.

4. Instrumentation and Data Recording
   (a) Error in writing down data and readings
   (b) Electrical resistance gages
      (i)   Incomplete bonding of adhesives
      (ii)   Chemical attack on plastics by adhesives
      (iii) Temperature compensation
      (iv)   Calibration errors
      (v)   Inherent recording instrument error
      (vi)   Gage factor error
      (vii) Transverse sensitivity
      (viii) Current heating effect on plastic materials
      (ix)   Gage stiffening of plastic materials
   (c) Displacements
      (i)   Judgment errors in smallest division of instrument
      (ii)   Support system of recording device not compatible with magnitude of displacements
      (iii) Calibration errors
      (iv)   Inherent recording instrument error
   (d) Pressure
      (i)   Meniscus corrections in a liquid manometer
      (ii)   Improper calibration and/or reading of pressure gages
5. Interpretation
   (a) Incorrect assumption for transformation of measured surface strains into stresses, e.g., assumption of plane strain instead of plane stress
   (b) Error in reduction of data

The above listing includes a wide variety of errors, some integration of which determines what is commonly referred to as *experimental error*. In a more specific sense, however, each of the errors listed above may be considered to fall into one of three general error categories: (1) blunders, (2) random errors, and (3) systematic errors. These are discussed in the next section.

9.3 TYPES OF ERRORS

9.3.1 Blunders

These are errors that have no place in scientific experiments. They are outright mistakes and should be eliminated by care and repetition of measurements. Examples of blunders would be:

1. Using incorrect logic in dimensional analysis
2. Misreading an instrument
3. Making a mistake in dimensional units
4. Mounting a strain gage in incorrect position

9.3.2 Random Errors

It is impossible to give a rigorous operational definition of randomness; however, the nature of the concept is associated with the fact that a random phenomenon is characterized by the property that its empirical observation under a given set of circumstances does not always lead to the same observed outcome but rather to different outcomes in such a way that there is *statistical* regularity among these different outcomes. In view of this vagueness, it is not surprising that several meanings have been advanced for random errors. The differences in such meanings are rather subtle, however, and one can think of a random error as the difference between a single measured value and the "best" value of a set of measurements whose variation is random. What constitutes the best value depends on one's purpose, but the best value here will always be taken as the arithmetic mean of all the actual trial measurements. It should be noted that the algebraic sign of a random error can be either positive or negative.

Random errors may arise in two rather different contexts. First, there are random phenomena associated with the statistical nature of the physical model or the property being measured. For example, the depth of 1000 W6 $\times$ 8.5 steel beams would not each be expected to equal the nominal value of 6.00 in. In fact the steel companies specify a tolerance of $\pm\frac{1}{8}$ in. so that one might expect to find a range of depths, perhaps the great majority lying between 5.9 and 6.1 in. but with an exceptional one falling outside these limits. Similarly, the yield stress in a certain portion of each of the 1000 beams would vary over a range of values, perhaps between 28,000 and 48,000 psi. Second, random errors may be introduced directly as a part of the measuring process. Examples of these errors would be: (1) variation inherent in estimating the smallest division on some measuring instrument, and (2) the fluctuation in apparent strain due to random supply voltage changes in an electrical resistance gage circuit.

9.3.3 Systematic Errors

Suppose now that the best value of the depth of the W6 × 8.5 beams is 5.99 in. Now someone comes along with an old ruler graduated in hundredths of an inch, but the ruler has been used so much that the ends have been worn very considerably. He measures the 1000 beams and finds a range of depths between 5.76 and 6.01 in. It is seen that, in addition to the inherent random error, an error that always has the same algebraic sign (in this case about 0.11 in.) has been inserted. Such an error is called a systematic error.

If the systematic error is always of constant magnitude, it merely shifts the entire range of values either up or down the scale. If it changes in magnitude during the course of the experiment, the relation of the measurements, one to another, is altered and little can be said. In the limit, as the changes become more and more chaotic, systematic error may be considered random.

Other examples of systematic error would be:

1. Improper bonding of electrical resistance strain gage
2. Support which offers moment restraint when a hinge is desired
3. Incorrect calibration of a measuring instrument
4. Use of radial pressure in place of vertical pressure
5. Effect of unknown residual stresses on the buckling of a compression element
6. Use of wrong E or v in converting strain ϵ to stress σ
7. Gage stiffening effects on plastics.

9.4 STATISTICS OF MEASUREMENTS

A rather extensive mathematical theory has been formulated that enables the engineer to make logical quantitative statements concerning the behavior of a structural system that is influenced by *random* fluctuations. It has already been stated that many of the errors involved in an experimental small-scale model study may be of a systematic nature, and hence the model results may not be amenable to statistical argument. Nevertheless, there are many experimental phenomena that are random, and the models engineer should certainly be aware of the basic techniques for the statistical treatment of random phenomena. A brief introduction is given below, and more complete treatments of this subject will be found in the works by Beers (1957), Parratt (1961), Wilson (1952), and Parzen (1960).

9.4.1 Probability Density Functions

Suppose that one had a six-sided die, and after throwing this die 1000 times it was found that the number of appearances of each side were as indicated in Table 9.1. These results could be plotted in another way, as is shown

TABLE 9.1   **Throws of a Six-Sided Die**

Side	Number of Appearances
1	181
2	158
3	162
4	167
5	170
6	162
	1000

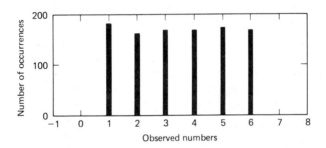

Figure 9.2   Histogram observations on 6-sided die.

in Fig. 9.2. Upon inspection of this figure it might be suspected that if the experiment had been performed 100,000 times instead of 1000, then the number of occurrences of the numbers 1, 2, 3, 4, 5, and 6 might differ by a relatively small number. In the limit when the number of occurrences approached infinity it would be reasonable to say that each of the numbers would appear one-sixth of the time. Such considerations lead directly to Fig. 9.3, where the *probability density function* (or probability mass function as it is sometimes called when the population is discretely distributed) for this experiment is shown. Two particular points should be noted with regard to Fig. 9.3: (1) the probability of any occurrence is never less than zero and (2) the summation of the probabilities of all possible occurrences equals 1. For a discrete system these two

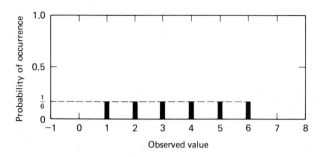

Figure 9.3   Probability density function: throws of a 6-sided die.

points can be described mathematically as:

$$0 \leq p_i \leq 1 \tag{9.1}$$

$$\sum_{i=-\infty}^{\infty} p_i = 1 \tag{9.2}$$

Of course, structural engineering problems seldom encompass situations wherein the possible outcomes are discretely distributed. For example, if one were making a measurement of surface strain on some particular structure, that strain is not by nature restricted to have a magnitude equal to some one of a number of discrete values. It may be that our measuring instruments are capable only of determining from among the discrete values such as 0.000254, 0.000255, 0.000256, or 0.000257 in./in., for example; but in fact the actual magnitude may have been closer to 0.0002554327 or 0.000255432756789 or even 0.0002554337567892742734, etc. This problem lies in the domain of a *continuously distributed* variable.

For example, if one took 20 measurements of surface strain under what seemed to be identical conditions one would not always obtain the same value. A typical set of measurements might be as indicated in Table 9.2, and as before,

TABLE 9.2   Measurements of a Certain Surface Strain

Trial	Measured Strain	Trial	Measured Strain
1	0.000255	11	0.000250
2	0.000250	12	0.000245
3	0.000240	13	0.000255
4	0.000260	14	0.000235
5	0.000255	15	0.000250
6	0.000250	16	0.000250
7	0.000240	17	0.000255
8	0.000250	18	0.000240
9	0.000245	19	0.000260
10	0.000270	20	0.000245

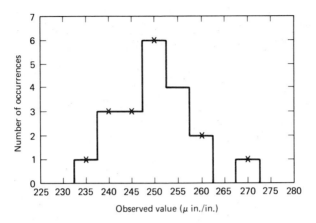

Figure 9.4 Histogram: observations of strain measurements.

this table can easily be transformed into a histogram, as shown in Fig. 9.4. In Fig. 9.4 the block type of diagram is used to indicate that these measurements come from some continuous system. If we now proceed to take more and more measurements and at the same time were able to continuously reduce the magnitude of our smallest interval (here 0.000005), the histogram would approach some kind of a smooth curve. The exact nature of this continuous curve depends upon the process that is generating the measurements. It is at this point that the experimental model engineer encounters a real difficulty. Seldom will the model engineer want to make a large number of measurements or tests; however, this very lack of a large number of tests makes it difficult to predict the true probability density function from which the measurements are being drawn. It has been found that many of the experimental measurements in science and engineering seem to follow a normal probability density function. If it can be reasonably assumed that a set of measurements has come from a population governed by the normal probability density function, then many statistical inferences can be drawn with regard to the phenomenon. On the other hand, it may be very difficult to establish the likelihood of similarity between an observed set of measurements and a normal probability density function. In these cases it may be useful to employ Chebyshev's inequality, which enables one to make certain deductions regardless of the true probability density function. The following two sections deal in turn with the normal probability density function and with Chebyshev's inequality.

9.4.2 Normal Probability Density Function

The most common continuous probability density function arising in the field of direct measurements is the normal (Gaussian) density function. Books on probability and statistics are full of discussions on the origin, derivation, applicability, and use of the normal density function; however, only certain results will be presented here.

The mathematical equation for the normal density function is

$$p(x) = \frac{1}{\sigma\sqrt{2\pi}} e^{-[(x-\mu)^2/2\sigma^2]} \tag{9.3}$$

where $x =$ random variable (observed strain in previous example).

   $p(x) =$ probability density of the particular value.

   $\mu =$ mean (or average) value for the distribution of the whole universe of measurements. Denoted as $\bar{X}$ when referring to the mean value of any finite sample.

   $\sigma =$ a measure of the dispersion about the mean value. Known as the standard deviation for the distribution of the whole universe of measurements. Denoted as S when referring to the standard deviation of any finite sample.

The mean value of any series of measurements is determined by

$$\bar{X} = \frac{\sum_{i=1}^{n} x_i}{n} \tag{9.4}$$

whereas the value of the standard deviation of a series of measurements is

$$S = \sqrt{\frac{\sum_{i=1}^{n} (x_i - \bar{X})^2}{n}} \tag{9.5}$$

In reality, S comes from the variance of a series of measurements. The variance is defined to be the average of the sum of the squares of the individual deviations from the mean value $\bar{X}$,

$$\text{Variance} = S^2 = \frac{\sum_{i=1}^{n} (x_i - \bar{X})^2}{n} \tag{9.6}$$

In passing from the sample mean and standard deviation to the mean and standard deviation of the entire population or universe of measurements, it is usually stated that the best values of universe mean and universe standard deviation are given by

$$\mu \doteq \bar{X} \tag{9.7}$$

$$\sigma \doteq \sqrt{\frac{n}{n-1}} S \tag{9.8}$$

The determination of these best values requires the introduction of some measure of what is best [Bacon (1953)]; the method of maximum likelihood provides such a measure and is generally accepted for all practical situations occurring in the measurements of engineering phenomena.

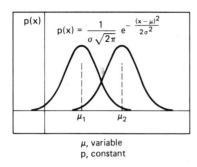
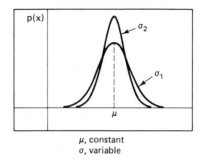

$$p(x) = \frac{1}{\sigma\sqrt{2\pi}} e^{-\frac{(x-\mu)^2}{2\sigma^2}}$$

μ, variable
p, constant

μ, constant
σ, variable

Figure 9.5 Parameter variation with normal density function.

Equation (9.3) involves the two parameters μ and σ. It is instructive to note the manner in which the probability density changes for variations in the two parameters and these variations are shown in Fig. 9.5. As was pointed out in Eqs. (9.1) and (9.2), we note that

$$0 \le p(x) \, dx \le 1 \tag{9.9}$$

$$\int_{-\infty}^{\infty} p(x) \, dx = 1 \tag{9.10}$$

If one had a random variable x that obeyed the normal law given by Eq. (9.3), one could determine the probability that x was less than some value, say t, merely by integrating the probability density function as follows:

$$p(x < t) = \Phi(t) = \int_{-\infty}^{t} \frac{1}{\sigma\sqrt{2\pi}} e^{-(1/2)[(x-\mu)/\sigma]^2} \, dx \tag{9.11}$$

If the change of variable $t = (x - \mu)/\sigma$ is made, Eq. (9.11) reduces to

$$\Phi(t) = \frac{1}{\sqrt{2\pi}} \int_{-\infty}^{t} e^{-(t^2/2)} \, dt \tag{9.12}$$

Equation (9.12), which is denoted as a cumulative probability distribution function in that it accumulates all the probability density from $-\infty$ up to t, can alternatively be set in several forms. One common form is

$$\text{Error function } t = \text{erf}\,(t) = \frac{1}{\sqrt{2\pi}} \int_{-t}^{t} e^{-(t^2/2)} \, dt \tag{9.13}$$

The probability density function, which is the parent of the cumulative probability distribution function, is shown in Fig. 9.6, and a very short table of numerical values for Eq. (9.12) is given in Table 9.3.

Of course, the question always arises as to whether a series of measurements has actually been drawn from a normal density function or for that matter any other mathematical representation of a density function. Thus, for example, one might wonder whether the series of measurements represented in

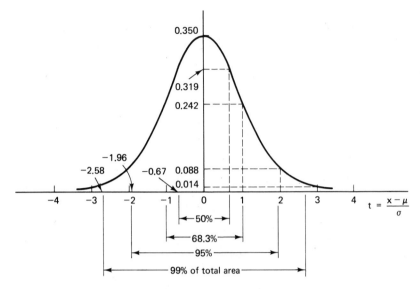

Figure 9.6 The normal density function.

TABLE 9.3   Area under Normal Density Function

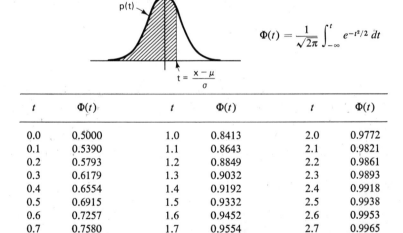

$$\Phi(t) = \frac{1}{\sqrt{2\pi}} \int_{-\infty}^{t} e^{-t^2/2}\, dt$$

t	$\Phi(t)$	t	$\Phi(t)$	t	$\Phi(t)$
0.0	0.5000	1.0	0.8413	2.0	0.9772
0.1	0.5390	1.1	0.8643	2.1	0.9821
0.2	0.5793	1.2	0.8849	2.2	0.9861
0.3	0.6179	1.3	0.9032	2.3	0.9893
0.4	0.6554	1.4	0.9192	2.4	0.9918
0.5	0.6915	1.5	0.9332	2.5	0.9938
0.6	0.7257	1.6	0.9452	2.6	0.9953
0.7	0.7580	1.7	0.9554	2.7	0.9965
0.8	0.7881	1.8	0.9641	2.8	0.9974
0.9	0.8159	1.9	0.9713	2.9	0.9981

the histogram of Fig. 9.4 could have come from a normal density function. Clearly, the histogram does not have the exactly symmetrical bell shape, but then since only 20 observations were made one might not expect to obtain a symmetrical histogram. On the other hand, a normal density function would allow for strains of the order of -5000.000000 or 230.246254, although the probability associated with such magnitudes would be small indeed. Physically our problem may convince us that there is absolutely no possibility of having negative (compressive) strains, just as there is absolutely no possibility of having the height of a human being be less than 0 feet 0 inches. Two questions must be answered with regard to whether one may use the normal density function to represent a set of observations:

1. What do we do about the tails of the normal density function where physically it may be known that no such tails could occur?
2. In a small sample how much variation from the symmetrical bell-shaped density function might be expected?

The first question is usually answered just by stating that the mathematical simplifications afforded by the normal density function are such that the discrepancy with regard to the tails is just to be neglected, although it should be mentioned that in certain problems such neglect may not be permissible. As far as the second question is concerned, one could calculate the skewness of the set of measurements, as shown by

$$\text{Skewness} = \frac{\sum\limits_{i=1}^{n}(x - \bar{X})^3}{ns^3} \qquad (9.14)$$

For the normal distribution the skewness would vanish; however, it would be most unlikely that this quantity would vanish for a small sample. Thus for the measurements of Fig. 9.4 one finds

$$\bar{X} = \frac{\sum\limits_{i=1}^{20} x_i}{20} = \frac{5000}{20} = 250 \ \mu\text{in./in.}$$

$$\sum\limits_{i=1}^{20}(x - \bar{X}) = 0 \qquad \text{(This is an identity.)}$$

$$\sum\limits_{i=1}^{20}(x - \bar{X})^2 = 1300 \ (\mu\text{in./in.})^2$$

$$\sum\limits_{i=1}^{20}(x - \bar{X})^3 = 3750 \ (\mu\text{in./in.})^3$$

$$S = \sqrt{\frac{\sum\limits_{i=1}^{20}(x - \bar{X})^2}{20}} = \sqrt{65} = 8.062 \ \mu\text{in./in.}$$

$$\text{Skewness} = \frac{\sum\limits_{i=1}^{20}(x - \bar{X})^3}{20S^3} = \frac{3750}{20(8.062)^3} = 0.36$$

The fact that the skewness $= 0.36$ does not establish that the measurements did not come from a normal distribution, but at the same time it does not establish that they did. Other checks such as the kurtosis and χ^2 test are useful when the sample is large, but the end result is that there are no good methods for establishing the likelihood of agreement for small sample results; and it comes down to the fact that the investigator must simply take a stand—perhaps on the basis of a histogram, or prior knowledge, or the skewness test, or blind faith—that the measurements either do or do not come from a normal density function.

Assuming that one is willing to accept that the universe of measurements from which the sample of Fig. 9.4 was drawn is in fact a normally distributed universe, Eqs. (9.7) and (9.8) can be used to estimate the best values of μ and σ as

$$\mu \doteq \bar{X} = 250 \ \mu\text{in./in.}$$

$$\sigma \doteq \sqrt{\frac{n}{n-1}} S = \sqrt{\frac{20}{19}} 8.062 = 8.27 \ \mu\text{in./in.}$$

Through the use of Fig. 9.6 or Table 9.3, one could then estimate that if a twenty-first measurement were to be taken, then with a probability of 0.95 the strain would lie between $\mu \pm 1.96\sigma = 250 \pm 1.96(8.27) = 250 \pm 16.2 \ \mu\text{in./in.}$

For several reasons, only a single probability density function, namely the normal density function will be presented in this chapter. It seems that the measurements of engineering phenomenon are fitted fairly well by a normal density function. In addition, the law is well known and tables that provide for quantitative predictions of the probability of a certain event are readily available; the law allows convenient equations for curve fitting, although such equations are not covered here; and it has the property that certain quantities derived from normally distributed quantities are themselves normally distributed.

However, as Parratt (1961) has stated rather well:

> a loud note of caution must be sounded in pointing out that the fit is typically not very good in the tails. Almost any set of trial measurements is generally bell shaped in the central region, and if interest in the statistics of the set is not very quantitative the normal approximation suffices. But if the interest is precise or if the tail regions are of special concern (as in situations where the probability of failure should be very low), a specific test of goodness of fit must be made and the reliability of the normal approximation judged accordingly.

9.4.3 Chebyshev's Inequality

Chebyshev's inequality states that the probability that any random measurement will deviate from the mean, in either direction, by more than $h\sigma$, where h is any positive constant, is less than $1/h^2$. This fact is valid regardless of the governing probability density function. Mathematically, it can be stated as

$$p[|x - \mu| \geq h\sigma] \leq \frac{1}{h^2}$$

or equivalently

$$p[|x - \mu| \leq h\sigma] \geq 1 - \frac{1}{h^2} \tag{9.15}$$

On the assumption that the distribution was normal it was found that $x_{21} = 250 \pm 16.2$ with a probability of 0.95. Let us see what would be predicted by Eq. (9.15):

$$p[|x_{21} - \mu| \leq h(8.27)] \geq 1 - \frac{1}{h^2} = 0.95$$

Thus $h = 4.47$ and $x_{21} = 250 \pm 4.47(8.27) = 250 \pm 37.0$. It is seen that the bounds given by the Chebyshev inequality are more conservative than the bounds that were obtained in accordance with the assumption that the strain measurements were governed by the normal density function. Figure 9.7 indicates the overall nature of the conservatism afforded by the use of the Chebyshev inequality and in particular shows the situation just considered.

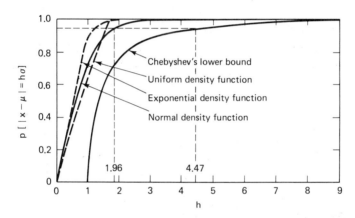

Figure 9.7   Chebyshev's inequality compared to other density functions.

If it is desired, one need not make a best estimate of the universe mean and standard deviation. Thus it is possible to take into account, through the Chebyshev inequality, the fact that the universe mean is not the same as the sample mean. This is done by considering the sample mean, $\bar{X}$, as a random variable, and setting

$$p[|\bar{X} - \mu| \leq \epsilon] \geq 1 - \left(\frac{\sigma_m}{\epsilon}\right)^2 \tag{9.16}$$

where σ_m = standard deviation of the mean
        ϵ = some positive constant greater than σ_m

If $\epsilon = k\mu$, the right-hand side can be expressed in terms of S and $\bar{X}$; and it is then possible to determine, for any desired probability, absolute bounds on the universe mean. Of course, this additional variation spreads the final obtained bounds in relation to those obtained by using Eq. (9.15) with best estimates of σ and μ.

In light of the fact that measured data may not agree with a normal density function in the tail region, it might be thought that the Chebyshev inequality would be useful in such a situation. Observation of Fig. 9.7 will reveal, however, that it is in these tail regions (i.e., regions where $p[|x - \mu| \le h\sigma] \doteq 1$) that the inequality becomes particularly conservative. In fact, it is perhaps so conservative that it may not be worthwhile.

9.5 PROPAGATION OF RANDOM ERRORS

In Section 9.4 the concern was with the statistics of measurements, i.e., with determining the nature of the probability density function from which the sample of measurements was drawn and further with determining in a probabilistic way the likely outcome of an additional, as yet unmeasured, outcome. There are many cases, however, in which it is not enough merely to know something about the measured phenomenon. For example, if one wanted to deduce experimentally information regarding the physical quantity known as stress, the customary procedure is to measure a strain and certain mechanical properties of the loaded material. Thus, if the material were linearly elastic a surface stress would be determined from the well-known formula

$$f_1 = \frac{E}{1 - v^2}(\epsilon_1 + v\epsilon_2)$$

which reduces in the case of uniaxial stress conditions to

$$f_1 = E\epsilon_1 \tag{9.17}$$

The question now arises as to whether it is possible to determine statistical relationships regarding f_1, where in fact we have no measurements of the quantity itself. Of course, the intent is to answer this question for a much more general class of situations than the simple product relationship in Eq. (9.17).

Suppose that one has a derived quantity that is related to the directly measured values of two random variables x and y. The functional relationship might have the general form

$$v = f(x, y) \tag{9.18}$$

If the deviations about the mean $\delta x = x - \bar{X}$ and $\delta y = y - \bar{Y}$ are small, then they might be considered as differentials, and the chain rule of partial differentiation may be utilized. With this assumption one obtains

$$v - \bar{v} = \delta v = \frac{\partial f}{\partial x}\delta x + \frac{\partial f}{\partial y}\delta y = \frac{\partial f}{\partial x}(x - \bar{X}) + \frac{\partial f}{\partial y}(y - \bar{Y}) \tag{9.19}$$

remembering that the equality is justified only because it is assumed that the higher-order terms can be neglected in view of the smallness of $\delta x = x - \bar{X}$

and $\delta y = y - \bar{Y}$. It should be noted that Eq. (9.19) implicitly sets $\bar{v} = f(\bar{X}, \bar{Y})$ in place of $\bar{v} = \sum_{i=1}^{n} v_i/n$. When the deviations are small, the two expressions for $\bar{v}$ are equivalent. For example, suppose that $v = xy$, then

$$v_i = x_i y_i = (\bar{X} + \delta x_i)(\bar{Y} + \delta y_i) = \bar{X}\bar{Y} + \bar{X}\,\delta y_i + \bar{Y}\,\delta x_i + \delta x_i\,\delta y_i$$

if $\delta x_i\,\delta y_i$ can be neglected in this expression then

$$v_i \doteq \bar{X}\bar{Y} + \bar{X}\,\delta y_i + \bar{Y}\,\delta x_i$$

$$\bar{v} = \frac{\sum_{i=1}^{n} v_i}{n} \doteq \frac{\sum_{i=1}^{n}(\bar{X}\bar{Y} + \bar{X}\,\delta y_i + \bar{Y}\,\delta x_i)}{n} = \frac{\sum_{i=1}^{n}\bar{X}\bar{Y}}{n} + \frac{\sum_{i=1}^{n}\bar{X}\,\delta y_i}{n} + \frac{\sum_{i=1}^{n}\bar{Y}\,\delta x_i}{n}$$

Now by definition

$$\sum_{i=1}^{n} \delta y_i = \sum_{i=1}^{n} \delta x_i = 0$$

so that $\bar{v} = \bar{X}\bar{Y}$.

It was shown in Section 9.4.2 that the standard deviation of a set of measured quantities is defined as

$$S = \sqrt{\frac{\sum_{i=1}^{n}(x_i - \bar{X})^2}{n}} = \sqrt{\frac{\sum_{i=1}^{n} \delta x_i^2}{n}}$$

According to Eq. (9.19), the standard deviation of v is

$$S_v = \sqrt{\frac{(\partial f/\partial x)^2 \sum_{i=1}^{n} \delta x_i^2 + 2(\partial f/\partial x)(\partial f/\partial y) \sum_{i=1}^{n} \delta x_i\,\delta y_i + (\partial f/\partial y)^2 \sum_{i=1}^{n} \delta y_i^2}{n}}$$

$$(9.20)$$

It should be noted that δx_i^2 and δy_i^2 are always positive, whereas $\delta x_i/\delta y_i$ may be either positive or negative. Consequently, no matter how large n is $\sum_{i=1}^{n} \delta x_i^2$ and $\sum_{i=1}^{n} \delta y_i^2$ are positive, but $\sum_{i=1}^{n} \delta x_i\,\delta y_i$ may take on a variety of values. In particular if x_i and y_i are completely *independent*, then in the limit as n becomes large one should expect $\sum_{i=1}^{n} \delta x_i\,\delta y_i \rightarrow 0$ since any $\delta x_i\,\delta y_i$ is just as likely to be positive as negative. If x_i and y_i are not independent, it becomes necessary to introduce the notion of correlation between x_i and y_i. Such correlation is measured by a quantity known as the correlation coefficient, which can vary between ± 1. When x and y are independent, the correlation coefficient vanishes.

The previous results can easily be generalized so that for *independent random* variables it can be said that

$$\bar{v} = f(\bar{X}_1, \bar{X}_2, \ldots, \bar{X}_n) \tag{9.21a}$$

$$S_v = \sqrt{\sum_{i=1}^{j} \left(\frac{\partial v}{\partial x_i}\right)^2 \delta x_i^2} \tag{9.21b}$$

In illustration of Eq. (9.21b), if

$$v = x_1 \pm x_2 \pm \ldots \qquad \text{then} \quad S_v = \sqrt{\delta x_t^2 + \delta x_2^2 + \ldots}$$

$$v = x_1 x_2 \qquad\qquad\quad \text{then} \quad S_v = \sqrt{\bar{X}_2^2\, \delta x_1^2 + \bar{X}_1^2\, \delta x_2^2}$$

$$v = x_1^a x_2^b \qquad\qquad\quad \text{then} \quad S_v = \sqrt{a^2 \bar{X}_1^{2(a-1)} \bar{X}_2^{2b}\, \delta x_1^2 + b^2 \bar{X}_1^{2a} \bar{X}_2^{2(b-1)}\, \delta x_2^2}$$

where it is noted that the partial derivatives are to be evaluated at the mean values $\bar{X}_1$ and $\bar{X}_2$, and consequently are constants. The validity of this procedure may be verified, since Eq. (9.19) was determined by taking only the first two terms in the Taylor series expansion about the mean values.

It should be noted that the derivation that led to Eq. (9.21) did not require a specification of the probability density functions of the independent random variables $x_1, x_2, \ldots, x_i$. However, having the knowledge of the mean and standard deviation of the derived variable does not imply knowledge of the probability density function of the derived variable even when the density functions of $x_1, x_2, x_3, \ldots, x_i$ are known. If one wants to know this additional information, then one must resort to the use of convolutions, or generating functions, or other less elementary techniques of the theory of probability. It may be stated here that if each of the independent random variables $x_1, x_2, \ldots,$ x_i are normally distributed, then it is true that the derived random variable of a sum or difference of $x_1, x_2, \ldots, x_i$ is also normally distributed. A similar statement cannot be made when the derived variable is a product, logarithm, square root, etc., of the variables $x_1, x_2, \ldots, x_i$. On the other hand, it is always possible to fall back on Chebyshev's inequality when one cannot easily determine the exact nature of the probability density function of the derived variable.

Example

Suppose that one were considering a prototype structure that was a simple-span prismatic beam (Fig. 9.8), the material of which is linearly elastic. The beam is subjected to a midspan load P, and it is of interest to determine the tensile stress in the bottom fibers of the midspan cross section.

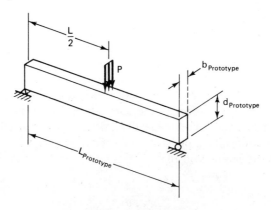

Figure 9.8 Prismatic beam.

Analytical solution. It is well known that the mathematical formulation of this problem is expressible as $f = Mc/I$, or

$$f = \frac{Mc}{I} = \frac{3}{2}\frac{PL}{bd^2} \tag{9.22}$$

It is seen that the stress is a function of four variables, three geometric and one loading. If these four variables are considered to be independent random variables with known means and known standard deviations, it would be possible to determine the mean and standard deviation of the derived quantity f by utilizing Eq. (9.21). Of course, it would be highly desirable to be able to readily obtain the probability density function of f, but such information cannot be obtained from the theory of error propagation embodied in Eq. (9.21). Thus Eq. (9.21) can be used in this example problem even if the random variables P, L, b, and d each have different types of probability density functions (e.g., uniform, normal, log normal, triangular). The equations yield no information regarding the type probability density function of f even in the case where P, L, b, and d all have the same type of probability density function (e.g., all normal). With these limitations in mind, one can write

$$\bar{f} = \frac{3}{2}\frac{\bar{P}\bar{L}}{\bar{b}\bar{d}^2} \tag{9.23}$$

$$\sigma_f = \sqrt{\left(\frac{3}{2}\frac{\bar{L}}{\bar{b}\bar{d}^2}\right)^2\sigma_P^2 + \left(\frac{3}{2}\frac{\bar{P}}{\bar{b}\bar{d}^2}\right)^2\sigma_L^2 + \left(\frac{3}{2}\frac{\bar{P}\bar{L}}{\bar{b}^2\bar{d}^2}\right)^2\sigma_b^2 + \left(\frac{3\bar{P}\bar{L}}{\bar{b}\bar{d}^3}\right)^2\sigma_d^2} \tag{9.24}$$

A quantitative probabilistic prediction regarding the stress at the midspan cross section could now be made through the use of the Chebyshev inequality.

Experimental solution. If one were to adopt an experimental investigation on a small-scale structural model as a means of determining the stress in the prototype beam, a dimensional analysis of the problem should be performed first. The physical quantity of interest is the stress, but to measure stress directly is not possible, and measurements of strain are usually taken. Then with knowledge of the elastic properties of the model material the strain measurements are transformed into stress. Equation (9.17) can be used to make the transformation in this plane stress problem. Thus it is necessary only that the model material be elastic and that we have some knowledge of the magnitude of the modulus of elasticity. Then

$$F(f, L, b, d, P) = 0$$

which according to Buckingham's theorem can be reduced to

$$\Phi\left(\frac{fL^2}{P}, \frac{L}{b}, \frac{L}{d}\right) = 0$$

or, in the solved form,

$$f = \frac{P}{L^2}\phi\left(\frac{L}{b}, \frac{L}{d}\right)$$

Of course, in this simple problem it is known that

$$\phi\left(\frac{L}{b}, \frac{L}{d}\right) = \frac{3}{2}\frac{L}{b}\left(\frac{L}{d}\right)^2 = \frac{3L^3}{2bd^2}$$

However, such information cannot be obtained from the dimensional analysis alone. Thus the model restrictions and model-to-prototype extrapolation is given by:

$$\left(\frac{L}{b}\right)_m = \left(\frac{L}{b}\right)_p \quad \text{and} \quad \left(\frac{L}{d}\right)_m = \left(\frac{L}{d}\right)_p \tag{9.25}$$

$$f_p = f_m \frac{P_p L_m^2}{P_m L_p^2} \tag{9.26}$$

Now actually the model strain is measured, perhaps by means of an electrical resistance strain gage. If the model is carefully constructed according to Eq. (9.25), and several measurements (say five) are made of the strain, there will be some variation among the individual measurements. This variation is certainly not due to variations in L, b, or d, and most likely not due to variations in P either. Thus the major part of the variation observed in the five measurements may be due to errors in the measuring system itself—a factor not present in the prototype structure. If a second model were constructed to duplicate the first model, five new strain measurements would again show some dispersion about a mean value; but in all likelihood the mean value would not be the same as the mean value obtained from the first model beam. If this procedure were repeated 10 times, a possible set of results might be as indicated in Fig. 9.9.

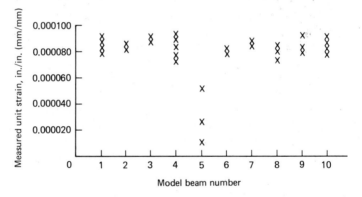

Figure 9.9 Possible results of model tests.

The question arises as to how such a set of data can and should be interpreted, keeping in mind that the real quantity of interest is the stress in the prototype beam. Further, is it possible to deduce quantitative probabilistic results with regard to the prototype stress in the same way as Eqs. (9.23) and (9.24) and Chebyshev's inequality allowed in the analytical solution? Although a closed form solution cannot be given to the first question, the following points should be mentioned:

1. The results indicated in Fig. 9.9 may be clouded with all three error types: blunders, systematic errors, and random errors. Equations (9.21) can be applied only to random phenomenon so every effort should be made to eliminate blunders and systematic errors. In this respect the results

obtained on model 5 are certainly suspect and perhaps should be eliminated.

2. Since it is felt that the variation within each model may be due largely to errors in the measuring system, a more realistic determination of the model strain could be obtained by taking the mean value and the standard deviation of the nine remaining means. Of course, there may be a systematic error present in all models (e.g., a consistent error in the setting of the gage factor dial on the strain indicator), which would lead to an equal error in each of the nine individual means. Such an error cannot be suspected merely by inspection of the experimental data.

3. In this particular problem the strain was measured only in order to be able to compute stresses. Therefore, to get the best possible value for the mean and standard deviation of the stress, test coupons of each model should be tested to determine the modulus of elasticity of the model material. After computing $\bar{f}_i = \bar{E}_i \bar{\epsilon}_i$ for each of the nine valid model beams, one could compute the mean of the means and the standard deviation of the means. Thus from Eqs. (9.7) and (9.8)

$$\bar{\bar{f}}_{\text{model}} \doteq \frac{\sum\limits_{i=1}^{9} \bar{f}_i}{9} \quad \text{and} \quad \sigma_{f \text{ model}} \doteq \sqrt{\frac{\sum\limits_{i=1}^{9} (\bar{f}_i - \bar{\bar{f}}_{\text{model}})^2}{8}} \qquad (9.27)$$

As to how one can use the information obtained in Eq. (9.27) to obtain quantitative probabilistic results with regard to the stress in the prototype, attention should be focused on the extrapolation equation deduced from the dimensional analysis. Equation (9.26) was deduced by setting

$$\left(\frac{fL^2}{P}\right)_{\text{model}} = \left(\frac{fL^2}{P}\right)_{\text{prototype}}$$

But Eq. (9.25) states that

$$\left(\frac{L}{b}\right)_{\text{model}} = \left(\frac{L}{b}\right)_{\text{prototype}} \quad \text{and} \quad \left(\frac{L}{d}\right)_{\text{model}} = \left(\frac{L}{d}\right)_{\text{prototype}}$$

so that Eq. (9.26) is certainly not a unique extrapolation equation. It could be written in several ways, for example,

$$f_p = f_m \frac{P_p L_m^2}{P_m L_p^2} = f_m \frac{P_p b_m^2}{P_m b_p^2} = f_m \frac{P_p L_m^5 d_p^3}{P_m L_p^5 d_m^3} = f_m \frac{P_p L_p b_m d_m^2}{P_m L_m b_p d_p^2} = \cdots \qquad (9.28)$$

Clearly all of the above expressions will yield the same result if they are used to extrapolate the observed mean values to obtain an approximation of the mean value of the prototype stress. On the other hand, each expression will lead to a different value of the standard deviation of the prototype stress. In fact, none of these expressions could lead to the result given by Eq. (9.23), which is known to be correct for this simple problem. Thus, we have seen that *while experimental results obtained from small-scale structural models can be used to predict the average or mean value of a particular physical quantity in a prototype structure, it is in general not possible to determine anything regarding the possible dispersion about this derived mean value.* In other words, even if the mean and standard deviation of f, P, L, b, and d were obtained by measurements on a series of model experiments and if the mean and standard deviation of P, L, b, and d in the prototype were known, it would be impossible to determine the standard deviation of the prototype stress unless the correct analytical formulation given by Eq. (9.22) were known—and in that case there is often no advantage to be gained in conducting a model study. Two further points are worthy of mention. First, in the simple beam problem that has just been discussed there is no ambiguity regarding what is meant by the depth d. As it related to this problem it was clearly the depth at the cross section under investigation. Similarly there was no indefiniteness involved in the definition of P, L, or b. It should be realized, however, that as soon as one enters into the field of statically indeterminate structures, real problems of definition arise. For example, where should the depth d be measured in a two-span beam, since the internal moment acting at the cross section of interest is some function of the depth existing along the entire beam? Or, going one step further into the woods, what is meant by the thickness of a small-scale, thin-shell model when that thickness varies in an irregular manner over the entire shell surface?

Finally, it would only be the rare occasion when 10 completely separate models would be fabricated and tested. In fact, perhaps the most serious immediate problem facing the experimental designer is that the time and cost involved in even a single model may be prohibitive. If only one model is constructed and tested, how are the results from that single test to be interpreted? What if in our simple beam problem we had tested only one model, and it had been beam number 5? It has been shown that the only advantage of fabricating and testing more than one model is to obtain a better approximation of the true model mean. If the mean is taken to be the result obtained from a single model, then it is particularly important that the investigator be convinced of the absence of blunders and major systematic errors in that single model study. Such systematic errors can enter into the model results through a variety of means, e.g., through the physical means of providing for the boundary supports, through incomplete bonding of a strain gage, through cementing a strain gage to a plastic material that is not resistant to the solvents in the cement, through switching circuits incorporating large switching resistances, through battery decay in a recording device, and through the use of radially applied loads in place of actual gravity loads. A partial check against many systematic errors can be obtained by building into the model certain internal checks. For example, it may be possible to provide for several static checks within the model. Such considerations are considered in more detail in the following section. Other indicators of blunder or systematic error are trends in the data, jumps in the data, a periodicity in the data, or a change in precision of the data; however, when only one model is to be studied, such indicators can seldom be used. Finally, it would be extremely useful to know that an extensive set of tests had been successfully carried out on a similar problem using techniques of fabrication, loading, and instrumentation similar to those proposed. Thus, the great background of experience underlying the use of electrical resistance strain gages on metallic materials leads one to have confidence in such results, whereas results obtained on foamed plastic materials might be suspect.

9.6 ACCURACIES IN (CONCRETE) MODELS

Following the above general discussion of errors in structural models, the remaining part of the chapter deals with those problems specifically related to concrete models. The topics discussed are:

1. Dimensional and fabrication accuracies
2. Accuracy-related material properties
3. Accuracy in testing and taking measurements
4. Accuracy in interpretation of test results

9.6.1 Dimensional and Fabrication Accuracies

Dimensional accuracy is an important aspect of modeling work and affects the fabrication of concrete models in two ways: first, in the making of forms and second, in the actual concreting process. There is always a question as to how accurately the fabrication process should be controlled. As geometrical scale becomes smaller, so do the absolute fabrication tolerances, if the model is to appropriately represent a physically larger specimen. In the case of shells, for example, beyond a certain scale it may just become impractical to construct an accurate mortar model. In general, a philosophy to maintain approximately the same degree of accuracy as in the prototype is desirable; a tolerance of $\pm 5\%$ is recommended to achieve sufficient accuracy. To achieve this:

1. The forms should be machined accurately from a material that maintains its dimensions with time and wet environments. Plexiglas is considered to be a very effective and relatively inexpensive material for making molds. Plastic-coated wooden forms may also be used to reduce costs. Aluminum sections can also be used for model forms, especially where stiffness is a problem. Aluminum works well in combination with Plexiglas but tends to be expensive.

2. Even after the forms are machined accurately, there may be some errors in dimensions during casting and screeding the material in the forms to obtain the required thickness. Although this does not pose a problem

for linear models (frames, beams, and columns), it is very important in spatial structures (slabs, shells).

The accurate measurement of cross sections and thicknesses, therefore, must be recorded and considered for the correlation between the model and the prototype behavior. Experience indicates that model dimensions tend to be slightly greater than design values, even when using accurately machined forms.

The dimensional accuracy in the fabrication of concrete models appears not only in the actual concreting process but also in the placement of the reinforcement. Compared with the large-scale models or prototype structures, extra care is required in this respect because of the flexibility of reinforcement and its response to the casting and vibration process during casting. In the case of underreinforced sections, wherein steel governs the failure of the structure, the accuracy will depend on the positioning of the reinforcement. Any change in position will directly influence the load capacity of the element.

Let us consider, for example, the concrete dimensions of two types of structures: linear and two-dimensional. In the case of a beam or column, screeding of the model in one direction during casting is easier than in the case of a slab or a shell, and will be reflected in the thickness measurements of these elements.

Typical tolerances obtained in cement mortar and gypsum mortar beams and columns are shown in Tables 9.4 to 9.7 [Harris, Sabnis, and White (1965)].

TABLE 9.4   Details of Mortar Beam Dimensions

Beam	Depth, Horizontal Cast			Depth, Vertical Cast		
	Designed	Actual	Ratio	Designed	Actual	Ratio
1	0.375	0.378	1.008	0.375	0.379	1.01
2	0.375	0.382	1.018	0.375	0.380	1.029
3	0.375	0.381	1.016	0.375	0.386	1.029
4	0.375	0.380	1.013	0.375	0.393	1.048
5	0.375	0.381	1.016	0.375	0.394	1.05
6	0.375	0.382	1.018	0.375	0.390	1.04
7	0.750	0.757	1.009	0.750	0.762	1.01
8	0.750	0.752	1.003	0.750	0.766	1.013
9	0.750	0.753	1.004	0.750	0.766	1.013
10	0.750	0.752	1.003	0.750	0.769	1.025
11	0.750	0.757	1.009	0.750	0.770	1.027
12	0.750	0.748	0.997	0.750	0.762	1.016
13	1.50	1.50	1.000	1.50	1.51	1.006
14	1.50	1.51	1.007	1.50	1.51	1.006
15	1.50	1.54	1.026	1.50	1.51	1.006
16	1.50	1.50	1.000	1.50	1.50	1.000

TABLE 9.5   Accuracy of Gypsum Beam Dimensions

Beam	Width			Depth		
	Designed, in.	Actual, in.	Percent Error	Designed, in.	Actual in.	Percent Error
1	1.000	1.030	3.0	0.500	0.49	−2.0
2	1.000	1.015	1.5	0.500	0.495	−1.0
3	1.000	1.032	3.2	0.500	0.497	−0.6
4	1.000	1.019	1.9	1.000	1.012	1.2
5	1.000	1.015	1.5	1.000	1.006	0.6
6	1.000	1.018	1.8	1.000	1.009	0.9
7	1.000	1.008	0.8	1.500	1.511	0.73
8	1.000	1.025	2.5	1.500	1.499	−0.07
9	1.000	1.024	2.4	1.500	1.515	1.0
10	1.000	1.044	4.4	2.000	2.014	0.7
11	1.000	1.025	2.5	2.000	2.007	0.35
12	1.000	1.040	4.0	2.000	2.012	0.60
13	1.000	1.021	2.1	3.000	3.012	0.4
14	1.000	1.047	4.7	3.000	3.000	0.0
15	1.000	1.035	3.5	3.000	3.012	0.40

TABLE 9.6   Details of Beam Dimensions

Beam	Width			Depth		
	Designed, in.	Actual, in.	Percent Error	Designed, in.	Actual, in.	Percent Error
B_{21}	0.6	0.615	2.5	1.1	1.128	2.56
B_{22}	0.6	0.595	−0.83	1.1	1.145	4.1
B_{31}	0.6	0.596	−0.66	1.1	1.138	3.43
B_{32}	0.6	0.610	1.67	1.1	1.139	3.55

TABLE 9.7   Details of Column Dimensions

Column	Width			Depth		
	Designed, in.	Actual, in.	Percent Error	Designed, in.	Actual, in.	Percent Error
C_1	1.0	0.981	−1.9	1.0	1.0	0.0
C_2	1.0	0.985	−1.5	1.0	1.0	0.0
C_3	1.0	0.959	−4.1	1.0	1.0	0.0
C_4	1.0	0.986	−1.5	1.0	0.994	−0.6
C_5	1.0	0.969	−3.1	1.0	0.999	−0.1
C_6	1.0	0.965	−3.5	1.0	0.995	−0.5

As shown in Table 9.4, the ratio of measured to desired values of depth for these modulus of rupture specimens ranged from 1.00 to 1.05, with an average of 1.02. The smallest beams, $\frac{3}{8}$ in. (10 mm) deep, were oversize by an average of 3%. Plexiglas forms were used for all model beams. Tolerances for gypsum mortar modulus of rupture specimens (Table 9.5) were of the same magnitude. Typical accuracies achieved with reinforced gypsum beams are given in Table 9.6. Desired cross-sectional dimensions were 0.6×1.1 in. (15×28 mm); average dimensions for four beams were 0.604×1.137 in. (15.4×29 mm), to provide values within 3% of the design values. Typical model column dimensions are given in Table 9.7. The six model columns had average dimensional errors of 2.6% in width and 0.2% in depth.

Another example of dimensional control from the work of Litle and Paparoni (1966), for various scales, is shown in Fig. 9.10. These models were singly reinforced and simply supported reinforced concrete beams constructed in the laboratory at MIT. The prototype beams themselves were being fabricated simultaneously in Venezuela by another group of model engineers. Data points for each of the six beams of each size as well as the average deviation from the intended dimensions are shown in Fig. 9.10. It is of interest to note

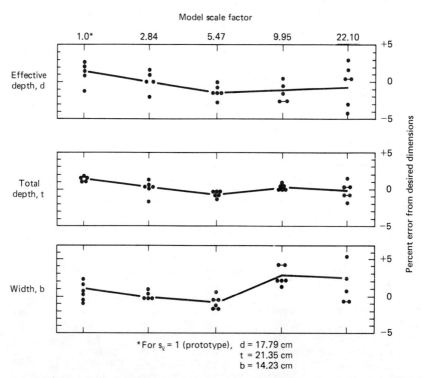

Figure 9.10 Fabrication control of model r.c. beams. [From Litle and Paparoni (1966).]

that, in this particular study, the "errors" in even the smallest models were of the same relative magnitude as those in the prototype beams.

Slab thickness is relatively easy to control provided accurate forms are used and reasonable care is taken in the screeding of the slab surface. Dimensional variations measured in several slab models are given in Table 9.8. It

TABLE 9.8   Model Slab Thickness Measurements

| Nominal Slab Dimensions, in. | Number of Points | Measured Thickness t, in. | | | |
		Maximum t	Minimum t	Average t	$\dfrac{\text{Measured } t}{\text{Design } t}$
$\frac{1}{2} \times 12 \ \times 12$	25	0.530	0.476	0.504	1.01
$0.75 \times 18 \ \times 18$	25	0.804	0.748	0.780	1.04
$0.77 \times 27.5 \times 27.5$	25	0.901	0.741	0.850	1.10*
$0.77 \times 27.5 \times 27.5$	25	0.862	0.762	0.822	1.07

*It was realized during the casting process that this slab was excessively thick.

should be noted that all slabs were cast by students who were engaged in model studies for the first time. It is seen that slab thicknesses of the order of $\frac{1}{2}$ in. (12 mm) may be achieved to an accuracy of approximately 5% even by inexperienced personnel using proper care. Thinner slabs, of the order of $\frac{1}{4}$ in. (6 mm) thick, are substantially more difficult to produce, as are curved shell sections. Several reinforced mortar plate and shell model studies having thickness of from 0.4 to 0.2 in. (10 to 5 mm) have been conducted successfully, as shown by Bouma et al. (1962), Lee (1964), Litle et al. (1967), White (1975), and others. The above fabrication limits have been reduced in the work by Harris (1967), where a series of reinforced mortar cylindrical shells having a design thickness of 0.131 in. (3.5 mm) were fabricated and tested to failure. Figure 9.11 shows a thickness contour map of one of these shells with two layers of reinforcement. The average measured thickness was 0.123 in. These studies indicate that thickness tolerances no greater than those found in Plexiglas sheets ($\pm 15\%$) can be achieved.

As was pointed out earlier, placement of model reinforcement is also important. Placement of reinforcement in a model can be done in a similar way to that in the prototype by the use of chairs (or using short pieces of wire) underneath. Because of the small size of models, very often holes are accurately drilled in the end blocks of the form and the reinforcement passed through to obtain exact placement. A slight tension placed on the model wires helps to maintain their correct alignment. An error of $\pm 5\%$ in positioning of reinforcement is considered acceptable. The accuracy of the process can be evaluated using actual test results as presented in Table 9.9. A series of three $1 \times 1.5 \times 10.5$ in. ($25 \times 38 \times 265$ mm) beams having three types of model wires and

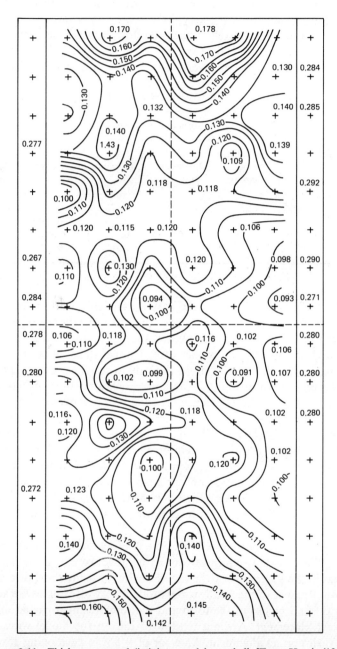

Figure 9.11 Thickness control (in.) in a model r.c. shell. [From Harris (1967).]

TABLE 9.9   Measured Positions of Reinforcement in Model Beams

Measured Cover to Center of Wire, in.

Type of Wire	Design Depth	Face 1 of Cut Wire 1	Face 1 of Cut Wire 2	Face 2 of Cut Wire 1	Face 2 of Cut Wire 2	Average No. 1	Average No. 2	C.G. Location	Percent Error, %
Plain	0.18	0.181	0.163	0.185	0.174	0.183	0.168	0.176	−2.2
Deformed	0.18	0.176	0.193	0.167	0.189	0.171	0.191	0.181	+0.5
Fabri-bond	0.18	0.181	0.202	0.181	0.204	0.181	0.203	0.192	+6.6

Average =   0.183   +1.6

reinforced identically were cast to measure the final position of the reinforcement after curing. These beams were cut along the center-line section using a diamond blade saw, and measurements for each wire were made from the bottom of the beam on both sides of the cut. The designed concrete cover from the bottom of the beam to the center line of the reinforcing wires was 0.18 in.; the average measured value was 0.183 in. (within less than +2%). These dimensions (Table 9.9) are in remarkably close agreement, showing that with good quality of work model specimens can be cast with accuracies comparable to those achieved in prototype beams. In the final analysis of these beams, the above measured discrepancies will cause approximately 2 to 3% error in the depth of steel or, in turn, on the lever arm, and will introduce error of similar magnitude in the calculation of loads. A correction is highly recommended where possible, and should be made by checking the position of reinforcement at the end of the test.

Some insight into the anticipated stress conditions will also help in determining the degree of accuracy; e.g., in the case of a column or tie where the stress condition is fairly axial or uniform, the location of longitudinal reinforcement in a corresponding reinforced concrete model column has a much smaller effect on the accuracy of the results compared with a similar situation of misplaced steel in a flexural specimen.

In small-scale models there is always the question about how accurately the fabrication process can be controlled. As geometric size becomes smaller and smaller, the absolute fabrication tolerances become less and less, if the model is to appropriately represent a physically larger specimen. While it is obvious that there becomes a size below which it just becomes impractical to construct accurate models, reinforced mortar models of plate- and shell-type structures have been adequately fabricated, and at reasonable cost, down to $\frac{1}{8}$-in. (3-mm) thicknesses, and beam-type models ranging down to about $\frac{1}{4}$ in. (6 mm) are possible.

9.6.2 Material Properties

Material properties are very important in the case of ultimate strength or inelastic models, in which the properties generally have to be identical to the prototype for investigating the true behavior of a structure. In case of elastic models (e.g., materials, PVC, Plexiglas, etc.), a number of similitude requirements can be relaxed without affecting the accuracy.

In an extensive study of elastic buckling of spherical shells [Litle (1964)] the models were vacuum-formed from flat sheets of polyvinylchloride (PVC). The variability of the modulus of elasticity and its sensitivity due to the manufacturing process were examined. In Fig. 3.12, test values of the modulus of elasticity from four specimens taken from each of 20 shell models are shown. Since the fabrication process in such materials involved heat forming, it was of interest to know the effect of annealing. By necessity, the vacuum forming process causes the material to stretch in going from the initial flat sheet to the contour of the model mold. Figure 3.13 presents Litle's (1964) results to illustrate the influence of the annealing temperature on the material's bending modulus.

On the other hand, when considering the behavior of structures at or near collapse, other material properties such as the tensile and compressive strengths, the ductility, and possible changes in the elastic constants become extremely significant. The ultimate compressive stress-strain curve can be modeled with a sufficient accuracy for any practical strength of prototype concrete (discussed in detail in Chapter 4) using either cement or gypsum mortar, as shown in Fig. 4.5. As was discussed in Chapter 2, the similitude requirement for modulus and for ultimate compressive strain are less satisfactory for concrete. A slight distortion of the strain scale will be a general condition of the modeling process, and corrections and allowances for this will have to be made for certain problems in which the total amount of strain is of prime importance in the overall behavior of the structure. Tensile strength (Fig. 4.21) and ultimate tensile strain tend to be higher than desired in the model material at a given compressive strength level, but the effects of this distortion can usually be accounted for in most model studies, without sacrificing accuracy to any significant degree.

Normally, it is not possible to simulate accurately the effects of shrinkage or creep in a microconcrete model. Although both model and prototype materials have similar shrinkage characteristics and identical construction techniques are used where possible, the effect of member size on shrinkage cannot be easily simulated. Similarly, the creep characteristics of microconcrete will qualitatively be similar to that of prototype concrete; only a limited amount of evidence exists to indicate the reliability of any quantitative similarities.

The mechanical strength properties are extremely sensitive to test procedures and imperfections in specimens, with the sensitivity increasing as the specimen size decreases. The errors associated with these properties can be classified as fabrication or testing errors.

In case of reinforcement, it was shown in Chapter 5 that careful choice of model reinforcing wires and suitable annealing processes may be used to produce deformed wire with surface properties similar to the prototype reinforcement. The desired accuracy of model tests can be achieved by modeling the stress-strain characteristics and the surface properties of reinforcement.

The properties of steel reinforcing can be controlled well in the factory. However, since the strength of steel is generally specified as "the minimum" only, actual strengths may vary considerably from batch to batch and from source to source. This will affect the accuracy of the final results of model tests, especially in the usual case of underreinforced members. Some of the remedies used to increase the accuracy of steel properties are:

1. Use reinforcement from the same batch and source. For model tests, it is a relatively minor problem since the quantities required are small and the batch is large enough to outlast the project needs.
2. Make frequent tests on samples of the specimens used to reveal all possible variations of strength properties.

The effect of steel strength variability can be considered in the strength calculations without any problem.

9.6.3 Accuracy in Testing and Measurements

Accuracy in testing is reflected by the accuracy of both the testing setup and the loading techniques. Scaling down of some of the items associated with testing can be quite involved, and proper attention must be given to such aspects. Even in a simple test setup, the reproduction of the support system for a beam by mere scaling might pose a problem, and sometimes an alternate scheme will work out much more accurately. The judgment on such variations will have to be made only with experience in the testing laboratory.

The success of a model study will also depend on the accuracy of the loading techniques. As discussed in Chapter 7, the prototype load should be represented as truly as possible. For example, a uniform loading in the prototype is simulated by discrete loads on the model for two reasons: First, the simplicity involved in applying them (such as whiffle-tree–type loading) and second, the feasibility. To show the accuracy of the discretization, a number of studies have been made. In a simple beam [see Litle, et al. (1972)] and two-hinged arch [see Pahl et al. (1964)], the effect of a uniformly distributed load replaced by a number of concentrated loads was studied. Figure 7.7 indicates the effect of discretizing the uniform load on the beam. Although the convergence of the moment coefficient k is obvious as the number of loading points is increased, only the even number of loads will give better accuracy. While considering the accuracy, other types of behavior (e.g., deflection coefficient) should also be considered for better representation of the prototype loading. Figure 7.8 presents a similar analysis of load discretization with reference to arch bending [Pahl et al. (1964)]. In this case, it becomes apparent that such a system may be entirely unsatisfactory unless a large number of points are used, and an alternate scheme should be worked out. In case of buckling experiments, the loading may produce totally acceptable results. Figure 7.9 from Kausel (1967) shows that the buckling load of an arch is not as badly degraded by load discretization as is the stress distribution. Whereas the 5- and 15-point load representations caused errors of 600 and 100%, respectively, in the maximum stress level in the shallow arch; these same loading patterns lead to errors of only 17 and 6%, respectively, in the magnitude of the buckling load.

In addition to the loading, instrumentation should also be considered in the accuracy of the results. The improper location will not only affect the measurements but will also influence their interpretation. Increased care is required in locating the measuring devices on the model structure as its size is reduced, particularly in the case where a strain gradient exists. Electrical resistance strain gages, which are so popularly used, should not only be located (both placing and direction) with extreme care but also should be mounted (i.e., bonded) properly. Improperly mounted gages may cause false strains and lead to erroneous conclusions.

Measurements in an experiment, whether on a model or a prototype, form an important aspect of accuracy. The success of an entire testing program will depend on how accurately measurements are made. Measurements range from loads, reactions, and deflections in simple tests to strains, curvatures, and cracking (both location and width) in ultimate strength models. The basic philosophy should be to achieve as much accuracy as possible but in no case less than that obtained or expected in a full-scale test. This means that a proper choice of instruments has to be made to obtain the best results. Most manufacturers specify the accuracy and range of applicability for their equipment, which should help the models engineer considerably in successfully achieving the desired degree of accuracy.

Loads are measured using dead weights or load cells. Good accuracy can be achieved using either method of measurements. Dead weights are particularly suitable for measurements in the case of slab- or shell-type structures, since a number of load cells will be required to achieve loading and may become extremely expensive. In case of a dead-weight system, a proper count should be kept while loading and unloading to minimize any error. Load cells are generally accurate up to 0.5% of the range, but accuracy in actual measurements will depend on the useful range; this is done by selecting the proper size load cells.

Accuracy of deflection measurements will depend on the type of instruments, such as dial gages or transducers (LVDTs). Accuracy of dial gages varies between 0.001 and 0.0001 in. (0.025 and 0.0025 mm), and some of the transducers can be read continuously up to an accuracy of 0.00001 in.

In the measurement of strains and curvatures, electrical resistance strain gages are probably the most widely used. The limitations on the accuracy should therefore be recognized fully, since it will depend on the material used in the structure as well as the gage itself. In plastic models, for example, the problems associated with gage stiffening and gage heating should be carefully examined. A certain time duration of between 5 and 15 min should be allowed to elapse before the readings are taken in the steady state; otherwise errors of the order of 15 to 50% can occur. Such errors, however, take place as a result of the material property rather than the strain gage itself. Electrical gages have been successfully used on models made of cement and gypsum mortar.

9.6.4 Accuracy in Interpretation of Test Results

Accurate and reliable test results and their interpretation are absolutely essential to a successful model technique. The experience so far on small-scale models indicates that direct ultimate strength models will predict the failure mode and ultimate load of a prototype structure within acceptable tolerance of the order of 10%. This degree of accuracy is completely dependent upon an intimate knowledge of the material properties of the model as was described earlier in this chapter. It is also important to realize that the results of several model tests compared with a single prototype test will normally show some difference because the prototype itself is subject to variations in strength; the single prototype might well be 10% over or under strength as compared with the entire population.

When interpreting an experimental result, it is desirable to compare it with some available theory. In case of difference between the two, the question arises as to whether an experimental or analytical model is a better predictor of the structural behavior. First, an attempt should be made to determine why the theory was not able to predict the test result. Perhaps there was a measurement error in the test, but in the absence of such improprieties, and in the absence of other experimental evidence to the contrary, one should depend on the experimental work and not on the theory if this is not very well established (especially in the case of concrete structures).

Serious difficulties are encountered in test results as well as in their interpretation only when the model material properties deviate from the prototype properties to such an extent that failure modes are changed; for example, this could happen if the tensile strength of model concrete were sufficiently high to prevent an expected diagonal tension failure. Bond-critical structures remain difficult to model with any degree of confidence, but it is believed that further development of deformed wires will markedly improve this situation.

9.7 OVERALL RELIABILITY OF MODEL RESULTS

One unfortunate aspect of model study programs is that there is sufficient time and money for only one or a few tests. In order to guard against loss of accuracy as a result of systematic errors in one or more of the techniques employed, procedures can usually be followed to detect any potential error sources before the results are properly interpreted. These procedures might include:

1. Calibration procedures prior to testing
2. Statics checks during and after testing
3. Symmetry checks
4. Repeatability checks
5. Comparisons with analytical predictions
6. Observation of trends in the data, e.g., linearity
7. Gross behavior observations

Calibration procedures prior to testing must be emphasized, they are an absolute necessity. Similar to full-scale testing, model test setups need debugging. Such calibration, similar to debugging of computer programs, can save both time and money and is helpful in obtaining good results and interpreting them properly.

Statics checks performed during and after testing can give the engineer confidence in the results, or in those cases where the check turns out to be negative, it is at least helpful to learn this fact as early as possible. It may even be possible to stop the test and correct the situation before the model structure and associated equipment are damaged. The provision for these checks must be built into the model study, as instrumentation must be positioned properly and must be adequate to provide the required data. *Symmetry checks*, like statics checks, can indicate the reliability of the test data.

Repeatability in any one test or (better yet) from one test specimen to another or (best of all) from one test to a completely different test using a different procedure can help the engineer's confidence.

Earlier, examples were cited to demonstrate the importance of various basic quantities and the degree of accuracy that can be achieved. It always is of interest for the engineer to have an "overall" reliability of the results obtained from the model tests. Using the same examples presented in earlier sections, the overall accuracy will be illustrated. In Tables 9.10 and 9.11, the accuracy of measurements from beam and column model studies are compared. The correlation for these $\frac{1}{10}$-scale models is shown to be very good. Results of these basic tests are presented in the following manner:

1. Comparison of ultimate load predictions
2. Comparison of load-deflection or moment-curvature (rotation) behavior

Table 9.10 compares the prediction values for beams from two model beams for each of the two prototypes. The ratio of values of loads in the model tests to those theoretically obtained is somewhat higher (average = 1.11 for four specimens compared with 1.07 for two specimens). Similar obser-

TABLE 9.10 Comparison of Behavior of Beams

Beam (1)	Quantity (2)	Model			Prototype				
		Theoretical (3)	Experiment (4)	Col. (4)/ Col. (3) (5)	Theoretical (6)	Model (7)	Test (8)	Col. (7)/ Col. (6) (9)	Col. (8)/ Col. (6) (10)
B_{21}	P^*	176	191.5	1.09	13.05	14.1	13.6	1.08	1.04
	M_y†	484	527	1.09	359	387	373	1.08	1.04
B_{22}	P	176	203	1.15	13.05	15.05	13.6	1.15	1.04
	M_y	484	557	1.15	359	414	373	1.15	1.04
B_{31}	P	269	286	1.06	24.9	26.5	27.4	1.07	1.10
	M_y	740	785	1.06	684	728	753	1.07	1.10
B_{32}	P	269	312	1.16	24.9	28.9	27.4	1.16	1.10
	M_y	740	855	1.16	684	794	753	1.16	1.10

$*P$ = lb for model and kips for prototype.

$†M_y$ = lb·in. for model and kip·in. for prototype.

TABLE 9.11 Comparison of Column Behavior

| | Model | | | | Prototype | | | |
Speci-mens (1)	Theoretical Load, lb. (2)	Test Load, lb. (3)	Ratio (3)/(2) (4)	Theoretical Load, kips (5)	Predicted from Model, kips (6)	Prototype Test, kips (7)	Ratio (6)/(5) (8)	Ratio (7)/(5) (9)
C_1	1020	1024	1.0	94	115	94	1.22	1.0
C_2	1006	1036	1.03	94	116	94	1.23	1.0
C_3	998	908	0.91	94	110	94	1.17	1.0
C_4	1020	1048	1.03	94	117.5	94	1.25	1.0
C_5	1060	1170	1.10	94	131	94	1.39	1.0
C_6	3820	3960	1.04	412	435	456	1.05	1.11

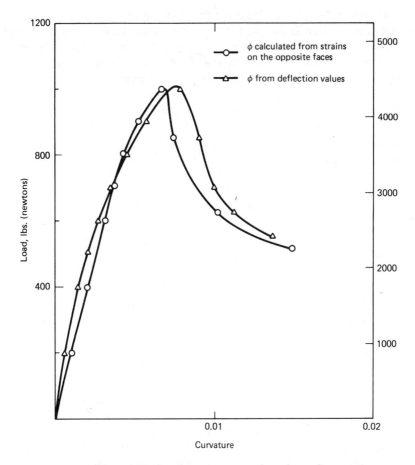

Figure 9.12   Load curvature curve for column C_1.

vation may be made from Table 9.11 for column load values. Dimensional accuracies of these specimens were discussed in Section 9.6.1. The accuracy of calculations of unit rotations computed from strain readings on opposite faces and dial gages is shown in Fig. 9.12. The values are quite close; some difference is due to the gage length used for the dial gage and strain gage measurements.

It may be concluded, based on this discussion, that the overall accuracy of the models is less (approximately 10%) compared with errors usually encountered in the basic dimensions, which were within $\pm 2\%$. The other error-causing factors, discussed previously, must interact and thus influence the overall accuracy.

9.8 INFLUENCE OF COST AND TIME ON ACCURACY OF MODELS

Cost and time are important to the extent that a project may not be carried out because of their limitations. Very often satisfactory solutions using model studies can be obtained if proper time and cost are allotted, in comparison to the other types of solutions. From the design engineer's point of view, even a very accurate solution is no good if the information cannot be obtained in a reasonable amount of time; ideally, the design engineer will always want the cheapest, most accurate solution in the shortest possible time. The research engineer is, on the other hand, generally, less concerned about the time factor and more concerned with accuracy and cost.

Similar to any other project, cost and time will undoubtedly affect accuracy of model results. Generally, the cost of an experimental structure or model will be more than the cost of a theoretical solution; but as was pointed out in the introduction of this book, a physical model is more useful and desirable, especially in the case of structures whose behavior is to be studied up to and including the ultimate load stage and where question of design criteria and loading are not resolved. Provided that the previous conditions of accuracy of modeling are satisfied, experimental results will be within an acceptable engineering range.

In the case of important or special structures, a test on a physical model is particularly helpful. In such projects, time will be an important aspect; but costs will play a secondary role to achieve the desired accuracy. The cost, among other things, will determine the scale factor. It is a general belief that the cost of a model will be reduced as the scale is reduced; this is true only to a certain extent.

9.9 SUMMARY

From the foregoing discussion on errors, accuracy, and reliability in model studies, the examples presented indicate that structural models can be used effectively for studying a wide range of problems. An introduction to the theory and propagation of random errors has been presented. The implications of different types of errors on the experimental results has been discussed. Although our knowledge of their reliability is by no means complete, techniques and materials are available today to apply physical modeling with confidence to many structural problems. For situations where a linear elastic solution is deemed to be satisfactory, structural models have little limitation. In the case of ultimate strength models, limitations on materials, fabrication, and loading will probably govern the optimum scale size.

It must be emphasized that confidence levels in model testing can best be established by comparing similar members or structures of different size. As is indicated by several examples, concrete models can predict effectively deflections, strains, modes of failure, and failure loads for beams, columns, slabs, and shell structures.

Modeling Applications and Case Studies

∿∿∿∿∿∿∿∿∿∿∿∿∿

10

10.1 INTRODUCTION

A large number of successful modeling studies has been reported in the literature in the past few decades. These fall into the categories of educational, research, and design models and are far too numerous to even list here. As an alternative, only a small sample of model applications are examined for the normal structural classifications encountered in civil engineering construction. Examples of typical design and research studies are illustrated with varying degrees of completeness.

Four case studies described briefly in Chapter 1 are discussed in greater detail in this chapter. These consist of two design applications and two research applications.

10.2 MODELING APPLICATIONS

10.2.1 Building Structures

10.2.1.1 Australia Square Tower

The tower structure (Fig. 10.1) consists essentially of a circular service core, composed of an outside and inside wall interconnected by 20 radial diaphragm walls forming cells that extend the full height of the tower. The circular core is encircled by a framework of 20 tapered columns connected with

Figure 10.1 The Australia Square Tower, Sydney, Australia. [After Gero and Cowan (1970).]

radial beams to the diaphragm wall opposite (Fig. 7.20). Dense concrete was used for the construction of the structure below the third floor, and lightweight concrete was used for the remainder of the tower.

The unusual tower shape made the basic structural actions far more complex to analyze than those of a conventional multistory frame. Moreover, the presently available sophisticated computer analysis programs were not available in the early sixties to analyze this tower, which is statically indeterminate to the 12,000th degree. The consulting engineers considered that dead-load stresses could be calculated with the needed accuracy; however, the stresses caused by wind loading needed further checking. Therefore, it was decided to build a $\frac{1}{30}$-scale model in the University of New South Wales Laboratories to study the behavior of the tower under a simulated wind loading.

Development of a suitable model construction material in the form of a polyester resin combined with varying amounts of calcite filler to simulate the properties of prototype concrete that varied along the lower height is described in Section 3.3.5.

Model Construction. The model construction consisted of assembly of repetitive mass-produced elements. The central core, along with its dividing walls and opening, was cast in heights of one floor at a time. The adjoining slabs were omitted in this case as it was impossible to prevent its warping. Each floor slab was cast in 20 equal sectors; all components (approximately 2200)

Figure 10.2 Model of Australia Square
Tower. [After Gero and Cowan (1970).]

were then glued together to form the completed model as shown in Fig. 10.2.*
Special attention was paid to the tensile strength of the glued joints, as it was
always lower than that of the solid calcite–filled resin. Different glues (epoxies,
Plastrenes, etc.) in combination with the various types of surface preparations
and joints were tested for strength.†

The design wind load was simulated by a static loading system consisting
of 28 concentrated loads that were combined into a single resultant using a
whiffle-tree lever system (Fig. 7.20). U-shaped steel bands [0.625 × 0.018 in.
(15.9 × 0.46 mm)] girdled the tower at 28 selected levels to transfer the load to
the model structure as the whiffle-tree system was moved (away from the model)
to produce the required loading.

The model was supported on a welded foundation base, consisting of a
6 × 6 ft (1.8 × 1.8 m) steel plate stiffened by a substantial two-way grid, which

*Each of the 20 columns was cast in 5 lifts. An accuracy of ±0.1 in. (2.5 mm) was
achieved in casting all components of the tower model.

†After several trials, it was decided to use Plastrene 97 glue (100 parts of Plastrene 97
with 50 parts of filler) along with butted joint faces that were mechanically cleaned with 20
grade silicon carbide.

was bolted down to 30-ft (9.1 m) long interconnected girders that formed the base joists. A stiff dial gage tower cantilevered from the foundation base (in the theoretical bending plane) to provide the reference axis for deflection measurements. This ensured coupled rigid body movement between the model and the reference axis. Small battery-operated electric motors were attached to each dial gage holder to eliminate by vibration any frictional effects that could have influenced the dial gage readings. This system proved quite reliable and was an excellent substitute for cumbersome hand tapping, which gave inconsistent results.

Instrumentation.   The objectives of the model analysis were to establish the deflection and stress conditions in certain parts of the structure that were due to a uniformly distributed horizontal loading action in a vertical plane. It was also required to find axial and shearing forces and bending moments in columns and beams, bending stress distribution in the core, and the shearing stress distribution in vertical planes above door openings and in horizontal planes between two door openings in the core. All stresses and internal actions were evaluated from measured strain data, and the deflections were measured using dial gages.

Over 500 linear and 45° rosettes (electrical resistance strain gages) with polyester backing were installed at critical regions using the polyester glue used in the model. A protective wax coating was applied, not so much to waterproof the gages, but to prevent temperature changes between the active and dummy gages resulting from air streams of different temperatures in the laboratory.

Testing.   The tower was tested in two stages. Stage A consisted of testing a free-standing core cantilevered from the base with all slabs attached, but before the column joints had been glued. Simple beam theory was used to calculate the bending and shearing stresses, which were compared with values determined from the experimental strain data. It must be noted that such a situation (that is, minus the columns) will not occur in the prototype. However, stage A testing provided a valuable check on the results and established confidence in the validity of the model.

The columns were then attached, and the model assumed the exact reduced shape of the prototype with all its complexity. The model was now ready for stage B tests, and the model test results could not be checked by theory. The loads were applied in four equal increments. The observed strains were corrected for time and temperature using the data in Fig. 3.6 before their conversion into stresses.

Data Evaluation.   All strains were multiplied by a factor E_R/E_0, where E_R is the reduced modulus at a given time and temperature (Fig. 3.7) and E_0 is the modulus of elasticity of Plastrene at a time interval of 30 min and a temperature of 75°F (24°C). This theoretically reduces all strain readings to the value at the standard time interval of 30 min and the standard temperature of 75°F (24°C). The corrected values of stresses and deflections at various points of the model were then converted to prototype quantities using the scaling factors in Table 2.2. Typical results for lateral deflection for both stages A and B are shown in Fig. 10.3.

Lateral deflections for 1 K/F horizontal UDL on structure

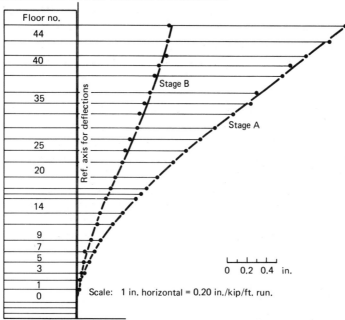

Figure 10.3 Lateral deflection of Australia Square Tower model under a lateral force of 1 k/ft (vertically). Stage A represents the structure with the central core and no external columns. Stage B results clearly show the propping effect of the section of the perimetric columns. [After Gero and Cowan (1970).]

10.2.1.2 Horizontal Joints in Concrete Masonry Bearing Wall Structures

Introduction.   Recent advances in reinforced masonry structures and the widespread use of precast, prestressed hollow-core floor slabs resting on masonry bearing walls have necessitated a closer examination of the structural behavior of these components. One of the problems of prefabricated construction is the additional attention that must be paid by the designer to the details in the connection areas to ensure that the various wall and floor elements have adequate structural continuity and can work together to accommodate the lateral as well as the gravity loadings. In order for masonry to realize its full potential as a competitive form of construction in high-rise buildings and buildings situated in more severe earthquake zones, the experimental behavior of such construction must be investigated. A direct small-scale modeling technique that has been successfully used in both reinforced and prestressed concrete structures was used as an economical alternative to full-scale testing at the Structural Models Laboratory, Drexel University, Philadelphia, Pennsylvania. Appropriately, a quarter scale was chosen as a first step in modeling the behavior of hollow-core concrete masonry structures using carefully constructed $\frac{1}{4}$-scale masonry blocks supplied by the National Concrete Masonry Association. These units are used to study the modeling technique of concrete masonry structures as described in Chapter 4. Attention has been focused on the physical properties of the constituent materials and the necessary tests that must be performed to ascertain the basic strength characteristics of the masonry units and masonry components. To this effect, both the hollow-core concrete masonry block and the mortar that is used to bind these units together to form masonry structures were studied in detail. Similitude requirements for modeling of masonry structures under static loading are presented in Chapter 2. It is important to understand from the outset that the masonry units used in structural components form a composite mechanical action system because the two materials, the masonry block and the masonry mortar, do not have identical mechanical characteristics. The model analyst, therefore, must have a good understanding of both components so that direct modeling will be achieved to a satisfactory level of confidence for the masonry composite to function under loading in the same manner as the prototype. Not only is one interested in elastic behavior of the composite system but also the behavior beyond cracking up to ultimate loading. It is this inelastic behavior requirement that complicates the modeling problem and forces one to adopt techniques in the modeling of masonry structures that parallel those used in the modeling of reinforced and prestressed concrete structures. Additional difficulties arise, however, because the masonry model specimens (walls, prisms, etc.) are not cast but must be fabricated using techniques such as those that have been evolved at Drexel University [Becica and Harris (1977), Harris and Becica (1978)].

Concrete Masonry Bearing Walls. Small wall specimens two blocks long by three blocks high were fabricated in running bond and tested in axial compression. The lower courses were laid in full mortar beds atop a $\frac{3}{16}$-in.-thick (5 mm) strip of aluminum. This allowed for transport of the specimens and eliminated the need for lower bearing surface capping. Another advantage of the full mortar bed is to simulate actual boundary conditions. The first series (series A), consisting of three tests, was used as a pilot study to determine typical small wall compressive strengths. The specimens of series B were instrumented with 4-in. SR-4 strain gages. These were placed at midheight at the center on both faces of the walls and wired in series, thus providing average readings. Series A specimens were capped and tested within 1 hr of their removal from the wet room. Series B specimens were removed from the wet room after 26

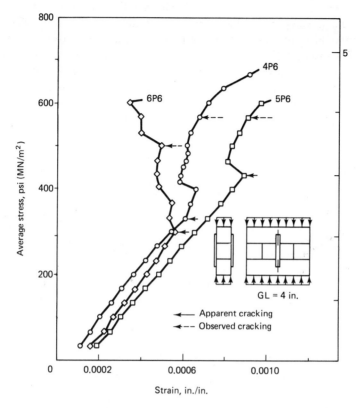

Figure 10.4 Compressive stress-strain curves for small masonry walls [After Harris and Becica (1978).]

days and allowed to dry for 4 days to allow for instrumentation. The stress-strain behavior of series B walls is plotted in Fig. 10.4. End splitting became visible for specimens 4P6 and 5P6 at 570 psi (3.9 MPa), which is 89 % and 95 % of f'_m for 4P6 and 5P6, respectively. First end splitting appeared at 500 psi or 85 % of f'_m for specimen 6P6. The discontinuities which appear at 63 % [400 psi (2.8 MPa)], 70 % [420 psi (2.9 MPa)], and 50 % [300 psi (2.1 MPa)] of f'_m for 4P6, 5P6, and 6P6, respectively, is attributed to tension increments being recorded as a result of face-shell buckling. These points, therefore, represent the failures of the interior web structures. In addition, cracking was audible at these load levels. The average secant modulus of elasticity of $0.5 f'_m$ was found to be 505,500 psi (3,500 MPa) for the three model specimens.

Horizontal Wall-to-Floor Joints.   In concrete masonry structures that utilize precast prestressed concrete hollow-core floor and roof slabs, the horizontal wall-to-floor joint plays a critical role. It must be able to accommodate the gravity and lateral loading without appreciable distress. An evaluation of horizontal wall-to-floor joint behavior under axially applied gravity loading is therefore the first step in developing model techniques for masonry with precast components. For this purpose the joint detail commonly found in low-rise masonry construction (Fig. 10.5) was used. Two unreinforced and two Durowall reinforced joints were tested to determine the strength and joint shortening for the full range of loading. Hollow-core precast slabs at quarter scale were used in these joints [Harris and Muskivitch (1977), Dow and Harris (1978)].

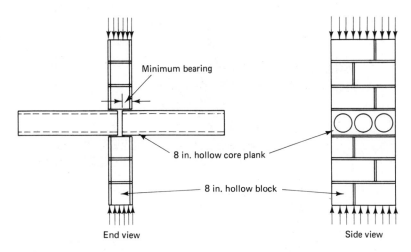

Figure 10.5 Interior horizontal joint detail. [After Harris and Becica (1978).]

Joints M/PCP-J-1 and M/PCP-J-2 were unreinforced and also served as pilot studies for this program. The fabrication of the joints was very similar to the small wall specimens described above. Wall elements consisted of 3 units in height and $1\frac{1}{2}$ units in width, thus giving a nominal 6×6 in. (150×150 mm) wall panel 2 in. (50 mm) thick. The hollow-core floor units were 6 in. (150 mm) wide, 6 in. (150 mm) in length, and 2 in. (50 mm) thick. A gage length of 8 in. (200 mm) across the joint was used to determine the joint shortening. Load vs. joint shortening for the unreinforced joints is shown in Fig. 10.6. Joint M/PCP-J-2 appears to be stiffer and stronger than joint M/PCP-J-1. The mode of failure of both joints was by end splitting of either the top or the bottom masonry wall panel. The results of the two unreinforced joints are summarized in Table 10.1. The compressive strength of the $\frac{1}{4}$-scale masonry units used was 1100 psi (7.7 MPa) as shown in column (2) of Table 10.1. Column (3) shows the compressive strength of the joint mortar and column (4) the compressive strength of two three-block prisms cast together with the joints. The ultimate joint load and joint strength on the gross area are given in columns (5) and (6), respectively, of Table 10.1. A compressive strength reduction of approximately 23% over the prism strength was shown by the average of the two unreinforced joints. This

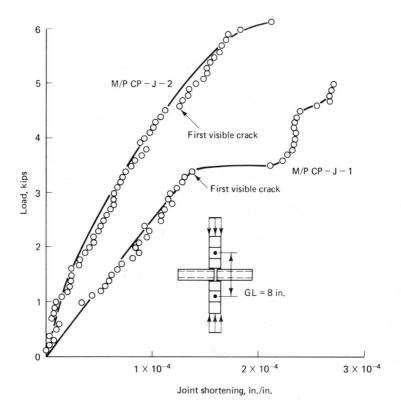

Figure 10.6 Compressive load shortening curves for models M/PCP-J-1 and M/PCP-J-2. [After Harris and Becica (1978).]

TABLE 10.1 Summary of Horizontal Joint Test Results

Joint Designation (1)	Compressive Strength Unit Masonry Gross A f', psi (MPa) (2)	Compressive Strength Mortar 1 × 2 in. (25 × 50 mm) Cyl. f'_b, psi (MPa) (3)	Compressive Strength Three-Course Prism Gross A f'_m, psi (MPa) (4)	Ultimate Joint Load P'_u, lb (kN) (5)	Joint Strength Gross A, psi (MPa) (6)
M/PCP-J-1	1100(7.6)	687(4.7)	671(4.6)	6140(42.3)	560(3.9)
M/PCP-J-2	1100(7.6)	687(4.7)	671(4.6)	5260(36.3)	479(3.3)
			Av.	5700(39.3)	519(3.6)
M/PCP-J-3R*	315(2.2)	1310(9)	191(1.3)	2925(20.2)	267(1.8)
M/PCP-J-4R*	315(2.2)	1310(9)	292(2)	2200(15.2)	201(1.4)
			Av.	2563(17.7)	234(1.6)

*R means wire-reinforced.

reduction was anticipated because the hollow-core slabs provide only partial support to the upper masonry wall panel (see Fig. 10.5)

The second series of horizontal joints tested were reinforced with model Durowall trusses made of 0.022-in. (0.6 mm)-diameter wire and 45° diagonals. The test set-up and instrumentation is shown in Fig. 10.7. Scaled model rein-

Figure 10.7 Horizontal joint test set-up. [After Harris and Becica (1978).]

forcement was placed between the second and third courses of the bottom wall panel and between the first and second courses of the upper wall panel to simulate field construction procedures. Companion three-block compression prisms with and without the Durowall reinforcement were also cast at the same time as the joints to determine the effect of the reinforcement in increasing the compressive strength. As can be seen from Fig. 10.7, the shortening of the upper and lower masonry wall panels was measured by both dial gages and strain gages to allow comparison of the results. The joints were loaded axially in 100-lb (0.45 kN) increments allowing adequate time for the ten dial gage and four strain gage readings to be made. The results of the load vs. axial shortening of the top and bottom wall panels and across the joint is shown in Fig. 10.8. Figure 10.8 shows that the stiffness of the upper and lower walls is higher than across the joint for both specimens. The results of the two reinforced joints are summarized in Table 10.1. As can be seen from Table 10.1, the type of model masonry units used in

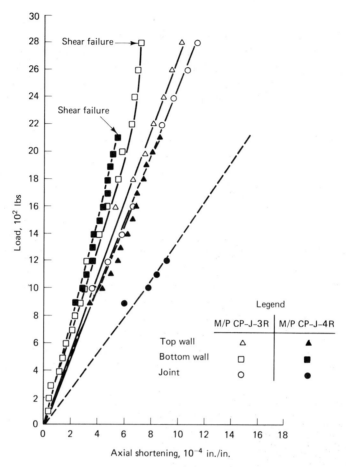

Figure 10.8 Compressive load shortening curves for joint M/PCP-J-3R and M/PCP-J-4R. [After Harris and Becica (1978).]

the two reinforced joints were of the very weak kind with an average compressive strength of only 315 psi (2.2 MPa) in six specimens.

The potential of ultimate strength models of concrete masonry structures has been clearly demonstrated. Studies of structural joint details or components are thus feasible and a necessary first step in the study of the behavior of three-dimensional concrete masonry structures.

10.2.2 Bridge Structures

10.2.2.1 Zarate-Brazo Largo Highway-Railway Bridge

The response of the elastic model of the Zarate-Brazo Largo Highway-Railway Bridge near Buenos Aires, Argentina, to static loads is presented in this section. Natural mode shapes and frequency measurements are described in Chapter 11. The prototype structures consist of two cable-stayed bridges over Parana de las Palmas and Parana Guazo rivers approximately 50 mi (80 km) from Buenos Aires. The 550-m long bridges have identical superstructures (central span of 330 m and two side spans of 110 m each) but different pile foundations.* The bridge has long prestressed concrete approach viaducts [Baglietto et al. (1976)]. The general layout of the bridge consists of the following elements (Fig. 10.9):

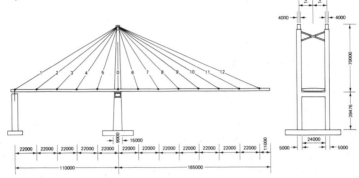

Figure 10.9   General layout of Zarate-Brazo Largo Bridges. [After Balietto et al. (1976).]

1. Four reinforced concrete piers. Both central piers have a rectangular, hollow cross section and are 105 m high above the rigid pile cap. Towers are connected by a concrete beam at the bridge deck level and are connected by a steel cross near the top, thus forming a statically indeterminate frame laterally. The bridge cables are anchored to steel caps at the tower tops. The remaining two piers are also statically indeterminate reinforced concrete frames laterally.

2. A fan-shaped cable system, emanating from the steel cap at the tower top and anchored to the steel deck at every 22 m. The cables consist of parallel 7-mm high-strength wires.

3. A steel deck (Fig. 10.10) consisting of:
   (a) Two longitudinal trapezoidal box beams that provide anchorage for the cables at its external face (22-m centers).
   (b) A 10-mm-thick plate deck stiffened by closed trapezoidal section ribs is connected to the two box beams by site welding. Transverse plate girders attached to the box beams at 22-m centers support the deck and resist bending moments due to the eccentric cable anchorages in the box beams. Transverse truss beams at 3.15-m spacings provide additional torsional stiffness.

---

*Each bridge accommodates four highway lanes, two sidewalks, and an asymmetric rail track. The asymmetrical position of the railway loads over a large span of 330 m is an unusual characteristic, which had to be considered in analysis.

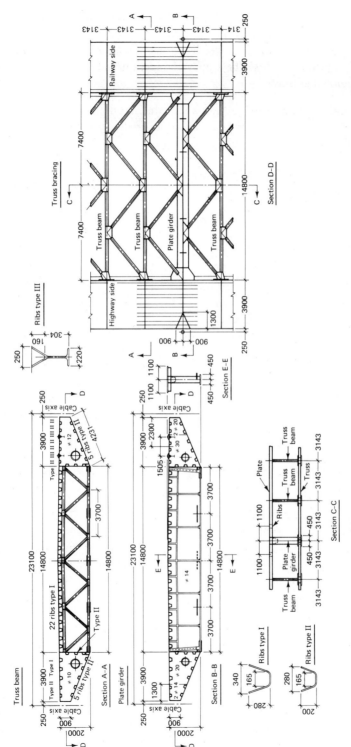

Figure 10.10 Typical cross section and truss bracing. [After Baglietto et al. (1976).]

(c) A truss bracing at the lower level of the plate girders and transverse truss beams to close the transverse section and to provide adequate torsional stiffness.

The cross section provides an adequate aerodynamic shape, and its torsional stiffness, along with the contribution of the suspension cables, contributes to the aerodynamic stability of the bridge. This was verified by wind tunnel tests.

A 100-mm-thick concrete pavement is provided on top of the steel deck. This concrete pavement does not interact with the steel deck; however, it assists in stiffening the deck locally.

The asymmetrical disposition of the railway load results in an asymmetrical structural design with the cables and the box beams on the railway side having twice the cross-sectional area of otherwise identical elements on the highway side.

The consulting engineers conducted a very careful examination of the computer-based design calculations for this complex structure. However, despite using the best available analysis and design tools, they decided to establish reliability of design assumptions and calculations by checking the assumed behavior by tests on a structural model. Although the design criteria were technically and economically valid, a number of problems required information that could be obtained only from suitable experimental work. The areas needing attention were: torsional stiffness of the highly indeterminate bridge deck, its interaction with the cables, their fatigue behavior and the influence of secondary cables under the action of highly eccentric high-speed railway loads, and evaluation of impact coefficient.

Similitude Criteria. The following considerations were made in designing the model:

1. Correct reproduction of the characteristics of the structure with respect to the established goals of the model
2. Limitations imposed by the availability of suitable materials, skilled personnel, laboratory space, and capacity of the testing equipment, etc.
3. Limited funds available for the project

As a compromise between all above considerations, a length scale factor of 33.3 was selected. The acceleration scale factor s_a was set equal to 1, because a correct simulation of the static and dynamic behavior of the bridge requires that the cable tension and therefore the geometrical coefficient be correctly reproduced. This can only be achieved by reproducing the dead-load effects on the model. The choice of a material immediately establishes the scale factors s_E and s_ρ, and thus a material cannot be chosen arbitrarily. Moreover, since $s_a = 1$,

$$s_E = s_L \cdot s_\rho \tag{10.1}$$

and any chosen material must satisfy Eq. (10.1). Commercially available steel sheets were used for the bridge deck model, thus making $s_E = 1$. The model was suitably ballasted to satisfy the condition $s_\rho = 1/33.3 = 0.03$.

The Model.   For reasons of practicality, the following modifications were necessary in the model:

1. The hollow cross section of the prototype piers was not reproduced. However, the flexural rigidity EI was correctly scaled.
2. The number and shape of the ribs were not reproduced. However, the axial rigidity EA was correctly scaled.
3. In some cases the deck was simulated by an aluminum sheet glued on a steel sheet, which distorted the model thickness. However, the axial rigidity EA was correctly simulated.
4. It was not possible to find exact scaled diameters of high-strength steel wires. However, the total elongation of the prototype cables was simulated by using two pieces of suitable lengths and diameters joined together.
5. The dead load was simulated by a series of lead weights distributed on the exterior and the interior of the bridge, arranged so as to allow passing of the train and to reproduce as accurately as possible the distribution of the masses, the position of the center of gravity, and the polar moment of inertia without modifying the overall stiffness of the structure.
6. Small steel cylinders were coaxially assembled on each cable to simulate its dead load without changing its axial stiffness.
7. The dimensions of cable anchorages, welding, and connections in the various parts of the model were significantly different from those of the prototype.

The above simplifications impose the following limitations on the model:

1. The model can satisfactorily simulate the overall behavior of the prototype. However, the local behavior may differ considerably, and this fact must be considered in any analysis.
2. As the limit of proportionality and the strength characteristics of the materials used and of the welds and the connections in the model are different from those of the prototype, the model results are valid only in the linear elastic range.
3. It was not possible to assess the similitude of damping achieved in the model. There were uncertainties about the damping characteristics of the prototype itself.

Details of bridge construction and preparation of the model for static and dynamic tests are shown in Figs. 10.11 to 10.14.

Figure 10.11 Welding of deck elements. [After Baglietto et al. (1976).] (Courtesy of ISMES, Bergamo, Italy.)

Figure 10.12 Concrete pier details and simulation of cable dead weight by cylindrical steel masses. [After Baglietto et al. (1976).] (Courtesy of ISMES, Bergamo, Italy.)

Figure 10.13 Simulation of dead load by lead masses. [After Baglietto et al. (1976).] (Courtesy of ISMES, Bergamo, Italy.)

Figure 10.14 Completed bridge model ready for tests. [After Baglietto et al. (1976).] (Courtesy of ISMES, Bergamo, Italy.)

Static Tests—Loading and Instrumentation. The following series of static tests were conducted on the model (using prototype dimensions):

1. Influence lines for the railway side. Concentrated unit load was placed in turn at every 22 m along the railroad axis.
2. Influence lines for the outside highway lane—repetition of item (1) above.
3. Railroad and highway load applied to the 110-m side span.
4. Railroad and highway load applied to the 330-m central span.
5. Railroad and highway loads 66-m long were placed along the bridge length to produce the maximum stress in the deck section.

The loads for static tests consisted of a series of steel cylinders of suitable weight that could be easily manipulated into the required positions using the available lifting equipment (Fig. 10.15). The measuring and data acquisition equipment were quite complex, with numerous electrical resistance strain gages and rosettes and 41 displacement transducers installed at a total of 258 measuring points on the bridge deck and the cables. All measurements were recorded by an automatic data acquisition system. The signals from the various measurement transducers, connected to a series of balancing circuits, were sent to an amplifier by means of an automatic switch. The analog signal from the amplifier was digitized by an analog-digital converter and stored in the memory of the

Figure 10.15 Loading equipment for static tests. [After Baglietto et al. (1976).] (Courtesy of ISMES, Bergamo, Italy.)

computer. The processed results were available as a printout, a punched tape, or a plot. A block diagram of the data acquisition system is shown in Fig. 10.16.

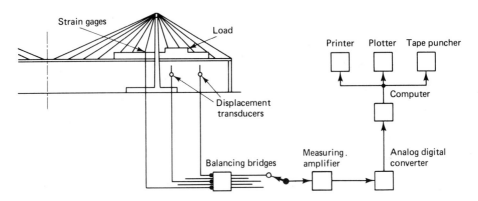

Figure 10.16   Block diagram—testing and measuring equipment for static tests. [After Baglietto et al. (1976).] (Courtesy of ISMES, Bergamo, Italy.)

Test Results.   All components of deck displacement and the cable tensions on both sides of the bridge were measured for the various loading conditions. Because of the asymmetry of the bridge cross section and the loading asymmetry, it was necessary to make measurements on both sides. Figure 10.17 shows the deck displacement and cable tension results for static loads on the highway and the railway sides. A total of 114 influence lines for vertical displacements and cable tension on both sides were determined. Strain distribution for 13 transverse sections were obtained from 161 strain readings for each loading condition. Some of this data is reported by Baglietto et al. (1976).

The model test results for vertical deflections showed a discrepancy of about 5 to 15% from the analytical calculations, and the variation for cable stresses was approximately 5 to 20%. These differences can be attributed to the fact that analytically determined influence lines were based on consideration of planar systems corresponding to the railway and highway sides, ignoring the deck torsional stiffness. However, the differences are small for all engineering purposes, and the slightly conservative analytical influence lines were considered adequate. The model test showed the importance of two planes of cables related to the deck torsional stiffness in redistributing the stresses by torsion. A comparison of analytical and experimental influence lines for stresses in a typical cable and for vertical displacements are shown in Fig. 10.18.

In summary, satisfactory agreement was noted between analytical and experimental results, with most of the data showing differences of less than 10%.

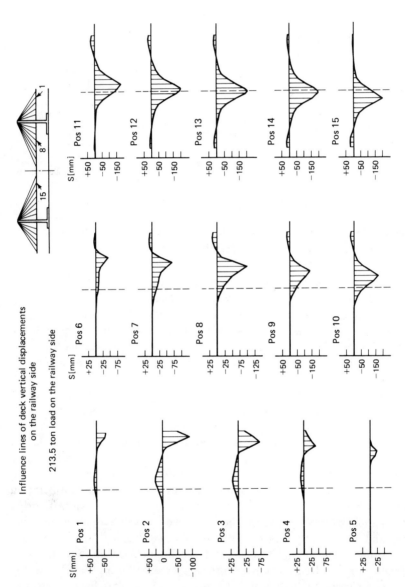

Figure 10.17 Typical influence lines for bridge deck displacement and cable tension. [After Baglietto et al. (1976).]

Influence lines of deck vertical displacements on the railway side

213.5 ton load on the railway side

S [mm]

Pos 1
Pos 2
Pos 3
Pos 4
Pos 5

Pos 6
Pos 7
Pos 8
Pos 9
Pos 10

Pos 11
Pos 12
Pos 13
Pos 14
Pos 15

Influence lines of cable tension of the highway side

213.5 ton load on the railway side

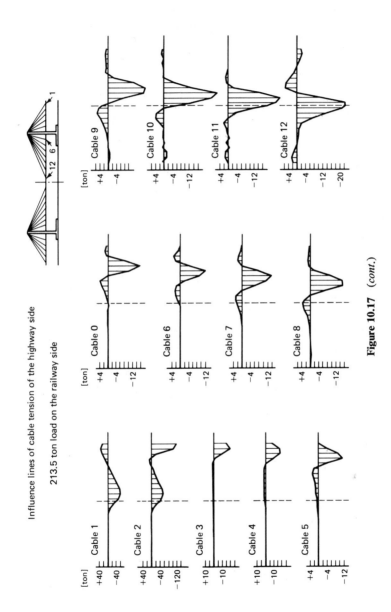

Figure 10.17 (*cont.*)

401

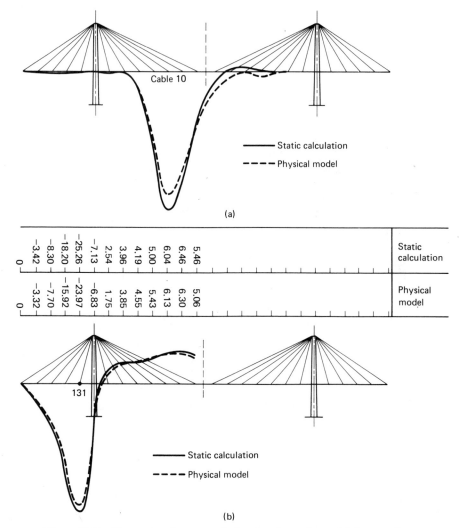

	Static calculation
	Physical model

(a)

| 0 | -3.42 | -8.30 | -18.20 | -25.26 | -7.13 | 2.54 | 3.96 | 4.19 | 5.00 | 6.04 | 6.46 | 5.46 | | | | | | | | | | | | | | | | Static calculation |
| 0 | -3.32 | -7.70 | -15.92 | -23.97 | -6.83 | 1.75 | 3.85 | 4.55 | 5.43 | 6.13 | 6.30 | 5.06 | | | | | | | | | | | | | | | | Physical model |

Figure 10.18 Comparison between calculated and experimental influences lines. (a) Influence lines for tension in cable no. 10. (b) Influence lines for displacement at deck point 131. [(After Baglietto et al. (1976).]

10.2.2.2 Curved Box Bridges

A number of extensive model studies of box and open girder bridges have been conducted in recent years. Clarke (1966) studied the behavior of curved two-span two-cell girders using three $\frac{1}{8}$-scale models constructed of rolled I beams and a concrete slab. A concrete model of a four-cell straight box girder at a scale of 1/3.78 was tested at the University of California, Berkeley, as reported by Davis et al. (1972). Roll and Aneja (1966), Culver and Christiano (1969), and Aneja and Roll (1971) used plastic models to study the behavior of straight and curved box and open girder bridges having single and continuous spans. Nicholls and Fuchs (1972) conducted small-scale model tests on concrete and steel composite construction of a single-cell horizontally curved box bridge. More recently, Godden and Aslam (1973) and Aslam and Godden (1975) used small-scale aluminum models to study the behavior of both skew and curved multicell box girder bridges. Of the several small-scale model investigations reported, those conducted at the University of Pennsylvania, Philadelphia, and the University of California, Berkeley, will be described in greater detail.

The experimental study of Roll and Aneja (1966), and Aneja and Roll (1971) consisted of testing Plexiglas models at approximately one-thirtieth scale of both a straight and a curved single-cell, single-span box beam bridge. The main purpose of their research was to obtain the experimental response of the models subjected to various typical loading conditions and measuring the corresponding strains at the inner and outer surfaces of the top deck, bottom flange, and webs at selected cross sections. The experimental data could then be used to check theoretical analyses of similar structures and make an attempt to develop a simple design method.

The dimensions of the curved bridge model is shown in Fig. 10.19. The

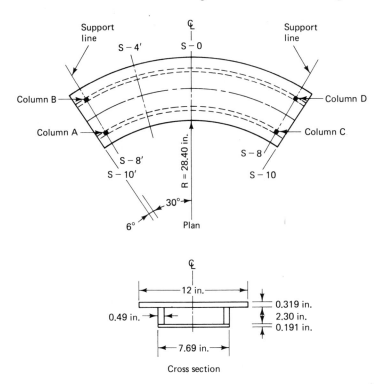

Figure 10.19   Dimensions of model. [After Aneja and Roll (1971).]

model, constructed from Plexiglas II UVA, was simply supported on pairs of Plexiglas columns, each column placed directly under a web. Each pair of columns lay on a radial transverse section, as shown in Fig. 10.19, so that the center-line span length was 36.0 in. (200 mm). The columns were prestressed to provide the large downward reactions required for the curved bridge under certain loading conditions. Details of the columns are shown in Fig. 10.20.

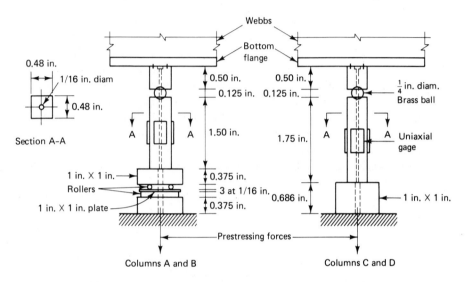

Figure 10.20 Column details. [After Aneja and Roll (1971).]

Average values of the material properties of the model are given in Table 10.2.

TABLE 10.2 Material Properties

Components (1)	Thickness, in.(mm) (2)	Modulus of Elasticity, lb/in.²(MPa) (3)	Poisson's Ratio (4)
Top flange	0.319(8.1)	519,700(3,583)	0.334
Bottom flange	0.191(4.9)	506,300(3,491)	0.323
Webs	0.490(12.4)	498,400(3,436)	0.353

After the components of the model were ready for final assembly, electrical resistance strain gage rosettes were attached to the external and internal surfaces of flanges and webs at the support, quarter-span and midspan sections designated S-8', S-4', and S-0, respectively.

At each gage section a total of 40 rosettes were used: 18 for the top flange (9 on each face), 10 for the bottom flange (5 on each face), and 6 for each web (3 on each face of each web). In all, there were 126 rosette gages and 16 uniaxial gages (4 on each column) to be read for any loading condition. All gages were foil gages that were temperature-compensated for plastic. The gages were attached to the specimens and the model components with Eastman 910 adhesive in a typical arrangement shown in Fig. 10.21.

Loading of the model was accomplished by means of hanging weights through small predrilled holes in the deck and bottom flange (Fig. 10.22).

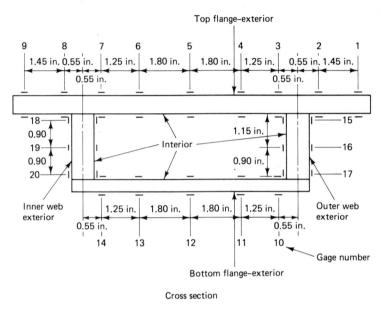

Cross section

Figure 10.21 Experimental data identifications for location of gages at gage section and model member surfaces. [After Aneja and Roll (1971).]

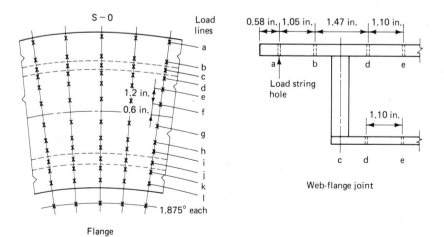

Figure 10.22 Typical loading points. [After Aneja and Roll (1971).]

Typical experimental results consisted of the tangential and radial stress distributions in the deck, webs, and bottom flange. A comparison with analytical results obtained from a finite element solution showed good agreement. Discrepancies between analytical and experimental results were attributed to shortcomings of the mathematical model by the authors in that the computer programs used for comparisons did not contain curved elements for the webs and a plate element with six degrees of freedom per node.

Aslam and Godden (1975) used a $\frac{1}{29}$-scale model made from aluminum plate to study the behavior of a four-cell horizontally curved box girder bridge. Their main objectives were (1) to produce accurate experimental data for checking the validity of computer analysis of such structures and (2) to determine the elastic behavior trends of such structures.

The basic glued model was provided with a midspan internal radial diaphragm that could be tightened or collapsed from outside to make it effective or inoperative. The model was cut back successively to study shorter spans. Thus, it was possible to get a large amount of test data from the same basic model. The instrumentation was common to all tests, and the accuracy of the strain gages was checked on the bridge components prior to assembling the model. The plan view of the model is shown in Fig. 10.23, and its cross section in Fig. 10.24a. Aluminum was selected for the model material in order to eliminate the time- and temperature-dependent properties usually experienced with plastics and to reduce the heating effect in reading strain gages. The alumi-

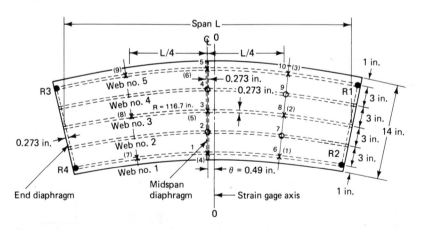

Deflection position x (deflection position no)
Load position o load position no
Reactions ● reaction no

Note: The span L is taken as the horizontal projection
of central web

Figure 10.23 Plan view of typical curved box-girder bridge model (1 in. = 25.4 mm). [After Aslam and Godden (1975).]

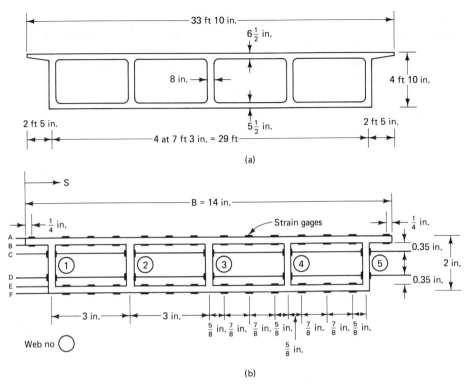

Figure 10.24 Cross section. (a) Prototype bridge. (b) Model and location of strain gages at section 0-0 (1 in. = 25.4 mm). [After Aslam and Godden (1975).]

num alloy used for the model was 6061 T-6 plate with $E = 10.2 \times 10^6$ psi (70,400 MPa) and $v = 0.332$.

The quantities measured in each test included the applied loading W, boundary reactions, deflections at nine points under outer and central webs (Fig. 10.23), and strains at Section 0-0. The load was applied by a small hydraulic jack, measured on a proving ring, and transferred to the model via two linear motion bearings and a steel ball to eliminate horizontal constraints.

The load cells were designed specifically to have large axial stiffness without sacrificing the required accuracy of reaction measurement. Linear variable differential transformers (LVDTs) with an accuracy of ± 0.0001 in. (0.0025 mm) were used to measure the deflections.

A total of 72 strain gages were located closely at Section 0-0 (Fig. 10.24b). All gages were two-element crosses except for four single-element gages used on overhangs. All strain gages were 0.125-in. (3-mm) foil gages. The elements were oriented tangentially and radially to the horizontal curvature at Section 0-0. The gages were placed closely to get a complete stress and moment distribution at Section 0-0 (Fig. 10.24b).

The equipment used for recording the data was a slow-speed scanner consisting of terminal boxes, a digital voltmeter, a portable computer, and a telex typewriter. Proving ring, load cells, LVDT, and all the strain gages were connected to the computer through the terminal boxes and digital voltmeter.

Experimental longitudinal plate forces and bending moments, radial plate bending moments, and transverse deflections compared very favorably to analytical results from computer programs developed at Berkeley. A comparison of radial distribution of experimental and theoretical vertical deflections for the model with the longest span are shown in Fig. 10.25. The distribution curves are essentially parallel and show similar trends. The test values are approximately 5% higher than the analytical results. It was found from this test program that the midspan diaphragm has a comparatively small influence on the response of the system.

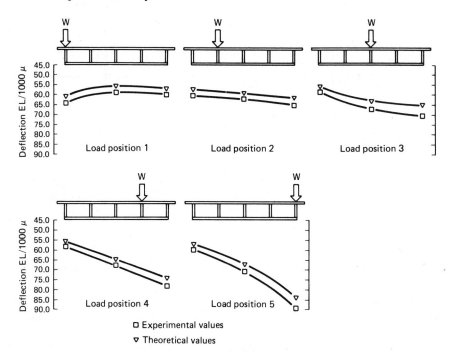

Figure 10.25   Radial distribution of vertical deflections at midspan for model 1A (experiment and theory). [After Aslam and Godden (1975).]

10.2.3 Special Structures

10.2.3.1 Prestressed Concrete Reactor Vessels (PCRV)

Prestressed concrete reactor vessels are complex structural forms that do not lend themselves easily to a completely analytical solution. For this reason, their development has been made possible only through extensive experimental programs using small-scale models. Table 10.3 summarizes some

TABLE 10.3 PCRV Scale Models

Organization	Test Item	Scale	Project	Number of Models	Test for*
French AEC	Head, PCRV	Not known	G-2, G-3	2	A, B, C
	Cylindrical PCRV	$\frac{1}{10}$	G-2, G-3	3	A, B, C
	Cylindrical vessels	I.D. 2 ft 6 in.(0.8m) I.H. 7 ft 6 in.(2.3m)	Safety studies	25	C, D
	Cylindrical vessel	Unavailable	G-2, G-3	2	A, B, C
Societe d'Etudes et d'Equipments d'Entreprises (SEEE), France	Cylindrical PCRV	$\frac{1}{6}$	EDF-3	3	A, B, C, D
	Cylindrical PCRV	$\frac{1}{10}$	EDF-3	1	T
	Cylindrical PCRV	$\frac{1}{5}$	EDF-4	2	A, B, C, T
	"Hot liner" vessel	Not known	General	1	A, B, C, T
Electricite de France (EDF), France	Cylindrical PCRV	$\frac{1}{5}$	Bugey I	2	A, B, C, T
	Two-layer cylinder	$\frac{1}{3}$	General	1	
Central Electric Research Laboratory England	Cylindrical PCRV	$\frac{1}{8}$	Oldbury	1	A, B, C, T
	Cylindrical PCRV	$\frac{1}{8}$	Pre-Oldbury	1	B, C
Sir Robert McAlpine & Sons, England	Cylindrical PCRV	$\frac{1}{7}$	Oldbury	1	A, B, C, T, D
Taylor Woodrow Construction Ltd.	Spherical PCRV	$\frac{1}{12}, \frac{1}{40}$	Wylfa	2	A, B, C
	Cylindrical PCRV	Not known	Wylfa	3	A, B, C
(TWC), England	Cylindrical PCRV	$\frac{1}{10}$	Hunterston B	1	A, B
	Heads, PCRV	$\frac{1}{24}$	Several	12	A, B, C
	Multicavity PCRV	$\frac{1}{10}$	Hartlepool	1	A, B, C
	Head PCRV	$\frac{1}{13}$	Ft. St. Vrain	2	A, B, C, D
Kier Ltd., England	Spherical PCRV	$\frac{1}{12}$	Wylfa	1	A, B, C, T

*A, Elastic response; B, design overpressure; C, failure; D, abnormal conditions; T, long-term creep and temperature.

TABLE 10.3 (cont.)

Organization	Test Item	Scale	Project	Number of Models	Test for*
Atomic Power Constr., England	Cylindrical PCRV	$\frac{1}{10}$	Dungeness B	1	A, B, C
	Cylindrical PCRV	$\frac{1}{26}$	Dungeness B	1	B, C
	Heads, PCRV	$\frac{1}{72}$	Dungeness B	1	B, C
	Heads, PCRV	$\frac{1}{24}$	Dungeness B	3	B, C
	Heads, PCRV	$\frac{1}{26}$	Dungeness B	2	B, C
Building Research Station, England	Cylindrical PCRV	$\frac{1}{10}$	Hinkley Pt B	1	T
	Cylindrical PCRV	$\frac{1}{20}$	Hinkley Pt B	4	T
Foulness, England	Cylindrical PCRV	$\frac{1}{20}$	Study by UKAEA Safety group	10 models to date (30 total) 40% pneumatic	C, D
General Atomic	Cylindrical PCRV	$\frac{1}{4}$	General	1	A, B, C
	Cylindrical PCRV	$\frac{1}{4}$	Ft. St. Vrain	1	A, B, C, D, T
	Multicavity PCRV	$\frac{1}{20}$	HTGR	1	A, B, C
Oak Ridge National Laboratory	Cylindrical PCRV	$\leq \frac{1}{5}$	General	4	A, B, C
University of Illinois	Cylindrical vessels		General		C, D
University of Sydney, Australia	Head, PCRV	$\frac{1}{20}$	General	4	C, D
Siemens, Germany	Cylindrical PCRV (prefabricated blocks)	$\frac{1}{3}$		1	A, B, C
Krupp, Germany	Cylindrical PCRV	$\frac{1}{5}$	Gas-cooled reactor	1	
ENEL/ISMES, Italy	Cylindrical PCRV	$\frac{1}{20}$	HTGR	1	
Ohbayashi-Gumi, Japan	Cylindrical PCRV	$\frac{1}{20}$	HTGR	1	A, B, C
	Multicavity PCRV		HTGR	1	A, B, C

*A, Elastic response; B, design overpressure; C, failure; D, abnormal conditions; T, long-term creep and temperature.

410

of the model testing activities on PCRV structures. As can be seen from Table 10.3, a wide variety of model studies have been conducted on this type of pressure vessel all over the world. Experimental studies in the United States have been reported by Corum et al. (1969) on an extensive program directed by the Oak Ridge National Laboratory. Their test program attempted to investigate the suitability and accuracy of small-scale models for determining certain behavioral aspects of the PCRV.

The study encompassed tests on three different size structures: a small concrete prototype vessel, two mortar models of the prototype, and an epoxy model of the prototype.

The prototype structure is a relatively small and simple prestressed concrete vessel. Shown in Fig. 10.26, the concrete prototype represents the top half of a cylindrical vessel closed with flat heads; its dimensions are on the order of one-tenth those of a full-size PCRV. The prototype has been tested to failure under internal hydraulic pressure, and the data obtained provided extensive information on the elastic and inelastic behavior of this type of prestressed pressure vessel as well as being used for direct comparison with the test results of the two smaller mortar models shown in Fig. 10.26. The mortar models were exact replicas of the concrete model at a geometrical scale of 1 to 2.75. The scale of the mortar models was chosen to produce a model structure that was at or near the minimum feasible size.

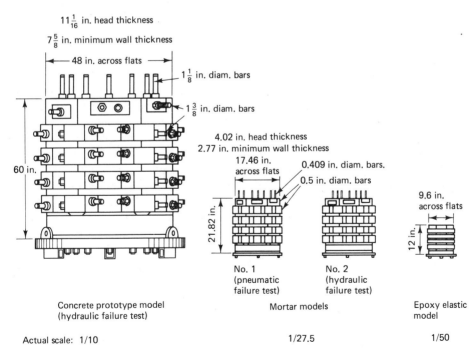

Figure 10.26 Prototype and model vessels. [After Corum et al. (1969).]

One of the mortar models was subjected to pneumatic pressure (nitrogen) and the other to hydraulic pressure (oil) in an attempt to distinguish any possible difference in failure behavior for the two pressurizing media.

The smallest model shown in Fig. 10.26 is a $\frac{1}{5}$-scale prestressed epoxy model that has been analyzed using strain gages. Data were obtained for both prestressing and pressure loading.

The design of the prototype had many features similar to those found in an actual PCRV, including thick head and walls, unbonded post-tensioned prestressing in three directions to counteract the effects of internal pressure, and a pattern of penetrations through the head. The only conventional reinforcing elements in the structure are the two concentric rings of steel mesh at the base and small pieces of mesh under the prestressing anchor plates in the head. The prototype was designed for 500-psi (3.5 MPa) internal pressure, which is representative of pressures met in actual gas-cooled reactors. A conventional concrete with nominal uniaxial compressive strength of 6000 psi (41 MPa) was used, and all prestressing elements were conventional Stressteel bars. A fiberglass-reinforced epoxy liner was applied to the interior cavity of the vessel to prevent permeation of the pressurization fluid into the concrete.

All geometrical features of the prototype were reduced by a scale factor of 2.75. Model materials were chosen to duplicate prototype material properties as closely as possible. The resulting mortar models can thus be considered as true models of the prototype, with strains and stresses theoretically identical in model and prototype at any given internal pressure level. The motrar mix proportions were 1:3:1 [cement, sand, $\frac{5}{16}$-in. (7.9 mm) aggregate] with a water-cement ratio of 0.517.

The models were prestressed prior to pressurization using conventional Stressteel jacking equipment. Six rams (Fig. 10.27) connected to a common hydraulic pressure source were used to prestress six tendons simultaneously. The prestressing was accomplished in two phases, with all tendons stressed to half the final load level in the first phase. In each phase, the axial tendons were stressed first, then the head tendons, the first band of circumferential tendons, the second band, etc. There was a total of 12 steps plus a final adjustment and checking of stress levels. No difficulties were met in achieving an accurate level of prestress in these very short tendons.

Both the concrete prototype and the second mortar model were tested hydraulically, as opposed to the pneumatic test of the first mortar model. Both were cycled twice to a pressure of 300 psi (2 MPa), once to 600 psi (4 MPa), once to 900 psi (6 MPa), and then to the maximum pressure dictated by liner leakage and the capacity of the hydraulic pumps used. The maximum pressure reached in the prototype by using a 60-gpm (gallons per minute) pump was 1130 psi (7.8 MPa). For the second mortar model, the maximum pressure reached with a 3-gpm pump was 1390 psi (9.6 MPa). For comparison, the maximum pneumatic pressure reached in the first mortar model was 965 psi (6.7 MPa). The only apparent evidence of failure was a sudden drop in pressure and the loud

Figure 10.27   Prestressing circumferential tendons in model PCRV. [After Corum et al. (1969).] (Reprinted by permission of the Oak Ridge National Laboratory, operated by Union Carbide Corporation under contract with the U.S. Dept. of Energy.)

hissing of escaping gas. The vessel was then filled with oil and pressurized to 1045 psi (7.2 MPa), which was more than 2.5 times the working pressure of 417 psi (2.9 MPa). In the case of the concrete prototype, visible leakage from the model was first observed at 1000 psi (6.9 MPa), and by the time 1100 psi (7.6 MPa) was reached, oil was spraying from the vessel. The corresponding pressures in the second mortar model were 1100 and 1200 psi (7.6 and 8.3 MPa), respectively.

A comparison has been made of the measured stresses in the three models at the 500-psi (3.5 MPa) design pressure. The predicted and measured meridional and circumferential stresses for the concrete prototype and for the two mortar models are shown in Fig. 10.28. The solid lines show the finite-element predictions for each surface, and the points show values measured with strain gage rosettes for each model. For the meridional stresses the agreement between theory and experiment for all three models is reasonably good. For the circumferential stresses, agreement is reasonably good except on the inside surface of the cylinder. The final cracking and deformation patterns of the mortar models are shown schematically in Fig. 10.29.

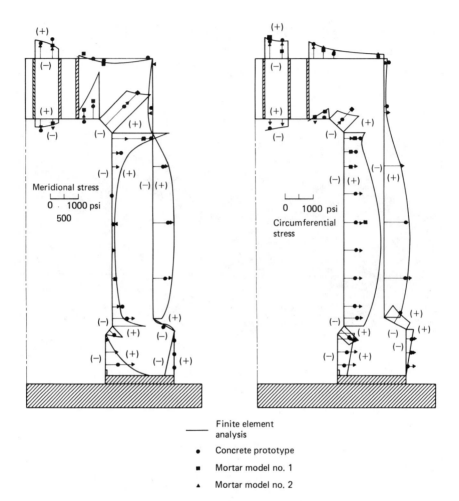

Figure 10.28 Measured and predicted meridional and circumferential stress distributions for an internal pressure of 500 psi. [After Corum et al. (1979).]

In addition to the model tests summarized in Table 10.3, an extensive experimental program was carried out at the University of Illinois, Urbana [Paul et al. (1969)], to study the structural response and modes of failure of PCRVs and to develop analytical procedures. Sixteen small-scale models were tested in their program. The test vessels were cylindrical, 3 ft 6 in. (1.1 m) in diameter and 3 ft 6 in. (1.1 m) long. One end was closed by a concrete slab (the test slab), and the other end was closed by a steel plate. Circumferential prestress was provided by wrapping wire continuously. Longitudinal prestress was provided by straight Stressteel rods or seven-wire strand. The thickness of the end

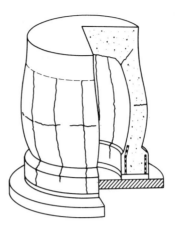

Figure 10.29 Schematic of final cracking pattern and deformation. (Courtesy Gulf General Atomic Inc.)

slab, which had no reinforcement in it, varied from 6 to 15 in. (150 to 381 mm). The wall thickness was either 5 or 7.5 in. (125 or 190 mm).

The majority of the vessels were lined with $\frac{1}{16}$-in. (1.6 mm) neoprene sheets and pressurized internally with nitrogen. Two were tested hydraulically. The measured internal pressures at failure ranged from 240 to 3690 psi (1.6 to 25.4 MPa).

Appa Rao (1975) has carried out a very comprehensive test program on a $\frac{1}{12}$-scale model of a secondary nuclear containment structure up to ultimate failure as a means of checking the linear and nonlinear analytical predictions for such structures.

Testing of physical models of PCRVs for boiling water reactors has been carried out at ISMES, Bergamo, Italy. Fumagalli and Verdelli (1976) report on one such model at one-tenth scale of a single-cavity cylindrical vessel prestressed longitudinally and circumferentially.

A $\frac{1}{20}$-scale model of a more complex multicavity PCRV for a 1000-MW (mega-watt) high-temperature, gas-cooled reactor system was constructed and tested by General Atomic, San Diego, California [Cheung (1969)]. The response of the model to increasing internal pressure was examined in four tests constituting two primary phases of testing. In the first phase, three tests were performed to demonstrate the behavior of the complete vessel under pressure. The maximum pressures achieved in the three tests were 1680, 1400, and 1350 psig [(11.6, 9.6, and 9.3 MPa) pounds per square inch, gauge], respectively. In the first test, separation of the head from the barrel section at the slip-plane occurred at a pressure of about 850 psig (5.9 MPa) (1.3 MCP, i.e., 1.3 times the main cavity pressure). At the maximum pressure of 1680 psig (11.6 MPa) (2.6 MCP), vertical radial cracks had formed through the steam generator cavities, and the liner and rebar steel spanning the cracks had reached yield. No sign of distress was observed at the top head. In the subsequent two pressurizations, the model with cracked barrel section was tested to a maximum pressure equivalent to 2.1

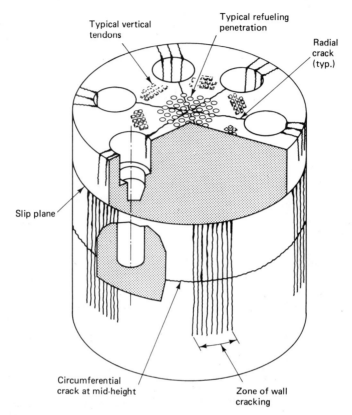

Figure 10.30 Crack pattern on 1/20 scale General Atomic model after all testing was completed. [After Cheung (1969).]

MCP. The final crack pattern is shown in Fig. 10.30 and is a predominant flexural mode of failure.

Takeda et al. (1973) reports on tests of $\frac{1}{20}$-scale single and multicavity PCRVs conducted by Ohbayashi-Gumi Ltd., Japan. The principal dimensions and layout of the $\frac{1}{20}$-scale PCRV model are shown in Fig. 10.31. The model was constructed from concrete with stress-strain characteristics similar to those of the concrete proposed for use in actual vessel construction. Model bonded reinforcement was provided in the form of high tensile deformed wire. Top head penetrations were simulated by seventeen 2.36-in.-diameter standpipes. Representative steel liners for the core and steam generator cavities and cross ducts were included in the model.

Vertical prestress in the model was applied by means of high tensile bars of 0.473- and 0.630-in. (12 and 16 mm) diameter. Circumferential prestressing was accomplished by 0.114-in. (2.9 mm) diameter, high-strength wire wound under tension around the surface of the model.

The model was adequately instrumented to record deflection and strain

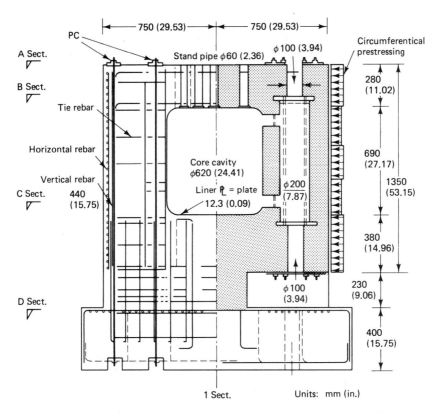

Figure 10.31   1/20th scale Ohbayashi-Gumi model. [After Cheung (1969).]

data during test. Over 300 transducers were included in the model for measurements of strain in the concrete, steel liners, representative standpipes, and bonded reinforcement. Deformation profiles of the model were measured by dial gages and linear potentiometers mounted on an independent reference frame. The main cavity and steam generator cavities together with cross ducts were pressurized hydraulically.

The development of top head and barrel deflections with internal pressure is shown in Figs. 10.32 and 10.33, respectively. The deflection response clearly illustrated the gradual mode of behavior as the model approached the failure pressure. The model was shown to have essentially elastic response to a pressure equivalent to 1.8 times the design main cavity pressure. The model demonstrated an ultimate load factor exceeding 4.

10.2.3.2 Plate and Shell Structures

Plate and shell structures are one form of construction where the modeling technique has played a primary role in its widespread use and general public acceptance. The complex nature of the structural action in such systems has

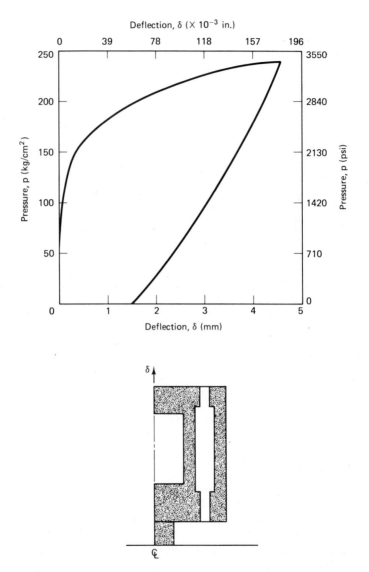

Figure 10.32 Pressure-deflection curve for top head of Ohbayashi-Gumi 1/20 scale model. [After Cheung (1969).]

been made simpler for the student, the researcher, and the designer by means of physical modeling. The modeling technique has been applied successfully for a wide range of problems from elastic behavior studies (to verify and expand the existing analytical techniques) to inelastic and nonlinear studies for research purposes and finally to the ultimate load behavior studies needed by the designer to investigate the structural integrity and safety of the whole

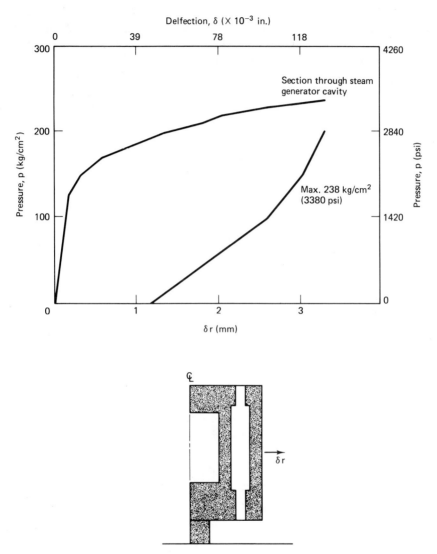

Figure 10.33 Pressure-deflection curve for mid height of barrel of Ohbayashi-Gumi 1/20 Scale model. [After Cheung (1969).]

system. In this section a brief summary of model studies from each of the above categories—educational, research, and design—is presented. Considerable information on model applications to plate and shell structures has been presented at various shell conferences [W. Olszak and Sawczuk (1964), Haas and Bouma (1961), Aas-Jakobsen (1958), Medwadowski (1964), Davies (1966) *Proceedings International Congress of the Application of Shells in Architecture*

(1967), *Proceedings RILEM International Symposium on Experimental Analysis of Instability Problems on Reduced and Full-Scale Models* (1971)], in the *Journal of International Association of Shell Structures*, and in the journals of the American Concrete Institute and the American Society of Civil Engineers. A brief discussion of some of these applications is presented in what follows.

The use of small-scale elastic plate and shell models for obtaining detailed information to aid in the formulation of more rigorous mathematical models has been championed at Princeton University with considerable success [Elms (1964), Billington and Mark (1965), Mark and Riera (1967), Mark and Billington (1969), and Hedgren and Billington (1967)]. To illustrate this type of research application, the study of a hyperboloid of revolution under axisymmetrical and concentrated edge loads is illustrated in Figs. 10.34 and 10.35. The model, studied by Elms (1964), was made from epoxy on a carefully constructed mandrel (Fig. 10.34) to a thickness tolerance of 0.4%. Loading of the model at the upper edge was accomplished using the loading frame shown in Fig. 10.35. Small strain gages were applied to the surface to measure strains in the highly stressed regions.

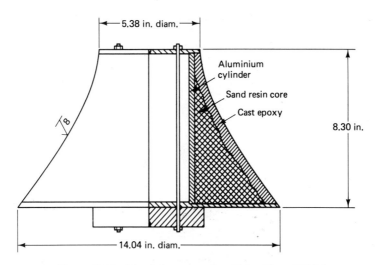

Figure 10.34   Mandrel construction. [After Elms (1964).]

Another powerful experimental technique was used in the study of a continuous folded-plate structure shown in Fig. 10.36. In this case a photoelastic technique using stress freezing was used to "lock" the stresses and deformations into the loaded model. The loading was done in an oven at about 300°F (140°C) for the particular epoxy used in the model. The oven temperature was then slowly lowered to ambient temperature, and the model unloaded. After the deflections were measured, the structure is sliced by careful sawing to be studied under a polariscope. The advantages of this technique are:

Figure 10.35 Uniform radial load applied to hyperboloid of revolution. [After Elms (1964).] (Courtesy of Prof. R. Mark, Dept. of Civil Eng., Princeton University, Princeton, New Jersey.)

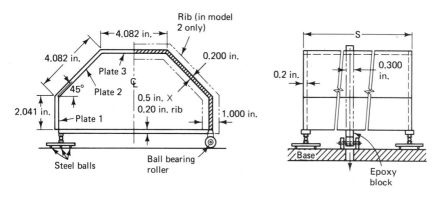

Model	Material	Lengths (see figure)	Total testing load vertical load in psi			Interior supports
			Plate 1	Plate 2	Plate 3	
1 A	Hysol epoxy resin CPS–4290	15.80 in.	0.009	0.009	0.109	Columns
1 B	Photolastic Inc. epoxy resin PL4	15.80 in.	0.009	0.009	0.109	Columns
2	Photolastic Inc. epoxy resin PL4	15.24 in.	0.009	0.009	0.129	Gable

Figure 10.36 Model geometry. [After Mark and Riera (1967).]

1. External strain measuring devices are eliminated.
2. Full-field observation enables one to study the behavior of a complete system.
3. Because the effective gage length of the measurements is nearly zero, complete stress analysis is feasible at discontinuities—openings, column capitals, stiffening ribs, etc.—even in very small scale models. Magnitudes of stress concentrations can be readily determined.
4. Simple loading systems may be used.
5. Modern photoelastic instrumentation is simple and inexpensive.
6. Internal stresses are obtained.

A limitation of stress freezing is the requirement that the model material be homogeneous, elastic, and subjected only to static loadings. Nevertheless, because almost all plate and shell structural design is based on forces and moments that are predicted by linear elastic analyses of assumed homogeneous structural systems, the use of homogeneous elastic models has wide application.

A comparison between theory and experiment of the transverse moments and deflections at the quarter span of the folded-plate model is shown in Fig. 10.37. The logical extension of the photoelastic technique to singly curved surfaces such as barrel vaults is shown in Fig. 10.38. The stress-freezing method was used to obtain deflections, membrane stresses, and bending stresses on cut slices. The effect of the end diaphram on the longitudinal bending stress observed from a longitudinal slice is shown in Fig. 10.39.

In order to be able to study the inelastic behavior and the ultimate strength of plate and shell structures made from reinforced and prestressed concrete, a

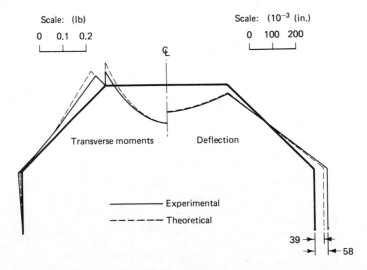

Figure 10.37 Transverse moments and deflections at $x = 3.81$, model 2. [After Mark and Riera (1967).]

Figure 10.38   Edge loads applied to shell model. [After Billington and Mark (1965).] (Courtesy of Prof. R. Mark, Dept. of Civil Eng., Princeton University, Princeton, New Jersey.)

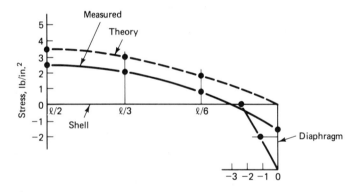

Figure 10.39   Longitudinal bending stress on outer surface along longitudinal slice. [After Billington and Mark (1965).]

different type of model than those described above is required. For such cases, we must resort to the type of model made from materials as close to the proto-type materials as possible (Chapters 4 and 5) and having the same internal construction as that used in the prototype structure. Some examples of ultimate strength model studies of shell structures are described below.

   One of the most extensive model studies dealing with the behavior of cylindrical shell roofs under gravity loading was undertaken at the Institute for Building Materials and Building Constructions, center for Applied Scientific Research, T.N.O., in the Netherlands [Bouma et al. (1961), Van Riel et al. (1958)]. A series of 11 circular cylindrical shells made from reinforced mortar

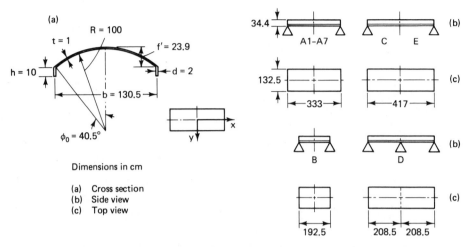

Figure 10.40 Dimensions of shell models. [After Bouma et al. (1961).]

and having the geometry shown in Fig. 10.40 at one-eighth scale were loaded to destruction by means of increasing uniform gravity loading. The objective was to study the effects of cracking and plasticity and amount and position of the reinforcement on the stress distribution and how this deviates from the theory of elasticity at higher loads. The amount of reinforcement and its distribution in both the shell and edge beams is shown for each of the shells tested in Fig. 10.41. Also shown by means of bar charts in Fig. 10.41 are the loading histories from initial cracking to ultimate failure of each of the models. Note from Fig. 10.41 that the ultimate carrying capacity of all models was greater than the design load by factors ranging from 2.7 to 4.5. The load vs. deflection charac-teristics of the models in series A are shown in Fig. 10.42 and indicate practically elastic behavior up to the design load. Beyond this load level the deviation from elastic behavior is dependent on the amount and the distribution of the reinforcement. Typical failed specimens with longitudinal yield lines in both the positive and negative transverse moment regions are shown in Fig. 10.43.

A series of smaller ($\frac{1}{24}$-scale) mortar shells having the same geometry as the shells in series A in Fig. 10.44 were studied by Harris (1967) and Harris and White (1972). The main objective of this study was to determine the feasibility of using very small scale "table top" models that required relatively little material and modest loading facilities to study the inelastic behavior of circular cylindrical shells. The $\frac{1}{24}$-scale models having a thickness of only 0.131 in. (3.3 mm) were made of reinforced mortar with similar properties to the larger $\frac{1}{8}$-scale models used as "prototype." Loading was accomplished in a scaled version of that used in the prototype shells. This consisted of a whiffle-tree arrangement with the load applied by means of pads to the outer surface of the shell [Fig. 7.16(b)]. The inner and outer surfaces of a failed shell model are shown in Fig. 10.44. The remarkably close correlation of the yield lines at failure of two small-scale

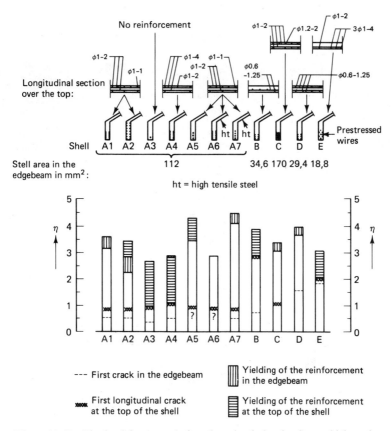

Figure 10.41 The load factors η (referred to the design load) at which various phenomena occurred. [After Bouma et al. (1961).]

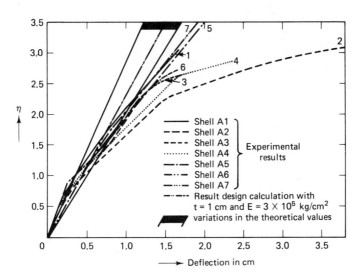

Figure 10.42 Shells of series A. Average deflection of edge beams at mid span. [After Bouma et al. (1961).]

Figure 10.43   Failure patterns of large model shells. [After Bouma et al. (1961).] (Courtesy of T.N.O., Delf, Holland.)

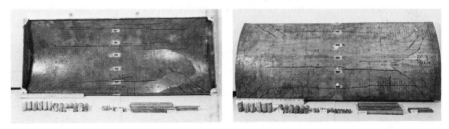

Figure 10.44   Inside and outside failure surfaces of small model. [After Harris and White (1972).]

models and their four times larger counterparts (prototype) are shown in Fig. 10.45.

Other shell forms have also been studied extensively by means of small-scale models. Representative of this work is the testing of a variety of hyperbolic paraboloid (hypars) shell models recently summarized by White (1975). Both plastic and reinforced mortar models were used to study such diverse questions as:

1. General behavior and failures modes

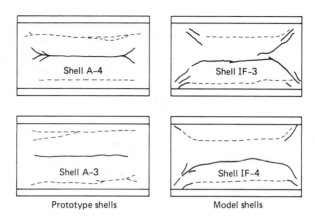

Figure 10.45 Comparison of large and small model failures. [After Harris and White (1972).]

2. Influence of size and location of edge members on shell stiffness, strength, and bending
3. Interaction of shell and edge members

A much larger inverted-umbrella hypar shell, tested at the Portland Cement Association Laboratories, was used to establish reliability between models and prototype. Table 10.4 gives a comparison of model and prototype hypar shell properties. Figure 10.46 gives the geometry and reinforcement of the prototype and the mortar and plastic small-scale models tested at Cornell University. A typical inverted-umbrella hypar mortar model shell during testing is shown in Fig. 7.16. From this investigation it can be concluded, that the tensile edge member, in the periphery of the shell, had high tensile stresses that led to severe

TABLE 10.4   Model Behavior Compared with PCA Prototype

Structure (1)	E, kips/in.^{2} (MN/m^2) (2)	Poisson's Ratio, ν (3)	Stress Scale, $S_\sigma = S_E$ (4)	Length Scale, S_L (5)
Prototype	3,610 (24,910)	0.18	—	—
Model 2-MA	3,210 (22,150)	0.18	1.11	6
Model 3-MA	2,870 (19,800)	0.18	1.26	6
Model 5-MB	3,320 (22,910)	0.18	1.09	6
Models 1-PA and 2-P	434 (2,995)	0.34	8.31	8

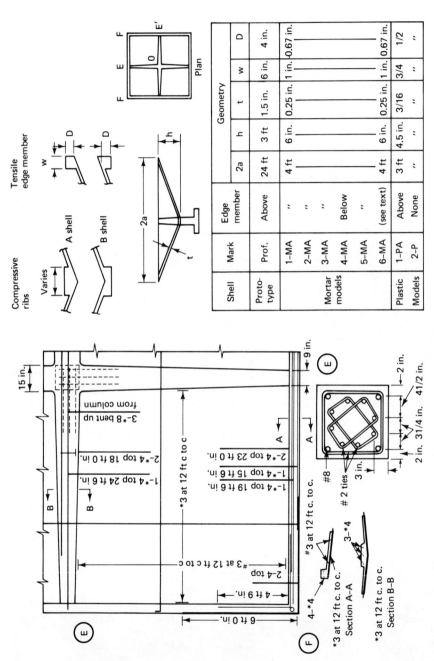

Figure 10.46 Shell geometry and reinforcing (1 in. = 25.4 mm; 1 ft = 0.305 m). [After White (1975).]

Shell	Mark	Edge member	2a	h	t	w	D
							Geometry
Proto-type	Prof.						
	1-MA	Above	24 ft	3 ft	1.5 in.	6 in.	4 in.
Mortar models	2-MA	"	4 ft	6 in.	0.25 in.	1 in.	0.67 in.
	3-MA	"					
	4-MA	Below					
	5-MA	"					
	6-MA	(see text)	4 ft	6 in.	0.25 in.	1 in.	0.67 in.
Plastic Models	1-PA	Above	3 ft	4.5 in.	3/16	3/4	1/2
	2-P	None	"	"	"	"	"

cracking at high overloads, thus establishing this as the "weak link" of the system. Also, it was found that a portion of the shell near the edge members picks up some of the edge member load and is effective in resisting these in conjunction with the edge members. Compressive edge members were found to be very essential for carrying unsymmetric live loads. It is recommended, from this study, that the inclined compressive ribs be placed under the shell while the horizontal tensile edge members be placed above the shell surface.

The combination of reinforced mortar models and plastic models is particularly useful in studying behavior of reinforced concrete shell structures. The mortar model is essential for determining post elastic response (Fig. 10.47), and the plastic model is much better suited for detailed studies of strain distributions and the interrelationships between shell and edge members at working load levels. Either type is adequate for predicting working load displacements.

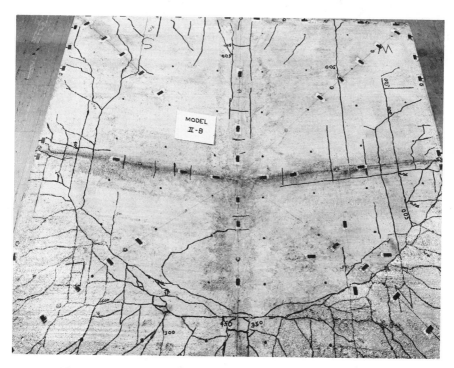

Figure 10.47   Model 4-MB after loading to failure. [After White (1975).]

In addition to stress distribution and strength determination studies of plate and shell structures, small-scale models have played an important role in studies of the instability of these complex three-dimensional structures. A recent survey of experimental studies of concrete shell buckling has been reported by Billington and Harris (1979). A similar review for folded-plate shells has been reported by Guralnick, Longinow, and Schwartz (1979). Only a brief summary of studies dealing with the buckling of spherical domes will be covered here.

The most extensive experimental buckling program on concrete spherical shells was conducted at the University of Ghent, Belgium, under the direction of Dr. Vandepitte from 1967 to 1976. Results from the testing of the first seven shells has been reported by Vandepitte and Rathe (1971), and a final report covering the testing of an additional 83 shells has been given by Weymeis (1977) in an internal report.

The shells used for this program were made of microconcrete. The concrete quality and the dimensions of the dome were chosen in such a manner that instability was to occur before the stresses in the concrete could attain the value of the ultimate compressive strength of the material, which was about 45 MPa. In all cases, the rise of the domes and the diameter of their perimeter were 193 mm and 1900 mm, respectively (Fig. 10.48). They were 7-mm thick, had a radius of curvature of 2431 mm, and were loaded in an upside-down manner with uniform radial pressure, using a hydraulic system. A steel ring beam support arrangement into which the domes were cast could also be prestressed during testing to minimize the edge stresses.

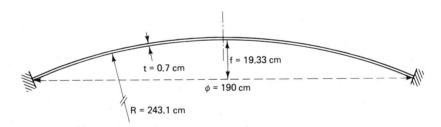

t = 0.7 cm

f = 19.33 cm

ϕ = 190 cm

R = 243.1 cm

Figure 10.48   Shell geometry. [After Vandepitte and Rathe (1971).]

From this very comprehensive experimental program the following conclusions were drawn by the investigators:

1. All domes failed in the same manner. A nearly circular disc is punched out of the dome, generally closer to the edge than to the center of the dome. This shows a clearly asymmetric failure pattern (Fig. 10.49).
2. The average value of the coefficient c in the buckling formula for a radially loaded spherical shell, given by

$$p = cE\left(\frac{t}{R}\right)^2 \tag{10.2}$$

where

$$E = \text{the modulus of elasticity}$$
$$t = \text{the thickness of the shell}$$
$$R = \text{the radius of curvature}$$

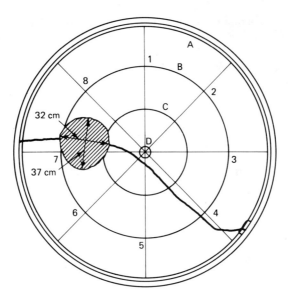

Figure 10.49 Failure mode of domes. [After Vandepitte and Rathe (1971).]

was found to be 0.542, with a standard deviation of 9.5%. This is somewhat higher than the theoretically determined values of c given in Table
10.5.

**TABLE 10.5   Analytically Determined Values
for c in Eq. (10.2)**

c	Reference
0.365	Von Karman and Tsien (1939)
0.34	Tsien (1942)
0.32	Mushtari and Surkin
0.312	Feodosiev
0.31	Vreedenburgh (1966)
0.178	Classical value ($v = 0.2$) (not taking finite changes of the geometry into account)

10.3 CASE STUDIES

10.3.1 Case Study A, TWA Hangar Structures

10.3.1.1 Introduction

This section describes the elastic and strength model studies used to
assist in the design of the roof structure of the Trans World Airlines hangars
at Kansas City [Guedelhoefer et al. (1972)]. (See also Chapter 1.)

After some discussion among the parties involved, the basic objectives of the study were defined as follows:

1. To obtain data for determination of wind forces on the prototype structure.
2. To determine stresses in the shell, edge beams, and center arch under various combinations of loadings in the design or service load range
3. To obtain information to serve as a basis for accurately predicting short-term and long-term deflections
4. To assess the ultimate strength and the factor of safety of the roof structure

Basic characteristics and applications of three elastic models A1, A2, and A3 and one ultimate strength model A4 will be briefly reviewed in this section.

10.3.1.2 Wind Effects Model A1 (Phase I)

A $\frac{1}{300}$-scale wooden site configuration model was built reproducing accurately the shape of the entire TWA hangar complex along with certain features of the adjoining open country terrain that were considered to be influential on the distribution of external wind pressures or suction on the hangar complex [Guedelhoefer et al. (1972)].

Four configurations were used, consisting first of the general shopping area and then successively adding hangars until the final configuration was obtained. Tufts of string mounted on the model established the air-flow patterns from photographs in a wind tunnel test. Pitot tubes were installed at various elevations in the vicinity of the model hangars to establish the wind profile. The surrounding structures did not create unusual wind patterns influencing the pressures on the model. The model was mounted on a rotating table and subjected to all wind directions in 30° increments.

10.3.1.3 Wind Effects Model A2 (Phase II)

Subsequently, a $\frac{1}{100}$-scale wooden model accurately simulating the hangar geometry details was constructed to develop more data regarding distribution of wind forces (Fig. 1.3). The exterior surface of the model was roughened to create turbulent effects expected in the prototype. The instrumentation consisted of Pitot tubes at 90 selected locations on the outside surface and five tubes in the interior of the structure. The phase II program consisted of the following three testing arrangements:

1. The shell construction phase completed and all doors and walls absent

2. The hangar building completely enclosed by side walls and all movable hangar doors closed

3. The complete structure with the enclosing walls but with all hangar doors fully open

The artificially generated wind profile represented the average of worst conditions as established from model A1. The wind tunnel could generate 70- to 90-mph (112–144 km/h) winds; all recorded data were corrected to the expected 100-yr return period wind velocity of 87 mph (139 km/h). Again, this model was subjected to all wind directions in increments of 30°. Similar to the observation in the smaller model, the surrounding structures were found to have no increasing effect on the pressures on the structure; in fact, the observed pressures showed a slight decrease.

10.3.1.4 Elastic Model A3

The data from the wind tunnel tests on models A1 and A2 were used to finalize the preliminary shell design and details of a $\frac{1}{50}$-scale elastic model of the hangar structure. The elastic model was built of fiberglass and tested to study the elastic behavior of the structure under dead loads, live loads, concentrated loads, and their combinations (Fig. 1.4).

Electrical resistance strain rosettes were installed at 50 stations to determine the principal stresses and the principal directions, the resulting maximum and minimum flexural and membrane stresses, and the orientation of the planes on which these stresses acted. Additional instrumentation was placed on the tie beam and on each of the two rear columns to monitor the loads in these elements.

The loading system initially consisted of 256 small containers of lead shot to represent the prototype dead load. Each lead shot container was suspended by a string passing through the shell and tied to a button that transmitted the load to the shell. This system caused large local bending stresses because the shell was extremely thin. Various other types of button and pad systems were tried in conjunction with the lead shot system, and all of these were considered unsatisfactory for representation of dead or live loads. Therefore, vacuum loading was adopted as an alternate loading system; the local bending stresses were noted to decrease considerably in the shell areas. Thus, dead load using lead shot containers was used for simulating the heavier edge beam and hangar door loads, and vacuum loading was used to reproduce the uniform portion of the dead load (Fig. 1.4).

The design provided for large crane, scaffolding, and other mechanical support equipment to be suspended from the shell. Figure 10.50 shows how the simulation of the concentrated loads was achieved in the model.

Figure 10.50   Rear dock loads on the elastic model (A3). [After Guedelhoefer et al. (1972).] (Courtesy of Wiss, Janney, Elstner & Associates, Inc., Northbrook, Illinois.)

10.3.1.5 Ultimate Strength Model A4

After the results of the preceding models were evaluated by the design engineers, and subsequent changes incorporated into the design, a microconcrete model was constructed to a scale of 1 to 10 (see Fig. 1.5). The primary purpose of this phase of the investigation was to duplicate to scale, as accurately as possible, the prototype structure, to perform load tests to determine the load-deflection response of the structure at various points, and to determine the ultimate strength, with special attention given to mode of failure and factor of safety. Pertinent dimensions of the model are shown in Figs. 10.51 and 10.52.

Several trial mixes were produced in the laboratory, and 2×4 in. (50×100 mm) cylinders were tested for agreement with prototype design specifications. Reinforcing bars were simulated by using individual wires of commercially available sizes and commercially available welded wire fabrics. Precise distortion is introduced into the model by adjusting the bar locations or spacing to provide for slight differences in yield strength. With a scale factor of 1 to 10, wire sizes were available so that reasonably accurate scaling of significant reinforcing bars was possible in most cases.

Some distortion was necessary in order to simulate the shell surface designated as surface number 2 (Fig. 10.51). In this region of the structure, a welded wire fabric of No. 15 gage wire was selected with spacing between wires chosen to provide the proper steel area in each of these directions. The design required reinforcing steel conforming to ASTM designation A432, with a minimum yield stress of 60,000 psi (414 MPa) throughout. The wire reinforce-

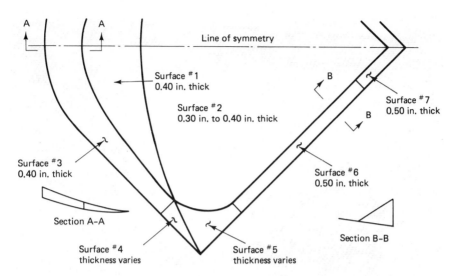

Figure 10.51   Shell surface numbering system. [After Guedelhoefer et al. (1972).]

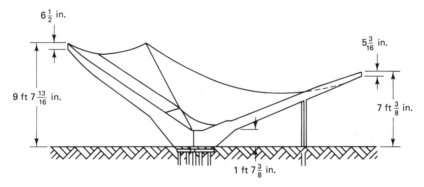

Figure 10.52   Elevation view of microconcrete model. [After Guedelhoefer et al. (1972).]

ment (used in the abutment regions, primarily) had a yield strength much higher than 60,000 psi (414 MPa); it was annealed moderately to reduce the yield strength. The most complex part of the model reinforcement was the front abutment wall, shown in the foreground of Fig. 10.53.

Because the shells of the model were to be posttensioned, supplementary investigations were undertaken to develop the necessary hardware and technique. Special auxiliary tests were conducted with the following objectives: to develop and test an anchorage system for the post-tensioning tendons; to develop and test a stressing system; to test a form coating material; and to test an air-springing system to assist in decentering the formwork.

Figure 10.53   Microconcrete model during construction. [After Guedelhoefer et al. (1972).] (Courtesy of Wiss, Janney, Elstner & Associates, Inc., Northbrook, Illinois.)

The actual sequence of casting the concrete was designed to be parallel to that anticipated for the prototype: footings, abutments, shell surfaces 1 and 2, vertical edge stiffeners, and then all other shell surfaces.

Screeding strips were used to control thickness in the 0.3-in. (7.6 mm) thick and 0.4-in. (10.2 mm) thick areas of the shells. The actual casting process was continuous and concluded in 6 hr. Vibration was accomplished by hand-held vibrating sanders.

When the forms were removed, deflections and strains due to model weight were measured. The deflections of the rear edge were imperceptible, while the front edge deflected only 0.05 in. (1.3 mm) (i.e., 0.91 in. (23.1 mm) in the prototype).

A series of front truss lateral load tests were conducted to determine the effect of wind loading against the hanging front enclosure walls. The forces acting on the walls from either wind pressure or suctions were distributed to a series of 14 trusses, which in turn delivered the loads to the shell structure. Various combinations of pressures and suctions statically simulating conditions observed with the wind models were studied. It was found that the edge members were carrying most of these forces directly to the abutments, with only nominal stresses observed in the shell areas.

Since the uniform loads were produced by vacuum techniques, the thickened edge beams required the application of additional superimposed loads to correct for the dead loads of these members. Accurately weighed sand bags were used to apply these loads.

Dead-load data were taken from the model in two stages: first, when superimposed dead load was applied to the edge beams, and second, when the uniform dead load was applied. As testing progressed, it was decided that useful information could be obtained if the superimposed dead load was added in two controlled stages. The first stage was to place the full superimposed load on the rear stiffener, during which time complete deflection and strain data were recorded. The second stage was to place the full superimposed load on the front stiffener and to monitor instrumentation.

The final test on this model consisted of loading to failure by the vacuum loading method. As the load was applied in increments, data were recorded at the following load levels: 44 psf (2.1 kN/m²), 94 psf (4.5 kN/m²), and 112 psf (5.4 kN/m²). At a vacuum of 25.5 in. of water, or 132.5 psf (6.3 kN/m²), the model failed. The actual total load at this point on the structure was 137.5 psf (6.9 kN/m²) uniform load, plus the superimposed load of the edge beam. The ultimate failure mechanism was a pure tension (membrane) failure of the shell reinforcement, allowing the main area of surface 2 to separate and "tear" perpendicular to the line of symmetry (Fig. 10.54).

It is interesting to note that the structure as a whole did not collapse,

Figure 10.54   Model after failure. [After Guedelhoefer et al. (1972).] (Courtesy of Wiss, Janney, Elstner & Associates, Inc., Northbrook, Illinois.)

even though the center portion of the rear shell surface had failed. In fact, later when it became necessary to remove the model from the laboratory, the curved edge members and the center arch were extremely difficult to break up, even with sledge hammers, which attests to the inherent strength of the system. Model displacement during loading and the load at collapse showed excellent agreement with predicted values. Structural performance characteristics of the prototype to date have been excellent, and checked well with the model.

10.3.2 Case Study B, Three Sisters Bridge

10.3.2.1 Introduction

The bridge, described briefly in Chapter 1, was to set several precedents, and thus it was decided that structural models should supplement the design calculations. Recognizing the complexity of the bridge with its curved side spans, slender webs, curved fascia, and very long main span, the Federal Highway Administration requested that a microconcrete model be constructed and tested. The objectives of constructing and testing the $\frac{1}{10}$-scale model were:

1. Confirmation that the design provides a structure complying with all service load criteria or indications where modifications are necessary or desirable
2. Determination that the design provides a structure with adequate strength. In addition, the construction of a relatively large scale model in concrete provided a better aesthetic appreciation of the proposed bridge.

Data obtained from the construction and testing of the laboratory model were:

1. Pier and abutment reactions
2. Stresses at selected locations during construction
3. Vertical deflections to provide information on both flexural and torsional stiffness at several cross sections of the bridge
4. Lateral stability of the slender webs and fascias
5. Stresses at selected sections as a result of dead load and uniform or concentrated live loads
6. Ultimate load, and identification of the element that governed strength
7. Shrinkage and creep deformations and the extent of cracking of the concrete model at all stages of construction and testing

10.3.2.2 Model Construction

Since the prototype bridge was symmetrical about the center of the main span, only one-half of the bridge was modeled, as shown in Figs. 10.55 and 1.7. Although the prototype is designed to be cast in place in segments, the model

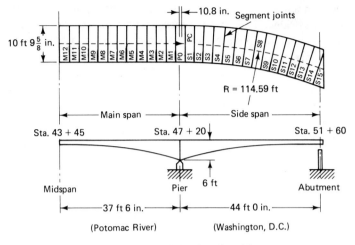

(a) Plan and elevation of model

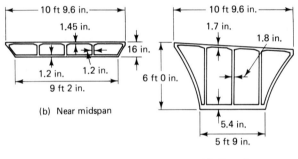

(b) Near midspan

(c) Near pier

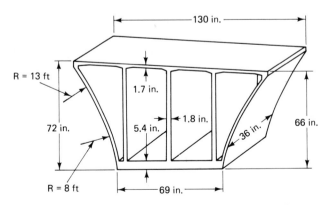

(d) Dimensions of precast model segment near pier

Figure 10.55 Model details and dimensions. [After Corley et al. (1972, 1975).]

[Corley et al. (1972, 1975)] was constructed of precast 3-ft. (0.91-m) segments that were sequentially grouted in position and post-tensioned together to form the complete bridge. This use of precast segments was strictly for convenience in the laboratory. To simulate the field construction, continuity of reinforcement was maintained across all joints. Dimensions of a model segment near the pier are shown in Fig. 10.55d. Longitudinal and transverse prestressing ducts were formed by greased steel rods encased in polyvinylchloride (PVC) tubing and placed in the deck. Continuity of the roadway sections was ensured by aligning longitudinal ducts with metal templates and by casting each new section against a previously cast section.

Using the cantilever method, the model bridge was constructed so that the superstructure was always heavier on the abutment side of the pier. Overturning of the partially completed bridge was prevented by a temporary support initially located near the pier in the side span and later moved to a position about two-thirds of the side span away from the pier. At all times after the first longitudinal tendons were stressed, a 3000-lb (1360-kg) weight representing the 300,000-lb (136-t) weight of construction equipment on the prototype, was kept near each end of the model. It was intended that the same cantilever construction procedure will be used for the prototype.

A microconcrete mix with proportions of 1 part Type III cement to about 5.25 parts Elgin fine aggregate was used to cast the segments. A water-cement ratio of 0.7 gave a slump of about 1 in. (25 mm). Properties of the model concrete used in the ultimate strength model are summarized in Table 10.6.

TABLE 10.6   Properties of Model Concrete at 28 Days

Location of Concrete	Test of 6 × 12 in. (150 × 300 mm) Cylinders*	Average Values, psi (MPa)	Standard Deviation, psi (MPa)	Coefficient of Variation, %
Segments	f'_c	5,870(40.5)	520(3.6)	8.9
Segments	E	3,350,000(23,100)	192,000(1,324)	5.7
Joints	f'_c	7,770(53.6)	595(4.1)	7.7

*f'_c = compressive strength; E = modulus of elasticity

Prestressing wire met the requirements of ASTM designation A421-65. Strengths obtained from tests of representative samples of wire used in the model are shown in Table 10.7.

10.3.2.3 Testing and Instrumentation

Eighteen load cells were used to measure applied forces; ten cells under the pier and two at the side span abutment measured reactions. Applied loads were monitored both with pressure cells in each hydraulic system and with six load cells distributed through the dead-load and live-load systems.

TABLE 10.7   Properties of Model Prestressing
Reinforcement

Diameter	Yield Stress at 1% Elongation, f_y, ksi (MPa)	Ultimate Strength, f'_s, ksi (MPa)
5 mm	264(1,820)	292(2,000)
0.25 in.	218(1,500)	254(1,750)

Electrical resistance strain gages were placed on eleven main deck tendons above the pier and on four soffit tendons at the calculated location of maximum positive moment. Tendon forces were also sensed with a load cell at each end of one long longitudinal deck tendon and with one load cell each on a soffit tendon, a web diagonal tendon, and a fascia tendon.

Strain gages were placed on the concrete surfaces at 360 locations. These included gages on nine heavily instrumented cross sections, two diaphragms, and on the soffit near the pier.

Deflections of the superstructure were sensed at 26 locations, including the quarter-points and midlengths of each span and the free end of the main span cantilever, using a linear potentiometer connected to the test floor.

At each load increment, loads, reactions, strains and deflections were recorded on printers and punched tape using a high-speed VIDAR digital data acquisition system. Recording of 400 channels of information with this equipment required about 40 sec. Rotation and web lateral deflection were continuously traced on oscillographic recorders. Selected load vs. rotation, load vs. vertical deflection, and load vs. web lateral deflection outputs were displayed continuously on x-y recorders with the y axis representing applied load.

Hydraulic loading equipment below the test floor was arranged to apply dead load (D) and live load (L) to the model. The two independent hydraulic systems used to apply these loads are shown schematically in Fig. 10.56.

The first step in the test sequence was to transfer the load from the springs to the hydraulic system and hold 1.0 times dead load of the prototype (1.0D), that is, a load 10 times the self-weight of the model. The initial readings were then taken.

10.3.2.4 Test Results

Performance during Construction.   Observed reactions, strains and deflections during construction of the bridge model were all within anticipated limits. The bridge model was observed to respond elastically as each new segment was erected and dead load was applied. In addition, no cracks attributed to applied load were found.

Performance under Design Service Load.   Under the design service load of $1.0D + 1.0(L + I)$ ($I =$ Impact), the microconcrete model was observed to

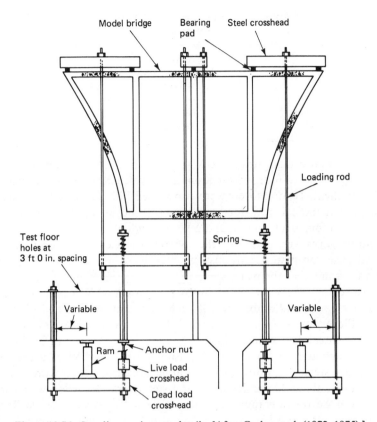

Figure 10.56 Loading equipment details. [After Corley et al. (1972, 1975).]

perform as anticipated. No cracks caused by applied load were found. Strains and deflections measured at critical locations were observed to be proportional to the applied load, indicating that the structure remained essentially elastic. A comparison of measured and calculated load deflection curves for midspan (end of cantilever) deflections is shown in Fig. 10.57. Calculated deflections were based on an uncracked section and measured material properties.

Behavior at service load was within the limits generally assumed in design. These experimental results indicate that the serviceability requirements implied by the AASHTO specifications are met by the design.

Performance under Design Ultimate Load.   After the service load tests, the bridge code design ultimate load of $1.5D + 2.5(L + I)$ was applied to the microconcrete model (Fig. 10.58). Under this extreme overload, the model was observed to safely carry the applied loads. Some inelastic strains and deflections were observed, and cracks occurred both in the negative moment region over the pier and in the positive moment region near the abutment. All the inelastic

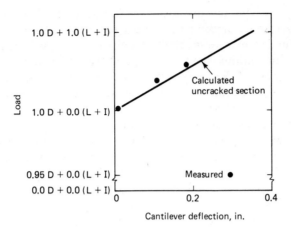

Figure 10.57 Load vs. deflection of cantilever end for service load test. [After Corley et al. (1972, 1975).]

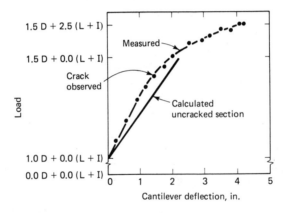

Figure 10.58 Load vs. deflection of cantilever end for ultimate load test. [After Corley et al. (1972, 1975).]

behavior observed under application of design ultimate load was within ranges anticipated.

After the overload was removed and the condition of $1.0D$ had been restored, all cracks were observed to have closed until they were barely visible to the naked eye. This behavior indicated, and measured strains confirmed, that the longitudinal prestressing tendons remained elastic under application of the design ultimate load.

Corley et al. (1972, 1975) concluded that the results of the structural tests carried out on the $\frac{1}{10}$-scale model (of the new Potomac River crossing, I-266) showed that under the application of service load, representing dead load of the prototype and live load plus impact under HS20-44 loading, no structural cracking occurred and the bridge remained essentially elastic. Consequently, the design met serviceability requirements implied by the AASHTO specifications. In addition, the results showed that the model safely carried the overload of $1.5D + 2.5(L + I)$ with only minor structural cracking. Consequently, it is concluded that the design also meets the ultimate strength required by AASHTO specifications.

Following completion of the design ultimate load test, several special tests were made to compare computed and observed behavior of the model. Results of concentrated load tests were used to obtain influence lines for the bridge. Similarly, analysis of data from torsion tests showed that the model had somewhat higher torsional stiffness than was used in the design.

After the concentrated load tests and torsion tests had been completed, the model was loaded to destruction (Fig. 10.59). Data obtained during this test showed that even under severe overload of $1.5D + 6.0(L + I)$, there was an additional margin of safety against instability of the fascias and webs. Figure 10.60 shows the pier region after the test to destruction.

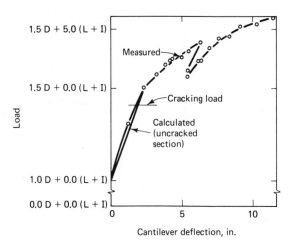

Figure 10.59 Load versus main-span deflection relationship during test to destruction. [After Corley et al. (1972, 1975).]

10.3.3 Case Study C, Multistory Reinforced Concrete Frames

10.3.3.1 Study of Joints under Cyclic Loading; C1 Models

Seven $\frac{1}{10}$-scale models of four full-scale prototypes of exterior beam-column joints for building [Hanson and Conner (1967)] were tested to determine the reliability of the models in predicting joint behavior under simulated seismic loadings [Chowdhury and White (1977)]. A typical model specimen is shown in

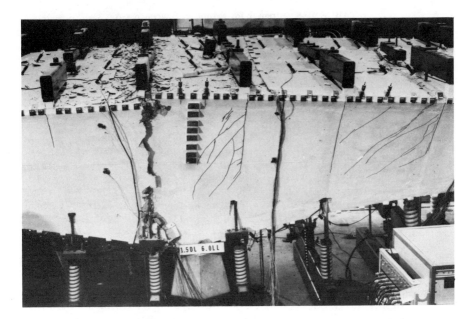

Figure 10.60 Pier region after test to destruction. [After Corley et al. (1972, 1975).] (Courtesy, Portland Cement Association, Skokie, Illinois.)

Fig. 1.9. Each prototype represented the exterior beam-column joints of a high-rise building between column inflection points and the beam inflection point for the condition of lateral loading.

Primary variables included (1) degree of joint confinement, (2) column axial load, and (3) concrete strength in the joint area. The load history for models and prototypes was simultaneous axial load in the column and a fully reversing bending action on the cantilever beam, as illustrated in Fig. 1.8b. The ductility is defined as the ratio of actual rotation over a gage length of one-half beam depth to the rotation that exists at the yield load over the same gage length.

The models were loaded in a small universal testing machine with an auxiliary hydraulic loading system for applying the beam loads (Fig. 7.21). The main instrumentation consisted of small electrical resistance strain gages (0.062-in., or 1.58-mm, gage length) on the reinforcing wires, and linear variable differential transformers (LVDTs) mounted to measure rotations, displacements, and joint shear distortion.

Three models of prototype specimen 1-A represent the most complete set of tests. In this design the steel percentages for the beam were $\rho = 1.9\%$ and $\rho' = 0.95\%$, and the column had 5.31% steel. The material strengths for the prototype and models Cla, Clb, and Clc are given in Table 10.8.

The yield strength of the beam steel of each of the models is 6% lower, and the applied model beam loads were scaled accordingly. In models Cla

TABLE 10.8 Strength for Prototype Specimen 1-A
and Models Cla, Clb, and Clc

Structure	Concrete Strength, psi (N/mm²)		Steel Yield Strength, ksi (N/mm²)		
	Beam	Column	Beam	Column	Ties
Prototype	3200	5320	47.8	69.8	52.8
Specimen 1-A	(22.1)	(36.7)	(329)	(481)	(364)
Model Cla	3670	5600	45.1	68.2	54.5
	(25.3)	(38.6)	(311)	(470)	(376)
Model Clb	3810	6160	45.1	68.2	54.5
	(26.3)	(42.5)	(311)	(470)	(376)
Model Clc	3150	4700	45.1	68.2	54.5
	(21.7)	(32.4)	(311)	(470)	(376)

and Clb the beam concrete strength was 15% and 19% higher than attempted, respectively, and it was 1.5% lower in model Clc. The column concrete in models Cla and Clb was also overstrength, while in model Clc it was about 9% understrength. Since model concrete tends to have higher tensile strength than desired, it is felt that the slight understrength of model Clc column concrete was beneficial in terms of modeling diagonal tension cracks in the column. Estimating the prototype concrete tensile strength at $7\sqrt{f'_c}$ and model concrete tensile strength at $9.5\sqrt{f'_c}$ (English units), model Clc column concrete is about 25% overstrength in tension and model Clb about 55% overstrength. Thus, it should be expected that model Clc would provide the best prediction of prototype specimen 1-A.

The moment-rotation responses of prototype specimen 1-A and model Clc are compared in Fig. 10.61 for load cycles 1, 2, 3, and 7. Cycles 4 to 6 were omitted for clarity purposes. The overall agreement is excellent; it is doubtful if two supposedly identical prototypes would have any better agreement.

Predicted and actual prototype rotations at the peak downward and upward loads at each cycle are shown in Fig. 10.62. The values for each of the three models are spread out horizontally at each cycle for clarity of plotting. Agreement between models and prototype is adequate over the entire loading history.

Vertical displacements at the end of the beam are compared in Fig. 10.63. Model Clc predicts the prototype behavior very well; the other models are adequate at low cycles and become increasingly less satisfactory as cycling progresses. Bond deterioration produces a large component of the total displacement, and it is believed that the higher-strength concretes in both the column and beam of these specimens produced improved bond behavior over both model Clc and the prototype. As discussed below, the two higher-strength models also had less distortion in the column immediately adjacent to the beam.

Development of major cracking in model Clc is compared with that of the

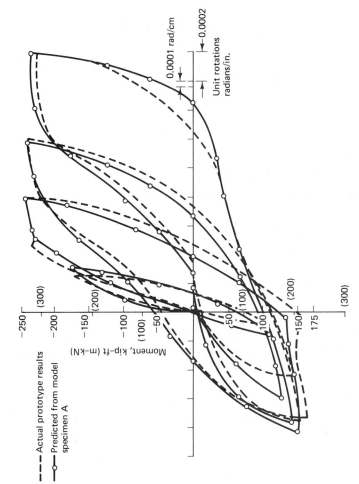

Figure 10.61 Comparison of predicted and actual moment-rotation response of prototype beam-column joint. [After Chowdhury and White (1977).]

447

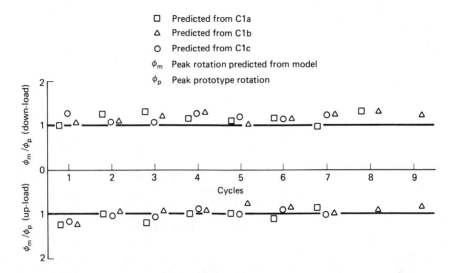

Figure 10.62 Comparison of predicted and actual peak rotation of prototype specimen 1-A. [After Chowdhury and White (1977).]

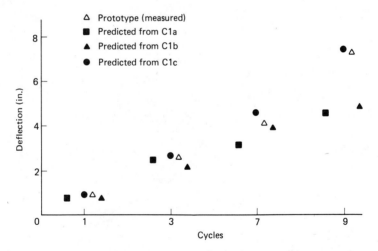

Figure 10.63 Comparison of predicted and actual peak deflections of prototype specimen 1-A at down-load of selected cycles. [After Chowdhury and White (1977).]

prototype in Fig. 10.64. The agreement is remarkable, and indicates that very complex behavior in reinforced concrete structures can be modeled if the model material properties meet the similitude requirements. Prediction of diagonal cracking in the joint region was not achieved with models Cla and Clb. It is felt that the overstrength of the column concrete (in tension) prevented the development of cracks.

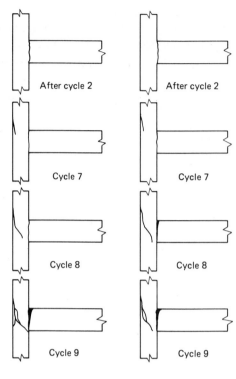

After cycle 2   After cycle 2

Cycle 7   Cycle 7

Cycle 8   Cycle 8

Figure 10.64 Qualitative major cracking patterns at peak downward load of
different cycles. [After Chowdhury and
White (1977).]

Cycle 9   Cycle 9

(a) Model specimen   (b) Prototype specimen I-A

Steel stresses in both beam and column were predicted with excellent
reliability by model C1c. The other models were less successful in predicting
column steel stresses; this is to be expected since the local concrete cracking and
resultant dowel bending stresses were not reproduced well in models C1a and
C1b.

The level of agreement between the other prototypes and their models
was generally good to excellent. The diagonal cracking of prototype specimen
II (low column load) was predicted by model C1d with the same degree of
success as that shown in Fig. 10.64 for prototype specimen 1-A.

Chowdhury and White (1977) concluded that:

1. The postcracking response of small-scale reinforced concrete models is
   highly sensitive to the properties of the model materials. In particular,
   the tensile strength of model concrete must be modeled accurately if
   diagonal cracking and other tension-dependent phenomena are to be
   reproduced.
2. Although small-scale structural models appear to have less cracks than
   their prototypes, both local moment-rotation characteristics and overall
   displacements can be modeled with excellent reliability. The bond charac-

teristics of the model reinforcing steel described in Section 5.8 appear to be adequate to model severe reversing load situations that are well beyond the initial yield stage of the reinforcing.

3. The availability of these techniques permits meaningful research studies of complex frame configurations at a small scale.

10.3.3.2 Multistory Frames, Models C2

In earthquake-resistant design, prediction of the available rotation capacity of the members is necessary to show that sufficient rotation is available at the critical sections so that the structure can resist collapse even when subjected to earthquakes of abnormal intensity. Chowdhury and White (1980) conducted an investigation of a three-story, two-bay frame that was selected as being representative of the high-rise building frame. It contains all possible types of joints of a multistory plane frame that need special detailing for earthquake-resistant, ductile design.

The principal structures of the investigation were two $\frac{1}{10}$-scale reinforced concrete models of a typical three-story, two-bay frame. One frame (frame C2a) was loaded with combined gravity load and unidirectional lateral load; the other (frame C2b) was subjected to combined gravity load and reversing lateral loads that simulated earthquake forces. Four beams with properties and dimensions similar to those of the frame beams were tested to obtain the actual moment-rotation curves for the frame members. Fixed base support conditions were used for both frames. The dimensions of frame C2a (unidirectional loading) and its members are shown in Fig. 1.10. The main reinforcing steel used in the beams and columns of the frame consisted of 0.159-in. (4 mm) diameter commercially deformed annealed bars. The beams were doubly reinforced with one bar at each of the four corners of the beam section. These bars ran straight through from one exterior column to the other exterior column, with anchorage bends inside the exterior columns. Plain annealed wire of 0.041-in. (1.04 mm) diameter was used for the stirrups, which were formed to very close tolerances on a machine built to manufacture stirrups rapidly.

The column longitudinal reinforcement consisted of eight bars; ties were deformed annealed wires of 0.0625-in. (1.6 mm) and 0.103-in. (2.6 mm) diameter. The details of the reinforcement are shown in Fig. 10.65. The concrete mix properties for the microconcrete used in the frames were 0.8:1:5 (water, cement, and sand by weight). The concrete had a uniaxial compressive strength of about 4000 psi (27.6 MPa) at the time of testing.

Frame C2a was loaded with the combined gravity design load and a monotonically increasing lateral load [Fig. 10.66(a)]. The design load was calculated in accordance with the ACI 318-71 Building Code. Gravity loads were simulated at two locations on each beam and were maintained constant by using gravity load simulators (Chapter 7) on each of the six beams. The top-story gravity loads were 88% of the lower-story loads. The monotonically

Figure 10.65   Details of reinforcement for frames C2a and C2b. [After Chowd-hury and White (1980).]

increasing lateral loads were applied at the level of each floor. Figure 1.10 shows the location of all loads.

Frame C2b (reversing lateral loads) was identical to frame C2a, with dimensions and reinforcement as shown in Figs. 1.10 and 10.65. The decision on a suitable loading history to simulate seismic effects is always difficult. The records of earthquake motions show that although the motions undergo a large number of reversals, the intensity is large in only a few cycles. With this point in mind, frame C2b was subjected to the design gravity load and nine cycles of lateral loads of varying intensity to simulate the seismic loading. Figure 10.66b shows the loading history for frame C2b, where the lateral load level is expressed as a ratio of the gravity load. Gravity loads were identical to those for frame C2a.

Fabrication and Instrumentation.   The fabrication of reinforcement and the casting of small-scale models form an important part of the modeling process, especially for earthquake-resistant ductile structural models. The use of closely spaced hoops at the critical sections in the latter type of structure produces very small clearances and thus complicates the casting procedure for the model specimens. This problem of casting was overcome by attaching the frame form laid in horizontal position to a vibrator table and shaking the entire assembly intermittently while placing the model concrete.

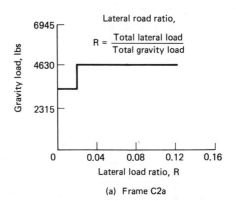

(a) Frame C2a

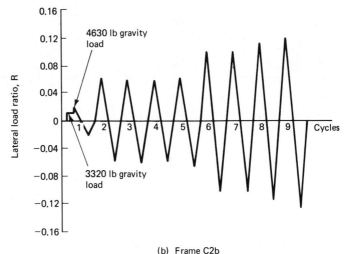

(b) Frame C2b

Figure 10.66 Load history for frames C2a and C2b. [After Chowdhury and White (1980).]

The frames were tested in a horizontal plane, with both the frame, its base, and the loading devices attached to a rigid steel table. Gravity loads were applied using hydraulic tension jacks acting through gravity load simulators, and lateral loads were applied at each floor with mechanical jacks. Each gravity load simulator had a lateral deflection range of ± 3 in. (75 mm).

Rotations in the beams were measured with pairs of linear variable differential transformers (LVDTs). Dial gages were used for measuring the lateral deflections of the frame and beam deflections. Concrete and steel strains were measured with $\frac{1}{2}$-in. (12.5 mm) and $\frac{1}{16}$-in. (1.6 mm) gage length electrical resistance strain gages, respectively.

Test Results—Monotonically Loaded Frame Model C2a.   The lateral load carried by the experimental frame C2a was 4% higher than the lateral load capacity predicted by theory. In designing the frame, the center-to-center distance between columns was taken as the span length, but actually the negative hinges were formed in the beam adjacent to the face of the columns where the moments were less than the theoretical maximum design moments. This factor, along with the fact that hinge lengths were greater than assumed in the analysis, contributed to the apparent increase in the load capacity of the test frame.

Details of the cracks in several critically stressed regions are shown in Fig. 10.67. The confinement of concrete in the joint area by closely spaced hoops effectively increases the concrete bearing capacity and provides sufficient anchorage to the beam-steel so that the desired ductility is ensured. The cores of the joint of the frame were provided with web reinforcement to carry the entire expected shear. Figure 10.67 also shows that there is no damage across the joint area of the columns.

The load-rotation (P-ϕ) characteristics of the bottom-story beam of the frame obtained from the experimental and analytical investigations are shown in Fig. 10.68. The abrupt change of the curves at the lateral load ratio of 0 and 0.0145 is due to the application of gravity loads in two stages at these two levels of lateral loads. These two applied gravity loads are equal to the code specified and to the ultimate design loads, respectively. A comparison between experimental and analytical results shows close agreement throughout the range of loading. The beam reached a rotational ductility factor μ_r of 24.

The lateral load ratio vs. lateral deflection curves for the frame are shown in Fig. 10.69. Although the curves are similar in shape to the load-rotation curves, the frame does not deflect laterally when only code-specified gravity load is applied with no lateral load. This is because of the symmetry of the loading and the frame geometry. But at a lateral load ratio of 0.0145, when the gravity load is increased to design gravity load, the lateral deflection increases as a result of the P-Δ effect. The experimentally determined curves show that at each story level there is some difference between the deflections at two sides of the frames, especially at higher loads. This difference in deflection is due to the deformation of the beams as produced by axial deformations, flexural and shear cracks, and transverse deflections of the beams. Of these factors, the wide flexural cracks in the beams contribute the most and is the primary source for this difference.

The frame attained a displacement ductility factor μ_d of 3 defined as the ratio of the lateral deflection under the specified load to the lateral deflection at the formation of the collapse mechanism of the frame. Thus the most critically loaded sections of the beams attained a rotational ductility factor of 8 times the displacement ductility factor of the frame.

Test Results—Cyclically Loaded Frame C2b.   Each of the six beams in frame C2b had four crack-forming regions (two positive and two negative moment regions). Details of the crack patterns at several hinging regions of the frame are shown in Fig. 10.70. At the peak of the first cycle, no flexural cracking was visible in the bottom-story interior beam-column joint (Fig. 10.70e). For

Figure 10.67 Cracking patterns at hinging regions, frame C2a. [After Chowdhury and White (1980).]

454

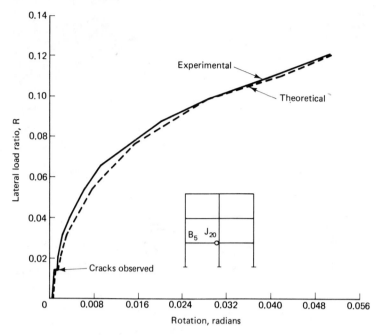

Figure 10.68 Load-rotation curves of bottom story beam B_5 of Frame C2a. [After Chowdhury and White (1980).]

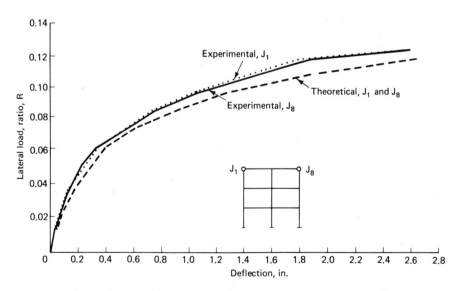

Figure 10.69 Load-deflection curves of frame C2a at the top story level (joints J_1 and J_8). [After Chowdhury and White (1980).]

Figure 10.70 Cracking patterns at hinging regions, frame C2b. [After Chowdhury and White (1980).]

(a)   (b)   (c)

(d)   (e)   (f)

the first half of the second cycle, when the lateral loads were applied from the left side of the frame, the top of the left beam yielded and a wide flexural crack was visible. On unloading, the crack narrowed but remained visible. In the second half of this cycle, cracks appeared in the right beam as a result of the yielding of the top beams, and the cracks in the left beam widened. At the completion of cycle 2, the cracks in both beams closed partially but remained visible. No new cracks developed in the next three cycles. In the subsequent cycles the number and dimension of cracks increased with the increase of load intensity.

The sequence of crack formation and crack widening in other critical sections of the frame was similar to that described above. First visible cracking at the bases of the columns was observed at the sixth cycle. Column base crack widths increased substantially in subsequent cycles.

The application of cylcic loads produces alternating diagonal tension forces in the joint regions of the frame. These forces may produce deterioration of reinforcement anchorage, yielding of shear reinforcement, additional yielding of beam steel, and shear failure of the joint itself. In order to prevent these inelastic deformations of the joint area, and to avoid shear failure and other nonductile behavior, proper reinforcement detailing must be followed in the joint areas and in other regions with high shear and anchorage forces. As shown in Fig. 10.70, no new cracks appeared in any joints in the frame designed and detailed according to SEAOC (1968) specifications.

Figure 10.71(a) shows the top-story lateral load ratio vs. the rotation of the bottom-story beam. The rotation could not be recorded beyond the peak load of cycle 8 because the total rotations beyond this load exceeded the rotation-measuring capacity of the LVDTs.

Cyclic loading of constant intensity causes some increase in rotation in consecutive cycles, but the consecutive increments of the rotation decreases with the increase of the number of cycles. However, the rotations increase as the intensity of the cycled load is increased. Figure 10.71b shows a comparison between the net rotation due to positive loading and the net recovery of rotation due to negative load at different cycles. The load-deflection curve for the lateral deflection of the frame is shown in Fig. 10.72.

Comparisons between Frames C2a and C2b—Effects of Cyclic Loads. Both frames C2a and C2b were subjected to the same maximum load intensity. The effect of cyclic loads on overall behavior will be discussed here through the use of load-rotation curves, load-deflection curves, and the deformations of the members.

Figures 10.67 and 10.70 show the cracking patterns of similar sections of frames C2a and C2b, respectively. From these figures, the cyclic loads are seen to increase the total deformations of the beams. These observed incremental deformations in the tension zones are believed to be caused by loss of bond and by additional cracking.

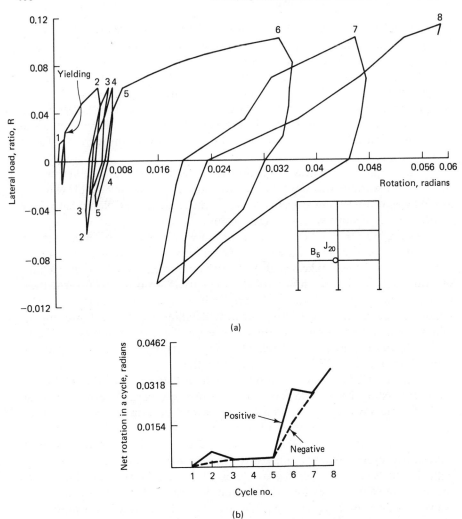

Figure 10.71   (a) Load-rotation curve of bottom story beam B$_5$ of frame C2b. [After Chowdhury and White (1980).] (b) Comparison of net rotation in different cycles of beam B$_5$ due to positive and negative lateral loads.

The lateral load ratio vs. the rotation of the bottom-story beam of frames C2a and C2b are shown in Fig. 10.73. There is good agreement between the two curves up to load cycle 6, when the beam reached about 80% of its load capacity. Beyond this point, the rotation of frame C2b increases more rapidly than in frame C2a. But unlike model results reported by Sabnis and White (1969), the strength and stiffness of the cyclically loaded frame did not reduce substantially at cyclic loads beyond 80% of the failure load level.

Figure 10.74 shows the lateral load vs. lateral deflection of the top story of the frames. Comparison between the curves of frame C2a and frame C2b

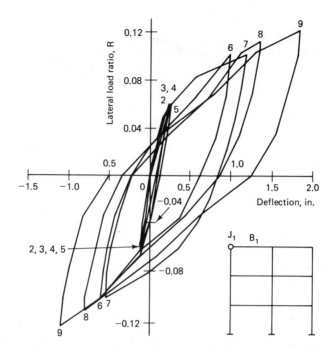

Figure 10.72 Load-deflection curve of frame C2b at the top story level (joint J_1). [After Chowdhury and White (1980).]

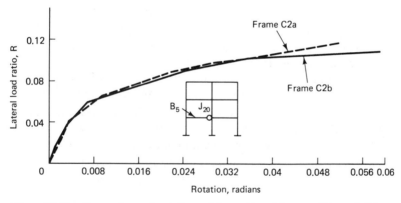

Figure 10.73 Comparison of rotation of beam B_5 of frames C2a and C2b. [After Chowdhury and White (1980).]

shows good agreement up to the ninth load cycle. The indispensable role of compression reinforcement and of closely spaced ties in providing toughness and ductility is apparent from these tests.

The following conclusions are based on the study of multistory frames subjected to this particular combination of loads conducted by Chowdhury and White (1980):

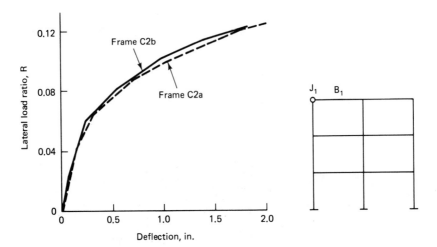

Figure 10.74 Comparison of deflection of frames C2a and C2b at the joint J_1. [After Chowdhury and White (1980).]

1. The three-story, two-bay, seismic-resistant, ductile-reinforced concrete frame can resist very strong cyclic and monotonically increasing lateral loads applied in the plane of the frame. The critical sections of the frame can reach a rotational ductility factor of 8 times the displacement ductility factor of the frame.

2. The stiffness of seismic-resistant ductile frames is not affected as long as the cyclic loads are less than 80% of the ultimate load capacity.

3. Although severe cyclic loads of more than 80% of the ultimate load capacity are found to increase the total deformations of the seismic-resistant frame, the strength and stiffness of this ductile frame do not reduce substantially, as is the case in ordinary reinforced concrete frames.

4. Properly designed web reinforcement is able to carry the entire shear force in the joint cores, preventing the formation of wide cracks in the column portion of the beam-column joint regions.

5. Small-scale microconcrete models can be used to investigate the complex inelastic behavior of reinforced concrete multistory frames under severe cyclic loads.

10.3.4 Case Study D, Precast Concrete Large-Panel Buildings

10.3.4.1 Horizontal Joints between Floor Slabs and Bearing Wall Panels, Models D1

The horizontal joint, which connects the load bearing wall panels to the horizontal floor slabs, is the most critical joint in large-panel structures. It must allow the wall and floor loading to be channeled to the foundations and in addition enable the structure to redistribute the gravity loads by cantilever action in case of loss of a bearing wall member.

A series of tests were conducted [Harris and Muskivitch (1977) and Dow and Harris (1978)] on $\frac{1}{4}$-scale models of an interior horizontal joint loaded in axial compression (Fig. 1.12). Thirty models tested consisted of two wall panels, two hollow-core floor panels, joint mortar, and dry pack material. The specimens were assembled from precast elements and are shown in Fig. 1.12.

Each specimen was instrumented with dial gages, which were used to measure wall and dry pack expansion and overall joint shortening, and strain gages, which were used to measure wall compressive strains. The model test results were compared to prototype tests conducted at the Portland Cement Association [Johal and Hanson (1978)]. Typical load vs. axial shortening for the walls and across the joint for model and prototype joints are shown in Fig. 10.75. The main variables in the model and prototype tests were the amount of wall reinforcement to prevent wall splitting and the strength of the joint grout. The ultimate strength of all model specimens, consisting of average values from three identical specimens, are compared with the full-scale proto- type tests in Fig. 10.76. As can be seen from Fig. 10.76, the correlation is very good. Typical crack patterns at failure of specimens with weak, intermediate- strength, and strong joint grout are shown in Fig. 10.77.

The authors concluded that a direct comparison of model and prototype results on joints of large-panel precast concrete buildings indicates that the models tested duplicated very well the ultimate load capacity and the mode of failure of the prototype. The load vs. deflection characteristics of models and prototype were also compared and found to correlate to within acceptable engineering limits. It was found that duplicate and triplicate model specimens are helpful in giving a statistical basis to the experimental results. Since the costs of fabricating and testing small-scale models is still relatively modest, when compared with prototype tests, the model solution is very attractive.

10.3.4.2 Cantilever Wall Component Assembly, Models D2

Tests were conducted [Harris and Muskivitch (1977)] on three $\frac{3}{32}$-scale (1/10.67-scale) models of a three-story exterior wall of a large-panel structure where one-half of the lowest floor was removed to simulate a missing panel caused by abnormal loading (Fig. 1.13). Two of the models tested (D2a and D2b) contained the same amount of transverse reinforcement in the horizontal joint [two $\frac{1}{16}$-in (1.6-mm), 19-wire, high-strength stainless steel cables] and a third model (D2c) with one-half of that amount (one such cable). Deflections, joint openings, and horizontal joint slip were measured and compared with test results obtained by Hanson (1975) at the Portland Cement Association on $\frac{3}{8}$-scale models.

The 19-wire, $\frac{1}{16}$ in. (1.6 mm) diameter stainless steel cable used as the main horizontal joint reinforcement had an area of 0.00215 in.^{2} (1.39 mm^2), a

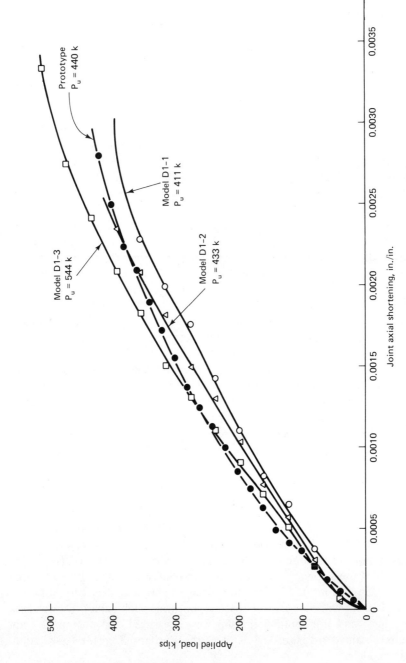

Figure 10.75 Load vs. axial shortening of models and prototype—joint B5. [After Harris and Muskivitch (1977).]

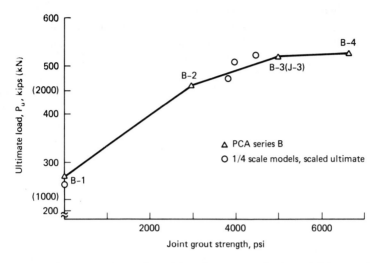

Figure 10.76 Comparison of model and prototype horizontal joint results. [After Harris and Muskivitch (1977).]

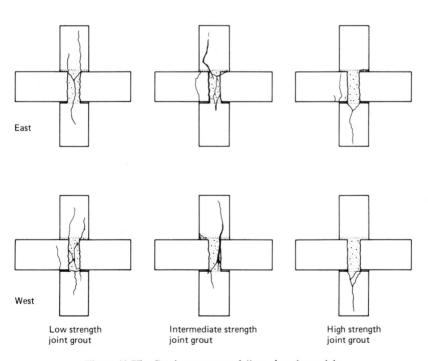

Figure 10.77 Crack patterns at failure, $\frac{1}{4}$ scale models.

0.2% offset yield stress of 246 ksi (1,700 MPa), and an ultimate strength of 273 ksi (1,880 MPa). Continuous-thread threaded steel rod was used for the vertical tie reinforcement. The area of the root of the thread was 0.00891 in.² (5.75 mm²), with a 0.2% offset yield stress of 80 ksi (552 MPa) and an ultimate strength of 85 ksi (586 MPa). The arrangement of these various ties and the instrumentation used in the model tests is shown in Fig. 1.13. Forms for the model components were fabricated using aluminum bars and Plexiglas sheets. Concrete used for the wall panels and floor stubs was made using Type III cement and Delaware Valley sand passing the U.S. Series No. 4 sieve and mixed in the proportion 0.6 : 1 : 3 (water, cement, sand by weight). The average compressive strength obtained from six 1 × 2 in. (25 × 50 mm) control cylinders for each mix was 5800 psi (40 MPa) at an average of 90 days. The grout used in casting the horizontal joint was made using Type III Cement and Delaware Valley sand passing the U.S. Series No. 16 sieve in a 0.5 : 1 : 1 mix by weight. The average grout com-

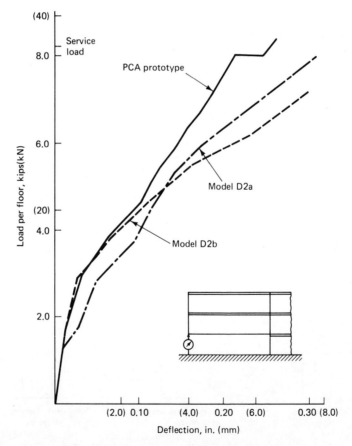

Figure 10.78   Load vs. deflection comparison of prototype and two extrapolated 1/10.67 scale models. [After Harris and Muskivitch (1977).]

pressive strength from six 1×2 in. (25×50 mm) cylinders was 5300 psi (36.5 MPa) at an average of 35 days.

A comparison of the $\frac{3}{8}$-scale PCA prototype test results with two identical models (D2a and D2b) at $\frac{3}{32}$-scale is shown in Fig. 10.78. The nonlinear behavior of the prototype cantilever wall, even at service load levels, is duplicated by the two small-scale models (Fig. 10.78). The PCA specimen was not loaded to failure in the two-story cantilever configuration because additional stories were added and tested to service load levels at each stage, with the final configuration being a five-story cantilever. Vertical joint openings, measured at top and bottom of each story level (Fig. 1.13), are compared for models and prototype in Fig. 10.79. A range of values obtained from several tests to service load levels for models D2a and D2b follow the prototype measured values very closely, as can be seen from Fig. 10.79. Model D2c, which had half the steel

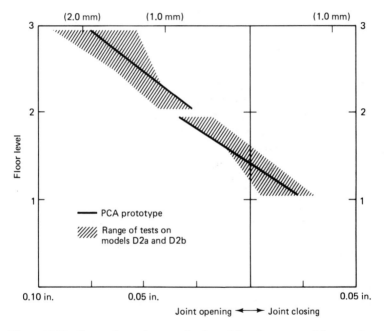

Figure 10.79 Comparison of extrapolated model and prototype joint openings. [After Harris and Muskivitch (1977).]

area in the transverse ties (one cable) as the previous two models, was loaded to ultimate failure, and the results are compared with the PCA tests in Fig. 10.80. As can be seen from Fig. 10.80, the results of the model and prototype are highly nonlinear even at low load levels, but considerable ductility existed in both model and prototype.

It was shown by this study that a methodology of fabricating and testing small-scale direct models of *precast concrete*, large-panel buildings down to

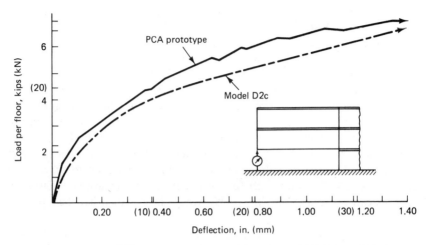

Figure 10.80 Load vs. deflection curves for model and prototype with less transverse tie reinforcement. [After Harris and Muskivitch (1977).]

scales of the order of one-tenth to one-fifteenth has been developed. These studies have shown that the small-scale modeling technique is a valid and useful method to study the behavior of three-dimensional assemblies of industrialized buildings under abnormal loading situations that may have a tendency to precipitate a progressive collapse.

10.3.4.3 Three-Dimensional Model D3

Using the techniques developed for the cantilevered wall model, a $\frac{3}{32}$-scale model of a six-story, three-bay cross-wall structure was constructed [Harris and Muskivitch (1980)]. The model configuration was chosen to have the characteristics of typical American-type cross-wall construction, although a specific structure was not modeled (Fig. 1.11). A model of six stories was chosen in light of the PCA recommendations [Schultz et al. (1977)] and constraints of overall model size that could be accommodated in the laboratory. The PCA recommendations indicate that the most stringent requirements for most of the ties occur in a damaged configuration containing 1 to 10 stories.

The model was fabricated using reinforced concrete wall panels and floor slabs. In addition, four steel frame removable wall panels replaced their concrete counterparts at four locations (Fig. 10.81). The purpose of using a steel frame as the removable wall panel (Fig. 10.82) was to facilitate its removal to simulate a damaged (missing) wall. The locations of these panels were chosen to represent four different damaged configurations (Fig. 10.81). The location of the three exterior removable wall panels in the constructed model is shown in Fig. 1.14.

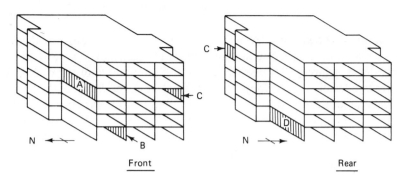

Wall	Identification	Model (prototype) damaged length
A	4th floor exterior long wall	31.125 in (27 ft 8 in.)
B	1st floor interior wall	30.875 in. (27 ft 5 in.)[a] 40.125 in. (35 ft 8 in.)[b]
C	4th floor exterior short wall	22.125 in. (19 ft 8 in.)
D	1st floor exterior wall	31.125 in. (27 ft 8 in.)

[a] with secondary support element

[b] with no secondary support element –
includes width of corridor – 8.25 in. (7 ft 4 in.)

Figure 10.81 Location of removable wall panels, model D3 [After Harris and Muskivitch (1980).]

Figure 10.82 Close-up view of in-place 1st floor exterior removable wall panel. [After Harris and Muskivitch (1980).]

Model Construction.   A major consideration in the design of the model was the design of the tie system (Fig. 10.83). The function of a tie system is to enable the structure to "bridge" over a wall failure and limit the extent of the damage to a localized area. Major factors involved in the design of the tie system are: the type of tie, the location of the tie within the structure (i.e., story level), the average story weight and height, and the length of wall panel damage. In most cases, ties consist of high-strength stranded cable because of its ability to develop large forces, and also to facilitate placement in joints. The criteria used in the design of the tie system were a combination of many of the recommendations proposed by the PCA with current local accepted practice.

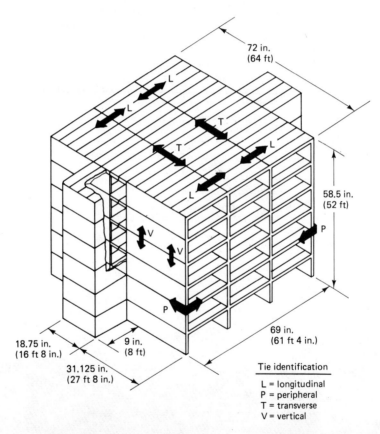

Tie identification

L = longitudinal
P = peripheral
T = transverse
V = vertical

Story height = 9.75 in. (8 ft 8 in.)
Floor slabs = 22.5 in. × 4.5 in. × 0.75 in. (20 ft × 4 ft × 0 ft 8 in.)
Wall panels = 31.875 in. × 9 in. × 0.75 in. (28 ft 4 in. × 8 ft × 0 ft 8 in.)

Figure 10.83   Three dimensional model D3 configuration showing tie system and model (prototype) dimensions (1 in. = 2.54 cm; 1 ft = 0.305 m). [After Harris and Muskivitch (1980).]

The vertical tie system (Fig. 10.83) provides continuous vertical continuity between successive wall lifts. In the model, two continuous-thread threaded steel rods were used in each wall panel to provide the story-to-story connection as shown in Fig. 10.84. Small square steel tubes along the bottom edge of the wall panels facilitated the use of hexagonal nuts to complete this connection. The location of the transverse tie, which provides tensile continuity in the plane of the wall, in the horizontal joints is also shown in Fig. 10.84.

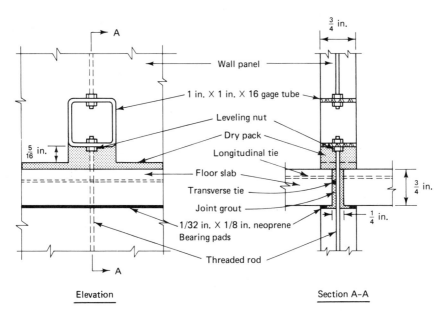

Figure 10.84   Model vertical tie connection detail. [After Harris and Muskivitch (1980).]

The longitudinal tie system ties the floor and roof elements together to prevent excessive debris loading. Stranded cable, grouted into the joints between adjacent floor slabs, provides the necessary tensile continuity and ductility to permit a suspension system to form above a damaged wall.

The peripheral tie (Fig. 10.83) is provided to establish a continuous tensile ring around the floor and roof systems of a large-panel (LP) structure. This tensile ring ensures the integral action of the system aiding the floor and roof to act as diaphragms. Requirements for the transverse and longitudinal ties will, in most cases, control peripheral tie design.

The model concrete used in the fabrication of the specimens in this study consists basically of Type III (high early-strength) Portland cement, local well-graded Delaware Valley sand, and enough water to create a workable mix. The proportions of sand, cement, and water vary with workability and strength requirements. Also, the ratio of the largest particle size in the sand to the smallest model dimension was kept at approximately 1 : 4. The strength vs. age curves for the model concrete used in casting the walls and floor slabs and the model grout used in filling all the construction joints are shown in Fig. 10.85.

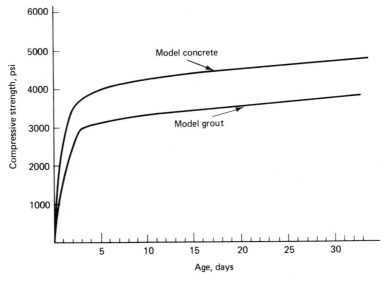

Figure 10.85 Compressive strength properties of model concrete and grout. [After Harris and Muskivitch (1980).]

Mild steel bars used for reinforcing the model wall panels and floor slabs were fabricated by passing small-diameter annealed steel wire through a knurling deformeter. This device forms the deformations in the wire that model the prototype reinforcing bar.

Continuous-thread threaded steel rod was used as tie reinforcing in locations where mechanical interconnection of elements was necessary. This rod was, therefore, used for vertical ties in the walls.

For modeling stranded prestressing cable, the use of small-diameter stainless steel cables, used in marine applications, was found to be most satisfactory.

The assembly of the model followed construction procedures that were similar to those used in prototype structures. Figure 10.86 shows a view of the model assembly after the introduction of the inner or compression whiffle tree and prior to the insertion of the loading beams. The loading beams are shown in Fig. 10.87 after the slab on the floor above has been grouted in place.

Instrumentation and Testing.   A number of displacement dial gages were placed at various locations to monitor floor slab deflections; overall axial, lateral, and support movements; and cantilevered wall deflections in the

Figure 10.86 Bird's eye view of model construction. [After Harris and Muski-vitch (1980).]

Figure 10.87 Sixth floor slabs grouted. Note fifth floor loading beams below. [After Harris and Muskivitch (1980).]

damaged configurations. The dial gage locations were chosen in an attempt to characterize the three-dimensional aspects of the behavior during each test.

All major bearing wall panels and several floor slabs near the simulated damaged regions of the model were instrumented with electrical resistance strain gages.

The load was applied to the three-dimensional model through a system of whiffle trees and loading beams (Figs. 7.16d to g) connected to a system of hydraulic tension jacks.

A series of tests were conducted on the intact structure and on the structure in each of its damaged configurations to determine its behavior when subjected to design load and overloads. The various damaged configurations were simulated through the removal of an entire wall panel, as shown in Fig. 10.88.

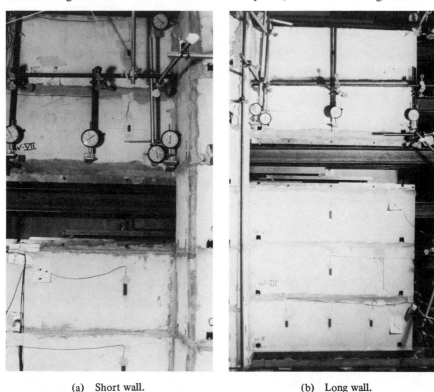

(a)   Short wall.   (b)   Long wall.

Figure 10.88   Fourth floor exterior wall panels removed during testing. [After Harris and Muskivitch (1980).]

The basic test procedure involved the loading of the model, through the hydraulic whiffle-tree system, in a number of increments up to a chosen load level. At each load increment, instrumentation readings were recorded. Because of similitude requirements, it was necessary to compensate for the dead-load effects of the prototype wall and floor panels through the application of additional load to the model floor system. The tie design load of the prototype was chosen to be equal to the dead load plus one-half the live load, or 90 psf (4.3 kN/m²). In the model, however, a total load ranging from 101 psf (4.8 kN/m²) to 166 psf (8.0 kN/m²) was applied to the floor system to accomplish the same net effect on the tie system. The value of this total applied load was a function of the particular damaged configuration under consideration.

Test Results.   One of the primary characteristics that defines the behavior of a damaged configuration is that of the values of cantilever end deflections. Load vs. deflection behavior of the tip of the cantilever for the three exterior wall–damaged configurations is shown in Fig. 10.89. From curves such as these, comparisons can be made that indicate to what extent the location and length of a damaged (missing) wall panel affects the overall cantilever behavior. For example, the effect of unsupported length of a damaged

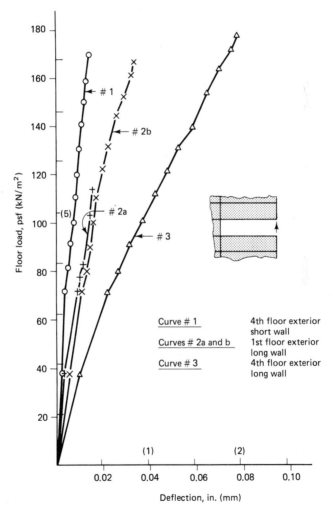

Figure 10.89   Load vs. cantilever end deflection curves for exterior damaged walls. [After Harris and Muskivitch (1980).]

wall panel for the same number (two) of cantilevered floors is shown by comparing curves 1 (short panel) and curve 3 (long panel) of Fig. 10.89. The effect of the number of cantilever floors above equal-length damaged wall panels is indicated by comparing curves 2a and b (five floors) with curve 3 (two floors) of Fig. 10.89.

An important characteristic is the overall deflection pattern of the cantilever wall assembly. Proper placement of instrumentation will provide the means through which rotations, deflections, vertical joint openings, and twisting of the wall assembly can be determined. Typical results from the tests on one of the damaged configurations are shown in Fig. 10.90. From these data, comparisons can be made as to the extent of the various overall deflection characteristics in each damaged configuration.

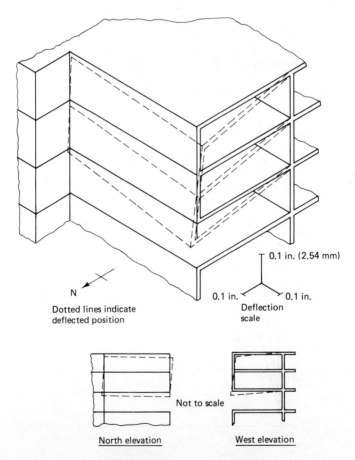

Figure 10.90 Overall deflection pattern of wall assembly above long exterior damaged wall panel. [After Harris and Muskivitch (1980).]

The extent of damage within a damaged structure can be determined through the examination of the deflection pattern of the floor system in the vicinity of the damaged wall.

Figure 10.91 shows the deflection pattern of the floor system in the vicinity of the missing first-floor interior wall. Large deflections occurred in the immediate area of the missing wall and in the adjacent floor slabs. The large floor-slab deflections were reduced to normal bending deflections in the region where the floor was supported by the rear wall. This demonstrates the apparent nature of large deflections being limited to the localized areas of damage. Figure 10.90 also indicates a relative absence of cantilever rotation, which is necessary to develop the tensile forces in the transverse ties. One recommendation of the PCA design criteria that was not included in the design of the model was the design criterion that requires the use of secondary support elements (integral columns or return walls). These elements, usually located at the innermost portion of the wall, provide the necessary vertical support to permit the cantilever mechanism to develop, but are not usually incorporated into current practice.

The need for these elements is demonstrated by the results of a second test on the first-floor interior damaged configuration. During this test, the innermost vertical member of the removable wall panel was left in position to simulate a secondary support element. A comparison of the tests on the first-floor interior damaged configuration with and without secondary support is shown in Fig. 10.92. This figure shows the dramatic effect that the provision of secondary vertical support has on both minimizing end deflections and enabling the structure to mobilize the cantilever mechanism and to sustain higher loads.

A major concern in any experimental program that involves structural models is how the results of the model tests can be used to predict the behavior of the prototype. Table 10.9 shows the values of the various model cantilever end deflections extrapolated to the prototype. Although there are no prototype test data available at present for comparisons, the ability of the structure to undergo what are sometimes very large prototype deflections was demonstrated in the model.

Harris and Muskivitch (1980) conclude from their tests on the three-D model that:

1. Various typical prototype construction details (i.e., materials, reinforcement, connections) of a three-dimensional, precast concrete, large-panel structure can be accurately modeled at $\frac{3}{32}$ scale.

2. The proper choice of the location of instrumentation will give data that will aid in the characterization of the behavior of a damaged structure.

3. The use of steel frame removable wall panels facilitated their removal from and replacement in the model.

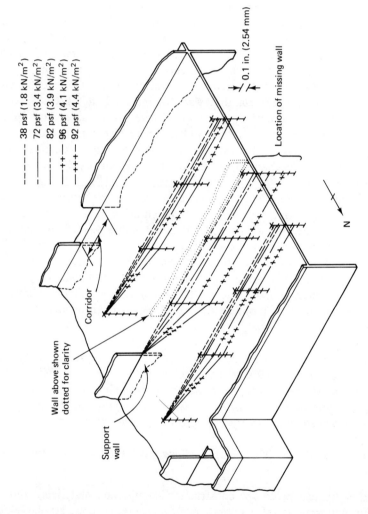

------- 38 psf (1.8 kN/m²)
———— 72 psf (3.4 kN/m²)
——— 82 psf (3.9 kN/m²)
+ + 96 psf (4.1 kN/m²)
+++ 92 psf (4.4 kN/m²)

0.1 in. (2.54 mm)

Location of missing wall

N

Corridor

Wall above shown
dotted for clarity

Support
wall

Figure 10.91 Floor system deflection pattern above missing interior wall panel with no secondary support. [After Harris and Muskivitch (1980).]

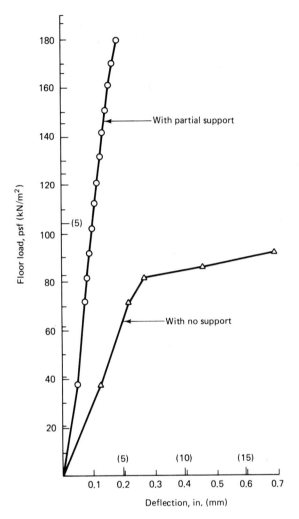

Figure 10.92 Comparison of load vs. deflection curves for first floor interior wall assembly with and without secondary support. [After Harris and Muskivitch (1980).]

4. The tie system, designed according to PCA recommendations [Schultz et al. (1977)] allowed the three exterior-damaged configurations to satisfactorily sustain the tie design load enabling the structure to bridge over local damage.

5. In some cases, where a whole interior panel is assumed to be removed, tests on the model indicate that serious problems may arise if there is an absence of secondary support elements for interior walls. The PCA recommendations call for the use of this type of element.

TABLE 10.9 Extrapolated Wall Assembly End Deflections

	1st Floor Exterior Wall	Damaged Configuration		4th Floor Exterior Long Wall	4th Floor Exterior Short Wall
		1st Floor Interior Wall			
		With Secondary Support	Without Secondary Support		
Tie design load, psf (kN/m²)	139(6.66)	108(5.2)	101(4.9)	111(5.3)	111(5.3)
Test load/tie design load	1.21	1.66	0.91	1.53	2.07
Model deflection at test load, in. (mm)	0.033(6.8)	0.172(4.4)	0.690(17.5)	0.014(0.36)	0.118(3)
Extrapolated prototype deflection, in. (mm)	0.35(8.9)	1.83(46.5)	7.36(187)	0.15(3.8)	1.26(32)

In addition, the test results show that the small-scale modeling technique is a valid and useful method to study the behavior of precast concrete, large-panel components and structures.

PROBLEMS

10.1. (a) Design a $\frac{1}{12}$-scale model for a singly reinforced concrete beam, using prototype and model properties given below:

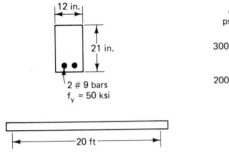
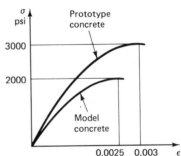

(b) Neglecting dead-load effects, and using elementary ultimate strength calculation methods, determine the maximum midspan concentrated load each beam can carry. Verify your model design by scaling the computed model capacity up to the prototype and comparing this with the calculated prototype capacity.

(c) If the deal-load effect is to be included, how much additional uniform load must be applied to the model?

(d) What are the requirements on the model concrete if shear capacity is to be modeled correctly?

(e) Assuming you are using two bars for the model steel, what is the bond stress at failure under part (b)? How are the bond stresses in model and prototype related?

(f) If the prototype beam also had two No. 9 bars as compression steel, would the same modeling of steel as used in part (a) be valid? Discuss.

10.2. You are asked to design a $\frac{1}{10}$-scale model reinforced concrete beam made from gypsum-based model concrete with the uniaxial compressive stress-strain curve shown below. Prototype material properties are also given.

(a) What other requirements would you place on model concrete properties, and why?

(b) What are the advantages and disadvantages of using gypsum-based model concretes as compared with cement-based microconcrete?

(c) Specify the required yield point of the model steel.

(d) What are the major problems met in modeling the postcracking and ultimate load behavior of reinforced concrete structures at this scale?

(e) The model beam deflects 1 in. (25 mm) at midspan just prior to failure. What is the corresponding prototype deflection?

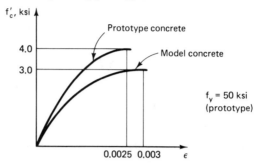

10.3. Design a reinforced concrete beam, either with single or double reinforcing, to fail in flexure or shear. Construct a model of the beam design and test it to failure, determining:

(a) Load-deflection curve

(b) Ultimate load capacity, including mode of failure

Compare the test results with the theoretical predictions and discuss *briefly*.

Ultracal (gypsum-based) model concrete with proportions of 1 part Ultracal, 1 part sand, and 0.31 parts water will produce a mix of about 3000 psi (21 MPa) in 24 hr. Rather than becoming involved with sealing the beam surfaces, it is suggested that you arrange the casting and testing of the beam within a 24- to 27-hr. time period, such as casting on Monday and testing Tuesday afternoon.

You may test the model steel in tension to determine its yield point.

10.4. The behavior of a prestressed concrete box girder bridge, made by post-tensioning 20-ft long segments together, is to be studied by model analysis in the laboratory. The tentative prototype structure design is:

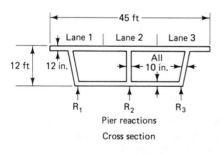

Pier reactions

Cross section

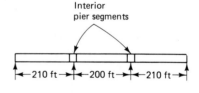

a) f'_c = 6000 psi at 28 days
wt = 145 lb/ft^3

b) Reinforcing steel is 60 ksi yield strength, $\frac{3}{4}$ in. ϕ

c) Prestress steel is $1\frac{1}{2}$ in. diam. strand with E = 25,000,000 psi and f_y = 140 ksi
Working (design) stress = 120 ksi

Each span is composed of 10 similar segments. The construction sequence is to place a segment at each interior pier and then add segments in each direction alternately, in a cantilever mode, with continuous prestressing. The prestressing elements are placed in pipe conduits cast into the concrete. Outer portions of each end span are constructed on falsework.

Critical *loading conditions* that are to be studied experimentally are:

(a) Dead-weight effects during the construction process.

(b) Full design live load, represented by a uniformly distributed surface load of 100 psf (5 kN/m²) on the entire top deck of the bridge.

(c) Partial live load of 100 psf (5 kN/m²) on one side lane only (either lane 1 or 3) to produce maximum torsional effects.

(d) Ultimate load capacity with a uniform loading on the full deck of the entire bridge.

(e) Slab-punching effects produced by a heavy wheel load of 18 kips (80 kN) over an area of 300 in.² (0.2 m²) [20 × 15 in. (500 × 375 mm) contact area of tire].

Quantities to be determined include:

(f) Concrete stresses at critical sections.

(g) Displacements.

(h) Changes in force in the post-tensioning prestress elements.

(i) Reactions of the bridge on its supporting piers.

(j) Failure capacity and failure mode under full uniform load.

(k) Slab-punching strength from concentrated wheel load.

10.5. A footbridge similar to the tubular truss bridge shown below is to be investigated for a possible instability failure under a uniformly distributed live load. The spans of the structure are as shown below; cross-sectional properties are also given. A36 steel, 3000-psi (21 MPa) concrete, and 40,000-psi (276 MPa) reinforcing steel are to be used in building the structure.

Assuming that the elastic buckling load is to be investigated experimentally on a model basis, describe in detail the model study you would undertake, including basic configuration (need the full structure be tested?), size of model, materials to be used, scaling relations, loading, method of applying load, instrumentation, and the actual method for determining the critical buckling load level. If resistance strain gages are to be used, specify their location and the types of gages to be employed, as well as any special circuitry you wish to use. Give justification for any distortion of true similitude. Comment on any unusual problems you feel this modeling will present, and discuss the confidence level you expect to achieve in the results.

Finally, make a time and cost estimate based on the following rates:

Engineer (yourself)	x/day	
Mechanician	$(\frac{2}{3}x)$/day	+75% overhead
Technician	$(\frac{1}{2}x)$/day	

Materials need not be estimated unless you wish to. Assume all testing equipment is available in the laboratory.

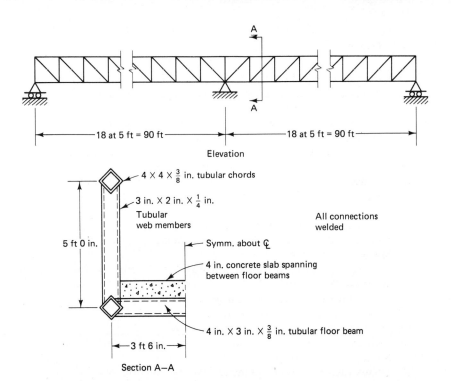

4 X 4 X $\frac{3}{8}$ in. tubular chords

3 in. X 2 in. X $\frac{1}{4}$ in.
Tubular
web members

All connections
welded

5 ft 0 in.

Symm. about ₵

4 in. concrete slab spanning
between floor beams

4 in. X 3 in. X $\frac{3}{8}$ in. tubular floor beam

3 ft 6 in.

Section A–A

18 at 5 ft = 90 ft

18 at 5 ft = 90 ft

Elevation

10.6. An existing underground storage tank (for liquids) is to be investigated for its strength to resist a high overpressure (blast loading) to be applied on the covering soil, at ground level. Since a full-scale test is not possible, it is proposed to study the strength of the structure with a model. The prototype tank is made of *reinforced concrete* and is detailed as shown below. The full reinforcing scheme is not shown here because a complete detailing of the model reinforcement is not wanted.

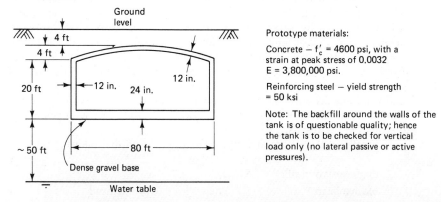

Ground level

4 ft

4 ft

20 ft

12 in.

12 in.

24 in.

~50 ft

80 ft

Dense gravel base

Water table

Prototype materials:

Concrete – f_c' = 4600 psi, with a strain at peak stress of 0.0032
E = 3,800,000 psi.

Reinforcing steel – yield strength = 50 ksi

Note: The backfill around the walls of the tank is of questionable quality; hence the tank is to be checked for vertical load only (no lateral passive or active pressures).

The structural engineer will be satisfied if the tank strength under *static* overpressure (that is, not dynamic) is known. The engineer wants to know the failure mode, which may be either strength-controlled or stability-controlled and the failure capacity, and would also like to know critical stresses at critical sections just prior to failure.

Outline a program for doing the model analysis for this situation. The program should include:

(a) Decision on basic type of model, and choice of scale factor.

(b) Choice of model materials. Two model concretes are available: one $(E_1) =$ 3,400,000 psi (23,443 MPa) with a strength of 3500 psi (24.1 MPa) and a strain at peak stress of 0.0025, and a second $(E_2) = 4,200,000$ psi (28,959 MPa) with a strength of 5500 psi (37.9 MPa) and a strain at peak stress of 0.0035. Model reinforcing is available in whatever strength and diameter you need.

(c) Fabrication techniques and a discussion of any special problems.

(d) Model loading method, describing specifically how you will apply the load and measure it. Give scaling relationships to relate model loads to the prototype.

(e) The necessary instrumentation to measure stresses, to detect any possible instability modes, and to determine the general deformations of the model. Types of strain gages and their lengths should be specified, along with the types of bridge circuits (single-arm, double-arm, or full-bridge) and how temperature compensation will be achieved.

(f) Desired and expected level of accuracy in your results.

(g) An estimate of the total person-days needed to complete the work, in terms of (1) engineer (yourself) and (2) technician (for construction, instrumentation, and testing).

10.7. A composite steel girder, concrete slab bridge, curved in plan, has been designed by an analytical approach that involves both elastic analysis and ultimate strength concepts. The tentative prototype structure is shown below:

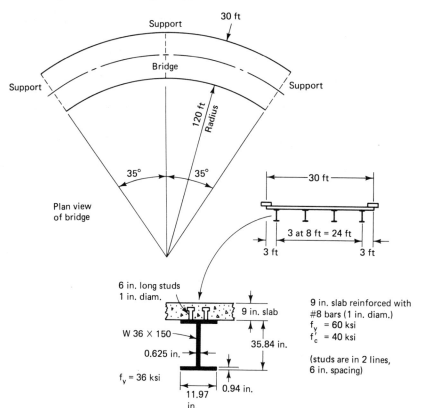

The shear connection between beam and slab is provided by metal studs welded to the upper flange of the beam. Such a connection can guarantee flexural strain compatibility between the two materials for loads appreciably beyond the design load. However, with inadequate, insufficient, or overstressed shear connectors, slippage of the slab relative to the beam will occur, thereby reducing the stiffness of the system and increasing stresses because the two elements then tend to act individually rather than as an integral unit. Thus the potential failure modes include yielding of the steel beams (bending, shear, torsion effects), concrete crushing from flexural compressive stress, localized concrete failure from bearing stresses produced by shear studs, flexural or shear yielding of the studs, and combinations of the above. Lateral motions of the curved girders could also be a problem.

In constructing composite structures, the formwork for the concrete slab is normally supported by the steel beams; thus the dead weight of the slab is carried by the beams alone, and only live-load stresses are resisted by the composite, integral section.

It is proposed to use *physical model analysis* to check on the overall validity of the analysis and to establish the behavior of the structure at overloads leading up to failure of the bridge. Questions to be resolved include:

(a) Accuracy of the analytical representation of the stiffnesses of the various elements (in particular the curved beams).

(b) Dead-load deformations produced by the weight of the fresh concrete acting on the steel girders.

(c) Girder reactions and maximum stresses (bending, shear, torsion) produced by dead load.

(d) Live-load influence lines for displacements, reactions, and peak stresses (within the range of behavior for normal live loads).

(e) Ultimate load capacity (in shear) of the shear transfer line between slab and girder (shear connectors).

(f) Failure mode and ultimate load capacity of the bridge subjected to a uniformly distributed loading that is meant to simulate maximum traffic.

Outline a program for doing the model analysis of this structure. The program should include:

(g) Decisions on basic type of model (or models, if you choose to study one or more of the problems with a special partial model of the structure).

(h) Choice of scale factor(s).

(i) Choice of model material(s).

(j) Fabrication techniques.

(k) Loading arrangements, including dead-weight effects.

(l) Instrumentation. Concentrate on specifying instrumentation at one section of the model and then indicate how many sections would have similar instrumentation. Types of strain gages (foil, epoxy-backed, or wire, paper-backed) and their gage lengths must be given. Locations of gages must be shown clearly, and the strain gage circuitry used should be defined.

Give brief justifications for your decisions. If a distorted model is used, discuss the effects of the distortion. Discuss the level of accuracy you desire and the level you actually expect to achieve. Comment on any unusual problems you feel the modeling will present. Finally, make a rough estimate of the time required for completing the project. Time should be in two categories: engineer (yourself) and mechanic/technician.

10.8. A reinforced concrete beam-column is to be modeled for ultimate load capacity. The model is to be loaded to failure by increasing P and M in a fixed ratio. Prototype material properties and model concrete properties are given below.

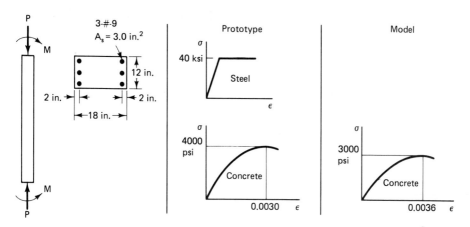

(a) What is the minimum size model you would recommend for this prototype? Discuss your reasons for your choice.
(b) Assuming a $\frac{1}{8}$-scale model, establish:
   (i) Required yield strength of the model steel
   (ii) Required areas of model steel
   (iii) Prediction factors for converting values of moment and axial load in the model to prototype values
(c) We cannot achieve true similarity with this model because the strain scale factor is not unity. Discuss the possible effects of this distortion on the model results. Will it tend to make the measured ultimate P and M higher or lower than they should be?
(d) Recommend a size of compressive cylinder for determining f'_c for the model material. Does the strength vary as a function of cylinder size?
(e) Sketch a suitable loading arrangement to achieve the conditions of combined axial load and equal end moments.
(f) Describe the instrumentation you would use for this study.

Structural Models for Wind, Blast, Impact, and Earthquake Loading

〰〰〰〰〰〰〰〰〰〰

11

11.1 INTRODUCTION

Physical modeling of structures with dynamic loadings has steadily developed in the past two decades [Baker et al. (1973), Hudson (1967), Schuring (1977), Castoldi and Casirati (1976), Krawinkler et al. (1978), and Harris (1982)]. Time-dependent loadings, because of their complex nature and effect on structures, have placed small-scale structural model techniques on a par with analytical techniques. The dynamic loadings of interest to the structural engineer range from wind- or traffic-induced elastic vibrations to blast and impact loadings that can cause considerable structural damage. Of special interest is the problem of earthquake loading, which, because of its widespread nature and potentially devastating effects, has assumed a greater importance in our highly urbanized society.

Dynamic modeling of structures is important in education, research, and design. In education, simple laboratory experiments demonstrate basic concepts of vibrations to undergraduate and graduate students. In the area of structural research, the small-scale dynamic model has proved to be a powerful tool in extending our knowledge and understanding of structural behavior in many complex situations where analytical techniques are inadequate. Also, a carefully constructed model aids the design of many dynamically loaded structures. In recent years, the quantity and the quality of the information obtained from the model test has increased as a result of improved instrumentation and data processing systems.

11.2 SIMILITUDE REQUIREMENTS

11.2.1 General

The dynamics of any structure is governed by an equilibrium balance of the time-dependent forces acting on the structure. These forces are the inertia forces that are the product of the local mass and acceleration, the resisting forces that are a function of the stiffness of the structure in the particular direction in which motion is occurring, and the energy dissipation of damping forces, whether material- or construction-related. In addition to these forces that produce dynamic stresses and deformation in the structure, there are certain types of massive structures in which gravity-induced stresses play an important role in dynamic situations and affect modeling. The similitude requirements that govern the dynamic relationships between the model and prototype structure depend on the geometric and material properties of the structure and on the type of loading (Fig. 11.1). These relationships can be derived as described in Chapter 2 using the Pi theorem. A summary of these

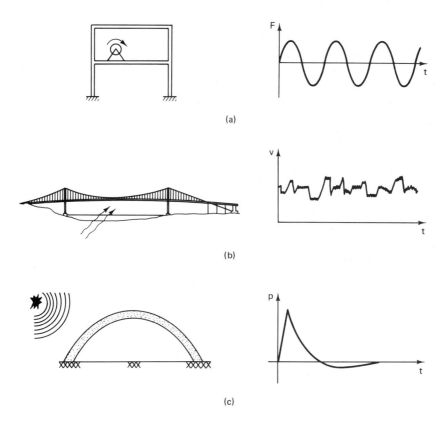

(a)

(b)

(c)

Figure 11.1 Various dynamic effects to be modeled.

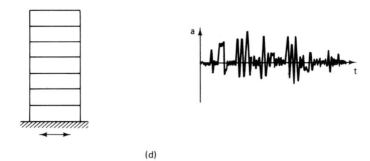

(d)

Figure 11.1 (*cont.*)

relationships for the most commonly encountered dynamic loads affecting civil engineering–type structures follows.

11.2.2 Vibrations of Elastic Structures

Vibration problems of elastic structures (Fig. 11.1a) are very common in civil engineering practice. Traffic-induced vibrations of bridges, wind-induced vibrations of tall buildings, towers, and chimneys, and water flow–induced vibrations of pilings and submerged structures are but a few of many such examples. These problems can be very conveniently studied by means of small-scale models. Consideration of the variables that govern the behavior of vibrating structures reveals that in addition to length (L) and force (F), which we considered in static loading situations (Chapter 2), we must now include time (T) as one of the fundamental quantities before we proceed with dimensional analysis.

As an example, consider an elastic structure of a homogeneous isotropic material whose vibration conditions are to be determined. A typical length in the structure is designated by l and a typical force by Q. The materials of both the model and prototype can be characterized by the material constants: the modulus of elasticity E, the Poisson's ratio v, and the mass density ρ. The important parameters to be determined from the structural vibration are the deflected shapes δ, the natural frequency f, and the dynamic stresses σ. The dimensions of the governing variables in both absolute and common engineering units are shown in Table 11.1. The acceleration due to gravity g must also be included in Table 11.1 since it is common to both model and prototype structures. The relationships that govern both model and prototype can be determined with the help of the Pi theorem (Chapter 2). For a true model, dimensionless parameters that govern the behavior can be shown to be:

$$\phi\left(\frac{\delta}{l}, \frac{\sigma}{E}, \frac{f^2 l}{g}, \frac{\rho g l}{E}, \frac{Q}{El^2}, v\right) = 0 \qquad (11.1)$$

TABLE 11.1   Dimensions of Governing Variables

Quantity	Dimensions	
	Absolute System	Engineering System
Length, l	L	L
Force, Q	MLT^{-2}	F
Modulus of elasticity, E	$ML^{-1}T^{-2}$	FL^{-2}
Poisson's ratio, ν		
Mass density, ρ	ML^{-3}	FT^2L^{-4}
Deflection, δ	L	L
Stress, σ	$ML^{-1}T^{-2}$	FL^{-2}
Frequency, f	T^{-1}	T^{-1}
Acceleration, g	LT^{-2}	LT^{-2}

If the deflections are to be of primary interest, the functional relationship becomes

$$\frac{\delta}{l} = \phi'\left(\frac{\sigma l^2}{Q}, \frac{f^2 l}{g}, \frac{\rho g l}{E}, \frac{Q}{El^2}, \nu\right) \tag{11.2}$$

and similarly if the stresses are of primary interest

$$\frac{\sigma l^2}{Q} = \phi''\left(\frac{\delta}{l}, \frac{f^2 l}{g}, \frac{\rho g l}{E}, \frac{Q}{El^2}, \nu\right) \tag{11.3}$$

From Eqs. (11.2) and (11.3) we can determine the dynamic characteristics of the structure by means of a model test by forcing the dimensionless terms on the right-hand side of Eqs. (11.2) and (11.3) to be identical in model and prototype. The deflections and stresses then become

$$\left(\frac{\delta}{l}\right)_m = \left(\frac{\delta}{l}\right)_p \quad \text{or} \quad \delta_p = \delta_m S_l \tag{11.4}$$

and

$$\left(\frac{\sigma l^2}{Q}\right)_m = \left(\frac{\sigma l^2}{Q}\right)_p \quad \text{or} \quad \sigma_p = \sigma_m S_E \tag{11.5}$$

In order to accomplish this, the second, third, and fourth terms in parentheses [Eqs. (11-2) and (11-3)] impose certain restrictions on the model design. The implied scale factors that govern these relationships are summarized in Table 11.2.

As can be seen from Table 11.2, the time scale is equal to $S_l^{1/2}$ for a true elastic model and it is equal to S_l in the case of a model where gravity loading effects can be neglected. The frequency of vibration of the model, which is inversely proportional to the period, will be $S_l^{-1/2}$ and S_l^{-1} in the above two cases, respectively. This means that the model will have higher frequencies. The shaking-table facility or the vibrator, which is exciting the model, must

TABLE 11.2 Similitude Requirements for Elastic Vibrations

| | | | Scale Factors | |
| | | | | |
Group (1)	Quantity (2)	Dimension (engineering units) (3)	Exact Scaling (4)	Gravity Forces Neglected (5)
Loading	Force, Q	F	$S_E S_l^2$	$S_E S_l^2$
	Gravitational acceleration, g	LT^{-2}	1	1
	Time, t	T	$S_l^{1/2}$	S_l
Geometry	Linear dimension, l	L	S_l	S_l
	Displacement, δ	L	S_l	S_l
	Frequency, f	T^{-1}	$S_l^{-1/2}$	S_l^{-1}
Material properties	Modulus, E	FL^{-2}	S_E	S_E
	Stress, σ	FL^{-2}	S_E	S_E
	Poisson's ratio, ν		1	1
	Density, ρ	FL^{-3}	S_E/S_l	Neglected

have a frequency and force output capability equivalent to that required for the highest important natural frequency of the model to be studied.

11.2.3 Fluidelastic Models

Wind loading forms another major design criterion for many civil engineering structures. Wind effects on buildings and structures of complex shapes are usually studied by means of wind tunnel testing of small-scale models. Two important modeling considerations must be addressed by the model analyst in studying wind effects on structures. The first consideration is a proper modeling of the wind loading through the scaling of the roughness, pressure, and velocity parameters of the flow, and the second is the scaling of the model structure itself. Tests on determining wind effects on buildings date to at least 1893, when Irminger made some experiments in Copenhagen. A report of this early work was presented by Irminger and Nokkentved (1930). Early work in the United States includes the tests on the Empire State Building rigid model that were conducted at the National Bureau of Standards by Dryden and Hill (1933) with no attempt to scale the velocity or the turbulence of the model flow. Tests on small-scale rigid building models were first subjected to boundary layer flow by Bailey and Vincent (1943). At about the same time, Jensen (1958) and Jensen and Franck (1965) established on a firm basis the need to use a thick turbulent boundary layer with upwind surface roughness scaled to the same scale as the building.

The modeling requirements for boundary layer type of winds can be easily obtained from dimensional considerations. Detailed derivations of the various similarity requirements are given by Cermak (1975) and Snyder (1972). Exact similarity requirements of the atmospheric boundary layer may be achieved as follows:

1. Topographic relief, terrain roughness, and surface temperature must be simulated.
2. The mean and fluctuating velocity and temperature fields of the approaching flow must be simulated.
3. Equality of the Reynolds, Rossby, Richardson, Prandl, and Eckert numbers need to be maintained in model and prototype atmospheres.

Because of these restrictions, exact similarity at a scale ratio other than 1 is not possible; however, if requirements (1) and (2) above are satisfied, complete similitude of all of the nondimensional parameters listed in (3) is not necessary. A discussion of the effects of relaxing some of the above requirements can be found in Cermak (1975), Snyder (1972), and Surry and Isyumov (1975).

The field of fluidelasticity covers a wide range of structural dynamics problems that result from the interaction of the structure and the fluid medium that is exciting it. Wind-induced problems range from the classical suspension bridge oscillations (Fig. 11.1b) and overhead power line galloping to buffeting of tall buildings and oscillations of tanks, stacks, and towers. Water- and wave-induced problems occur in the oscillations of harbor and offshore structures. A summary of scale factors for fluid leastic models [Scanlan (1974), Scanlan (1973), Isyumov (1976), and Cermak (1977)] are given in Table 11.3 for the

TABLE 11.3   Similitude Requirements for Fluidelastic Models

| | | | Scale Factors | |
| | | | Reynolds No. Neglected | Froude No. Neglected |
Group (1)	Quantity (2)	Dimension (3)	(4)	(5)
Loading	Force, Q	MLT^{-2}	$S_\rho S_l$	$S_\rho S_l^{-1}$
	Pressure, q	$ML^{-1}T^{-2}$	$S_\rho S_l$	$S_\rho S_l$
	Gravitational acceleration, g	LT^{-2}	1	1
	Velocity, v	LT^{-1}	$S_l^{1/2}$	S_v
	Time, t	T	$S_l^{1/2}$	$S_l S_v^{-1}$
Geometry	Linear dimension, l	L	S_l	S_l
	Displacement, δ	L	S_l	S_l
	Frequency, ω	T^{-1}	$S_l^{-1/2}$	$S_v S_l^{-1}$
Material properties	Modulus, E	$ML^{-1}T^{-2}$	$S_\rho S_l$	$S_\rho S_l$
	Stress, σ	$ML^{-1}T^{-2}$	$S_\rho S_l$	$S_\rho S_l$
	Poisson's ratio, ν		1	1
	Mass density, ρ	ML^{-3}	S_ρ	S_ρ

case of negligible Reynolds number effect [column (4)] and negligible effect due to gravity stresses (or neglect of Froude's number) [column (5)].

Recent use of improved higher-strength materials and construction techniques has resulted in the development of generally lighter and more flexible structures. The presently widespread use of welded steel and prestressed concrete structures are some examples where not only the generally more flexible construction has led to increased fluidelastic action but also the lower damping characteristics of these materials has added to fluid structure interaction. The resulting greater sensitivity to dynamic excitation by wind action has prompted an increased use of wind tunnel simulations for providing design information.

11.2.4 Blast and Impact, Load Modeling

Impulsive type loading due to external or internal blast effects, or impact of moving objects on structures, is considered an abnormal type of loading that has a relatively low risk of occurrence. However, the fact that it may occur at all in certain types of structures requires knowledge of the ability to treat its effects on their dynamic response. For these reasons we will consider in this chapter the modeling of blast effects either external or internal to the structure.

11.2.4.1 External Blast on Structures

Because of the difficulty and expense of conducting studies dealing with blast loading (Fig. 11.1c) at full scale and the enormous difficulties in performing analytical studies, the need for model experiments in this particular area has developed from the very beginning [Baker (1973)]. The first scaling law to be developed for the blast phenomenon itself has been the *Hopkinson* or *cube-root scaling law*. This law states that self-similar blast (shock) waves are produced at identical scaled distances when two explosive charges of similar geometry and the same explosives but of different size are detonated in the same atmosphere; thus

$$L_m = \frac{L_p}{\bar{E}^{1/3}} = \frac{L_p}{W^{1/3}}$$

where

$$L_m = \text{length in model}$$
$$L_p = \text{length in prototype}$$
$$\bar{E} \text{ (or } W) = \text{the energy of the prototype explosion}$$

This type of modeling is sometimes referred to as *replica* modeling, and the pressure amplitude and loading in such modeling will be similar in form as in the prototype. The pressure amplitude and velocity of the pressure wave are the same. The duration of the pulses will be scaled by the length scale.

The most general and useful modeling techniques used in the design and analysis of structures to predict the blast loads are those associated with the ability to predict the elastoplastic as well as the elastic behavior and the ability to study with confidence the mode of failure of the structure. These techniques are, however, very restrictive on the choice of model materials and methods of fabrication. Extensive research has been conducted at MIT [Antebi et al. (1960), Antebi et al. (1962), Harris et al. (1962), Smith et al. (1963), and Harris et al. (1963)] and several government facilities to develop and utilize such techniques in the study of nuclear blast effects on various structures. Although these studies were conducted for external blast effects on structures, the same methodology can be applied to the case of other gaseous explosions.

For a geometrical scale factor of S_l and when the energy of the blast source is scaled according to the cube-root law and the model is made from components of the same strength properties with the same manufacturing details and supports as the prototype, then the scaling laws that must relate the model to the prototype structure are given in Table 11.4. Since specific weight

TABLE 11.4   Summary of Scale Factors for Blast Loading

| | | | Scale Factors | |
| | | | True Replica Model | Gravity Forces Neglected, Prototype Material |
Group (1)	Quantity (2)	Dimension (3)	(4)	(5)
Loading	Force, Q	F	$S_E S_l^2$	S_l^2
	Pressure, q	FL^{-2}	S_E	1
	Time, t	T	S_l	S_l
	Gravitational acceleration, g	LT^{-2}	1	Neglected
	Velocity, v	LT^{-1}	1	1
Geometry	Linear dimension, l	L	S_l	S_l
	Displacement, δ	L	S_l	S_l
	Strain, ϵ		1	1
Material	Modulus of elasticity, E	FL^{-2}	S_E	1
	Stress, σ	FL^{-2}	S_E	1
	Poisson's ratio, ν		1	1
	Mass Density, ρ	FT^2L^{-4}	S_ρ	1

scales the same as the length scale, we cannot, in general, provide a model material with the same strength properties and yet S_l times denser than the prototype. Thus, the effects of gravity stresses cannot be modeled through a choice of a denser material in most direct model studies, unless we resort to preloading techniques such as those discussed in Chapters 2 and 7.

11.2.4.2 Internal Blast Effects

The nature of gaseous explosions in domestic surroundings has been studied extensively after the progressive collapse at Ronan Point [Rashbash (1969), Stretch (1969), Alexander and Hambly (1970), and Slack (1971)]. The 100-psi theoretical maximum pressure that can be reached in a completely confined explosion in a maximum concentration of most gaseous types is rarely realized [Rashbash (1969)]. Most gaseous explosions are less concentrated, and venting of the burning gases drastically reduces the theoretical peak pressures. Most experimentally induced explosions have peak pressures in the range of $\frac{1}{20}$ to $\frac{1}{10}$ of the theoretical maximum.

Venting area and covering can be simulated in dynamic models using scaled-down window and door areas and coverings found in actual construction. An expression, which shows clearly the effect of venting and developed in the Netherlands by Dragosavic (1973), is also sometimes used in analytical work. Venting can also occur through weak partitions between rooms, although more tests are needed to quantify these effects. The effects of turbulence created as the gas expands from one room into another may create higher pressures in the second room. In the particular problem of interior gaseous explosions, the computed energy release of typical prototype explosions can be scaled, and since the length scale will be dictated by the size of model to be used, the model explosion energy is defined. The model testing can be carried out in a small strong chamber, and the model charges will follow the cube-root scaling law for the prototype gaseous explosion. Extensive verification of Hopkinson's law (Section 11.2.4.1) has been reported [Baker (1973)]. More sophisticated blast scaling laws such as Sach's [Baker (1973)] can be employed in the event that Hopkinson's law proves inadequate for the particular problem at hand.

11.2.5 Earthquake Modeling of Structures

Earthquake loading (Fig. 11.1d) is an important design consideration in many civil engineering structures because of its potential catastrophic nature. In practical structures it requires that the inelastic behavior of the structure be mobilized in order for the design to be economical. Such considerations, however, impose severe restrictions on the possible choice of materials for model testing. A summary of the scale factors obtained from similitude considerations [Krawinkler et al. (1978)] for earthquake loading is given in Table 11.5. True replica models [column (4)] satisfy the Froude and Cauchy scaling requirements, which imply simultaneous duplication of inertial, gravitational and restoring forces. Unfortunately, however, such models are practically impossible to build and test because of the severe restrictions imposed on the model material properties, especially the mass density. Alternate scaling laws shown in columns (5) and (6) of Table 11.5 have been shown to adequately simulate the behavior of the structure. Considerable success has been achieved in the testing of

TABLE 11.5 Summary of Scale Factors for Earthquake Response of Structures

(1)	(2)	Dimension (3)	Scale Factors		
			True Replica Model (4)	Artificial Mass Simulation (5)	Gravity Forces Neglected Prototype Material (6)
Loading	Force, Q	F	$S_E S_l^2$	$S_E S_l^2$	S_l^2
	Pressure, q	FL^{-2}	S_E	S_E	1
	Acceleration, a	LT^{-2}	1	1	S_L^{-1}
	Gravitational acceleration, g	LT^{-2}	1	1	Neglected
	Velocity, v	LT^{-1}	$S_l^{1/2}$	$S_l^{1/2}$	1
	Time, t	T	$S_l^{1/2}$	$S_l^{1/2}$	S_l
Geometry	Linear dimension, l	L	S_l	S_l	S_l
	Displacement, δ	L	S_l	S_l	S_l
	Frequency, ω	T^{-1}	$S_l^{-1/2}$	$S_l^{-1/2}$	S_l^{-1}
Material properties	Modulus, E	FL^{-2}	S_E	S_E	1
	Stress, σ	FL^{-2}	S_E	S_E	1
	Strain, ϵ	—	1	1	1
	Poisson's ratio, ν	—	1	1	1
	Mass density, ρ	$FL^{-4}T^2$	S_E/S_l	*	1
	Energy, EN	FL	$S_E S_l^3$	$S_E S_l^3$	S_l^3

$*(g\rho l/E)_m = (g\rho l/E)_p$.

reduced-scale structures and structural components [Mirza et al. (1979)] on shaking tables where additional material of a nonstructural nature has been added to simulate the required scaled density of the model. The similitude laws for artificial mass simulation are shown in column (5). Both lumped mass and distributed mass simulation has been reported. A third type of scaling law, shown in column (6), applies to the case where gravity stresses can be neglected in the structural behavior and where the same materials are used in both model and prototype to enable the testing to proceed to failure.

11.3 MATERIALS FOR DYNAMIC MODELS

11.3.1 Dynamic Properties of Steel Structures

Experimental evidence indicates that the strength behavior of most materials depends on the rate of strain, especially at high strain rates during testing. In steel these changes are shown in Fig. 11.2 for the case of ASTM A36

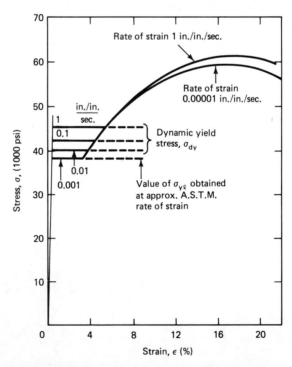

Figure 11.2 Effect of rate of strain on stress-strain curve for structural steel. [After Norris et al. (1959).]

structural steel. The effects of increasing rate of strain can be summarized as follows:

1. The yield stress increases to some dynamic value (σ_{yd}).
2. The yield point strain (ϵ_y) increases.
3. The modulus of elasticity (E) remains constant.
4. The strain at which strain hardening begins (ϵ_{st}) increases.
5. The ultimate strength increases slightly.

The most important effect that will influence the design of steel structures to resist dynamic loads is the increase in the yield stress. In Fig. 11.3 [Shaw (1962)]

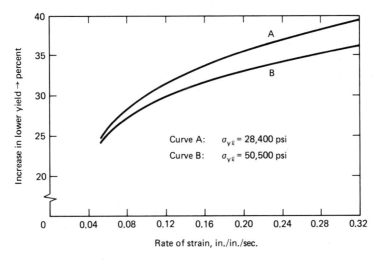

Figure 11.3   Increase of lower yield point of steel with strain rate. [After Norris et al. (1959).]

the percentage increase in yield stress is given as a function of the rate of strain for two steels of different static yield stress. It is evident from the figure that the increase in dynamic yield point is greater for steels with lower static yield point, as is shown by the higher slopes of curve A. Figure 11.4 shows the dynamic yield stress as a function of the time required to reach that value of stress (σ_{yd}) for ASTM A7 steel. From this curve, values of design yield stress could be found if the time to reach yield stress in a particular structure is known.

Useful data on the effect of strain rate on smooth and knurled steel wire suitable for model reinforcement has been obtained by Staffier and Sozen (1975) and is shown in Fig. 11.5. Additional strain rate effects on steel tension specimens has been reported by Krawinkler et al. (1978), and Mills et al. (1979), with similar increases in yield strength with strain rate.

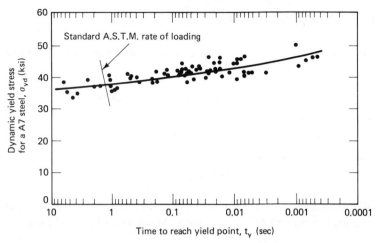

Figure 11.4 Effect of rate of strain on yield stress. [After Norris et al. (1959).]

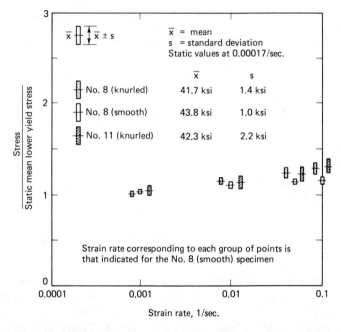

Figure 11.5 Ratios of standard deviation and mean lower yield stress to static mean lower yield stress at different strain rates for no. 8 and no. 11 gage black annealed wire. [After Staffier and Sozen (1975).]

11.3.2 Reinforced Concrete

11.3.2.1 General

For modeling of concrete structures subjected to inelastic deformations it is not possible to use anything but concrete-like materials such as cement mortar or "microconcrete" and gypsum mortar. The static properties of model concrete and model reinforcement have been studied more extensively [Harris et al. (1966), Harris et al. (1970), Sabnis and White (1969), Johnson (1962), Little and Paparoni (1966)] than the dynamic properties. Strain rate effects on model concretes have been reported in connection with blast and impact loading effects [Harris et al. (1963), Ferrito (1969)], and cyclic reversed loading tests, albeit quasistatic, have been conducted in connection with earthquake loading [Chowdhury and White (1977), Yeroushalmi and Harris (1978)].

11.3.2.2 Effect of Strain Rate on the Unconfined Compressive Strength

A series of 2 × 4 in. (50 × 100 mm) cylinders of microconcrete with a mix of water-cement-sand of 0.9 : 1 : 4.5 was tested at increased strain rates in unconfined compression. The strain was measured by two SR-4-type A-3 strain gages placed on opposing generators. The test setup using a 10-k (44.5 kN) Instron machine is shown in Fig. 11.6, with a two-channel Sanborn recorder for strains and load. A total of 44 specimens were tested at four rates of strain ranging from 10^{-5} to 10^{-2} in./in./sec (mm/mm/sec). The results are shown in

Figure 11.6   Test set-up. [After Harris et al. (1963).]

TABLE 11.6 Strain Rate Effects on Unconfined Compressive Strength 2 × 4 in. (50 × 100 mm) Cylinders

Average Strain Rate, in./in./sec (mm/mm/sec)	Head Speed of Instron Machine, in./min (mm/min)	$f'_c = 1730$ psi (11.9 MPa)		$f'_c = 1860$ psi (12.8 MPa)		$f'_c = 2270$ psi (15.7 MPa)		Average $\dfrac{f'_{cd}}{f'_c}$
		No. Specimens	$\dfrac{f'_{cd}}{f'_c}$	No. Specimens	$\dfrac{f'_{cd}}{f'_c}$	No. Specimens	$\dfrac{f'_{cd}}{f'_c}$	
1×10^{-5}	0.02(0.5)	5	1.040	7	1.023	2	0.968	1.021
1×10^{-4}	0.20(5)	4	1.109	7	1.116	3	1.153	1.122
1×10^{-3}	2.0(50)	4	1.206	3	1.199	3	1.160	1.187
1×10^{-2}	20.0(500)	3	1.367	0		3	1.303	1.335

Table 11.6 as the ratio of dynamic unconfined compressive strength to the static value. The average values of f'_{cd}/f'_c vs. the average rate of strain are plotted in Fig. 11.7. A comparison with similar results of ordinary concrete shows a higher rate of increase of f'_{cd}/f'_c for the microconcrete over the range tested. This is partly due to the size effect when testing smaller specimens of microconcrete, as discussed in Chapter 6.

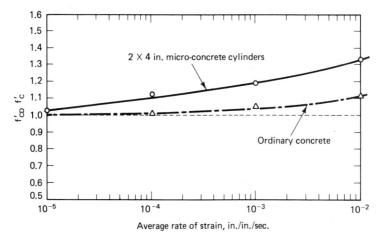

Figure 11.7 Effect of increased strain rate on the unconfined compressive strength.

An empirical relation was obtained from the mean values of Fig. 11.7 by passing a cubic parabola through these points [Harris et al. (1963)].

11.4 LOADING SYSTEMS FOR DYNAMIC MODEL TESTING

11.4.1 Vibration and Resonant Testing

Elastic vibration studies can be performed on full-scale structures using a variety of techniques [Hudson (1967)] to excite the natural modes and frequencies of the structure. These methods tend to be expensive, and alternate procedures using scaled models are therefore widely utilized.

Although free-vibration measurements are sometimes easily accomplished by pulling on the structure, then releasing quickly, and measuring the free motions of the structure, most vibration tests in the laboratory are performed by forcing the structure to vibrate in one of its natural modes. This is accomplished by the use of mechanical or electromagnetic oscillators or by placing the model on a shaking table. A setup for studying the natural modes and frequencies of a cantilever model plate is illustrated in Fig. 11.8. Using a small-capacity electromagnetic shaker, the plate can be made to oscillate sinusoidally or in

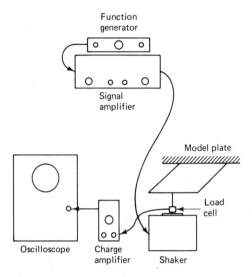

Figure 11.8 Vibration test set-up.

other types of programmed periodic motions. The plate structure is forced to oscillate with the shaker through a small coupling rod to which a load cell is attached (Fig. 11.8). By changing the frequency of the forcing signal, the plate is forced to vibrate at different frequencies. The frequencies that correspond to the natural frequencies of the plate will result in zero applied force through the load cell. This condition can be observed on the oscilloscope and becomes therefore the point at which the natural mode shape and frequency are determined.

11.4.2 Wind Tunnel Testing

Fluidelastic studies of structural models are performed in a variety of wind tunnels. At present, such facilities are available with the capability of simulating the characteristics of the atmospheric boundary layer which is essential for determining wind effects on buildings and other structures. These facilities can be divided into three basic types:

1. Long wind tunnels (of the order of 30 m) in which the boundary layer develops naturally over a rough floor. Examples of such tunnels are the environmental wind tunnel of Colorado State University (Fig. 11.9) and the University of Western Ontario Boundary Layer Wind Tunnel (Fig. 11.10) whose boundary layer development is shown schematically in Fig. 11.11.

2. Wind tunnels with passive devices such as grids, fences or spires which are used to generate a thick boundary layer. The flow then passes over a short section of roughness arrays. An example of such a tunnel is that of the National Aeronautical Establishment, Ottawa, Canada.

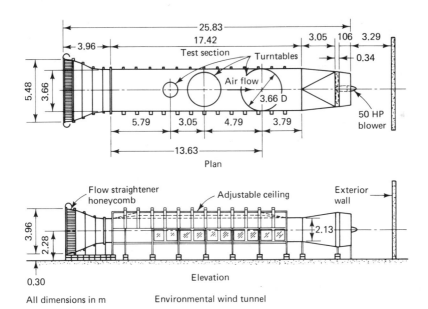

Figure 11.9 Wind tunnel for physical modeling of flow around buildings—Fluid Dynamics and Diffusion Laboratory, Colorado State University. [After Cermak (1977).]

Figure 11.10 Upstream view of boundary layer wind tunnel with model of a rectangular building in foreground. [After Davenport and Isyumov (1968).] (Courtesy of Prof. A. G. Davenport, Director, Boundary Layer Wind Tunnel Laboratory, The University of Western Ontario, London, Ontario, Canada.)

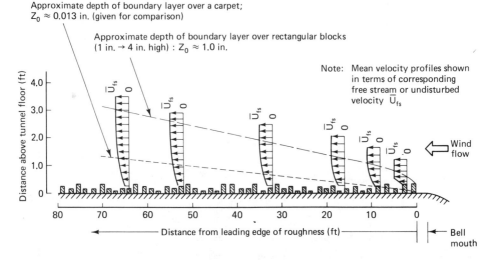

Figure 11.11 Development of boundary layer over typical tunnel surfaces. [After Davenport and Isyumov (1968).]

3. Wind tunnels with active devices such as jets or machine-driven shutters or flaps. Many tunnels used for aeronautical research fall into this category.

Wind effects on tall buildings require the determination of the shears, moments and deformations along the height caused by the integrated effect of the wind pressures as well as the wind effect on glass panels and cladding. By introducing instrumented small scale models of the structure and its surroundings into the test section of the wind tunnel, all of the above quantities can be determined experimentally.

An example of a rigid model of a building mounted in the wind tunnel is shown in Fig. 11.12. Further examples of the testing techniques used in wind tunnel studies are presented by [Davenport and Isyumov (1968), Cermak (1977), Simiu and Scanlan (1978), and Sachs (1978)].

11.4.3 Shock Tubes and Blast Chambers

Laboratory studies of external blast loading effects have been conducted successfully in "shock tube" facilities and specially constructed blast chambers. A shock tube consists of a straight, usually uniform, section separated into a high-pressure and a low-pressure portion by means of a diaphragm (Fig. 11.13a). By quickly opening the diaphragm—by bursting, a compression wave is first propagated into the low-pressure region, followed by a rarefaction wave into the high-pressure region. The compression wave rapidly propagates into a shock wave as it progresses down the tube as illustrated in Fig. 11.13. A scale

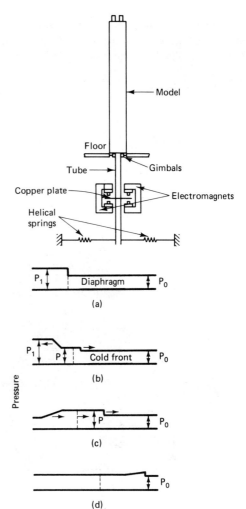

Figure 11.12 Rig for linear-mode models [After Scruton (1968).]

Figure 11.13 Pressure distribution in shock tube at different times. [After Norris et al. (1959).]

model placed in the shock tube would therefore be subjected to the shock wave for a period of time prior to the shock's eventual alteration by the interaction of the rarefaction with the closed high-pressure end of the tube and the compression wave with the open end of the tube.

Special concrete bunker-like facilities have also been used to simulate blast effects on small-scale models [Norval and Cohen (1970)]. A typical facility constructed at Drexel University consists of an 8.5 × 10-ft (2.6 × 3 m) box-like bunker having 2-ft (0.6 m) thick concrete walls. A blast deflector and masonry stack direct the blast and detonation products to the roof, where they are safely dispersed. The model is loaded by a blast wave from a scaled charge exploded inside the chamber. Instrumentation consists of pressure, deflection, strain, and acceleration pickups, and a quartz glass window provides access to the chamber for instrumentation cables or high-speed photography. Charges of up to 1 lb (0.45 kg) of TNT have been detonated without difficulty in this particular facility.

11.4.4 Shaking Tables

Shaking tables come in a variety of sizes and capabilities. Some are hydraulically actuated, and others, such as the 6.5×10 ft (2×3 m) table at ISMES, Italy [Castellani et al. (1976)], are actuated by electromagnetic shakers as shown in Fig. 11.14. A classification of various shaking tables (Table 11.7)

Figure 11.14   Lay-out of the test. [After Castellani et al. (1976).] (Courtesy of ISMES, Bergamo, Italy.)

indicates that the smaller facilities are more suited to testing scale models. The model to be tested under simulated earthquake motions is usually bolted to the moving table as shown in Fig. 11.14, which illustrates the testing of a model dam and its surroundings. The moving table is supported with various techniques indicated in the last column of Table 11.7.

A testing and measuring system for the sinusoidal and random vibration of model structures used at the National Civil Engineering Laboratory, Lisbon, Portugal [Borges and Pereira (1970)], is shown in Fig. 11.15. The experimental data from such tests can be plotted during the test or recorded on magnetic tape for later processing.

TABLE 11.7 Classification of Various Shaking Tables*

Location (1)	Dimensions, ft(m) (2)	Payload Limit, lb(kN) (3)	a_{max}, g (4) Horizontal	Vertical	d_{max}, ±in. (5) Horizontal	Vertical	f_{max}, Hz (6)	Type of Support (7)
Small (<10 ft) (<3 m)								
Stanford Univ.	5 × 5 (1.6 × 1.6)	5,000(22.2)	5	—	2.5	—	50	Roller bear.
Univ. of Calgary	4.5 × 4.5(1.3 × 1.3)	2,000(9)	20	—	3	—	—	—
ISMES, Italy	10 × 6.5 (3 × 2)	300(1.3)	100	—	—	—	800	Oil film
Drexel University	4 × 6 (1.2 × 1.8)	2,000(8.9)	3.6	—	0.25	—	2000	Roller bear.
Medium (10–30 ft.) (3–9 m)								
Univ. Calif., Berkeley	20 × 20 (6 × 6)	100,000(444.8)	1.5	1.0	5	2	15	Air press.
Univ. Illinois	12 × 12 (3.6 × 3.6)	10,000(44.5)	7	—	4	—	100	Flex. supp.
Corps of Engineers	12 × 12 (3.6 × 3.6)	12,000(53.4)	34	60	2.2	1.8	200	—
Wyle Lab, Huntsville, Ala.	17 × 11 (5.5 × 3.5)	9,500(42.3)	8	8	3	3	500	—
Large (>30 ft) (>9 m)								
National Research Center, Japan	50 × 50 (15 × 15)	1,000,000(4,448)	0.6	1.0	1.2	—	16	—
Berkeley (proposed)	100 × 100 (30 × 30)	4,000,000(17,792)	0.6	0.2	6	3	—	—

*After Krawinkler et al. (1978).

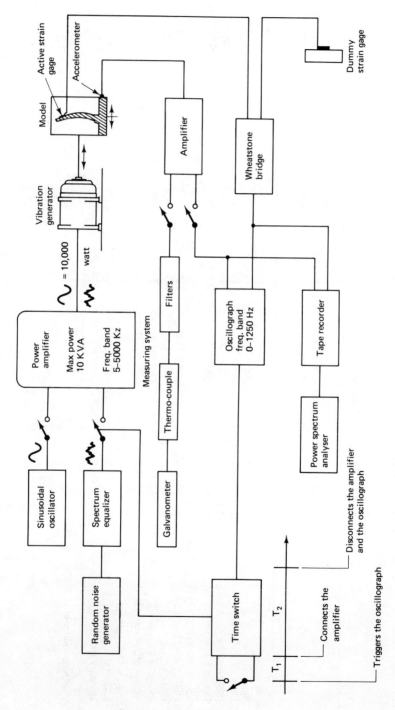

Figure 11.15 Testing and measuring systems. [After Borges and Pereira (1970).]

11.5 EXAMPLES OF DYNAMIC MODELS

11.5.1 Natural Modes and Frequencies

11.5.1.1 Buildings

The determination of natural vibration modes and frequencies of tall buildings is a necessary first step to an analysis involving any dynamic loads. It is during this phase of analysis or testing in which the dynamic characteristics of the structure are studied. The effect of dynamic loads, such as wind or earthquake, would be a tendency to excite the fundamental modes of the structure.

Sometimes the complexity of the structure and its mass distribution require that experimental verification of the analytical procedures be carried out. Scale model testing can easily be used to carry out this step. An example of such a study [Castoldi and Casirati (1976)] is the determination of the natural vibration characteristics of the Parque Central high-rise building in Caracas, Venezuela. This study, which preceded the seismic testing, was carried out at ISMES, Bergamo, Italy. An elastic 1 : 40 scale model was made using an epoxy resin as shown in Fig. 11.16. The first four mode shapes and frequencies

Figure 11.16 Parque Central building model. [After Castoldi and Casirati (1976).] (Courtesy of ISMES, Bergamo, Italy.)

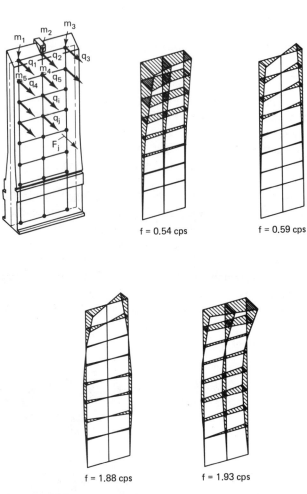

f = 0.54 cps   f = 0.59 cps

f = 1.88 cps   f = 1.93 cps

Figure 11.17 Parque Central building. Stiffness matrix determination and vibration modes for transverse excitation. [After Castoldi and Casirati (1976).]

are shown in Fig. 11.17, where we note that the torsional stiffness of the structure influences the vibration modes.

11.5.1.2 Bridges

An extensive static and dynamic model study [Baglietto et al. (1976)] was conducted at ISMES for the design of the Zarate-Brazo Largo Bridges, Buenos Aires. These identical twin bridges are braced steel box and cable stayed, carrying both highway and eccentrically placed railroad traffic. This particular design arrangement was one of the main reasons of concern, which prompted a 1 : 33.3 scale model study as the design work was progressing and

Figure 11.18 View of the completed model, Zarate-Brazo Largo Bridges. [After Baglietto et al. (1976).] (Courtesy ISMES, Bergamo, Italy.)

not as a final check or afterthought to the design process. A view of the complete model is shown in Fig. 11.18, with all simulated masses added and ready for vibration testing. A schematic of the method used in vibrating the model by means of four electromagnetic shakers is shown in Fig. 11.19. The data control

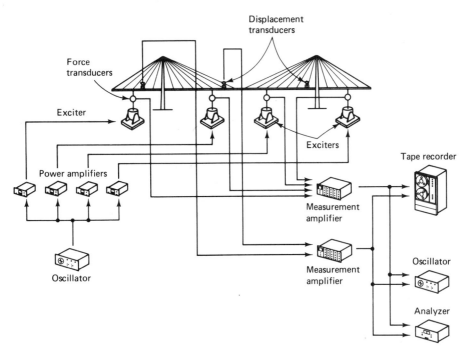

Figure 11.19 Block diagram of excitation and measurement system to determine vibration modes by means of concentrated forces. [After Baglietto et al. (1976).]

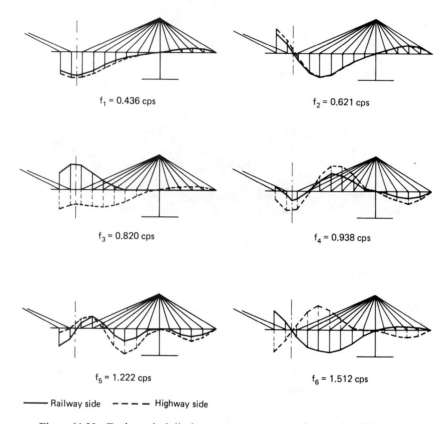

$f_1 = 0.436$ cps $f_2 = 0.621$ cps

$f_3 = 0.820$ cps $f_4 = 0.938$ cps

$f_5 = 1.222$ cps $f_6 = 1.512$ cps

——— Railway side   — — — Highway side

Figure 11.20   Deck vertical displacements at resonance frequencies. [After Baglietto et al. (1976).]

and recording equipment is also shown in this figure. Some typical results of the lowest resonant frequencies and corresponding mode shapes are shown in Fig. 11.20 for the case with the deck loaded in the middle with the mass corresponding to two locomotives. A comparison of the analytically predicted natural frequencies with those obtained from the model vibration tests is made in Table 11.8. A close agreement of the results is indicated for both the dead-load and live-load cases.

Several extensive model studies [Williams and Godden (1976) and Godden and Aslam (1978)] of the dynamic response of bridges to earthquake loading have been conducted at the University of California at Berkeley in the last few years. The models used in these studies were subjected to harmonic input as well as seismic inputs on the 20 × 20 ft (6 × 6 m) shaking table in the Earthquake Simulator Laboratory. The description of a $\frac{1}{30}$-scale ultimate strength model and its seismic behavior is given in Section 11.5.4.2.

TABLE 11.8   Comparison of Model Results with Analytical Predictions

Dead loads, frequencies

		Frequency	
Mode	Shape	Dynamic Calculated, Hz	Structural Model, Hz
1	Symmetrical bending	0.44	0.472
2	Antisymmetrical bending	0.58	0.587
3	Symmetrical torsional	0.81	—
4	Symmetrical bending	0.88	0.945
5	Antisymmetrical bending	1.01	—
6	Antisymmetrical flexotorsional	1.09	—

Bridge with railway load in the central 330 meters (1,000 ft)

		Frequency	
Mode	Shape	Dynamic Calculated, Hz	Structural Model, Hz
1	Symmetrical bending	0.42	0.436
2	Antisymmetrical bending	0.55	0.621
3	Symmetrical torsional	0.71	0.820
4	Symmetrical bending	0.82	0.938
5	Antisymmetrical flexotorsional	0.94	—
6	Antisymmetrical bending	1.04	1.222

11.5.1.3 Special Structures

Many types of structures other than those described in Fig. 11.1 undergo dynamic loading during their useful operation. One class of such structures are aerospace vehicles. A vibration evaluation is a first step in their design for dynamic loads. Dynamic models play a very important role in the design and analysis of these usually complex structures. As an example, we will illustrate the $\frac{1}{10}$-scale model of a portion of the Apollo simplified shell structure consisting of the conical lunar adapter (SLA), the short cylindrical instrumentation unit (IU), and the stiffened cylindrical shell of the S-IVB rocket forward skirt shown in Fig. 11.21. This model, simulating all the important axial and bending stiffness parameters, was vibrated through the four lunar module (housed within the conical adapter) attachment points by means of a small electrodynamic vibrator, seen in Fig. 11.21. The lunar module was simulated by means of a cruciform structure and additional mass so that the inertial forces at the attachment points, interacting with the conical shell, were dynamically modeled. In

Figure 11.21 Model of simplified Apollo shell shown driven by a 30 lb. maximum force shaker at the +Z apex fitting. [After Harris (1968).] (Courtesy Grumman Aerospace Corp., Bethpage, New York.)

addition, the mass and center of gravity of the command module on top of the conical shell was properly simulated. Mode shape surveys of the simplified shell model were made using a hand-held accelerometer.

A comparison of the model and prototype lowest mode test results [Harris (1968)] is shown in Fig. 11.22 in an exploded view of the conical adapter. The conical shell vibrates in basically one circumferential wave everywhere except in the vicinity of the four lunar module attachment points at the adapter circumferential stiffening ring, where it vibrates essentially in five waves.

11.5.2 Aeroelastic Model Studies of Buildings and Structures

11.5.2.1 Wind Effects on Buildings

For studies of wind effects on buildings, the model analyst is usually interested in both the mean deflection and sway of the structure as a whole and the magnitudes of the induced local pressures. The first effect is related to the design of the structural framing, and the second to the design of the skin (cladding and glass). Studies of aeroelastic and of rigid models of structures in representative wind flow can be performed to yield the information required to design both the structure and the cladding. A rigid model for such wind studies is shown in Fig. 11.10 and an aeroelastic semirigid model, spring mounted at the base, is shown schematically in Fig. 11.12.

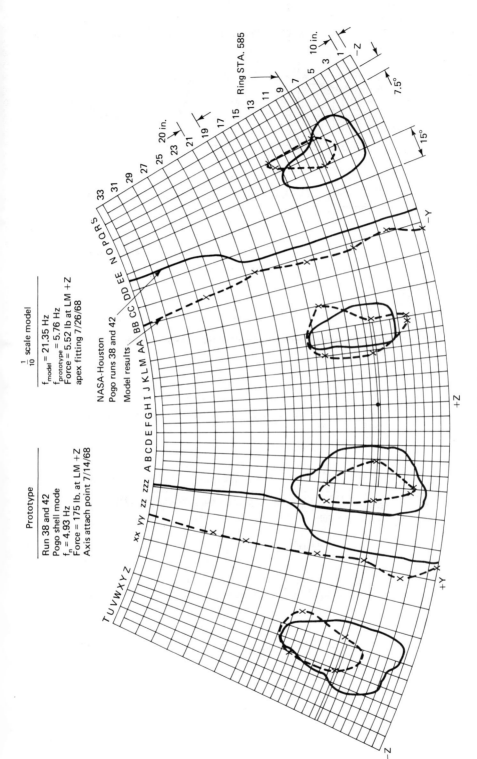

Figure 11.22 Comparison of the model and prototype mode surveys at the lowest natural frequency on developed shell. [After Harris (1968).]

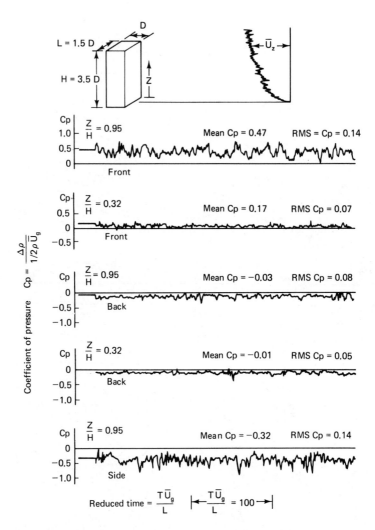

Figure 11.23 Exterior pressures on a rectangular building in boundary layer flow representative of wind over a large city. [After Davenport and Isyumov (1968).]

An illustration of the main effects of fluctuating pressure distributions on the exterior of a model of a rectangular tall building tested by Davenport and Isyumov (1968) is shown in Fig. 11.23. The model was situated in a scaled replica environment of a large city. In Fig. 11.23 the oscillograph records of pressures at front, back, and side of the building at two different elevations are plotted as functions of the reduced time. These results show that the pressure distributions are similar to full-scale measurements on tall buildings. According to the authors:

the complexity of fluctuating pressures, which in this case is further aggravated by the presence of surrounding buildings, emphasizes the difficulties associated in arriving at loads from a knowledge of velocity and stresses the advantages gained from obtaining the required structural response parameters from aero-elastic models.

A distinct advantage of the modeling of wind effects on structures is the ability to study the interference effects of groups of buildings and obstructions in the terrain in the immediate vicinity of a proposed structure. Such an investi-gation was made at the National Physical Laboratory, Teddington, England (Fig. 11.24), in an attempt to estimate the amount of sway in typhoon winds of a

Figure 11.24 A grouping of model tower blocks of octagonal section. [After Scruton (1968).] (Courtesy of University of Toronto Press, Toronto, Canada and the National Physical Laboratory, Teddington, England.)

proposed grouping of tower blocks. The buildings (Fig. 11.24) were of basically octagonal section and had a height-width ratio of 4.25. The effect of turbulence is shown in Fig. 11.25 for a single tower tested in isolation, which shows a significant increase of excitation of the turbulent flow over that of smooth flow. Two towers in line with the wind direction and spaced at 1.78 D apart (Fig. 11.26) show larger amplitudes for the downstream tower and somewhat larger amplitudes in turbulent than in smooth winds. The dependence of the response of the leeward tower (tower B, Fig. 11.26) on the spacing ratio S/D, is shown in Fig. 11.27 for smooth flow. The optimum spacing was found to be 2.75 D for smooth airflow and 2 D for turbulent airflow. A conclusion drawn from this study is that because of the very rough surface of the towers, turbulence had only a small effect on the vortex excitation when there was mutual inter-ference between a pair of towers.

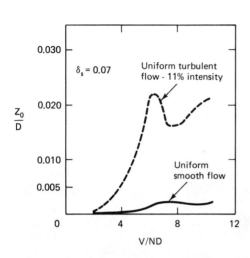

Figure 11.25 Variation of cross-wing amplitude with wind speed for an isolated tower block of octagonal section H/D = 425 immersed in smooth and turbulent flow. [After Scruton (1968).]

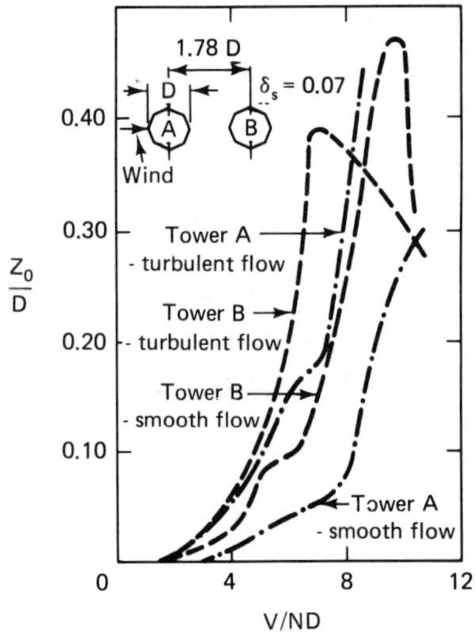

Figure 11.26 Amplitudes of oscillation (crosswind) of the towers of a twin-tower block configuration (H/D = 425) in smooth and turbulent airflow. [After Scruton (1968).]

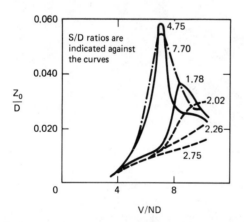

Figure 11.27 Variation of amplitude of leeward tower of a twin tower block configuration with the spacing (S) of the towers smooth airflow. [After Scruton (1968).]

11.5.2.2 Lions' Gate Bridge

Typical of aeroelastic model studies are the extensive experiments [Irwin and Schuyler (1977)] that were performed on a 1 : 110 scale full aeroelastic model of the Lions' Gate Bridge, Vancouver, British Columbia. The general configuration and a proposed modification to this 41-year-old bridge, using cantilever sidewalks to extend the three-lane roadway, are shown in Fig. 11.28. The full

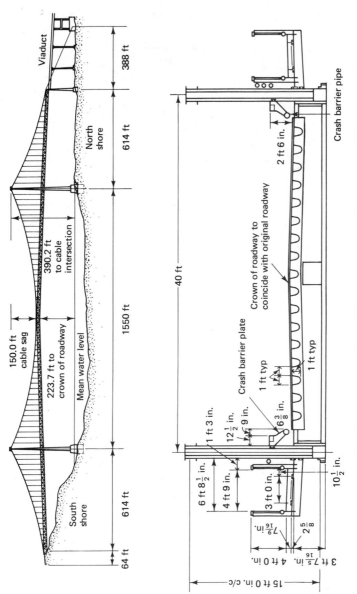

Figure 11.28 Lions' Gate Bridge with modified cross section. [After Irwin and Schuyler (1977).]

519

aeroelastic model that was tested in simulated turbulent flow and also in smooth flow using the National Aeronautical Establishment of Canada 30 × 30 ft (9 × 9 m) wind tunnel is shown in Fig. 11.29. Of prime concern in the program was

Figure 11.29 Aeroelastic model of the Lions' Gate Bridge in the 30-ft. × 30-ft. wind tunnel. [After Irwin and Schuyler (1977).] (Courtesy National Research Council Canada, Ottawa, Canada.)

the determination of the aerodynamic characteristics of the proposed changes as well as those of the present bridge. Two sectional models at $\frac{1}{24}$ scale (Irwin and Wardlaw [1976]), one of the present configuration and the other of the revised design, as well as a 1 : 110 sectional model of the new version were tested in smooth flow under this general investigation. In smooth flow, the full-scale model was in good agreement with the behavior of the $\frac{1}{110}$ scale and $\frac{1}{24}$ scale sectional models. The presence of turbulence had a large effect on the critical velocity for flutter, raising it to a value above the velocity range tested. The results of these investigations imply that conventional methods of determining the stability of bridges using sectional models in smooth flow may lead to conservative (low) estimates of the critical velocity. Comparison of the experimental and analytical results for the vertical modes of vibration is shown in Fig. 11.30. Excellent agreement is found between the two. Comparisons of the more complex torsional/lateral vibration modes is shown in Fig. 11.31.

11.5.3 Blast Effects on Protective Structures

In order to evaluate the effects of high blast loading on structures, a series of spherical shells at 1 : 25 scale were constructed at MIT [Smith et al. (1963)] and then field-tested under blast wave loading conditions at overpressures ranging from 20 to 80 psi (0.14 to 0.56 MPa). These were replica models of

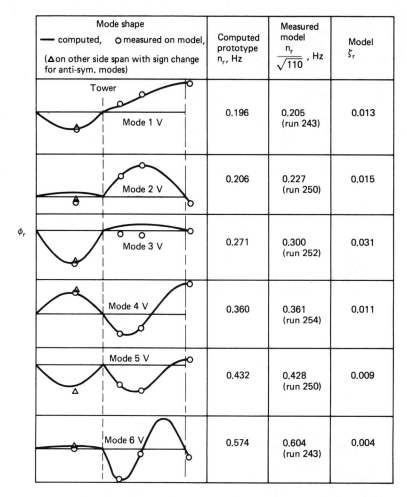

Mode shape — computed, ◯ measured on model, (△ on other side span with sign change for anti-sym. modes)	Computed prototype n_r, Hz	Measured model $\dfrac{n_r}{\sqrt{110}}$, Hz	Model ζ_r
Tower — Mode 1 V	0.196	0.205 (run 243)	0.013
Mode 2 V	0.206	0.227 (run 250)	0.015
Mode 3 V	0.271	0.300 (run 252)	0.031
Mode 4 V	0.360	0.361 (run 254)	0.011
Mode 5 V	0.432	0.428 (run 250)	0.009
Mode 6 V	0.574	0.604 (run 243)	0.004

Figure 11.30 Vertical modes of model compared with computed prototype modes. [After Irwin and Schuyler (1977).]

full-scale reinforced concrete domes tested in an above-ground nuclear explosion. Details of the model shell construction and the method of anchoring the edge ring to the base support are shown in Fig. 11.32.

The wire meshes used for reinforcement were fabricated on a wooden form (Fig. 11.33) of the same geometry as the domes, using a soldering technique. The form, ready for casting and using a rotating scribe to control the thickness of the shell, is shown in Fig. 11.34. Field testing using a charge of 100 t (90.8 Mg) of TNT was accomplished by arranging the 12 models at increasing distances from ground zero (Fig. 11.35) so that the overpressures experienced would be as shown in Table 11.9, from a low value of 20 psi (0.14

Figure 11.31 Torsional/lateral modes of model compared with computed modes. [After Irwin and Schuyler (1977).]

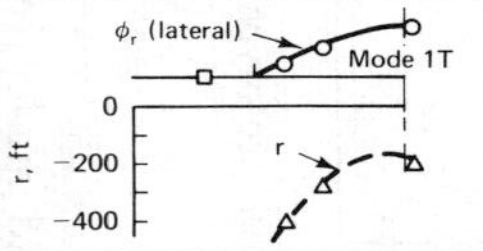

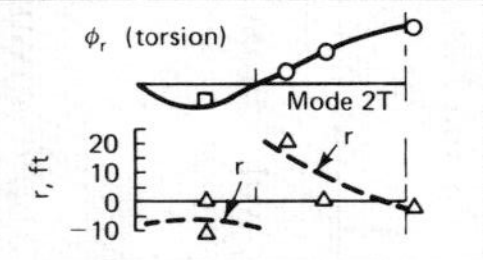

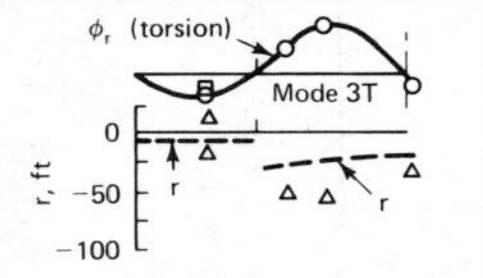

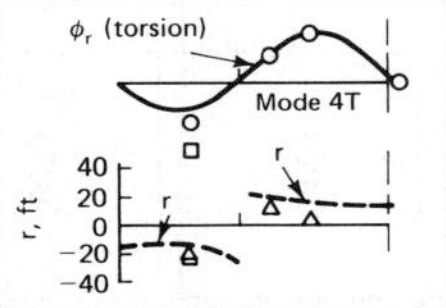

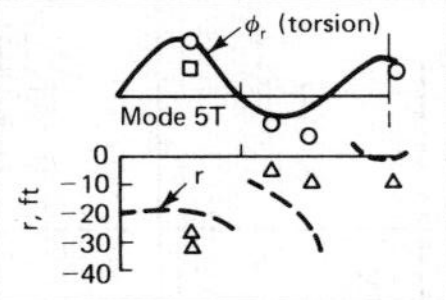

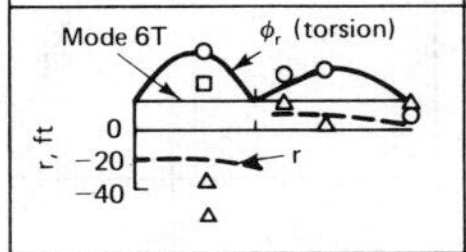

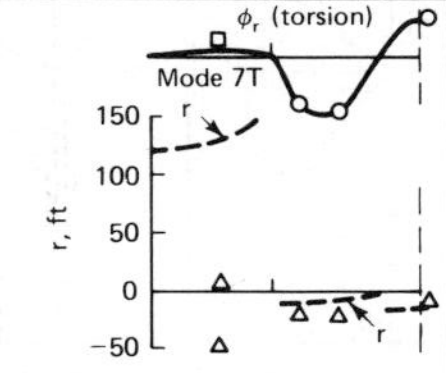

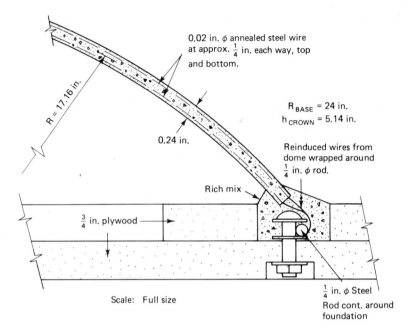

Figure 11.32 Part section through dome model showing base ring. [After Smith et al. (1963).]

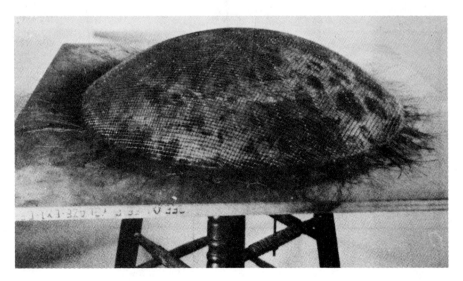

Figure 11.33 Model reinforcement. [After Smith et al. (1963).]

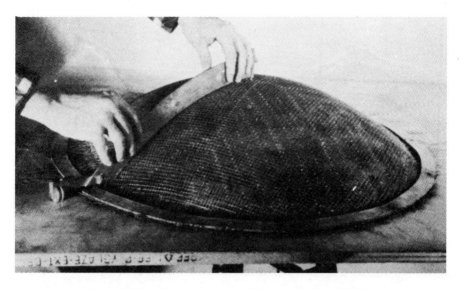

Figure 11.34   Casting technique. [After Smith et al. (1963).]

Figure 11.35   View of model installations looking away from ground zero. [After Smith et al. (1963).]

MPa) to a high of 80 psi (0.56 MPa). Typical photographs of a dome model subjected to an overpressure of approximately 40 psi (0.28 MPa) before and after the test are shown in Figs. 11.36 and 11.37, respectively. The mode of failure of the model domes was very similar to that experienced by the prototype domes.

TABLE 11.9 Summary of Model Distance-Pressure Values

Expected Overpressure, psi (MPa)	Dome Number	Distance from GZ, ft (m)	Overpressure Probably Experienced, psi (MPa)
20(0.14)	8	400(121.9)	21.4(0.15)
	9		
30(0.21)	12	330(100.6)	32.5(0.22)
	7		
35(0.24)	11	310(94.5)	37.5(0.26)
	6		
40(0.28)	10	290(88.4)	43.5(0.30)
	5		
60(0.42)	3	250(76.2)	61(0.42)
70(0.48)	4	230(70.1)	74(0.51)
	2		
80(0.56)	1	220(67)	82(0.57)

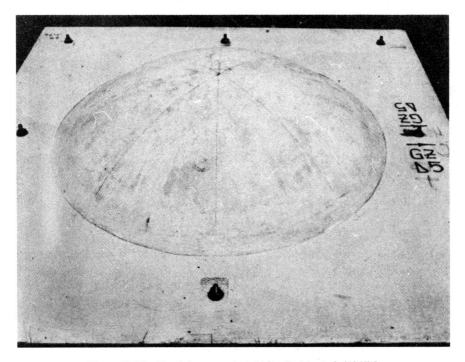

Figure 11.36 40 psi dome preshot. [After Smith et al. (1963).]

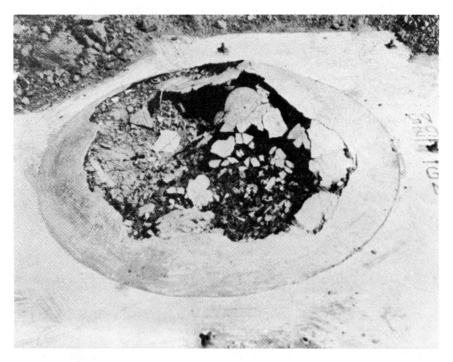

Figure 11.37   40 psi dome postshot. [After Smith et al. (1963).]

11.5.4 Earthquake Simulation of R/C Frames and Bridges

11.5.4.1 Reinforced Concrete Frames

A two-story reinforced concrete frame, representing a segment of a small office building, was studied on the 20×20 ft $(6 \times 6$ m$)$ shaking table of the Earthquake Engineering Research Center, University of California, Berkeley [Hidalgo and Clough (1974), Clough and Bertero (1977)]. This was a relatively large-scale model $(\frac{7}{10})$ chosen such that normal reinforcing and fabrication procedures could be used. Additional concrete blocks were added for ballast, and lateral bracing was provided to constrain the building against any lateral and torsional motions, as shown in Fig. 11.38.

Instrumentation installed to record the response of the test structure included accelerometers and displacement gages at each story level and on the shaking table, strain gages on the reinforcing bars at the column bases, and relative rotation measuring devices at the ends of columns and girders. In addition, the midcolumn moments, shears and axial forces were measured directly by means of transducers.

The input earthquake was one horizontal component of the Taft, California, earthquake of July 1952 with no vertical components. The excitation was applied with successively increasing intensities, starting with a peak ac-

Figure 11.38 Two-story R/C frame on shaking table. [After Clough and Bertero (1977).] (Courtesy of Prof. R. W. Clough, UC, Berkeley, Earthquake Engineering Research Center.)

celeration of 0.07 g in the first run and reaching a maximum run having a peak acceleration of 0.44 g. The first run produced only elastic response, but the maximum intensity test caused considerable damage. Typical test records are shown in Fig. 11.39. They depict the input table displacements and the resulting average column shears measured during the maximum test.

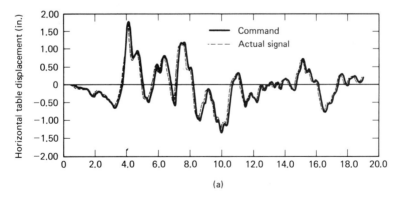

Figure 11.39 Shaking table motion and building response. (a) Table displacements: command vs actual.

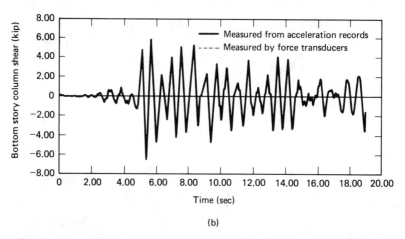

Figure 11.39 (*cont.*) (b) First-story column shear force. [After Clough and Bertero (1977).]

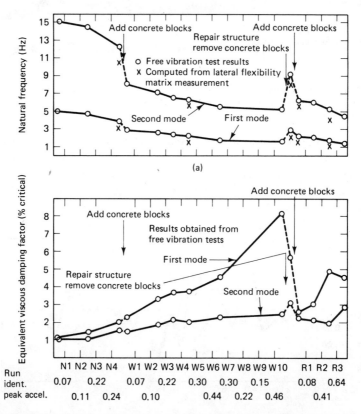

Figure 11.40 Changes of vibration properties during test sequence. (a) Frequencies vs test number. (b) Damping ratio vs test number. [After Clough and Bertero (1977).]

After each test, the free vibration frequency and damping of the test structure were determined by suddenly releasing a 1000-lb (6850 kN) horizontal force applied at the first-story level. The successive changes of these properties, plotted in Fig. 11.40, demonstrate the extent of damage inferred by the drop in natural frequencies and the increase in damping done to the structure in each test run. A comparison of the top-story displacement measured during the most intense test run with a nonlinear response analysis is shown in Fig. 11.41.

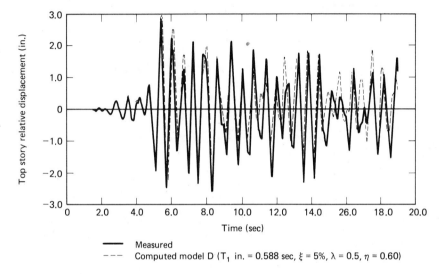

——— Measured
– – – Computed model D (T_1 in. = 0.588 sec, ξ = 5%, λ = 0.5, η = 0.60)

Figure 11.41 Correlation of top story displacement: measured vs. computed. [After Clough and Bertero (1977).]

11.5.4.2 Reinforced Concrete Bridges

The design, construction, and seismic testing of a model of a long, curved, reinforced concrete overcrossing structure was conducted at the University of California, Berkeley [Williams and Godden (1976)], to investigate the general dynamic behavior of such structures and to provide experimental data for comparison with theoretical solutions. A schematic drawing of the $\frac{1}{30}$-scale model, which represents only a portion of the total structure, is shown in Fig. 11.42. The multicell box girder section of the prototype structure was modeled using an equivalent rectangular solid section to simplify the model construction. Mild steel deformed bars were used for the model longitudinal reinforcement, and plain mild steel wires for temperature effects, shrinkage, and shear reinforcement (Fig. 11.43). The model concrete utilized was a specially designed high-strength [8000 psi (55 MPa)] low-shrinkage mortar. Special attention was given to the fabrication of the two expansion joints (Fig. 11.42). An equivalent-

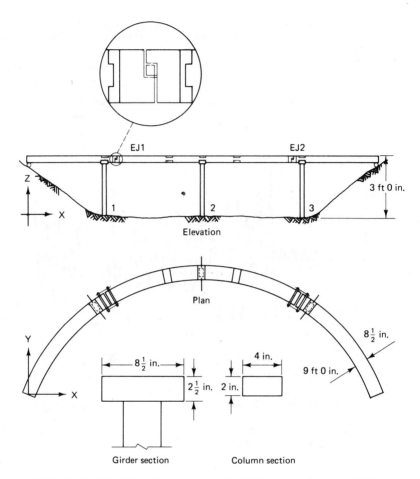

Figure 11.42 Schematic of test model. [After Williams and Godden (1976).]

strength section was used for the columns with the longitudinal reinforcement butt welded to steel plates at both ends of the model column (Fig. 11.44). Since the model and prototype materials had similar densities, external weights were placed on the model to preserve the effective mass density required by dynamic similitude (Table 11.5). Care was taken to ensure that the added lead ingots used did not change appreciably the section stiffness or the system damping of the structure.

A large number of preseismic tests were conducted on the model components and on the assembled model to determine its most important dynamic characteristics. The seismic-type tests were conducted on the Berkeley 20 × 20 ft (6 × 6 m) shaking table by subjecting the model to prescribed table motions

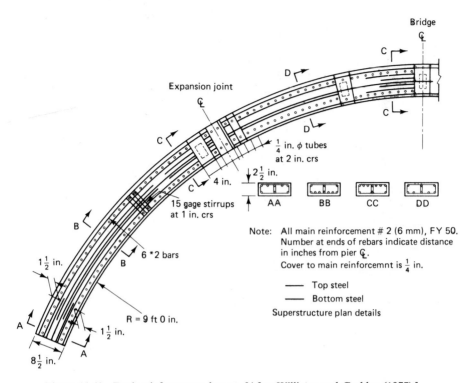

Figure 11.43 Deck reinforcement layout. [After Williams and Godden (1977).]

of increasing intensity. The measured dynamic response included displacements in the two horizontal directions at the expansion joints as well as the gap motion of the joint measured at both the inside and outside edges of the deck.

The model bridge was designed as a small representative structure on which trends of behavior could be studied and correlation studies with theory could be made. No attempt was made to model a specific prototype. For this reason, the mechanisms of failure are more significant than the maximum ground accelerations needed to produce these. Although the joints were supposedly overdesigned, they did suffer major damage but no catastrophic collapse. Typical results of the symmetric seismic response to an artificially generated table motion with a peak horizontal component of 0.48 g on the model are shown in Fig. 11.45.

The extensive seismic testing of the bridge model indicated that the most critical locations of the structure where damage occurs and hence where ductility is required are the expansion joints and the bases of the long columns. The joints are very susceptible to damage caused by multiple impacting in both the torsional and translational modes of response.

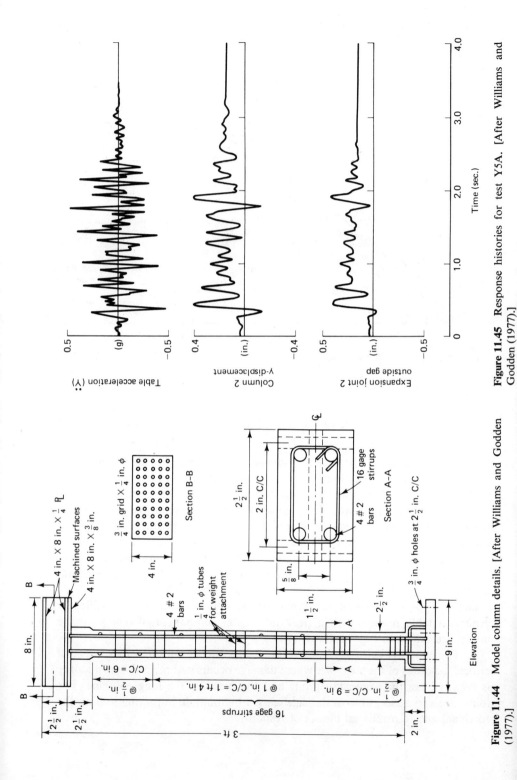

Figure 11.45 Response histories for test Y5A. [After Williams and Godden (1977).]

Figure 11.44 Model column details. [After Williams and Godden (1977).]

11.6 CASE STUDY, WIND TUNNEL TESTS
OF THE TORONTO CITY HALL

11.6.1 The Problem

Aeroelastic response of the two crescent-shaped towers, one approximately 290 ft (88.4 m) and the other 225 ft (68.6 m) high, of the new Toronto City Hall depends on the prevailing wind direction, and it may exhibit the flow characteristics of a diffuser, a nozzle, a semicylinder, or a complete cylinder. The pressure on the shell surface can vary considerably with the wind direction and can be steady or unsteady (that is, oscillating) for some specific wind directions. The unusual shape of the structure renders it weak against torsional loads. This along with the tower height led the engineers to suspect that the standard wind design pressures specified by the City of Toronto Building Code would not be applicable, and it was decided to conduct wind tunnel tests on a small-scale model of the structure at the Institute of Aerophysics of the University of Toronto.

Although the scale of the models was much smaller than desirable (no larger wind tunnels were immediately available), it was felt that despite the uncertainty associated with the scale effect it was preferable to base the design on these results rather than on the building code. The nature of the pressure distributions obtained gave the engineers sufficient confidence to make some changes in the design of the shell-like structure. To verify the aerodynamic characteristics on which to base the structural design, a second model, corresponding to the revised design, was also built and tested.

11.6.2 Test Program

Both models were tested in the 36 $\times$ 42 in. (914 $\times$ 1067 mm) test specimen of the University of Toronto subsonic wind tunnel. The original $\frac{1}{276}$-scale model, built using mahogany, was mounted on a circular plywood base plate, beveled and graduated at the outer edge (Fig. 11.46). Three horizontal rows of 20 static pressure taps each, spaced evenly along the arc of the tower, were installed on both the inside and outside walls of the two towers. A plastic strip, 1 in. (25 mm) wide, consisting of 20 pressure tubes was mounted horizontally and flush with the wall at the desired level for the full width of the tower. Each tap was formed by drilling a hole from outside. The upstream ends of pressure holes were plugged with pins, while the downstream ends were connected by a strip of tubes to a multiple manometer board. The second model ($\frac{1}{296}$-scale) was equipped with only two pressure tap levels since tests on the original model showed little variation of pressure with height.

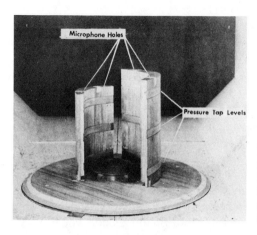

Figure 11.46 Solid mahogany wind model at 1/276 scale of the Toronto City Hall in the subsonic wind tunnel. [After Dau (1961).] (Courtesy of the University of Toronto Institute for Aerospace Studies, Downsview, Ontario, Canada.)

11.6.3 Testing Techniques

11.6.3.1 Steady Pressure Measurements

Although the maximum speed possible in the tunnel was in excess of 200 fps (61 m/sec.), the very large drag of the model resulted in a maximum wind speed of 175 fps (53.3 m/sec.). The objective was to obtain as high a Reynold's number as possible. A typical test consisted of rotating the model slowly in the airstream through a predetermined 60° sector while observing the pressure readings. A similar procedure was repeated for the remaining five 60° sectors. The wind direction that caused maximum suction on the outside of the tower was chosen for each 60° sector. This direction was maintained constant for all pressure readings on a given tower within each 60° sector. For a number of wind directions, pressure readings were taken in the two corners of each tower at a height of 1.5 in. (38 mm) from the base.

11.6.3.2 Unsteady Pressure Measurements

Unsteady pressures resulting from periodic vortex shedding from the building were measured at four pressure taps on the outside wall located 4 in. below the top edge. Vertical holes 4 in. deep and $\frac{5}{8}$ in. (15.9 mm) in diameter were drilled into the tower at required positions. Each hole was equipped with a microphone registering pressure through a small (horizontal) drill hole. To ensure undisturbed air flow over the wall, the outside opening of the drill hole was closed with a tape that was subsequently punctured with a pin. The microphone was connected to a pen recorder, a sound-level meter, and an oscilloscope for immediate visual observation.

Seven pen recordings were made for two stations for two wind directions with the wind speed varying from 75 to 160 fps (22.9 to 48.8 m/s). Sound-level readings in decibels were simultaneously taken on the sound-level meter. The vortex shedding frequency was calculated from the number of times the pen recording trace crossed the referenced line divided by twice the elapsed time.

11.6.3.3 Flow Visualization Tests

Flow visualization tests were conducted in a small smoke tunnel with a 1 × 12 in. (25.4 × 305 mm) test section. This precluded testing of a model properly scaled for height; instead a third $\frac{1}{960}$-scale, two-dimensional model was used. The test consisted of taking still photographs of the streamline pattern around the model at 15° intervals in the wind direction, the wind speed being about 6 fps (Fig. 11.47). In addition, movie pictures were taken of the streamline patterns, while the model was rotated through 360° to observe the unsteady flow associated with vortex shedding at certain wind directions.

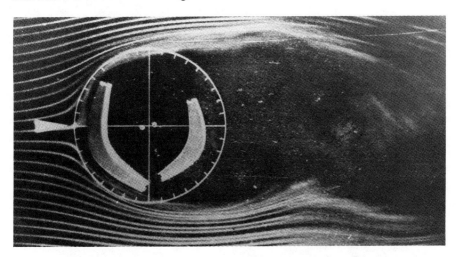

Figure 11.47   Unsteady flow around a distorted two-dimensional 1/960 scale model showing vortex shedding. [After Dau (1961).] (Courtesy of the University of Toronto Institute for Aerospace Studies, Downsview, Ontario, Canada.)

11.6.4 Conclusions

This investigation shows that the steady pressure can be predicted with reasonable accuracy from wind tunnel model studies, since viscous effects are negligible except in the regions of separated flow. In such cases, the pressure coefficients for the prototype are likely to be higher than those for the model.

Dau (1961) observed that the steady pressure distribution on the outside (convex) wall resembled that on a circular cylinder, with some suction peaks attaining a value of over twice the wind dynamic pressure. The resulting pressure distribution on the outer face causes torsional loads on the tower, while the pressure distribution on the inside (concave) contributes mainly to bending loads.

The wind tunnel test results were converted into design pressures using an assumed wind velocity distribution varying from 110 mph (180 km/h) at the tower top to 60 mph (96.6 km/h) at the bottom. The external design wind pressures were as high as 31 psf (1.5 kN/m²), and suctions as high as 72 psf (3.5 kN/m²). These high values, along with the unusual pressure distributions found from these tests, produced torsional and bending loads far in excess of those expected from standard design assumptions, thus proving the important value of the model study.

Because of the large difference in the model and prototype scale, it was not possible to formulate any conclusions regarding the amplitude and frequency of unsteady pressures caused by vortex shedding on the prototype structure except that vortex shedding may occur for two wind directions.

11.7 SUMMARY

A brief summary of some of the various dynamic models used in civil engineering practice has been presented. Types of loading effects on structures that have been considered include vibrations of elastic structures, wind effects, blast loading, and impact and earthquake loading. The similarity conditions in modeling structures loaded by the above dynamic effects have been presented. Material requirements for dynamic testing of model structures are given, and various types of loading facilities for testing such structures are described. Several case study illustrative examples of dynamic models from a variety of structural design situations are presented. Dynamic models of structures are shown to be powerful tools in many situations where the analytical techniques needed are lacking and yet practical designs must be generated.

PROBLEMS

11.1. Use the Pi theorem to develop an expression for the natural frequency of a freely vibrating fixed-ended beam of prismatic cross section.

11.2. It is desired to build and test a Plexiglas model of a large cast steel flywheel having a heavy rim and radial spokes. Establish the similitude conditions. For a $\frac{1}{20}$-scale model, what are the stress and velocity scales for a prototype that has an angular velocity of 50 rad/sec?

11.3. The natural frequency of a steel tuning fork is 200 vibrations per second. What is the natural frequency of a $\frac{1}{3}$-scale aluminum model of the tuning fork? The unit weight of aluminum is 36% that of steel, and its E value is $\frac{1}{3}$ that of steel. (A tuning fork is a freely vibrating elastic system.)

11.4. The idealized dynamic loading on a bridge is given by a forcing function that varies periodically with time. The exciting forces are fully specified by the frequency n and the maximum value of force F_0; for the prototype bridge, $n = 3$ Hz (cycles per second) and $F_0 = 10,000$ lb (44.5 kN). Determine similitude and scaling relations for a model bridge with a geometric scale factor $S_L = 30$. Both bridges are made of steel. Is it necessary to add additional mass (weights) to the bridge to satisfy similitude? Explain carefully.

11.5. For the beam of Problem 11.4, develop an expression for the maximum stress due to a concentrated load P at midspan, the beam's self-weight, and a uniform temperature rise of ΔT. Assume elastic action only.

11.6. The torque on an airplane propeller depends only on the diameter d of the propeller, its angular velocity ω, the velocity of advance v, and the mass density m and viscosity μ of the air. Using dimensional analysis, find an expression for the torque, and show that if the effect of viscosity can be neglected then the torque is proportional to air density. (Courtesy of W. Godden, University of California, Berkeley.)

11.7. In order to study the performance of a high-speed train on its track, a $\frac{1}{10}$-scale model is made of the complete system. The track is both curved and banked (transverse slope) in places. The model train is true-scale, and its effective density is found to be twice that of the prototype. The model is to be used to measure the forces exerted by the train on the track and to study the tendency for the train to overturn on corners. Specify the following model ratios required to simulate prototype behavior:

Track: **(a)** horizontal radius of curvature, r **(b)** transverse slope, ϕ
           (c) coefficient of friction, μ.

Train: **(d)** velocity, v **(e)** acceleration, a

Forces: **(f)** centrifugal force on the track, F_c at points of curvature
           (g) axial forces on track due to acceleration, F_a

Will this model correctly simulate the tendency for the train to overturn? (Courtesy of W. Godden, University of California, Berkeley.)

11.8. The structure of a tension roof system is composed of a series of freely hanging cables. Details of the prototype cables are as follows:

Span: 100 ft (30 m)
Diameter: 1 in. (25 mm)
Elastic modulus: 30×10^6 psi (200,000 MPa)
Specific weight: 500 lb/ft³ (8,015 kg/m³)

It is required to study the aerodynamic properties of the roof in a steady wind. For this purpose it is proposed to use a 10-ft (3 m) span model in a wind tunnel. The model will be made of aluminum wires [$E = 10 \times 10^6$ psi (68,950 MPa), specific weight $= 167$ lb/ft³. (2,677 kg/m³)]. It is known from prior experience on structures of this type that drag forces are not important.

(a) What is the required diameter of the model wires?

(b) Does the self-weight of the model wire have to be artificially increased? Explain.

(c) Flutter in the model is observed at a wind velocity of 10 mph (16 km/h). At what wind velocity would you expect the same phenomenon in the prototype?

(d) This flutter has a measured frequency of 5 cps in the model. At what frequency will the prototype vibrate? (Courtesy of W. Godden, University of California, Berkeley.)

11.9. It is required to study the response of a 400-ft (122 m) high concrete dam [properties of concrete: $E = 3.5 \times 10^6$ psi (24,133 MPa), specific weight = 150 pcf (2,405 kg/m³)]. For this purpose a 4-ft (1.2 m) high model is made in gypsum plaster (material properties: $E = 3.5 \times 10^5$ psi (24,133 MPa), specific weight = 100 pcf (1,603 kg/m³). Strain gages are mounted on the surface of the model.

(a) If it is required that the measured strains on the model are to be the same magnitude as those on the prototype, what is the required pressure in psi at the base of the model?

(b) Then, what is the ratio of top displacements of model and prototype?

(c) The free-standing model (that is, without back pressure) is put on a shaking table to determine its natural frequencies, and it is found that the first natural frequency of the model is 20 cps. What is the predicted first natural frequency of the prototype?

(d) It is required to predict the dynamic stresses in the prototype as the ground shakes in harmonic motion at 2 cps at a maximum acceleration of 0.5 g. In order to have the same strains in model and prototype, at what frequency would you shake the model, and at what acceleration? (Courtesy of W. Godden, University of California, Berkeley.)

11.10. A beam 6 in. (152 mm) wide and 12 in. (0.3 m) deep, made of timber with $E = 1.2 \times 10^6$ psi (8,274 MPa), is simply supported on an 8-ft (2.4 m) span and is to carry a moving vertical concentrated load. A model of the beam is constructed with a length scale of 20, but the beam is represented by a $\frac{1}{8}$-in. (3.2 mm) diameter steel rod.

(a) Neglecting shear effects, establish prediction equations for deflection, and for compressive stress on the top of the beam.

(b) Establish prediction equations for shearing deflection and horizontal shearing stresses at the neutral surface.

Appendices

Dimensional Dependence
and Independence

∿∿∿∿∿∿∿∿∿∿∿∿∿∿∿

A

In Chapter 2 the conditions for dimensional dependence or independence between a group of physical quantities were stated without proof. The basis for these statements will be presented here. First, it will be necessary to discuss the functional form of the dimensions of physical quantities. Then two different methods by which the results of the dimensional conditions can be obtained will be discussed.

A.1 THE FORM OF DIMENSIONS

Suppose that one is considering some n mechanical quantities. It is clear that the measures of these quantities can be established through the three *fundamental* dimensional units: force, length, and time. From a dimensional point of view then, these n quantities are known to be functions of force, length, and time.*

$$X_1 \doteq D_1(F, L, T)$$
$$X_2 \doteq D_2(F, L, T)$$
$$\cdots$$
$$X_n \doteq D_n(F, L, T)$$

$$(A.1)$$

*The symbol $\doteq$ will be used where dimensional equivalence is meant and the symbol $=$ will be left for those equations where numerical equivalence is also maintained.

It is of interest to ask whether there are any restrictions that should be placed upon the functions $D_1, D_2, \ldots, D_n$. Of basic importance in the concepts of dimensions as they have been formulated is that the relative magnitude of the numbers expressing the magnitudes of any two physical quantities remains unchanged when the size of the dimensional units is changed (e.g., the statement that John's car is traveling twice as fast as Joe's car does not depend upon whether speed is measured in feet per second, miles per hour, or centimeters per second). There cannot be any dispute over this fact: human beings simply established their measuring system on this premise. The important thing is to note that this fact places a definite restriction on the nature of the functional relationship $D_1, D_2, \ldots, D_n$. *In fact, $D_1, D_2, \ldots, D_n$ is restricted to be in the form of products of powers of the dimensional units.*

   Proof.

$$\text{John's speed} = D(aF, bL, cT)$$

$$\text{Joe's speed} = D(pF, qL, rT)$$

where a, b, c, and p, q, r are numbers that indicate the magnitude of the F, L, T dimensional units.

   Now we have stated that

$$\frac{D(aF, bL, cT)}{D(pF, qL, rT)}$$

has absolute significance regardless of the size of the dimensional units, thus if the size of the force unit is changed to be $1/x$ as large, the length unit $1/y$ as large, and the time unit $1/z$ as large, then it must still be true that:

$$\frac{D(aF, bL, cT)}{D(pF, qL, rT)} = \frac{D(axF, byL, czT)}{D(pxF, qyL, rzT)}$$

or

$$D(axF, byL, czT) = D(pxF, qyL, rzT)\frac{D(aF, bL, cT)}{D(pF, qL, rT)}$$

Now after applying the chain rule of partial differentiation with respect to x and remembering that F, L, and T represent dimensional units and are carried along only for completeness,

$$aF\frac{\partial D(axF, byL, czT)}{\partial axF} = pF\frac{\partial D(pxF, qyL, rzT)}{\partial pxF}\frac{D(aF, bL, cT)}{D(pF, qL, rT)}$$

Letting aF, bL, and cT vary while holding pF, qL, rT fixed, one gets when $x = y = z = 1$.

$$\frac{aF\dfrac{\partial D(aF, bL, cT)}{\partial aF}}{D(aF, bL, cT)} = \frac{pF\dfrac{\partial D(pF, qL, rT)}{\partial pF}}{D(pF, qL, rT)} \qquad \text{Constant}$$

or

$$\frac{\partial D(aF, bL, cT)}{D(aF, bL, cT)} = k_1 \frac{\partial aF}{aF}$$

This is a very special partial differential equation that can be treated as an ordinary differential equation, if it is kept in mind of course that the arbitrary constant of the ordinary differential equation becomes an arbitrary function for the partial differential equation. Thus the integration yields

$$\ln D(aF, bL, cT) = k_1 \ln aF + \ln G(bL, cT)$$

or

$$D(aF, bL, cT) = G(bL, cT)(aF)^{k_1}$$

This process could be repeated, except that one would differentiate partially with respect to y and then z. The final result then becomes

$$D(aF, bL, cT) = \text{constant } (aF)^{k_1}(bL)^{k_2}(cT)^{k_3}$$

or

$$D(aF, bL, cT) = \text{constant } (F)^{k_1}(L)^{k_2}(T)^{k_3} \qquad \text{(A.2)}$$

In this way it is seen that the functional form of the dimensions of any physical quantity is necessarily a product. Naturally such a conclusion depends directly upon the way in which dimensions were first defined and used, and so it may be more satisfying just to note that the dimensions of all physical quantities do occur in the form of a single product.

The concepts dealing with dimensional dependence and independence can now be approached and rigorously deduced from either of two points of view. The first approach, which has been used since the early 1900s, leads to the conclusion that a set of physical quantities are dimensionally independent (or dependent) if the determinant formed from the powers of the fundamental units does not vanish (or does vanish). The second approach, suggested by Pahl in 1962, considers the fact that if a set of functional relationships are independent (or dependent) with respect to certain arguments, then the Jacobian of these relationships with respect to the arguments does not vanish (or does vanish). Both approaches lead to the identical result, but since they are basically different it may be of interest to consider both.

A.2 METHOD I: THE NUMERIC METHOD

From the results of Section A.1 the dimensional form of mechanical physical quantities is known to be:

$$X_1 \doteq F^{a_1}L^{b_1}T^{c_1}$$
$$X_2 \doteq F^{a_2}L^{b_2}T^{c_2}$$
$$\dots$$
$$X_n \doteq F^{a_n}L^{b_n}T^{c_n}$$

$$\text{(A.3)}$$

A necessary and sufficient condition that r physical quantities be dependent (or independent) with respect to dimensions is that the determinant formed from the exponents of the "fundamental" dimensions of the quantities should vanish (or should not vanish).

The proof of the necessary portion can be effected in the following way. Assume that some three mechanical quantities, say X_1, X_3, and X_6, are dimensionally dependent. That is, assume that there exists some constants k_1, k_2, and k_3 (not all zero) for which

$$X_1^{k_1} X_3^{k_2} X_6^{k_3} = 1 \tag{A.4}$$

It follows that

$$(F^{a_1} L^{b_1} T^{c_1})^{k_1} (F^{a_3} L^{b_3} T^{c_3})^{k_2} (F^{a_6} L^{b_6} T^{c_6})^{k_3} \doteq 1$$

$$F^{(a_1 k_1 + a_3 k_2 + a_6 k_3)} L^{(b_1 k_1 + b_3 k_2 + b_6 k_3)} T^{(c_1 k_1 + c_3 k_2 + c_6 k_3)} = 1 \tag{A.5}$$

Now in order for Eq. (A.5) to hold, the exponents of F, L, and T must all equal zero. Thus it is necessary that

$$a_1 k_1 + a_3 k_2 + a_6 k_3 = 0$$
$$b_1 k_1 + b_3 k_2 + b_6 k_3 = 0 \tag{A.6}$$
$$c_1 k_1 + c_3 k_2 + c_6 k_3 = 0$$

In order for a nontrivial solution of Eqs. (A.6) to exist, it is necessary that the determinant of the coefficient matrix vanish. That is,

$$\begin{vmatrix} a_1 & a_3 & a_6 \\ b_1 & b_3 & b_6 \\ c_1 & c_3 & c_6 \end{vmatrix} = 0 \tag{A.7}$$

On the other hand, if this determinant did not vanish, then the trivial solution $k_1 = k_2 = k_3 = 0$ would be the only solution and there would be a contradiction to the initial assumption. The sufficiency condition is obtained by reversing steps.

One might wonder whether or not if would be possible to have a situation wherein the number of dimensionally independent quantities exceeded the number of fundamental dimensions. If one had four quantities and only three dimensions, Eqs. (A.6) would contain four unknowns in only three equations. Such a situation would lead to the possibility that three of the unknowns could only be expressed in terms of the fourth, and infinitely many solutions exist. In fact, it is just this sort of reasoning which can be applied to a group of n physical variables. With n variables and only r dimensions, there would be an $(n - r)$-fold infinity of solutions to the set of Eqs. (A.6). In other words, there would be $(n - r)$ infinity sets of the k's that would satisfy the associated Eq. (A.4), thus making the product dimensionless. Accordingly, it would never be possible to find more independent quantities than there would be dimensions.

Since only r quantities can be dimensionally independent, it is of interest to know how the remaining $(n - r)$ quantities depend upon the dimensionally independent ones. This dependence can be obtained in the following way. First, Eq. (A.4) can be written to include all n terms.

$$X_1^{k_1} X_2^{k_2} \ldots X_r^{k_r} \ldots X_n^{k_n} \doteq 1 \qquad (A.8)$$

Equations (A.6) now take the form

$$a_1 k_1 + a_2 k_2 + \ldots + a_r k_r + \ldots + a_n k_n = 0$$
$$b_1 k_1 + b_2 k_2 + \ldots + b_r k_r + \ldots + b_n k_n = 0$$
$$\ldots$$
$$r_1 k_1 + r_2 k_2 + \ldots + r_r k_r + \ldots + r_n k_n = 0$$

This set of r equations in n unknowns can only be solved for r of the k's in terms of the remaining $(n - r)$ arbitrary k values. Thus

$$a_1 k_1 + a_2 k_2 + \ldots + a_r k_r = -a_{r+1} k_{r+1} - \ldots - a_n k_n$$
$$b_1 k_1 + b_2 k_2 + \ldots + b_r k_r = -b_{r+1} k_{r+1} - \ldots - b_n k_n$$
$$\ldots \qquad\qquad (A.9)$$
$$r_1 k_1 + r_2 k_2 + \ldots + r_r k_r = -r_{r+1} k_{r+1} - \ldots - r_n k_n$$

where it is assumed that the first r numbered quantities are dimensionally independent (the quantities can always be renumbered to make this be so). Now, a particularly easy way to determine the dimensionally dependent relationship would be to set all but one of the $(n - r)$ arbitrary values in Eqs. (A.9) equal to zero and then let that remaining one equal 1. Values of k_1 through k_n could then be obtained. Substituting all of the values back into Eq. (A.8), a dimensionless product is obtained. Another dimensionless product can be obtained by starting again with Eqs. (A.9), and letting the value of some other one of the $n - r$ arbitrary k's be 1. In the end, $(n - r)$ such dimensionless products can be obtained. These dimensionless products have often been referred to as *numerics*. It is for this reason that the title of this section is Method I: The Numeric Method.

It will not be proved here, but it is a fact that there are only $(n - r)$ independent dimensionless products that can be formed from a set of n variables. These $(n - r)$ independent products are by no means unique, but they do correspond indirectly to the fact that Eqs. (A.9) have, in general, an $(n - r)$-fold infinity of solutions. One such set of independent dimensionless products is obtained by the procedure outlined in the previous paragraph. Such a special set of products is known as a *complete set* and could be used in the application of Buckingham's theorem.

A.3 METHOD II: THE FUNCTIONAL METHOD

From the results of Section A.1 the dimensional form of physical quantities is known to be:

$$X_1 \doteq F^{a_1} L^{b_1} T^{c_1}$$

$$X_2 \doteq F^{a_2} L^{b_2} T^{c_2}$$

$$\ldots$$

$$X_n \doteq F^{a_n} L^{b_n} T^{c_n}$$

It is shown in the calculus that *whenever r functions of r variables are functionally dependent, then the Jacobian of the functions with respect to these variables vanishes identically. Conversely, if the Jacobian does not vanish, then r functions are independent.* The Jacobian is defined as:

$$\frac{\partial(f_1, f_2, \ldots, f_r)}{\partial(m_1, m_2, \ldots, m_r)} = \begin{vmatrix} \dfrac{\partial f_1}{\partial m_1} & \dfrac{\partial f_1}{\partial m_2} & \cdots & \dfrac{\partial f_1}{\partial m_r} \\ \dfrac{\partial f_2}{\partial m_1} & \dfrac{\partial f_2}{\partial m_2} & \cdots & \dfrac{\partial f_2}{\partial m_r} \\ \cdot & \cdot & & \cdot \\ \cdot & \cdot & & \cdot \\ \cdot & \cdot & & \cdot \\ \dfrac{\partial f_r}{\partial m_1} & \dfrac{\partial f_r}{\partial m_2} & \cdots & \dfrac{\partial f_r}{\partial m_r} \end{vmatrix} \qquad (A.10)$$

In the case of dimensional relationships, one attempts to find any r of the functions such that the Jacobian of those r functions with respect to the r fundamental dimensions does not vanish. Suppose that it is found that

$$\frac{\partial(X_1, X_3, X_6)}{\partial(F, L, T)} = \begin{vmatrix} \dfrac{\partial X_1}{\partial F} & \dfrac{\partial X_1}{\partial L} & \dfrac{\partial X_1}{\partial T} \\ \dfrac{\partial X_3}{\partial F} & \dfrac{\partial X_3}{\partial L} & \dfrac{\partial X_3}{\partial T} \\ \dfrac{\partial X_6}{\partial F} & \dfrac{\partial X_6}{\partial L} & \dfrac{\partial X_6}{\partial T} \end{vmatrix} \neq 0$$

Then these three quantities are dimensionally independent. It is readily seen that the Jacobian determinant bears a strong resemblance to the determinant of the exponents that arose in the discussion of Method I. In fact, the two determinants are identical except that the Jacobian determinant may contain F's, L's, and T's as well as numbers. That is, where the Method I determinant contained a 2, the Method II determinant might contain $2FT/L$.

Once a set of dimensionally independent quantities has been found, it is then necessary to obtain a complete set of dimensionless products. A very convenient way of obtaining a complete set (and, in fact, the set which is obtained is exactly the one which is obtained by the suggested procedure of Method I) requires knowledge of another theorem of calculus. Namely, if it given that

$$X_1 \doteq D_1(F, L, T)$$
$$X_3 \doteq D_3(F, L, T)$$
$$X_6 \doteq D_6(F, L, T)$$

$$\frac{\partial(X_1, X_3, X_6)}{\partial(F, L, T)} \neq 0$$

then there exist *unique* functions

$$F \doteq f_1(X_1, X_3, X_6)$$
$$L \doteq f_2(X_1, X_3, X_6) \qquad (A.11)$$
$$T \doteq f_3(X_1, X_3, X_6)$$

With Eqs. (A.11) one can return to Eqs. (A.3) and substitute into the equations for $X_2, X_4, X_5, X_7, X_8, \ldots, X_n$ and thus determine the dimensions of $X_2, X_4,$ $X_5, X_7, X_8, \ldots, X_n$ in terms of $X_1, X_3,$ and X_6. These $(n - r)$ expressions can be transformed into dimensionless products. They will in fact be a complete set of dimensionless products.

A.4 ILLUSTRATIVE EXAMPLE

To illustrate how each method can be applied, suppose that one is considering the following set of physical variables:

Quantity	Units
X_1 Length	L
X_2 Force	F
X_3 Mass	$FL^{-1}T^2$
X_4 Stress	FL^{-2}
X_5 Strain	1
X_6 Acceleration	LT^{-2}
X_7 Displacement	L
X_8 Poisson's ratio	1
X_9 Modulus of elasticity	FL^{-2}

A.4.1 Method II Solution

Step 1. Search for three quantities for which the Jacobian of the three dimensional functions with respect to $F, L,$ and T does not vanish.

Try $X_1, X_3,$ and X_7; then

$$\frac{\partial(X_1, X_3, X_7)}{\partial(F, L, T)} = \begin{vmatrix} \dfrac{\partial X_1}{\partial F} & \dfrac{\partial X_1}{\partial L} & \dfrac{\partial X_1}{\partial T} \\[2mm] \dfrac{\partial X_3}{\partial F} & \dfrac{\partial X_3}{\partial L} & \dfrac{\partial X_3}{\partial T} \\[2mm] \dfrac{\partial X_7}{\partial F} & \dfrac{\partial X_7}{\partial L} & \dfrac{\partial X_7}{\partial T} \end{vmatrix} = \begin{vmatrix} 0 & 1 & 0 \\[2mm] T^2/L & -FT^2/L^2 & 2FT/L \\[2mm] 0 & 1 & 0 \end{vmatrix} = 0$$

so that X_1, X_3, and X_7 are dimensionally dependent. Clearly X_1 and X_7 cannot both be chosen.

Try X_1, X_3, X_5; then

$$\frac{\partial(X_1, X_3, X_5)}{\partial(F, L, T)} = \begin{vmatrix} \dfrac{\partial X_1}{\partial F} & \dfrac{\partial X_1}{\partial L} & \dfrac{\partial X_1}{\partial T} \\[2mm] \dfrac{\partial X_3}{\partial F} & \dfrac{\partial X_3}{\partial L} & \dfrac{\partial X_3}{\partial T} \\[2mm] \dfrac{\partial X_5}{\partial F} & \dfrac{\partial X_5}{\partial L} & \dfrac{\partial X_5}{\partial T} \end{vmatrix} = \begin{vmatrix} 0 & 1 & 0 \\[2mm] T^2/L & -FT^2/L^2 & 2FT/L \\[2mm] 0 & 0 & 0 \end{vmatrix} = 0$$

so that X_1, X_3, and X_5 are dimensionally dependent. At first glance this result is surprising; however, it is seen that $X_5 \doteq (X_1)^0 (X_3)^0$ (or that the dimensions of X_5 equal the dimensions of X_1 to the power 0 times the dimensions of X_3 to the power of 0). Such an occurrence implies the linear dependence just as strongly as if the exponents had been 1 and 2. In a certain respect this is a degenerate case, but it does point out that a dimensionless quantity can never be dimensionally independent from another group of quantities. In a similar manner one can prove that X_1, X_2, and X_4 are linearly dependent, etc.

Try X_1, X_6, and X_9; then

$$\frac{\partial(X_1, X_6, X_9)}{\partial(F, L, T)} = \begin{vmatrix} 0 & 1 & 0 \\[2mm] 0 & 1/T^2 & -2L/T^3 \\[2mm] 1/L^2 & -2F/L^3 & 0 \end{vmatrix} = -\frac{1}{L^2}\frac{2L}{T^3} = -\frac{2}{LT^3} \neq 0$$

so that X_1, X_6, and X_9 are dimensionally independent quantities.

Step 2. Since X_1, X_6, and X_9 are dimensionally independent, the functional relationships can be inverted and unique solutions for F, L, and T in terms of X_1, X_6, and X_9 can be obtained.

$$L \doteq X_1$$

$$T \doteq \sqrt{\frac{X_1}{X_6}} \qquad\qquad (A.12)$$

$$F \doteq X_1^2 X_9$$

Step 3. These in turn can be substituted into the remaining dimensional relationships to yield

$$X_2 \doteq F \qquad\qquad\qquad \doteq X_1^2 X_9$$

$$X_3 \doteq FL^{-1}T^2 \doteq X_1^2 X_9 \frac{1}{X_1} \frac{X_1}{X_6} \doteq \frac{X_1^2 X_9}{X_6}$$

$$X_4 \doteq FL^{-2} \qquad \doteq X_1^2 X_9 \frac{1}{X_1^2} \qquad \doteq X_9 \qquad\qquad \text{(A.13)}$$

$$X_5 \doteq 1 \qquad\qquad\qquad\qquad \doteq 1$$

$$X_7 \doteq L \qquad\qquad\qquad\qquad \doteq X_1$$

$$X_8 \doteq 1 \qquad\qquad\qquad\qquad \doteq 1$$

Therefore, the complete set of dimensionless products is

$$\frac{X_2}{X_1^2 X_9} \doteq \frac{X_3 X_6}{X_1^2 X_9} \doteq \frac{X_4}{X_9} \doteq X_5 \doteq \frac{X_7}{X_1} \doteq X_8 \doteq 1 \qquad \text{(A.14)}$$

A.4.2 Method I Solution

Step 1. Search for three quantities for which the determinant formed from the exponents of the fundamental dimensions does not vanish. Try X_1, X_3, and X_7; then

$$\Delta = \begin{vmatrix} a_1 & b_1 & c_1 \\ a_3 & b_3 & c_3 \\ a_7 & b_7 & c_7 \end{vmatrix} = \begin{vmatrix} 0 & 1 & 0 \\ 1 & -1 & 2 \\ 0 & 1 & 0 \end{vmatrix} = 0$$

Try X_1, X_3, and X_5; then

$$\Delta = \begin{vmatrix} a_1 & b_1 & c_1 \\ a_3 & b_3 & c_3 \\ a_5 & b_5 & c_5 \end{vmatrix} = \begin{vmatrix} 0 & 1 & 0 \\ 1 & -1 & 2 \\ 0 & 0 & 0 \end{vmatrix} = 0$$

Try X_1, X_6, and X_9, then

$$\Delta = \begin{vmatrix} a_1 & b_1 & c_1 \\ a_6 & b_6 & c_6 \\ a_9 & b_9 & c_9 \end{vmatrix} = \begin{vmatrix} 0 & 1 & 0 \\ 0 & 1 & -2 \\ 1 & -2 & 0 \end{vmatrix} = -2 \neq 0 \qquad \text{(A.15)}$$

so that X_1, X_6, and X_9 are dimensionally independent quantities.

Step 2. It must be possible to determine the dimensional relationship of all remaining quantities in terms of the dimensions of X_1, X_6, and X_9. This can be done in the same manner as was done in the Method II solution. Another procedure that can be made to lead to the identical result follows.

Any product π of the nine physical quantities has the form:

$$\pi = X_1^{k_1} X_2^{k_2} X_3^{k_3} X_4^{k_4} X_5^{k_5} X_6^{k_6} X_7^{k_7} X_8^{k_8} X_9^{k_9},$$

Regardless of the values of the constants $k_1, k_2, \ldots, k_9$, the corresponding dimension of π is

$$\pi \doteq [L]^{k_1}[F]^{k_2}[FL^{-1}T^2]^{k_3}[FL^{-2}]^{k_4}[1]^{k_5}[LT^{-2}]^{k_6}[L]^{k_7}[1]^{k_8}[FL^{-2}]^{k_9}$$

which can be rearranged to

$$\pi \doteq F^{(k_2+k_3+k_4+k_9)}L^{(k_1-k_3-2k_4+k_6+k_7-2k_9)}T^{(2k_3-2k_6)}$$

If it is now demanded that π be dimensionless, then each of the exponents of F, L, and T must vanish.

$$
\begin{aligned}
k_2 + k_3 + k_4 \qquad\qquad\qquad\;\; + k_9 &= 0 \\
k_1 \qquad - k_3 - 2k_4 \quad + k_6 + k_7 \quad - 2k_9 &= 0 \qquad \text{(A.16)} \\
2k_3 \qquad\qquad - 2k_6 \qquad\qquad\quad &= 0
\end{aligned}
$$

Here are three equations in nine unknowns (of course, k_5 and k_8 can just be considered to have zero coefficients). Our result in Eq. (A.15) guarantees that this set of equations has a sixfold $(9 - 3 = 6)$ infinity of solutions; i.e., we can choose six of the k's arbitrarily and then solve for the remaining three. Suppose we choose $k_2 = 1$ and $k_3 = k_4 = k_5 = k_7 = k_8 = 0$ and then solve for k_1, k_6, and k_9. Equations (A.16) reduce to

$$
\begin{aligned}
k_9 &= -1 \\
k_1 + k_6 - 2k_9 &= 0 \\
- 2k_6 &= 0
\end{aligned}
$$

which have the unique solutions of $k_1 = -2$, $k_6 = 0$, $k_9 = -1$. Thus

$$\pi \doteq \frac{X_2}{X_1^2 X_9} \doteq 1$$

the same as the first of Eqs. (A.14). The remaining five expressions in Eqs. (A.14) can be obtained in exactly the same manner, by letting $k_3 = 1$, then $k_4 = 1$, etc., while holding the other of k_2, k_3, k_4, k_5, k_7, and k_8 equal to zero.

It should be pointed out that Eqs. (A.14) do not represent anything unique. There is an infinite number of ways that Eqs. (A.14) could be written, corresponding to the fact that the six arbitrary k constants can be chosen in an infinity of ways. The particular choice of k's that has been made above will prove to be the most useful.

A Note on the Use
of SI Units in Structural
Engineering
∿∿∿∿∿∿∿∿∿∿∿∿∿∿
B

The International System of Units (Système International d'Unités), commonly called SI, is being adopted around the world as a uniform measurement system. Since the usage of SI in engineering and scientific circles is proceeding rather rapidly, it will soon be essential that the modern civil engineer be experienced in using the SI system.

The units adopted here for structural engineering work have been arrived at by careful study of SI references and using ASCE, AISC, AISI, and other technical recommendations. The SI user is urged to study the available references in the literature (e.g., ASTM, ASCE Standards) for a more general and complete treatment of this important topic.

The SI system differs from the MKS (meter-kilogram-second) system in the units of force and stress. In the MKS system the unit of force is the kilogram force (kgf), while in SI it is a newton, which is explained below. Stress in the old metric system was expressed in units such as kilogram force per square centimeter (kgf/cm²), while in SI it is newtons per square meter (N/m²), known as pascal.

The basic and derived SI units for various categories of measurement are discussed in the following sections. A summary of pertinent SI units and conversion factors from U.S. Customary units is given at the end of the discussion. SI values for typical easily recognized quantities are also listed.

B.1 GEOMETRY

The length units, the meter (m) and the millimeter (mm), will be used for geometrical quantities. The use of millimeter units for section modulus and moment of inertia (second moment of area) does involve large numbers for the majority of common structural shapes; this problem is met by listing steel section properties as (section modulus 10^3 mm^3) and (inertia 10^6 mm^4).

B.2 DENSITIES, GRAVITY LOADS, WEIGHTS

The standard of mass in SI is the kilogram (kg), equal to the mass of the international prototype of the kilogram (about 2.2 lb mass). This use of kg must not be confused with the old metric force called a kg or kgf.

Gravity loads exert forces on structures, and the conversion from mass to force becomes essential. In SI the force unit is a newton (N), about $\frac{1}{5}$ lb. It is the force required to accelerate one kilogram mass by one meter per second squared; a kilogram of mass exerts 9.80665 N on its support point. Load expressed in both kilogram and newton units will thus differ by the factor 9.80665 (9.8 for practical purposes).

B.3 FORCE, MOMENT, STRESS, AND OTHER STRESS RESULTANTS

The preceding section defines the newton (N) as the basic SI measure of force. The kilonewton (1000 N or kN) is about $\frac{1}{5}$ of a kip and will be used widely in structural design. These force units are combined with meters to express loadings, bending and twisting moments, and other quantities involving length and force.

The basic stress unit in SI is the newton per meter squared (N/m^2), called the pascal (Pa). This is a very small unit (1 psi = 6895 Pa) that becomes practical only when used with a large prefix (k or M). The most convenient SI stress unit for structures is 1 000 000 Pa, the megapascal, or MPa, which is identical to 1 MN/m^2 (or 1 N/mm^2), and approximates $\frac{1}{7}$ ksi. Since 1 ksi = 6.895 MPa, the modulus of steel as an example will be 200,000 MPa (or N/mm^2) in SI units.

Surface loadings and soil pressures have the units of pressure or stress, but the common usage will dictate their expression in kilonewtons per meter squared (kN/m^2). Surface loads in particular are well expressed in kilonewtons per meter squared because their effects must be converted into kilonewtons during structural analysis.

Moment is expressed in meter-newtons (m·N) or meter-kilonewtons (m·kN). These units are convenient since a meter-newton is close in value to a foot-pound (ft·lb) and a meter-kilonewton is close to a foot-kip.

B.4 MISCELLANEOUS (ANGLES, TEMPERATURE, ENERGY, POWER)

Plane angles are still measured in radians (rad) and solid angles in steradians (sr) in the SI system. Temperature in SI should be in degrees Celsius (C in old centigrade scale), but temperature in Kelvin (K) is also permissible, and the two are interchangeable for temperature gradients since $1°C = 1$ K. Energy is expressed in joules (J): a joule is a newton-meter (N·m), and work is in watts (W), which is one joule per second (J/sec).

B.5 SI SYSTEM STANDARD PRACTICE

There are several simple rules to be observed in using the SI system:

1. Preferred prefixes are to be selected from the following table, in which each prefix is a multiple of 1000:

Prefix	Symbol	Multiplication Factor
giga	G	10^9
mega	M	10^6
kilo	k	10^3
—	unit	1
milli	m	10^{-3}
micro	μ	10^{-6}
nano	n	10^{-9}

2. Use prefixes in the numerator only, except for kilogram, the base unit of mass. Thus the stress unit of newton per millimeter squared (N/mm²) is not recommended, rather meganewton per meter squared (MN/m²).

3. Separate digits in groups of three, counting from the decimal sign. Do not separate with commas since the comma is used for the decimal point in many countries. Examples:

$$1,234.57 = 1\ 234.57$$

$$0.58729 = 0.587\ 29$$

$$4789 = 4\ 789 \text{ or } 4789$$

4. Abbreviations of compound units, such as for moment, are written with a centered dot to indicate multiplication, such as m·kN. If the unit is spelled out, a hypen or space is used (meter-kilonewton). Abbreviations of compound units that are divided are always written with a slash; thus kilogram per meter is abbreviated kg/m.

U.S. Customary Units, SI Units, and Conversion Factors
(for converting U.S. to SI) in Structural Engineering

Property	U.S. Customary		Conversion Factor		SI Units
Overall geometry					
Spans	ft	×	0.3048*	=	m
Displacement	in.	×	25.4*	=	mm
Surface area	ft^2	×	0.0929	=	m^2
Volume	ft^3	×	0.0283	=	m^3
	yd^3	×	0.765	=	m^3
Structural properties					
Cross-sectional dimensions	in.	×	25.4*	=	mm
Area	in.^{2}	×	645.2	=	mm^2
Section modulus, volume	in.^{3}	×	16.39	=	10^3 mm^3†
Moment of inertia (second moment of area)	in.^{4}	×	0.4162	=	10^6 mm^4†
Material properties					
Density	lb/in.^{3}	×	27 680	=	kg/m^3
	lb/ft.^{3}	×	16.03	=	kg/m^3
Modulus and stress values	psi	×	0.006895	=	MPa
	ksi	×	6.895	=	MPa

*Exact.

†AISC uses this style in SI units.

553

	Mass Units				Force Units			
Loadings								
Concentrated loads	kip	× 0.4536	=	Mg*	kip	× 4.448	=	kN
Self weight (density)	lb/ft³	× 16.03	=	kg/m³	lb/ft³	× 0.1571	=	kN/m³
Line loads (linear density)	k/ft	× 1488	=	kg/m	k/ft	× 14.59	=	kN/m
Surface loads	lb/ft²	× 4.882	=	kg/m²	lb/ft²	× 0.0479	=	kN/m²
	k/ft²	× 4882	=	kg/m²	k/ft²	× 47.9	=	kN/m²
Stresses, moments								
Stress	psi	× 6895	=	Pa				
	ksi	× 6.895	=	MPa (MN/m² or N/mm²)				
Moment, torque	ft·lb (or lb·ft)	× 1.356	=	m·N (or N·m)				
	ft·k (or k·ft)	× 1.356	=	m·kN (or kN·m)				
Miscellaneous								
Velocity	fps	× 0.3048	=	m/s				
Energy	ft·lb force	× 1.356	=	N·m = J				
Temperature	$t_C^\circ = (t_F^\circ - 32)(5/9)$							
	$t_k = t_F^\circ + 273.15$							
Linear expansion coefficient	°F⁻¹	× 1.8	=	°C⁻¹ or K⁻¹				

*Exact.

Typical Values

Property	U.S. Customary	Approximate SI
Water density	62.4 lb/ft^3	1000 kg/m^3 = 1 t/m^3
Concrete density	150 lb/ft^3	2400 kg/m^3
Steel modulus E	29,000,000 psi	200 000 MPa
Concrete modulus E	3,500,000 psi	24 000 MPa
Allowable steel stress	25 ksi	170 MPa
Design live load	100 psf	5 kN/m^2

*This appendix is based on a similar one in White, Gergely and Sexsmith, "*Structural Engineering*," New York: John Wiley & Sons.

References

AAS-JAKOBSEN, A. ET AL., eds. (1958), *Proceedings of the Second Symposium on Concrete Shell Roof Construction*, July 1–3, 1957, Teknisk Ukeblad, Oslo, Norway.

ABELES, P. W. (1966), "Cracking and Bond Resistance in High Strength Reinforced Concrete Beams, Illustrated by Photoelastic Coating," *J. Am. Concr. Inst.* (November), pp. 1265–78.

ABRAMS, D. A. (1922), "Flexural Strength of Plain Concrete," *Proc. Am. Conc. Inst.*, 18, pp. 20–50.

ACI Ad Hoc Committee on Structural Models (1970), *Models for Concrete Structures*, R. N. White, Chairman, ACI SP-24, American Concrete Institute, Detroit, Mich., 495 pp.

ACI Committee 444 on Models of Concrete Structures (1972), *Preprints of the Symposium on Models of Concrete Structures*, American Concrete Institute, Detroit, Mich., 180 pp.

ACI Committee 531 (1970), "Concrete Masonry Structural Design and Construction," *Proc. Am. Concr. Inst.*, 67, no. 5 (May), pp. 380–403.

ADASZKIEWICZ, M. (1970), "Behavior of Continuous Reinforced Concrete Beams under Uniform Loads," Master's of Engineering thesis, Structural Concrete Series, McGill University, Montreal.

AKAZAWA, T. (1953), "Tension Test Method for Concrete," Bulletin No. 16, International Association of Testing and Research Laboratories for Materials and Structures, Paris, November.

ALAMI, Z. X AND FERGUSON, P. M., (1963), "Accuracy of Models Used in Research on Reinforced Concrete," *Proc. Am. Conc. Inst.*, 60, no. 11 (November) pp. 1643–1663.

ALDRIDGE, W. W. (1966), "Ultimate Strength Tests of Model Reinforced Concrete Folded Plate Structures," Ph.D. thesis, The University of Texas, Austin.

ALDRIDGE, W. W. AND BREEN, J. E. (1970), "Useful Techniques in Direct Modeling of Reinforced Concrete Structures," *Models for Concrete Structures*, ACI SP-24, American Concrete Institute, Detroit, Mich., pp. 125–40.

ALEXANDER, K. M. AND WARDLAW, J. (1960), "Dependence of Concrete Aggregate Bond Strength on Size of Aggregate," *Nature*, 187, 230.

ALEXANDER, S. J. AND HAMBLY, C. E. (1970), "Design of Structures to Withstand Gaseous Explosions," Parts 1 and 2, *Concrete*, 4, no. 2 (February), and *Concrete*, 4, no. 3 (March).

American Concrete Institute (1977), "ACI-318-77 Building Code Requirements for Reinforced Concrete," American Concrete Institute, Detroit, Mich.

American Concrete Institute Committee 444, Models of Concrete Structures (1979), "Models of Concrete Structures—State-of-the-Art," *Concr. Int. Des. Constr.*, 1, no. 1 (January), 77–95.

American Society for Testing and Materials (1978), ASTM A36-77a, "Standard Specification for Structural Steel," *Annual Book of Standards*, Part 4, Philadelphia, pp. 135–38.

American Society for Testing and Materials (1978), ASTM A242-75, "Standard Specification for High-Strength Low-Alloy Structural Steel," *Annual Book of Standards*, Part 4, Philadelphia, pp. 216–17.

American Society for Testing and Materials (1978), ASTM A416-74, "Standard Specification for Uncoated Seven-Wire Stress-Relieved Strands for Prestressed Concrete," *Annual Book of Standards*, Part 4, Philadelphia, pp. 378–82.

American Society for Testing and Materials (1978), ASTM A421-77, "Standard Specification for Uncoated Stress-Relieved Wire for Prestressed Concrete," *Annual Book of Standards*, Part 4, Philadelphia, pp. 383–86.

American Society for Testing and Materials (1978), ASTM A440-77, "Standard Specification for High Strength Structural Steel," *Annual Book of Standards*, Part 4, Philadelphia, pp. 389–90.

American Society for Testing and Materials (1978), ASTM A615-76a, "Standard Specification for Deformed and Plain Billet-Steel Bars for Concrete Reinforcement," *Annual Book of Standards*, Part 4, Philadelphia, pp. 579–84.

American Society for Testing and Materials (1978), ASTM A616-76, "Standard Specification for Rail Steel Deformed and Plain Bars for Concrete Reinforcement, *Annual Book of Standards*, Part 4, Philadelphia, pp. 585–88.

American Society for Testing and Materials (1978), ASTM A617-76, "Standard Specification for Axle Steel Deformed and Plain for Concrete Reinforcement," *Annual Book of Standards*, Part 4, Philadelphia, pp. 589–93.

American Society for Testing and Materials (1978), ASTM C190, "Tensile Strength of Hydraulic Cement Mortars," *Annual Book of Standards*, Part 13, Philadelphia, pp. 185–90.

American Society for Testing and Materials (1978), "ASTM D638-77a, Standard Test Method for Tensile Properties of Plastics," *Annual Book of Standards*, Part 35: *Plastics—General Test Methods*, Philadelphia, pp. 220–235.

American Society for Testing and Materials (1978), "ASTM D695-77, Standard Test for Compressive Properties of Rigid Plastics," *Annual Book of Standards*, Part 35, *Plastics—General Test Methods*, Philadelphia, pp. 267–73.

American Society for Testing and Materials (1978), "ASTM D790-71, Standard Test Methods for Flexural Properties of Plastics and Electrical Insulating Materials," *Annual Book of Standards*, Part 35: *Plastics—General Test Methods*, Philadelphia, pp. 319–28.

American Society for Testing and Materials (1975), "Standard Specifications for Brick and Applicable Standard Testing Methods for Units and Masonry Assemblages," ASTM, Philadelphia, May.

ANEJA, I. K. AND ROLL, F. (1971), "Model Analysis of Curved Box-Beam Highway Bridge," *ASCE, J. Struc. Div.* 97, no. ST 12, 2861–78.

ANTEBI, J., SMITH, H. D. AND HANSEN, R. J. (1960), "Study of the Applicability of Models for the Investigation of Air Blast Effects on Structures," Technical Report to the Defense Atomic Support Agency, Department of Civil Engineering, Massachusetts Institute of Technology, October.

ANTEBI, J., SMITH, H. D., SHARMA, H. D. AND HARRIS, H. G. (1962), "Evaluation of Techniques for Constructing Model Structural Elements," Research Report R62-15, Department of Civil Engineering, Massachusetts Institute of Technology, May.

APPA RAO, T. V. S. R. (1975), "Behavior of Concrete Nuclear Containment Structures Up to Ultimate Failure with Special Reference to MAPP-1 Containment," *Symposium on Structural Mechanics on Reactor Technology*, B.A.R.C., Bombay, India.

ARONI, S. (1959), "Pullout Resistance of Square Twisted and Plain Round Bars," Dept. of Civil Eng. Rept. (unpublished), Univ. of Melbourne, Australia.

ASLAM, M. AND GODDEN, W. G. (1975), "Model Studies of Multicell Curved Box-Girder Bridges," *ASCE, J. Eng. Mech. Div.*, 101, no. EM3 (June), 207–22.

Associate Committee on the National Building Code (1977), "National Building Code of Canada, 1977," National Research Council of Canada, Ottawa, p. 374.

BAGLIETTO, E., CASIRATI, M., CASTOLDI, A., DEMIRANDA, F. AND SAMMARTINO, R. (1976), "Mathematical and Structural Models of Zarate-Brazo Largo Bridges," Report No. 85, ISMES–Istituto Sperimentale Modelli e Strutture, Bergamo, Italy, September, 46 pp.

BAILEY, A. AND VINCENT, N. D. G. (1943), "Wind Pressure on Buildings Including the Effects of Adjacent Buildings," *J. Inst. Civ. Eng.*, London, vol. 20, 243–75.

BAKER, L. R. (1972), "Manufacture and Testing of Model Brickwork Wind Panels," *Proceedings of Structural Models Conference*, Sydney, Australia, sponsored by School of Architectural Science, University of Sydney.

BAKER, W. E. (1973), *Explosions in Air*, University of Texas Press, Austin.

BAKER, W. E., WESTINE, P. S. AND DODGE, F. T. (1973), *Similarity Methods in Engineering Dynamics—Theory and Practice of Scale Modeling*, Hayden Book Company, Inc., Rochelle Park, N.J.

BALINT, P. S. AND SHAW, F. S. (1965), "Structural Model of Australia Square Tower in Sydney," *Archit. Sci. Rev.*, Sydney, 8, 136–49.

BATCHELOR, B. (1972), "Materials for Model Structures at Queen's University," Dept. of Civil Eng. Rept. (unpublished), Queen's University, Kingston, Ontario.

Battle Memorial Institute (19—), *Prevention of Failure of Metals under Repeated Stress*, John Wiley & Sons, Inc., New York.

BEAUJOINT, N. (1960), "Similitude and Theory of Models," *RILEM Bull., Paris*, no. 7 (June).

BECICA, I. J. AND HARRIS, H. G. (1977), "Evaluation of Techniques in the Direct Modeling of Concrete Masonry Structures," Structural Models Laboratory Report No. M77-1, Department of Civil Engineering, Drexel University, Philadelphia, June.

BEERS, Y. (1957), *Introduction to the Theory of Error*, Addison-Wesley Publishing Company, Inc., Reading, Mass.

BEGGS, G. E. (1932), "An Accurate Mechanical Solution of Statically Indeterminate Structures by Use of Paper Models and Special Gages," *Proc. Am. Concr. Inst.*, 18, 58–82.

BENJAMIN, J. R. AND WILLIAMS, H. A. (1958), "The Behavior of One Story Brick Shear Walls," *ASCE, Journal of the Structural Division*, 84, no. ST4 (July), 1723–21–1723–30.

BEST, C. C. (1967), "Testing Microconcrete Structural Models," presented to the Joint British Committee for Stress Analysis at a meeting entitled "Model Testing Techniques, the Collection and Interpretation of Data," University College, London, June 28–29, 1967.

BILLINGTON, D. P. AND HARRIS, H. G. (1979), "Test Methods for Concrete Shell Buckling," presented at the ACI Symposium on Concrete Shell Buckling, ACI Annual Convention, October 29–November 2, Washington, D.C., ACI Special Publication SP-67, *Concrete Shell Buckling*, E. P. Popov and S. J. Medwadowski Editors, American Concrete Institute, Detroit.

BILLINGTON, D. P. AND MARK, R. (1965), "Small Scale Model Analysis of Thin Shells," *Proc. Am. Concr. Inst.*, 62, no. 6 (June), 673–88.

BLACKMAN, J. S., SMITH, D. M. AND YOUNG, L. E. (1958), "Stress Distribution Affects Ultimate Tensile Strength," *J. American Concrete Institute*, 55, 675–84.

BORGES, J. F. AND LIMA, J. A. E. (1960), "Crack and Deformation Similitude in Reinforced Concrete," *RILEM Bull. (Paris)*, New Series no. 7 (July), pp. 79–90.

BORGES, J. F. AND PEREIRA, J. (1970), "Dynamic Model Studies for Designing Concrete Structures," Paper SP-24-10, *Models for Concrete Structures*, ACI SP-24, American Concrete Institute, Detroit, Mich.

BOUMA, A. L. ET AL. (1962), "Investigations on Models of Eleven Cylindrical Shells Made of Reinforced and Prestressed Concrete," *Proceedings, Symposium on Shell Research* (Delft, 1961), Wiley-Interscience, New York, and North-Holland Publishing Company, Amsterdam, pp. 70–101.

BRADSHAW, R. R. (1963), "Some Aspects of Concrete Shell Buckling," *Proc. Am. Concr. Inst.*, 60, no. 3 (March), 316–28.

BREDSDORFF, P. K. AND KIERKEGAARD-HANSEN, P. (1959), discussion of paper by J. S. Blackman et al., "Stress Distribution Affects Ultimate Tensile Strength," *J. American Concrete Institute*, 55 (June), 1421–26.

BRIDGMAN, P. W. (1922), *Dimensional Analysis*, Yale University Press, New Haven, Conn.

BROCK, G. (1959), "Direct Models as an Aid to Reinforced Concrete Design," *Engineering (London)*, 187 (April), 468–70.

BROMS, B. B., (1965), "Crack Width and Crack Spacing in Reinforced Concrete Members," *Proc. Am. Conc. Inst.*, 62, no. 10 (October) p. 1237.

BROWNIE, R. D. AND McCURICH, L. H. (1967), "Measurements of Strain in Concrete Pressure Vessels," Paper 52, Group I, *Conference on PCRV*, London, March.

BUCKINGHAM, E. (1914), "On Physically Similar Systems," *Phys. Rev.*, London, 4, no. 345.

BULL, A. H. (1930), "A New Method for the Mechanical Analysis of Trusses," *Civil Engineer*, 1, no. 3 (December), 181–183.

BURTON, K. T. (1963), "A Technique Developed to Study the Ultimate Strength of P/C Structures by the Use of Small Scale Models," M.S. thesis, Cornell University, Ithaca, N.Y.

BUYUKOZTURK, O. (1970), "Stress-Strain Response and Fracture of a Model of Concrete in Biaxial Loading," Ph.D. thesis, Cornell University, Ithaca, N.Y., June.

BUYUKOZTURK, O., NILSON, A. H. AND SLATE, F. O. (1971), "Stress-Strain Response and Fracture of a Concrete Model in Biaxial Loading," *Proc. Am. Concr. Inst.*, 68, no. 8 (August), 590–99.

CARLSON, R. W. (1937), "Drying Shrinkage of Large Concrete Members," *J. American Concrete Institute*, 33 (January–February), 327–36.

CARNIERO, F. L. L. B. AND BARCELLOS, A. (1953), "Concrete Tensile Strength," Bulletin No. 13, International Association of Testing and Research Laboratories for Materials and Structures, Paris, March.

CARPENTER, J. E. (1963), "Structural Model Testing—Compensation for Time Effect in Plastic," *J. PCA Res. Dev. Labo.* Skokie, Ill., 5, no. 7, 47–61.

CARPENTER, J. E., MAGURA, D. D. AND HANSON, N. W. (1964), "Structural Model Testing Techniques for Models of Plastic," *J. PCA Res. Dev. Lab.*, 6, no. 2, 26–47. Also, Development Department Bulletin D76, Portland Cement Association, Skokie, Ill.

CARPENTER, J. E. AND MAGURA, D. D. (1965), "Structural Model Testing—Load Distribution in Concrete I-Beam Bridges," Development Department Bulletin D94, Portland Cement Association, Skokie, Ill.

CASTELLANI, A., CASTOLDI, A. AND IONITA, M. (1976), "Numerical Analysis Compared to Model Analysis for a Dam Subject to Earthquakes," Report No. 83, ISMES—Istituto Sperimentale Modelli e Strutture, Bergamo, Italy, September, reprint from the *Proceedings of the Fifth International Conference on Experimental Stress Analysis*, Udine, Italy.

CASTOLDI, A. AND CASIRATI, M. (1976), "Experimental Techniques for the Dynamic Analysis of Complex Structures," Report No. 74, ISMES–Istituto Sperimentale Modelli e Strutture, Bergamo, Italy, February.

CERMAK, J. E. (1975), "Applications of Fluid Mechanics to Wind Engineering," A Freeman Scholar Lecture, *J. Fluids Eng.*, ASME, 97, no. 1 (March), 9–38.

CERMAK, J. E. (1977), "Wind-Tunnel Testing of Structures," *J. Eng. Mech. Div.*, ASCE, 103, no. EM6 (December), 1125–40.

CHAO, N. D. (1964), "Ultimate Flexural Strength of Prestressed Concrete Beams by Small Scale Models," unpublished M.S. thesis, Cornell University, Ithaca, N.Y.

CHEUNG, K. C. (1974), "PCRV Design and Verification," Report No. GA-A12821 (GA-LTR-8), General Atomic Co., San Diego, Calif.

CHOWDHURY, A. H. (1974), "An Experimental and Theoretical Investigation of the Inelastic Behavior of Reinforced Concrete Multistory Frame Models Subjected to Simulated Seismic Loads," Ph.D. dissertation, Cornell University, Ithaca, N.Y.

CHOWDHURY, A. H. AND WHITE, R. N. (1971), "Inelastic Behavior of Small Scale Reinforced Concrete Beam-Column Joints under Severe Reversing Loads," Report No. 342, Department of Structural Engineering, Cornell University, Ithaca, N.Y. October, 135 pp.

CHOWDHURY, A. H. AND WHITE, R. N. (1980), "Multistory Reinforced Concrete Frame Models under Simulated Seismic Loads," presented at the Symposium on the Response of Buildings to Lateral Forces, 442, ACI Convention, Toronto, Canada, April; *See also* ACI Publication SP-63, *Reinforced Concrete Structures Subjected to Wind and Earthquake Forces*, American Concrete Institute, Detroit, pp. 275–300.

CHOWDHURY, A. H. AND WHITE, R. N. (1977), "Materials and Modeling Techniques for Reinforced Concrete Frames," *Proc. Am. Concr. Inst.*, 74, no. 11, (November) 546–51.

CLARKE, B. C. (1966), "Testing of a Model Curved Steel Girder Bridge," *AISC Eng. J.*, 3, no. 3 (July), 106–112.

CLARK, L. A. (1971), "Crack Similitude in 1: 3.7 Scale Models of Slabs Spanning One Way," Technical Report, Cement and Concrete Association (London), March.

CLOUGH, R. W. AND BERTERO, V. V. (1977), "Laboratory Model Testing for Earthquake Loading," *ASCE J. Eng. Mech. Div.*, 103, no. EM6 (December), Proc. Paper 13444, 1105–1124.

CORLEY, W. G., CARPENTER, J. E., RUSSELL, H. G., HANSON, N. W., CARDENAS, A. E., HELGASON, T., HANSON, J. M., AND HOGNESTAD, E. (1975), "Construction and Testing of 1/10-Scale Micro-Concrete Model of New Potomac River Crossing," 1–266 (RDO31.01E), Portland Cement Association, Skokie, Ill.

CORLEY, W. G., RUSSELL, H. G., CARDENAS, A. E., HANSON, J. M., CARPENTER, J. E., HANSON, N. W., HELGASON, T., AND HOGNESTAD, E. (1972), "Ultimate Load Test of 1/10-Scale Micro-Concrete Model of New Potomac River Crossing, I-266," *J. Prestressed Concr. Inst.*, 1971, 16, no. 6 (November–December), 70–84.

CORUM, J. M. AND SMITH, J. E. (1970), "Use of Small Models in Design and Analysis of PCRV's," ORNL-4346, Oak Ridge National Laboratory, Oak Ridge, Tenn.

CORUM, J. M., WHITE, R. N. AND SMITH, J. E. (1969), "Mortar Models of Prestressed Concrete Reactor Vessels," ASCE *J. Struct. Div.*, 95, no. ST2 (February), Proc. Paper 6419, 229–248.

COWAN, H. J. ET AL. (1968), *Models in Architecture*, Elsevier Publishing Company, New York, 228 pp.

COWAN, H. J. AND LYALIN, I. M. (1965), *Reinforced and Prestressed Concrete in Torsion*, Edward Arnold (Publishers) Ltd., London.

CRANSTON, W. B. (1965), "Tests on Reinforced Concrete Frames: Pinned Portal Frames," Cement and Concrete Association (London), Technical Report TRA/392, August.

CULVER, C. G. AND CHRISTIANO, P. P. (1969), "Static Model Tests of Curved Girder Bridge," ASCE *J. Struct. Div.*, 95, no. ST8 (August), Proc. Paper 6712, 1599–1614.

DALLY, J. W. AND RILEY, W. F. (1978), *Experimental Stress Analysis*, McGraw-Hill Book Company, New York, 520 pp.

DANIELS, H. E. (1945), "The Statistical Theory of Strength of Bundles of Threads," *Proc. of Royal Society*, Series A, 183, pp. 405–435.

DAU, K. (1961), "Wind Tunnel Tests of the Toronto City Hall," UTIA Technical Note No. 50, Institute of Aerophysics, University of Toronto, Toronto.

DAVENPORT, A. G. AND ISYUMOV, N. (1968), "The Application of the Boundary Layer Wind Tunnel to the Prediction of Wind Loading," *Proceedings, International Research Seminar on Wind Effects on Building and Structures*, vol. 1, University of Toronto Press, Toronto, pp. 201–30.

DAVENPORT, A. G. AND ISYUMOV, N. (1972), "A Study of Wind Effects for Federal Reserve Bank Building, New York," Report No. BLWT-8 of the Boundary Layer Wind Tunnel Laboratory, University of Western Ontario, London, Canada.

DAVENPORT, A. G., ISYUMOV, N., FADER, D. AND BOWEN, C. P. (1969), "A Study of Wind Action on a Suspension Bridge during Erection and on Completion," Report No. BLWT-3 of the Boundary Layer Wind Tunnel Laboratory, University of Western Ontario, London, Canada.

DAVENPORT, A. G., ISYUMOV, N. AND JANDALI, T. (1971), "A Study of Wind Effects for the Sears Project," Report No. BLWT-5 of the Boundary Layer Wind Tunnel Laboratory, University of Western Ontario, London, Canada.

DAVIDENKOV, N. et al, (1947), "The Influence of Size on the Brittle Strength of Steel, *Jour. of Appl. Mech.*, 14, (March) pp. 63–67.

DAVIES, R. M., ed. (1967), *Space Structures*, International Conference on Space Structures, University of Surrey (1966), Blackwell Scientific Publications, Ltd. Oxford.

DAVIS, R. E. ET AL. (1972), "Model and Prototype Studies of Box Girder Bridge," ASCE *J. Struct. Div.*, 98, no. ST1 (January), Proc. Paper 8631, 165–85.

DER, T. J. AND FIDLER, R. (1968), "Model Study of the Buckling Behavior of Hyperbolic Shells," *Proc. Inst. Civ. Eng.* (London), 41, 105–118.

DOBBS, N. AND COHEN, E. (1970), "Model Techniques and Response Tests of Reinforced Concrete Structures Subjected to Blast Loads," Paper SP-24-17, *Models for Concrete Structures*, ACI SP-24, American Concrete Institute, Detroit, Mich.

DOVE, R. C. AND ADAMS, P. H. (1964), *Experimental Stress Analysis and Motion Measurements*, Prentice-Hall, Inc., Englewood Cliffs, N.J., 515 pp.

DOW, B. N. AND HARRIS, H. G. (1978), "Use of Small Scale Direct Models to Predict the Response of Horizontal Joints in Large Panel Precast Concrete Buildings," Structural Models Laboratory Report No. M78-3, Department of Civil Engineering, Drexel University, Philadelphia, June, 128 pp.

DRAGOSAVIC, I. M. (1973), "Structural Measures Against Natural Gas Explosions in High-Rise Blocks of Flats," *Heron*, 19, no. 4, Department of Civil Engineering, Technological University, Delft, The Netherlands, 3–51.

DRYDEN, H. L. AND HILL, G. C. (1933), "Wind Pressure on Model of the Empire State Building," *J. Res. Nat. Bur. Stand.*, 10, 493–523.

EFSEN, A. AND GLARBO, O. (1956), "Tensile Strength of Concrete Determined by Cylinder Splitting Test," *Beton og Jernbeton (Copenhagen)*, no. 1, pp. 33–39.

ELMS, D. G. (1964), "The Stress Distribution in a Shell of Negative Curvature Subjected to Distributed and Concentrated Edge Loads," Ph.D. thesis, Civil Engineering Department, Princeton University, Princeton, N.J., January.

ENEY, W. J. (1939), "New Deformeter Apparatus," *Eng. News Rec.*, 122, no. 7 (February 16), 221.

EZRA, A. A. (1962), "Similitude Requirements for Scale Model Determination of Shell Buckling under Impulsive Pressure," NASA TN-1510, NASA, Washington, D.C., December, pp. 661–70.

FAM, A. R. M. (1973), "Static and Free Vibration Analysis of Curved Box Bridges," Ph.D. thesis, McGill University, Montreal.

FATTAL, S. G. AND CATTANEO, L. E. (1974), "Evaluation of Structural Properties of Masonry in Existing Buildings," NBSIR 74-520, National Bureau of Standards, Washington, D.C., July.

FAZIO, P. (1972), "Failure Modes of Folded Sandwich Panel Roofs," *ASCE J. Struct. Div.*, 98, no. ST5, 1085–1104.

FAZIO, P. AND PALUSAMY, S. (1972), "Data Acquisition System for Large Scale Structural Testing," *Proceedings of Structural Models Conference*, Sydney, Australia, sponsored by School of Architectural Science, University of Sydney.

FERRITO, J. (1969), "Dynamic Tests of Model Concretes," Technical Report R-650, Naval Civil Engineering Laboratory, Port Hueneme, Calif., November.

FIALHO, J. F. L. (1960 and 1962), "The Use of Plastics for Making Structural Models," *RILEM Bull.*, New Series No. 8 (September 1960) and Technical Paper No. 184, Laboratorio Nacional de Engenharia Civil, Lisbon, Portugal, 1962.

FIORATO, A. E., SOZEN, M. A. AND GAMBLE, W. L. (1970), "An Investigation of the Interaction of Reinforced Concrete Frames with Masonry Filler Walls," University of Illinois, Civil Engineering Studies, Structural Research Series No. 370, Urbana, Ill., November.

FISHBURN, CYRUS C. (1961), "Effect of Mortar Properties on Strength of Masonry," NBS Monograph 36, National Bureau of Standards, Washington, D.C., November 20.

FRUEDENTHAL, A. M. (1968), "Statistical Approach to Brittle Fracture," in *Fracture*, vol. 2, ed. H. Liebowitz, Academic Press, New York, chap. 6, pp. 591–619.

FUMAGALLI, E. (1973), *Statical and Geomechanical Models*, Springer-Verlag, New York, 182 pp.

FUMAGALLI, E., VERDELLI, G. (1976), "Research on PCPV for BWR—Physical Model as Design Tool—Main Results," Report No. 86, ISMES, Istituto Sperimentale Modelli e Strutture, Bergamo, Italy, September.

Fuss, D. S. (1968), "Mix Design for Small Scale Models of Concrete Structures," Report No. R-564, Naval Civil Engineering Laboratory, Port Hueneme, Calif.

Gergely, P. (1969), "Splitting Cracks along the Main Reinforcement in Concrete Members," Cornell University Report to the Bureau of Public Roads, U.S. Department of Transportation, April.

Gergely, P. and Winter, G. (1972), "Experimental Investigation of Thin-Steel Hyperbolic Paraboloid Structures," *ASCE J. Struct. Div.*, 69, no. ST10, 2165–79.

Gero, J. E. and Cowan, H. J. (1970), "Structural Concrete Models in Australia," *Models of Concrete Structures*, ACI, SP-24, American Concrete Institute, Detroit, Mich., pp. 353–386.

Geymeyer, H. G. (1967), "Strain Meters and Stress Meters for Embedment in Models of Mass Concrete Structures," Technical Report No. 6-811, U.S. Corps of Engineers, Vicksburg, Miss., March, 60 pp.

Gilkey, H. J. (1961), "Water-Cement Ratio versus Strength—Another Look," *Proc. Am. Concr. Inst.*, 57, no. 4 (April), 1287–1312.

Glücklich, J. and Cohen, L. J. (1968), "Strain Energy and Size Effects in a Brittle Material," *Mater. Res. Stand.*, 8, 17–22.

Godden, W. G. and Aslam, M. (1975), "Model Studies of Skew Multicell Girder Bridges," *ASCE J. Eng. Mech. Div.*, 99, no. EM1 (February), Proc. Paper 9576, 201–222.

Godden, W. G. and Aslam, M. (1978), "Dynamic Model Structures of the Ruck-a-Chucky Bridge," preprint of the paper presented at the ASCE Spring Convention, Pittsburgh, April.

Gonnerman, H. F. (1925), "Effects of Size and Shape of Test Specimen on Compressive Strength of Concrete," *Am. Soc. Test. Mater., Proc.*, 25, 237–50.

Gottschalk, O. (1927), "Use of Models in the Solution of Indeterminate Structures," *J. Franklin Inst.*, March.

Grieb, W. E. and Werner, G. (1962), "Comparison of Tensile Splitting Strength of Concrete with Flexural and Compressive Strengths," *Public Roads*, 32, no. 5 (December), 97.

Griggs, P. H. (1971), "Buckling of Reinforced Concrete Shells," *ASCE J. Eng. Mech. Div.*, 97, no. EM3, 687–700.

Guedelhoefer, O. C., Moreno, A. and Janney, J. R. (1972), "Structural Models of Hangars for Large Aircraft," *Proceedings of the Symposium of the ACI Canadian Chapter on Models in Structural Design*, Montreal, pp. 71–109.

Haas, A. M. and Bouma, A. L., eds. (1961), *Shell Research, Proceedings of the Symposium on Shell Research*, Delft, August 30–September 2, 1961, North-Holland Publishing Company, Amsterdam; Interscience Publishers, Inc., New York.

Hansen, T. C. and Mattock, A. H. (1966), "Influence of Size and Shape of Member on the Shrinkage and Creep of Concrete," *ACI J. Am. Concr. Inst.*, 63 (February), 267–90.

Hanson, N. W. (1975), "Interim Report, Task 4—Experimental Cantilever Test—Phase 1," *Design and Construction of Large Panel Concrete Structures*, prepared for the Office of Policy Development and Research, Department of Housing and Urban Development, by Portland Cement Association, Skokie, Ill., December.

HANSON, N. W. AND CONNER, H. W. (1967), "Seismic Resistance of Reinforced Concrete Beam-Column Joints," *ASCE J. Struct. Div.*, 93, no. 2, 533.

HANSON, N. W. AND CURVITTS, O. A. (1965), "Instrumentation for Structural Testing," *J. PCA Res. Dev. Lab.*, May, pp. 24–39.

HARRIS, H. G. (1967), "The Inelastic Analysis of Concrete Cylindrical Shells and Its Verification Using Small Scale Models," Ph.D. thesis, Cornell University, Ithaca, N.Y.

HARRIS, H. G. (1968), "Simplified Apollo Shell One-Tenth Scale Model," Lunar Module Report LED-520-50, Grumman Aerospace Corporation, Bethpage, N.Y., October 1.

HARRIS, H. G. (1982), Editor, *Dynamic Modeling of Concrete Structures*, Publication SP-73, American Concrete Institute, Detroit, 242 pp.

HARRIS, H. G. AND BECICA, I. J. (1977), "Direct Small Scale Models of Concrete Masonry Structures," paper presented at the 2nd Annual ASCE Engineering Mechanics Division Specialty Conference, North Carolina State University, Raleigh, N.C., May 23–25, 1977, published in *Advances in Civil Engineering through Engineering Mechanics*, ASCE, New York, pp. 101–04.

HARRIS, H. G. AND BECICA, I. J. (1978), "The Behavior of Concrete Masonry Structures and Joint Details Using Small Scale Direct Models," *Proceedings of the North American Masonry Conference*, University of Colorado, Boulder, Colo., August 14–16.

HARRIS, H. G. AND MUSKIVITCH, J. C. (1977), "Report 1: Study of Joints and Sub-Assemblies—Validation of the Small Scale Direct Modeling Techniques," *Nature and Mechanism of Progressive Collapse in Industrialized Buildings*, Office of Policy Development and Research, Department of Housing and Urban Development, Washington, D.C., October, 165 p., also Department of Civil Engineering, Drexel University, Philadelphia, 19104.

HARRIS, H. G. AND MUSKIVITCH, J. C. (1980), "Models of Precast Concrete Large Panel Buildings," *ASCE J. Struct. Div.*, 106, no. ST2 (February), Proc. Paper 15218, 545–65.

HARRIS, H. G., PAHL, P. J. AND SHARMA, S. D. (1962), "Dynamic Studies of Structures by Means of Models," Report R63-23, Department of Civil Engineering, Massachusetts Institute of Technology, Cambridge, Mass.

HARRIS, H. G., SABNIS, G. M. AND WHITE, R. N. (1966), "Small Scale Direct Models of Reinforced and Prestressed Concrete Structures," Report No. 326, Department of Structural Engineering, Cornell University, Ithaca, N.Y. September, 362 pp.

HARRIS, H. G., SABNIS, G. M. AND WHITE, R. N. (1970), "Reinforcement for Small Scale Direct Models of Concrete Structures," paper No. SP-24-6, *Models for Concrete Structures*, ACI SP-24, American Concrete Institute, Detroit, Mich., pp. 141–58.

HARRIS, H., SCHWINDT, R., TAHER, I. AND WERNER, S. (1963), "Techniques and Materials in the Modeling of Reinforced Concrete Structures under Dynamic Loads," Report R63-54, Department of Civil Engineering, Massachusetts Institute of Technology, Cambridge, Mass., December, also NCEL-NBY-3228, U.S. Naval Civil Engineering Laboratory, Port Hueneme, Calif.

HARRIS, H. G. AND WHITE, R. N. (1972), "Inelastic Behavior of Reinforced Concrete Cylindrical Shells," ASCE *J. Struct. Div.*, 98, no. ST7 (July), Proc. Paper 9074, 1633–53.

HARRIS, P. J. (1964), "The Analysis of Axially Symmetric Spherical Shells by Means of Finite Differences," Ph.D. thesis, McGill University, Montreal, July, 168 pp.

HASELTINE, B. A. AND FISHER, K. (1973), "The Testing of Model and Full-size Composite Brick and Concrete Cantilever Wall Beams," *Proc. Br. Ceram. Soc.*, no. 21, (April), Load Bearing Brickwork (4), pp. 243–60.

HEDGREN, A. W., JR. AND BILLINGTON, D. P. (1967), "Mortar Model Test on a Cylindrical Shell of Varying Curvature and Thickness," *Proc. Am. Concr. Inst.*, 64, no. 2 (February), 73–83.

HENDRY, A. W. (1964), *Elements of Experimental Stress Analysis*, Pergamon Press, London, 193 pp.

HENDRY, A. W. AND MURTHY, C. K. (1965), "Comparative Tests on Third and Sixth Scale Model Brickwork Piers and Walls," *Proc. Br. Ceram. Soc.*, July.

HERCEG, D. E. (1976), *Handbook of Measurement and Control*, Revised Edition, Schaevitz Engineering, Pennsauken, NJ.

HETENYI, M. (1954), *Handbook of Experimental Stress Analysis*, McGraw-Hill Book Company, New York, 1076 pp.

HIDALGO, P. AND CLOUGH, R. W. (1974), "Earthquake Simulator Study of a Reinforced Concrete Frame," Report No. EERC-74-13, Earthquake Engineering Research Center, University of California, Berkeley, December.

HOGNESTAD, E., HANSON, N. W. AND MCHENRY, D. (1955), "Concrete Stress Distribution in Ultimate Strength Design," *Proc. Am. Concr. Inst.*, 52, no. 12 (December), 455–79.

HOGNESTAD, E., HANSON, N. W., KRIZ, L. B. AND KURVITS, O. A. (1959), "Facilities and Test Methods of PCA Structural Laboratory," Development Department, Bulletin D33, Portland Cement Association Research and Development Laboratories, Skokie, Ill.

HSU, C. T. (1969), "Investigation of Bond in Reinforced Concrete Models," M. Eng. thesis, Structural Concrete Series No. 14, McGill University, Montreal, Canada, April.

HSU, C. T. (1974), "Behavior of Structural Concrete Subjected to Biaxial Flexure and Axial Compression," Ph.D. thesis, McGill University, Montreal, August.

HSU, T. T. C. AND SLATE, F. O. (1963), "Tensile Bond Strength between Aggregate and Cement Paste or Mortar," *Proc. Am. Concr. Inst.*, 60, no. 4 (April), 465–86.

HSU, T. T. C., SLATE, F. O., STURMAN, G. M. AND WINTER, G. (1963), "Microcracking of Plain Concrete and the Shape of the Stress Strain Curve," *Proc. Am. Concr. Inst.*, 60, no. 2 (February), 209–24.

HUDSON, D. E. (1967), "Scale Model Principles," in *Shock and Vibration Handbook*, ed. Harris and Crede, McGraw-Hill Book Company, New York, chap. 27.

HUGHES, B. P. AND CHAPMAN, G. P. (1965), "Direct Tensile Tests for Concrete Using Modern Adhesives," *RILEM Bull. (Paris)*, New Series, no. 26, (March), p. 77.

IPSEN, D. C. (1960), *Units, Dimensions, and Dimensionless Numbers*, McGraw-Hill Book Company, New York.

IRMINGER, J. O. V. AND NOKKENTVED (1930), "Wind Pressure on Buildings," *Ingenioer-vidensk. Skr.*,

IRWIN, H. P. A. H. AND SCHUYLER, G. D. (1977), "Experiments on a Full Aeroelastic Model of Lions' Gate Bridge in Smooth and Turbulent Flow," Laboratory Technical Report No. LTR-LA-206, National Aeronautical Establishment, Ottawa, October 18.

IRWIN, H. P. A. H. AND WARDLOW, R. L. (1976), "Sectional Model Experiments on Lions' Gate Bridge, Vancouver," Laboratory Technical Report No. LTR-LA-205, National Aeronautical Establishment, Ottawa, October.

ISBERNER, A. W. (1969), "Properties of Masonry Cement Units," in *Design Engineering and Construction with Masonry Products*, ed. Dr. Franklin B. Johnson, Gulf Publishing Company, Houston, pp. 42–50.

ISHAI, O. (1961), "Influence of Sand Concentration on the Deformations of Beams under Low Stresses," *Proc. Am. Concr. Inst.*, 58, 11 (November), 611–23.

ISSEN, L. (1966), "Scaled Models in Fire Research on Concrete Structures," *J. PCA Res. Dev. Lab.*, 8, no. 3 (September), 10–26.

Istituto Sperimentale Modelli e Strutture (1972), *ISMES 1961–1971*, Bergamo, Italy, p. 193.

ISYUMOV, N. (1972), "Wind Tunnel Methods for Evaluating Wind Effects on Buildings and Structures," *International Symposium on Experimental Mechanics*, University of Waterloo, Canada.

ISYUMOV, N. (1976), "Modelling of Wind Effects on Structures and Buildings," lecture notes.

JANNEY, J. R., BREEN, J. E. AND GEYMAYER, H. (1970), "Use of Models in Structural Engineering," *Models for Concrete Structures*, ACI SP-24, American Concrete Institute, Detroit, Mich. pp. 1–18.

JENSEN, M. (1958), "The Model-Law for Phenomena in Natural Wind," *Ingenioreen-* (international edition), vol. no. 4, 121–28.

JENSEN, M. AND FRANCK, N. (1965), *Model-Scale Tests in Turbulent Wind*, Parts I and II, The Danish Technical Press, Copenhagen.

JOHAL, L. S. AND HANSON, N. W. (1978), "Horizontal Joint Tests," Supplemental Report B, *Design and Construction of Large Panel Concrete Structures*, prepared for the Office of Policy Development and Research, Department of Housing and Urban Development, by Portland Cement Association, Skokie, Ill., April.

JOHNSON, A. E., JR., AND HOMEWOOD, R. H. (1961), "Stress and Deformation Analysis from Reduced Scale Plastic Model Testing," *Proc. Soc. Exp. Stress Anal.*, 18, no. 2.

JOHNSON, A. I. (1953), "Strength, Safety and Economical Dimensions of Structure," Bulletin No. 1, Royal Institute of Technology, Division of Building Studies and Structural Engineering, Stockholm, Sweden.

JOHNSON, A. N. (1926), "Concrete in Tension," *Am. Soc. Test. Mater. Proc.*, 26, part 2, 441.

JOHNSON, I. (1959), "An Optical Method for Studying Buckling of Plates," *Proc. Soc. Exp. Stress Anal.*, 16, no. 2, 145–52.

JOHNSON, R. P. (1962), "Strength Tests on Scaled-Down Concretes Suitable for Models, with a Note on Mix Design," *Mag. Concr. Res.*, 14, no. 40 (March).

KAAR, P. H. (1966), "High Strength Bars as Concrete Reinforcement, Part 8, Similitude in Flexural Cracking of T-Beam Flanges," *J., PCA Res. Dev. Lab.*, 8, no. 2 (May), 2–12.

KADLECEK, V. AND SPETLA, Z. (1977), "How Size and Shape of Specimens Affect the Direct Tensile Strength of Concrete," *Tech. Dig. (Prague)*, 9, 865–72.

KALITA, U. C. AND HENDRY, A. W. (1970), "An Experimental and Theoretical Investigation of the Stresses and Deflections in Model Cross-Wall Structures," *Proceedings of Second International Brick Masonry Conference*, Stoke-on-Trent, April.

KANDASAMY, E. G. (1969), "Stress-Strain Characteristics of Gypsum Plaster-Sand Mixes under Direct and Flexural Compression," M.S. thesis, North Carolina State University, Raleigh, N.C.

KAPLAN, M. F. (1963), "Strains and Stresses of Concrete at Initiation of Cracking and Near Failure," *Proc. Am. Concr. Inst.*, 60, no. 7 (July), 853–80.

KATZOFF, S. (1963), "Similitude in Thermal Models of Spacecraft," NASA Technical Note D-1631, April.

KAUSEL, E. (1967), "Teoría General de la Estabilidad Elástica de Arcos Planos," Master's of Engineering thesis, Universidad de Chile, Santiago, Chile, September, pp. 91–92.

KAWASHINA, K., AND PENZIEN, J. (1976), "Correlative Investigations on Theoretical and Experimental Behavior of a Model Bridge Structure," EERC Report, EERC-78-18, Earthquake Engineering Research Center, University of California, Berkeley, July.

KEMP, G. (1971), "Simply Supported, Two Way Prestressed Concrete Slabs under Uniform Load," Master's of Engineering thesis, Structural Concrete Series No. 71-4, McGill University, Montreal, August.

KHOO, C. L. AND HENDRY, A. W. (1973), "Strength Tests on Brick and Mortar under Complex Stresses for the Development of a Failure Criterion for Brickwork in Compression," Proc. Br. Ceram. Soc., No. 21 (April), Load Bearing Brickwork (4), pp. 51–66.

KINNEY, G. F. (1957), *Engineering Properties and Applications of Plastics*, John Wiley & Sons, Inc., New York, 278 pp.

KINNEY, J. S. (1957), *Indeterminate Structural Analysis*, Addison-Wesley Publishing Company, Inc., Reading, Mass., 655 pp.

KORDINA, K. (1964), "The Influence of Creep on the Buckling Load of Shallow Cylindrical Shells—Preliminary Tests," in *Non-Classical Shell Problems*, North-Holland Publishing Company, Amsterdam, pp. 602–8.

KRAWINKLER, H., MILLS, R. S., MONCARZ, P. D. ET AL. (1978), "Scale Modeling and Testing of Structures for Reproducing Response to Earthquake Excitation," The John A. Blume Earthquake Engineering Center, Department of Civil Engineering, Stanford University, Stanford, May.

KUPFER, H., HILSDORF, H. K. AND RUSCH, H. (1969), "Behavior of Concrete under Biaxial Stresses," *Proc. Am. Concr. Inst.*, 66, no. 8 (August), 656–66.

LABONTE, L. R. S. (1971), "An Investigation of Anchorage Zone Behavior in Prestressed Concrete Containments," Master's of Engineering thesis, Structural Concrete Series No. 71-6, McGill University, Montreal, September.

LANGHAAR, H. L. (1951), *Dimensional Analysis and Theory of Models*, John Wiley & Sons, Inc., New York.

LEE, S. T. (1964), "Behavior of Microconcrete Flat Plate Structures," M.S. thesis, Department of Civil Engineering, Massachusetts Institute of Technology, Cambridge, Mass., March.

LEET, K. M. (1966), "Study of Stability in the Hyperbolic Paraboloid," *ASCE J. Eng. Mech. Div.*, 92, no. EM1, 121–42.

LEVER, A. E. AND RHYS, J. A. (1968), *The Properties and Testing of Plastic Materials*, International Scientific Series, C. R. C. Press, Cleveland, Ohio, p. 445.

LIM, S. N., SYAMAL, P. K., KHAN, A. Q. AND NEMEC, J. (1968), "Development Length in Pullout Tests," McGill University, Montreal, January.

LIN, M. S. AND POPOV, E. P. (1969), "Buckling of Spherical Sandwich Shells," *Exp. Mech.*, 9, no. 10, 433–40.

LITLE, W. A. (1963), "Reliability of Shell Buckling Predictions Based upon Experimental Analysis of Plastic Models," Technical Publication T63-7, Department of Civil Engineering, Massachusetts Institute of Technology, Cambridge, Mass.

LITLE, W. A. (1964), *Reliability of Shell Buckling Predictions*, Research Monograph No. 25, Massachusetts Institute of Technology, Cambridge, Mass., p. 149.

LITLE, W. A., COHEN, E. AND SOMMERVILLE, G. (1970), "Accuracy of Structural Models," *Models for Concrete Structures*, ACI SP-24, American Concrete Institute, Detroit, Mich. pp. 65–124.

LITLE, W. A., FORCIER, F. J. AND GRIGGS, P. H. (1970), "Can Plastic Models Represent the Buckling Behavior of Reinforced Concrete Shells?" *Models for Concrete Structures*, ACI SP-24, American Concrete Institute, Detroit, Mich. pp. 265–87.

LITLE, W. A. AND FOSTER, D. C. (1966), "Fabrication Techniques for Small Scale Steel Models," Department of Civil Engineering, Masachusetts Institute of Technology, Cambridge, Mass., October.

LITLE, W. A. AND HANSEN, R. J. (1963), "The Use of Models in Structural Design," *J. Boston Soc. of Civ. Eng.*, 50, no. 2, 59–94.

LITLE, W. A., HANSEN, R. J. ET AL. (1965), Notes of the Special Summer Course on Structural Models, Massachusetts Institute of Technology, Cambridge, Mass.

LITLE, W. A. AND PAPARONI, M. (1966), "Size Effect in Small Scale Models of Reinforced Concrete Beams," *Proc. Am. Concr. Inst.*, 63, no. 11 (November), 1191–1204.

LITLE, W. A. ET AL. (1967), "A Study of Cylindrical Shell Buckling," *IASS International Congress on the Application of Shell Structures in Architecture*, Mexico City, September.

LOBO FIALHO, J. F. (1970), "Static Model Studies for Designing Reinforced Concrete Structures," ACI SP-24, *Models for Concrete Structures*, American Concrete Institute, Detroit, Mich., pp. 215–50.

LOH, G. (1969), "Factors Influencing the Size Effects in Gypsum Mortar," M.S. thesis, Cornell University, Ithaca, N.Y., September.

LORD, W. D. (1965), "Investigation of the Effective Width of the Slab of Reinforced Concrete T-Beams," M.S. thesis, Cornell University, Ithaca, N.Y.

LUTZ, L. A. AND GERGELY, P. (1967), "Mechanics of Bond and Slip of Deformed Bars in Concrete," *Proc. Am. Concr. Inst.*, 64, 11, (November), pp. 711–21.

MAISEL, E. (1978), "Reinforced and Prestressed Microconcrete Models," *Proceedings of Joint Institution of Structural Engineers/Building Research Establishment Seminar on Reinforced and Prestressed Microconcrete Models*, Garston, England, May.

MALHOTRA, V. M. (1969), "Effect of Specimen Size on Tensile Strength of Concrete," Report of Dept. of Energy, Mines and Resources, Ottawa, Canada, June, 9 pp.

Manufacturing Chemists Association (1957), "Technical Data on Plastics," Washington, D.C.

MARK, J. W. AND GOLDSMITH, W. (1973), "Barium Titanate Steam Gages," *Proc. Exp. Stress Analysis*, 13, no. 1, pp. 139–150.

MARK, R. AND BILLINGTON, D. P. (1969), "Photoelastic Model Analysis of Concrete Storage Tanks," ASCE *J. Struct. Div.*, 95, no. ST5 (September), 1939–51.

MARK, R. AND RIERA, J. D. (1967), "Photoelastic Analysis of Folded-Plate Structures," *Jour. of the Eng. Mech. Div.*, ASCE, 93, no. EM4, Proc. Paper 5387 (August), pp. 79–93.

MARTIN, I. (1971), "Full Scale Load Test of a Prestressed Folded Plate Unit," *Proc. Am. Concr. Inst.*, 68, no. 12, 937–44.

MASTRODICASA, A. G. (1970), "Size Effects in Models of Reinforced Concrete Slabs," Research Report No. R70-57, Massachusetts Institute of Technology, Cambridge, August.

MATTOCK, A. H. (1964), "Rotational Capacity of Hinging Regions in Reinforced Concrete," ACI SP12, *Flexural Mechanics of Reinforced Concrete*, American Concrete Institute, Detroit, Mich, pp. 143–82.

MCCONCHIE, R. E. AND SCHMIDT, L. C. (1972), "The Structural Behaviour of a Hyperbolic Cooling Tower under Static Loadings," *Proceedings of the Structural Models Conference*, The University of Sydney, Sydney, Australia.

MCDERMOTT, J. F. (1968), "Single-Layer Corrugated Steel Sheet Hypars," *ASCE Struct. Div.*, 94, no. ST6, 1279–94.

MEDWADOWSKI, S. J., ed. (1964), *Proceedings of the World Conference on Shell Structures*, San Francisco, 1962, National Academy of Sciences, Washington, D.C., 690 pp.

MEININGER, R. C. (1968), "Effect of Core Diameter on Measured Concrete Strength," *J. Mater.*, 3, no. 1 (March), 320–36.

MIKHAIL, M. L. AND GURALNICK, S. A. (1971), "Buckling of Simply Supported Folded Plates," *ASCE J. Eng. Mech. Div.*, 97, no. EMS, 1363–80.

MILLS, R. S., KRAWINKLER, H. AND GERE, J. M., "Model Tests on Earthquake Simulators Development and Implementation of Experimental Procedures," Report No. 39, The John A. Blume Earthquake Engineering Center, Department of Civil Engineering, Stanford University, Stanford, Calif., June 1979, 272 pp.

MIRZA, M. S. (1967), "An Investigation of Combined Stresses in Reinforced Concrete Beams," Ph.D. thesis, Department of Civil Engineering and Applied Mechanics, McGill University, Montreal, March.

MIRZA, M. S. (1969), "Direct Models in McGill Structural Concrete Investigations," Structural Concrete Series No. 25, McGill University, Montreal, November.

MIRZA, M. S. (1972), "Heat Treatment of Deformed Steel Wires for Model Reinforcement," Structural Concrete Series No. 72-4, McGill University, Montreal, March.

MIRZA, M. S. (1978), "Reliability of Structural Models," *Proceedings of the Joint Institution of Structural Engineers/Building Research Establishment Seminar on Reinforced and Prestressed Microconcrete Models*, Garston, England, May.

MIRZA, M. S., ed. (1972), *Structural Concrete Models (Materials, Instrumentation, Correlation)*, *A State-of-the-Art Report*, Department of Civil Engineering and Applied Mechanics, McGill University, Montreal, 232 pp.

MIRZA, M. S. ET AL. (1972), "Materials for Direct and Indirect Structural Models," *Structural Concrete Models, A State-of-the-Art Report*, Department of Civil Engineering and Applied Mechanics, McGill University, Montreal, pp. 1-78.

MIRZA, M. S., HARRIS, H. G., AND SABNIS, G. M. (1979), "Structural Models In Earthquake Engineering," *Proceedings of the Third Canadian Conference on Earthquake Engineering*, June 4-6, Montreal, Quebec, Canada, vol. 1, pp. 511-549.

MIRZA, M. S., LABONTE, L. R. S. AND McCUTCHEON, J. O. (1972), "Size Effects in Model Concrete Mixes," presented at the ASCE National Convention, Cleveland, April.

MIRZA, M. S. AND McCUTCHEON, J. O. (1971), "Bond Similitude in Reinforced Concrete Models," paper presented to the National Structural Engineering Meeting of the ASCE, Baltimore, April.

MIRZA, M. S. AND McCUTCHEON, J. O. (1978), "Direct Models of Prestressed Concrete Beams on Bending and Shear," *Build Int.*, London, England, 7, no. 2, (March 1974), 99-125. Reprinted in the *Proceedings of the Joint Institution of Structural Engineers/Building Research Establishment Seminar on Reinforced and Prestressed Microconcrete Models*, Garston, England, May.

MIRZA, M. S., WHITE, R. N. AND ROLL, F. (1972), "Materials for Structural Models," *Proceedings of the ACI Symposium on Models of Concrete Structures*, Dallas, American Concrete Institute, Detroit, Mich., March, pp. 19-112.

MIRZA, M. S., WHITE, R. N., ROLL, F. AND BATCHELOR, B. DEV. (1972), "Materials for Direct and Indirect Structural Models," in *Structural Concrete Models, A State-of-the-Art Report*, Department of Civil Engineering and Applied Mechanics, McGill University, Montreal, pp. 1-78.

MOAKLER, M. W. AND HATFIELD, L. P. (1953), "The Design and Construction of a Deformeter for Use in Model Analysis," Master of Civil Engineering thesis, Rensselaer Polytechnic Institute, Troy, N.Y.

MOHR, G. A. (1971), "Stiffened Brick Walls," M. Eng. Sc. thesis, Civil Engineering, University of Melbourne, Melbourne, Australia.

MORRISON, J. L. M. (1940), "The Yield of Mild Steel with Particular Reference to the Effect of Size of Specimen," *J. Proc. Inst. Mech. Eng.*, 142, no. 3 (January), 193-223.

MUFTI, AFTAB A. (1969), "Matrix Analysis of Thin Shells Using Finite Elements," Ph.D. thesis, McGill University, Montreal, July, 141 pp.

MURPHY, C. K. AND HENDRY, A. W. (1966), "Model Experiments in Load Bearing Brickwork," in *Building Science*, vol. 1, Pergamon Press, London, pp. 289-98.

MURPHY, G. (1945), "Graphical Solution of Principal Strains," *J. Appl. Mech.*, 12, A209.

MURPHY, G. (1950), *Similitude in Engineering*, The Ronald Press Company, New York.

MUSKIVITCH, J. C. AND HARRIS, H. G. (1979), "Behavior of Large Panel Precast Concrete Buildings under Simulated Progressive Collapse Conditions," *Proceedings, International Symposium, Behavior of Building Systems and Building Components*, Vanderbilt University, March 8–9, Nashville, Tenn.

MUSKIVITCH, J. C. AND HARRIS, H. G. (1979), "Report 2: Behavior of Precast Concrete Large Panel Buildings under Simulated Progressive Collapse Conditions," *Nature and Mechanism of Progressive Collapse in Industrialized Buildings*, Office of Policy Development and Research, Department of Housing and Urban Development, Washington, D.C., January, 207 pp., also Department of Civil Engineering, Drexel University, Philadelphia.

MOUSTAFA, S. E. (1966), "A Small Scale Model Study of a Prestressed Concrete Slab," unpublished M.S. thesis, Cornell University, Ithaca, N.Y.

NARROW, I. AND ULLBERG, E. (1963), "Correlation between Tensile Splitting Strength and Flexural Strength of Concrete," *Proc. Am. Concr. Inst.*, 60, no. 1, 27.

NEBRASKA, J. E. AND SUR, L. M. (1963), "Behavior of Miniature Prestressed Concrete Beams," Report of a project sponsored by the N.S.F. Undergraduate Research Program, Department of Civil Engineering, University of Illinois, Urbana, Ill., June.

NEILSEN, K. E. C. (1954), "Effect of Various Factors on the Flexural Strength of Concrete Test Beams," *Mag. Concr. Res. (London)*, 15, 105–114.

NEVILLE, A. M. (1959), "Some Aspects of the Strength of Concrete," Three parts, *Civ. Eng. Public Works Rev.*, 54, no. 639 (October); no. 640 (November); no. 641 (December).

NEVILLE, A. M. (1963), *Properties of Concrete*, I. Pitman and Sons Ltd., London.

NEVILLE, A. M. (1966), "A General Relation for Strengths of Concrete Specimens of Different Shapes and Sizes," *Proc. Am. Concr. Inst.*, 63, 1095–1110.

NEWMAN, K. (1965), "The Structure and Engineering Properties of Concrete," *Proceedings of the International Symposium on the Theory of Arch Dams*, Southampton, April, Pergamon Press, New York.

NEWMAN, K. (1965), "Concrete Control Tests as a Measure of the Properties of Concrete," *Proceedings of the Symposium on Concrete Quality*, November, Cement and Concrete Association, United Kingdom.

NEWMAN, K. (1965), "Concrete Systems," in *Complex and Heterophase Materials*, Elsevier Publishing Company, New York, chap. 8.

NICHOLLS, J. I. AND FUCHS, P. (1972), "A Comparison of Test Results and a Computer Analysis of a Single Cell Horizontally Curved Composite Box Bridge," Department of Civil Engineering, University of Washington, Seattle.

NORRIS, C. H., HANSEN, R. J., HOLLEY, M. J., JR., BIGGS, J. M., NAMYET, S. AND MINAMI, J. K. (1959), *Structural Design for Dynamic Loads*, McGraw-Hill Book Company, New York.

NORRIS, C. H. AND WILBUR, J. B. (1976), *Elementary Structural Analysis*, McGraw-Hill Book Company, New York.

OLSZAK, W. AND SAWCZUK, A., eds. (1964), *Non-Classical Shell Problems, Proceedings Symposium Warsaw*, September 2–5, North-Holland Publishing Company, Amsterdam.

ORR, D. M. F. AND BREEN, J. E. (1972), "A Rapid Data Acquisition and Processing System for Structural Model Usage," *Proceedings of Structural Models Conference*, sponsored by School of Architectural Science, University of Sydney, Sydney, Australia.

PAHL, P. J. (1963), "Confidence Levels for Structural Models," Report No. 63–05, Department of Civil Engineering, Massachusetts Institute of Technology, Cambridge, February.

PAHL, P. J. AND SOOSAAR, K. (1964), "Structural Models for Architectural and Engineering Education," Report No. R64-3, Department of Civil Engineering, Massachusetts Institute of Technology, Cambridge, Mass., 269 pp.

PANG, C. L. (1965), "Reliability of Models in the Analysis of Prestressed Concrete Beams in Flexure," Master's of Engineering thesis, McGill University, Montreal, April.

PARRATT, L. G. (1961), *Probability and Experimental Errors in Science*, John Wiley & Sons, Inc., New York.

PARZEN, E. (1960), *Modern Probability Theory and Its Applications*, John Wiley & Sons, Inc., New York.

PAUL, S. L. ET AL. (1969), "Strength and Behavior of Prestressed Concrete Vessels for Nuclear Reactors," Civil Eng. Series, Structural Research Series No. 346, University of Illinois, Urbana, Ill. July.

PERRY, D. C. AND LISSNER, H. R. (1962), *The Strain Gage Primer*, McGraw-Hill Book Company, New York, 332 pp.

PIPPARD, A. J. (1947), *The Experimental Study of Structures*, Edward Arnold, p. 29.

POPOVICS, S. (1967), "Relations between Various Strengths of Concrete," *Highway Research Record*, No. 210, National Academy of Sciences, Washington, D.C.

PREECE, B. W. AND DAVIES, J. D. (1964), *Models for Structural Concrete*, C. R. Books Ltd., London, England, 252 pp.

PRICE, W. W. (1951), "Factors Influencing Concrete Strength," *Proc. Am. Concr. Inst.*, 47 (February) 417.

Magnaflux Corporation (1965), "Principles of Stress-Coat-Operating Manual," Chicago,

Proceedings Congress of the International Association for Shell Structures, (1966), Leningrad, TSNIS, Moscow, USSR.

Proceedings International Congress of the Application of Shells in Architecture, (1967), Mexico City, International Association of Shell Structures, Madrid, Spain.

Proceedings International Association for Shell Structures International Colloquium on Progress of Shell Structure (1969), Madrid, Spain, September.

Proceedings RILEM International Symposium on Experimental Analysis of Instability Problems on Reduced and Full-Scale Models (1971), Buenos Aires, Argentina, September 13–18, Instituto Nacional de Technologia Industrial.

RANDALL, F. A. AND PANARESE, W. C. (1976), *Concrete Masonry Handbook*, Portland Cement Association, Skokie, Ill.

RAO, C. V. S. K. (1972), "Some Studies on Statistical Aspects of Size Effects on Strength and Fracture Behavior of Materials and Fracture Resistant Design," Ph.D. thesis, Indian Institute of Technology, Kanpur, India, June, 225 pp.

RASBASH, D. J. (1969), "Explosions in Domestic Structures; Part I: The Relief of Gas and Vapour Explosions in Domestic Structures," *Struct. Eng. (London)*, 47, no. 10 (October).

RAWLINGS, B. AND BURGMANN, J. B. (1969), "Tests on Small Scale Folded Plate Metal Roofs," Report of the Department of Civil and Structural Engineering, University of Sheffield, Sheffield, England.

Raytheon Company (1962), "A Discussion of Module Welding Techniques," presented for General Dynamics Corporation, Raytheon Company, Waltham, Mass., August.

RICHARDS, C. W. (1958), "Effect of Size on the Yielding of Mild Steel Beams," Preprint 75, ASTM, 61st Annual Meeting, June.

RICHARDS, C. W. (1954), "Size Effect in the Tension Test of Mild Steel Beams," *Am. Soc. Test. Mater. Proc.*, 54, 995.

ROCHA, M. (1961), "Determination of Thermal Stresses in Arch Dams by Means of Models," *RILEM Bull.* (Paris), no. 10, p. 65.

ROCHA, M. (1971), "Model Study of Structures in Portugal," Technical Translation O. TT-970 (by D. Sinclair), National Research Council of Canada, Ottawa, Canada, p. 30.

ROCHA, M. AND BORGES, J. F. (1961), "Photographic Method for Model Analysis of Structures," *Proc. Soc. Exp. Stress Analy.*, 8, no. 2, 129–42.

ROHM AND HAAS (1959), "Fabrication of Plexiglas," Bulletin No. PL-383-C, Philadelphia.

ROLL, F. (1968), "Materials for Structural Models," *ASCE J. Struct. Div.*, 94, no. ST6, 1353–82.

ROLL, F. AND ANEJA, I. K. (1966), "Model Tests of Box-Beam Highway Bridges With Cantilevered Deck Slabs," paper presented at the October 17–21, ASCE Transportation Engineering Conference held at Philadelphia, (Preprint No. 3905).

RONTSCH, G. (1966), "Model Tests for Investigating the Stability of Folded-Plate Structures," International Association for Shell Structures, Bulletin No. 25, pp. 29–55.

ROSS, A. D. (1944), "Shape, Size and Shrinkage," *Concr. Constr. Eng.*, 38 (August), 193–199.

ROSS, A. D. (1946), "The Effect of Creep on Instability and Indeterminacy Investigated by Plastic Models," *Struct. Eng.* (London), 24, no. 8 (August), 413–28.

ROWE, R. E. (1960), "Works on Models in the Cement and Concrete Association," *J. PCA Res. Dev. Lab.*, 2, no. 1, 4–10.

RUGE, A. C. AND SCHMIDT, E. O. (1939), "Mechanical Structural Analysis by the Moment Indicator," *Trans. ASCE*, 104, also *Proc. ASCE*, 65, no. 1 (January), 161–70 and no. 6 (June), 1037–40.

RUIZ, W. (1966), "Effect of Volume of Aggregate on the Elastic and Inelastic Properties of Concrete," M. S. thesis, Cornell University, Ithaca, N.Y.

RUSCH, H. (1954), "Specimen Size and Apparent Compressive Strength," *Proc. Am. Concr. Inst.* 51, 803 pp.

RUSCH, H. (1959), "Physikalische Fragen der Betonprufung," (Physical Problems in the Testing of Concrete), *Zem.-Kalk-Gips*, 12, no. 1, Cornell University, Ithaca, New York, (Translation by G. M. Sturman).

RUSCH, H. (1964), "Zur Statistichen Quilitaskontrolle de Beton," (On the Statistical Quality Control of Concrete), *Materialpruefung*, 6, no. 11 (November).

SABNIS, G. M. (1967), "Investigation of Reinforced Concrete Frames Subjected to Reversed Cyclic Loading Using Small Scale Models," Ph.D. thesis, Cornell University, Ithaca, N.Y., June.

SABNIS, G. M. (1969), "Behavior of Over-Reinforced Concrete Beams: A Small Scale Model Approach," *Indian Concr. J.*, 43, no. 1 (January), 13–24.

SABNIS, G. M. AND ARONI, S. (1971), "Size Effects in Material Systems, The State-of-the-Art," Paper No. 12, in *Structure, Solid Mechanics and Engineering Design*, The Proceedings of the Southhampton 1969 Civil Engineering Materials Conference. Editor, M. Te'eni, Wiley Interscience, 131–42.

SABNIS, G. M. AND MIRZA, M. S. (1979), "Size Effects in Model Concretes?" ASCE *J. Struct. Div.*, 105, no. ST6 (June), 1007–20.

SABNIS, G. M. AND ROLL, F. (1971), "Importance of Scaled Compressive Strength Cylinders in Shear Resistance of Reinforced Concrete Slabs," *Proc. Am. Concr. Inst.*, 68, no. 3 (March), 218–21.

SABNIS, G. M. AND WHITE, R. N. (1967), "A Gypsum Mortar for Small-Scale Models," *Proc. Am. Concr. Inst.*, 64, no. 11 (November), 767–74.

SABNIS, G. M. AND WHITE, R. N. (1969), "Behavior of Reinforced Concrete Frames under Cyclic Loads Using Small Scale Models," *J. Am. Concr. Inst.*, 66, no. 9, September, 703–15.

SABNIS, G. M. (1980), "Size Effects in Material Systems and Their Impact on Model Studies: A Theoretical Approach," Proc. of SECTAM X Conference, Knoxville, Tenn., pp. 649–68.

SACHS, P. (1978), "Wind Tunnel Techniques," in *Wind Forces in Engineering*, (2nd ed.), Pergamon Press, New York. chap. 5 p. 400.

SAHLIN, S. (1971), *Structural Masonry* (1st ed.), Prentice-Hall, Inc., Englewood Cliffs, N.J.

SALMON, C. G. AND JOHNSON, J. E. (1971), *Steel Structures—Design and Behavior*, Intext Educational Publishers, Scranton, Pa.

SARGIN, M. (1971), "A Stress-Strain Relationships of Concrete and the Analysis of Structural Concrete Sections," Study No. 4, Solid Mechanics Division, University of Waterloo, Waterloo, Canada.

SCANLAN, R. H. (1973), "Dynamic Similitude in Models," *Proceedings*, EDF Conference sur l'Aero-Hydo-elasticite, CEA-EDF, Chatou, France, September 9 1972, Eyrolles, Paris, pp. 787–828.

SCANLAN, R. H. (1974), "Scale Models and Modeling Laws in Fluid-Elasticity," Preprint No. 2247, ASCE National Structural Engineering Meeting, Cincinnati, Ohio, April 22–26.

SCHNEIDEWIND, R. AND HOENICKE, E. C. (1943), "A Study of the Chemical, Physical and Mechanical Properties of Permanent Mold Gray Cast Iron," *Am. Soc. Test. Mater. Proc.*, 42, 622–35.

SCHULTZ, D. M., BURNETT, E. F. P., AND FINTEL, M. (1977), "Report 4: A Design Approach to General Structural Integrity," *Design and Construction of Large Panel Concrete Structures*, Office of Policy Development and Research, Department of Housing and Urban Development, Washington, D.C., October.

SCHURING, D. J. (1977), *Scale Models in Engineering—Fundamentals and Applications*, Pergamon Press, New York.

SCRUTON, C. (1968), "Aerodynamics of Structures," *Proceedings, International Research Seminar on Wind Effects on Buildings and Structures*, vol. 1, University of Toronto Press, pp. 115–61.

SEAOC (1968), *Recommended Lateral Force Requirements and Commentary*, Seismology Committee, Structural Engineers Association of California.

SELF, M. W. (1975), "Structural Properties of Load-Bearing Concrete Masonry," in *Masonry: Past and Present*, ASTM STP 589, American Society for Testing and Materials, pp. 235–54.

SHAW, W. A. (1962), "Static and Dynamic Behavior of Portal-Frame Knee Connections," U.S. Naval Civil Engineering Laboratory, Port Hueneme, Calif., May.

SIDEBOTTOM, O. M. AND CLARK, M. E. (1954), "The Effect of Size on the Load-Carrying Capacity of Steel Beams Subjected to Dead Loads," ASME Meeting, Milwaukee, Wis.

SIMIU, E. AND SCANLAN, R. H. (1978), "The Wind Tunnel as a Design Tool," in *Wind Effects on Structures: An Introduction to Wind Engineering*, John Wiley & Sons, Inc. New York, chap. 9, 458 p.

SINHA, B. P. (1967), "Model Studies Related to Load Bearing Brickwork," Ph.D. thesis, University of Edinburgh, Scotland.

SINHA, B. P. AND HENDRY, A. W. (1969), "Racking Tests on Story-Height Shear-Wall Structures with Openings, Subjected to Precompression," in *Designing Engineering and Constructing with Masonry Products*, ed. by F. B. Johnson, Gulf Publishing Co., Houston, Texas, pp. 192–99.

SINHA, B. P., MAURENBRECHER, A. H. P. AND HENDRY, A. W. (1970), "Model and Full Scale Tests on a Five-Story Cross-Wall Structure under Lateral Loading," *Proceedings of Second International Brick Masonry Conference*, Stoke-on-Trent, April, pp. 201–8.

SLACK, J. H. (1971), "Explosions in Buildings—The Behavior of Reinforced Concrete Frames," *Concrete*, 5 (April), 109–14.

SMITH, H. D., CLARK, R. W. AND MAYOR, R. P. (1963), "Evaluation of Model Techniques for the Investigation of Structural Response to Blast Loads," Report R63-16, Department of Civil Engineering, Massachusetts Institute of Technology, Cambridge, Mass. February.

SNYDER, W. H. (1972), "Similarity Criteria for the Application of Fluid Models to the Study of Air Pollution Meteorology," *Boundary Layer Meteorol.* 3, no. 1 (September).

Society of Experimental Stress Analysis (1965), "Manual on Experimental Stress Analysis," Westport, Conn., p. 67.

SOOSAAR, K. (1963), "Systematic Errors in the Loading of Stress Models of Shells," Report T63-6, Dept. of Civil Eng., Massachusetts Institute of Technology, Cambridge, MA.

SPOKOWSKI, R. (1972), "Finite Element Analysis of Reinforced Concrete Members," M. Eng. thesis, Structural Concrete Series No. 72–2, McGill University, Montreal, February.

STAFFIER, S. R. AND SOZEN, M. A. (1975), "Effect of Strain Rate on Yield Stress of Model Reinforcement," Civil Engineering Studies, Structural Research Series No. 415, University of Illinois, Urbana, Ill., February.

STAFIEJ, A. P. (1970), "Bond Similitude in Reinforced Concrete Models," B. Sc. (Hon.) thesis, Department of Civil Engineering and Applied Mechanics, McGill University, Montreal, April.

STEPHEN, R. M. AND BOUWKAMP, J. G. (1975), "New General Purpose Component Test System," presented at the ASCE National Structural Engineering Convention, New Orleans, La. (Preprint No. 2475), p. 27.

STEVENS, L. K. (1959), "Investigations on a Model Dome with Arched Cut-Outs," *Mag. Concr. Res. (London)*, 11, no. 31 (March), 3–14.

STRETCH, K. L. (1969), "Explosions in Domestic Structures; Part 2: The Relationship between Containment Characteristics and Gaseous Reactions," *Struct. Eng.* (London), 47, no. 10 (October).

STRUMINSKY, E. S. (1971), "Low Cycle Fatigue Study of Fiberglass Reinforced Plastic Laminates," M. Eng. thesis, McGill University, Montreal.

SUBEDI, N. K. AND GARAS, F. K. (1980), "Bond Characteristics of Small Diameter Bars Used in Microconcrete Models," in *Reinforced and Prestressed Microconcrete Models*, ed. by F. K. Garas and G. S. T. Armer, The Construction Press, New York.

SURREY, D. AND ISYUMOV, N. (1975), "Model Studies of Wind Effects—A Perspective on the Problems of Experimental Technique and Instrumentation," *Int. Congress on Instrumentation in Aerospace Simulation Facilities*, Ottawa.

SURREY, D., KITCHEN, R. B. AND DAVENPORT, A. G. (1976), "Design Effectiveness of Wind Tunnel Studies for Buildings of Intermediate Height," Preprints of the ASCE National Structural Engineering Conference, Madison, Wis.

SWANN, R. A. (1970), "Test on a Prestressed Micro-Concrete Model of a Three Cell Box Beam Bridge," *Models for Concrete Structures*, ACI SP-24, American Concrete Institute, Detroit, Mich., pp. 333–51.

SWARTZ, S. E., MIKHAIL, M. L. AND GURALNICK, S. A. (1969), "Buckling of Folded-Plate Structures," *Exper. Mech.*, 9, no. 6, 269–74.

SWARTZ, S. E. AND ROSEBRAUGH, V. H. (1972), "Experiments with Elastic Folded Plate Models," *ASCE J. Eng. Mech. Div.*, 98, no. EM3, 531–38.

SYAMAL, P. K. (1969), "Direct Models in Combined Stress Investigations," M. Eng. thesis, Structural Concrete Series No. 17, McGill University, Montreal, July.

TABBA, M. M. (1972), "Free Vibrations of Curved Box Girders," Master's of Engineering thesis, McGill University, Montreal.

TAHER, I. (1963), "A Study of Bond Characteristics in Wire Reinforced Specimens," M.S. thesis, Massachusetts Institute of Technology, Cambridge, August.

TAKEDA, T., ET AL. (1973), "Pressure Tests of PCRV Models," Ohbayashi-Gumi Report OTN. TY. 48100, December, presented at the 7th FIP/PCI International Conference, New York, May 1974.

TASUJI, M. E. (1976), "The Behavior of Plain Concrete Subject to Biaxial Stress," Research Report No. 360, Department of Structural Engineering, Cornell University, Ithaca, N.Y., May.

TASUJI, M. E., SLATE, F. O., AND NILSON, A. H. (1978), "Stress-Strain Response and Fracture of Concrete in Biaxial Loading," *Proc. Am. Concr. Inst.*, 75, no. 8 (August), 306–12.

TIMOSHENKO, S. AND GOODIER, J. N. (1951), *Theory of Elasticity* (2nd ed.), McGraw-Hill Book Company, New York.

TSUI, S. H. AND MIRZA, M. S. (1969), "Model Microconcrete Mixes," Structural Concrete Series No. 23, McGill University, Montreal, November.

TUCKER, J., JR. (1941), "Statistical Theory of the Effect of Dimensions and Method of Loading on the Modulus of Rupture of Beams," *Am. Soc. Test. Mater. Proc.*, 41, 1072–88.

TUCKER, J., JR. (1945), "Effect of Dimensions of Specimens upon the Precision of Strength Data," *Am. Soc. Test. Mater. Proc.*, 45, 592–959.

University of Sydney and the Institution of Structural Engineers, Australia (1972), *Proceedings of the Structural Models Conference*, H. J. Cowan, Chairman, University of Sydney, Sydney, Australia.

VOGT, H. (1956), "Considerations and Investigations on the Basic Principle of Model Tests in Brickwork and Masonry Structures," Library Communication No. 932, Building Research Station, Garston, Watford, England.

WEIBULL, W. (1939), "A Statistical Theory of the Strength of Materials," *Royal Swedish Proc.*, vols. 151–152, Stockholm, Sweden.

WELLER, T. AND SINGER, JR. (1970), "Experimental Studies on Buckling of Ring-Stiffened Conical Shells under Axial Compression," *Exper. Mech.*, 10, np. 11, 449–57.

WEYMEIS, G. (1977), "Report on the Investigation of Instability of Concrete Domes Conducted from 1967–76 at the University of Ghent in Belgium" (in Dutch), Rijks Universiteit Laboratorium voor Modelonderzoek, Grote Steenweg Noord 12, Swijnaarde, Belgium.

WHITE, I. G. AND CLARK, L. A. (1978), "Bond Similitude in Reinforced Microconcrete Models," *Proceedings of Joint Institution of Structural Engineers/Building Research Establishment Seminar on Reinforced and Prestressed Microconcrete Models*, Garston, England, May.

WHITE, R. N. (1972), *Structural Behavior Laboratory—Equipment and Experiments*, Report No. 346, Department of Structural Engineering, Cornell University, Ithaca, N.Y.

WHITE, R. N. (1975), "Reinforced Concrete Hyperbolic Paraboloid Shells," *ASCE J. Struct. Div.*, September, pp. 1961–82.

WHITE, R. N. (1976), "High Strength Model Concrete Mixes," unpublished report, Department of Structural Engineering, Cornell University, Ithaca, N.Y.

WHITE, R. N. AND SABNIS, G. M. (1968), "Size Effects in Gypsum Mortar," *J. Mater.*, 3, no. 1 (March), pp. 163–77, ASTM, Philadelphia, Pa.

WILBUR, J. B. AND NORRIS, C. H. (1950), "Structural Model Analysis," in *Handbook of Experimental Stress Analysis*, ed. M. Hetenyi, John Wiley & Sons, Inc., New York, chapter 15, pp. 663–99.

WILLIAMS, D. AND GODDEN, W. G. (1976), "Experimental Model Studies on the Seismic Response of High Curved Overcrossings," Report No. EERC 76–18, College of Engineering, University of California at Berkeley, Berkeley, June.

WILLIAM, K. J. AND WARNKE, E. P. (1974), "Constitutive Model for the Triaxial Behavior of Concrete," *Proceedings of the Seminar on Concrete Structures Subjected to Triaxial Stresses*, Istituto Sperimentale Modelli e Strutture (ISMES), Bergamo, Italy, May.

WILSON, E. B. (1952), *An Introduction to Scientific Research*, McGraw-Hill Book Company, New York.

WRIGHT, P. J. F. AND GARWOOD, F. (1952), "The Effect of the Method of Test on the Flexural Strength of Concrete," *Mag. Concr. Res. (London)*, 11, 67–76.

YATES, J. G., LUCAS, D. H. AND JOHNSTON, D. L. (1953), "Pulse-Excitation of Resistance Strain Gauges for Dynamic Multi-Channel Observation," *Proc. Soc. Exp. Stress Anal.*, 11, no. 1.

YEROUSHALMI, M. AND HARRIS, H. G. (1978), "Behavior of Vertical Joints between Precast Concrete Wall Panels under Cyclic Reversed Shear Loading," Structural Models Laboratory, Report No. M78–2, Department of Civil Engineering, Drexel University, Philadelphia, March, 100 pp.

YOKEL, F. Y., MATHEY, R. G. AND DIKKERS, R. D. (1971), "Strength of Masonry Walls under Compressive and Transverse Loads," Building Science Series 34, National Bureau of Standards, Washington, D.C., March.

YORULMAZ, M. AND SOZEN, M. A. (1968), "Behavior of Single-Story Reinforced Concrete Frames with Filler Walls," University of Illinois, Civil Engineering Studies, Structural Research Series No. 337, Urbana, Ill., May.

ZIA, P., WHITE, R. N. AND VAN HORN, D. A. (1970), "Principles of Model Analysis," *Models for Concrete Structures*, ACI SP-24, American Concrete Institute, Detroit, Mich., pp. 19–39.

Index